Montserrat Filella (Ed.)
Antimony

Also of interest

Hazardous Substances.
Risks and Regulations
Thomas Schupp, 2020
ISBN 978-3-11-061805-1, e-ISBN 978-3-11-061895-2

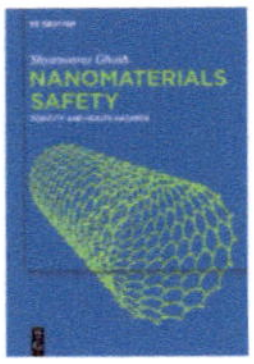

Nanomaterials Safety.
Toxicity And Health Hazards
Shyamasree Ghosh, 2019
ISBN 978-3-11-057808-9, e-ISBN 978-3-11-057909-3

Environmental Toxicology
Luis M. Botana (Ed.), 2018
ISBN 978-3-11-044203-8, e-ISBN 978-3-11-044204-5

Handbook of Rare Earth Elements.
Analytics
Alfred Golloch (Ed.), 2017
ISBN 978-3-11-036523-8, e-ISBN 978-3-11-036508-5

Antimony

Edited by
Montserrat Filella

DE GRUYTER

Editor
Dr. Montserrat Filella
University of Geneva
Department F.-A. Forel
Boulevard Carl-Vogt 66
CH-1205 Geneva
Switzerland
montserrat.filella@unige.ch

ISBN 978-3-11-066534-5
e-ISBN (PDF) 978-3-11-066871-1
e-ISBN (EPUB) 978-3-11-066545-1

Library of Congress Control Number: 2021931443

Bibliographic information published by the Deutsche Nationalbibliothek
The Deutsche Nationalbibliothek lists this publication in the Deutsche Nationalbibliografie; detailed bibliographic data are available on the Internet at http://dnb.dnb.de.

Cover image: Kesu01/iStock/Getty Images Plus
Typesetting: Integra Software Services Pvt. Ltd.
Printing and binding: CPI books GmbH, Leck

www.degruyter.com

Preface

The aim of this book is to give an account of current research and understanding about a chemical element, antimony. The book is addressed to all who share an interest in any aspect of the element.

Antimony has for centuries attracted the interest of many people in a number of different ways, and still does. The history of antimony applications is very long and unusually fascinating. The history of humankind follows the history of the exploitation and use of natural resources and, in particular, of metals and metalloids. However, not many chemical elements have been used in such a variety of applications as has antimony, and probably none has ended up being so ubiquitously present in our everyday life. This book has been designed to acquaint the reader with some of the main aspects of mineralogy, exploitation, biogeochemical behavior, and effects of the use of antimony. Surprisingly, in spite of the extensive scientific literature, no similar books have been published so far. This work aspires to fill that gap.

Any chemical element forms a multifaceted subject. I have chosen a very broad title for this book in order to provide a general umbrella allowing the discussion of different subjects around the element, but obviously the book does not contain 'all that is known' on antimony. Although I have tried to adopt a comprehensive approach, in managing the compromise between subjects to be treated and keeping the book within a reasonable size, some subjects have been deliberately left out. This text is admittedly idiosyncratic and its contents inevitably reflect my own background and interests as an environmental chemist. I admit that I regret the absence of chapters devoted to aspects such as the discussion of the chemical characteristics that make this element so versatile in its applications, or the best analytical methods to apply. I hope a future work will cover them. It has also been necessary to arrive at a compromise between depth of coverage of included subjects and book length, with many viewpoints given only limited space. The authors have devoted much effort to this aspect. I hope that, in spite of these limitations, readers will gain an appreciation of the scope of the subject and will consequently be adequately prepared to understand the growing literature on this interesting element.

The book comprises 16 chapters. It opens with a presentation of historical aspects of the use of antimony (Chapter 1). Then, mineralogy (Chapter 2), deposits (Chapter 3), and anthropogenic flows and stocks (Chapter 4) are discussed. The prominent situation of China in the current mining and production of the element is examined in Chapter 5. The ensuing three chapters are devoted to different aspects related to antimony in soils: diffuse soil contamination (Chapter 6), soil-plant transfer (Chapter 7), and shooting range contamination (Chapter 8). They are followed by a critical appraisal of the journey of antimony across the aquatic continuum (Chapter 9) and the examination of the fate of radioactive antimony in the environment (Chapter 10). Biomethylation and biovolatilization of the element is reviewed (Chapter 11), followed by a discussion of the molecular mechanisms of

https://doi.org/10.1515/9783110668711-202

antimony transport and detoxification (Chapter 12). The use of pentavalent antimonials in the treatment of human leishmaniasis is considered in the next chapter (Chapter 13) before moving to human exposure (Chapter 14), ecotoxicology (Chapter 15), and current regulations (Chapter 16).

As mentioned, there are gaps in the book that are my responsibility, but it is important to realize that there are also many gaps in our knowledge, which will be apparent in the text. I hope that this book will help to motivate future researchers to fill these gaps and also make readers aware of the need to embrace uncertainty when dealing with any scientific subject. Scientific knowledge by definition continuously evolves and our knowledge of antimony is no exception. Helping readers face up to the 'known unknowns' about antimony is also an objective of the book.

While preparing this book I have been very conscious of how much I owe to many friends and colleagues for their help in stimulating my efforts to understand this element over the years. Particular mention should be made of the three conferences on antimony that took place in 2005 (Heidelberg, organized by Bill Shotyk, Michael Krachler, and Bin Chen), 2011 (Jena, co-organized by Jurajz Majlan and myself), and 2015 (Leipzig, co-organized by late Birgit Daus and myself). I am deeply appreciative of the many hours of stimulating discourses and interactions in these conferences. I think they have played a key role in creating a thriving 'antimony community,' some of which members are contributing authors of this book.

Personally, I am also indebted to Nelson Belzile for introducing me to the charms of the element. I am most grateful to many colleagues who have reviewed individual chapters and have given useful advice. Credit for the creation of this volume is, of course, primarily due to its authors.

Contents

Gary Patterson

Chapter 1
A brief history of antimony

1.1 Introduction

Although antimony has been known for more than 10,000 years, its presence among the metals is more recent. The seven metals of antiquity included gold, silver, copper, iron, tin, lead, and mercury [1]. So where would we find antimony? The surface of the earth is filled with minerals. Silicates and carbonates are everywhere. And most elements can be found as oxides. But, there is a class of minerals that were important in antiquity: the sulfides. One of the most common is “fool’s gold” (FeS_2), iron pyrites (Fig. 1.1).

Fig. 1.1: Remarkable specimen of iron pyrites ($Fe(II)S_2$) in its cubic form (on marl, a calcium carbonate mineral) (Wikimedia Commons).

Gary Patterson, Carnegie Mellon University, Science History Institute, 5000 Forbes Avenue, Pittsburgh, PA 15213, U.S.A. gp9a@andrew.cmu.edu

https://doi.org/10.1515/9783110668711-001

Antimony occurs most frequently as the mineral stibnite (Sb_2S_3) (Fig. 1.2).

Fig. 1.2: Stibnite crystals in the Hillman Hall of Minerals, Pittsburgh, PA (by permission from the Carnegie Museum of Natural History).

But, one form in which stibnite was used in antiquity was as a mascara, a black eyelash pigment. In Egypt, it was called "kohl" (Fig. 1.3).

Fig. 1.3: Black stibnite powder (Sb_2S_3) (by permission of Reade International).

This substance is mentioned in connection with the infamous biblical character Jezebel, in II Kings 9:30. The Greek word for eye paint (in the Septuagint) is *εστιμμισατο τους οφθαλμους*. Stibnite powder was used throughout the Mediterranean world. It is mentioned by Pliny [2]. It was not the only black pigment used as a mascara (galena, PbS_2, lead sulfide was cheaper and more abundant) [3].

Ancient chemists were fond of roasting minerals. They heated marble ($CaCO_3$) and produced lime (CaO). They heated cinnabar (HgS) and produced liquid mercury (Hg). When they heated stibnite in air, the sulfur was replaced by oxygen to produce antimony(III) oxide, Sb_2O_3 (Fig. 1.4). It was used as a medicine, both orally, as a diaphoretic, and externally, as an astringent [4, 5].

Fig. 1.4: White antimony oxide (Sb_2O_3) (by permission of Industrial Metal Powders).

Although there is no manuscript evidence for the production of antimony metal in the Egyptian world, a remarkable glass vessel colored yellow by antimony exists in the Louvre in Paris (Fig. 1.5). It was discovered at Canosa, Italy, in 1895.

Fig. 1.5: Antimony-colored skyphos dated circa 200 BCE (Louvre Museum item MNC2200, by permission through Art Resource, NY, photographer, Herve Lewandowski).

While iron oxide (Fe_2O_3) can be reduced to the metal with charcoal, antimony oxide yields antimony carbonate ($Sb_2(CO_3)_3$) under these conditions. However, this substance is unstable when heated and produces liquid antimony metal, and carbon dioxide and oxygen gases.

$$Sb_2(CO_3)_3(s) \rightarrow 2\ Sb(l) + 3\ CO_2(g) + (3/2)\ O_2(g)$$

The ancient artisans would have had no clue about why this worked, but they used the process to produce cast objects. Pure antimony is very brittle, so most such objects would have broken. And there were better uses for antimony. But, the existence of the pure metal in antiquity is definitely documented.

Egypt developed a sophisticated artisanal chemical industry and produced materials remarkable for their colors and properties, even today. One of the most celebrated Egyptian chemists was named Zosimus, and a few of his reputed works exist today, as papyrus manuscripts. He practiced around 300 CE [6]. He referred to stibnite as "our lead" in one of his recipes.

Many manuscripts from the period from 800 CE to 1400 CE still exist. Antimony is mentioned in alchemical treatises, the most famous of which is *The Summa perfectionis of Pseudo-Geber* [7]. This thirteenth-century manuscript is a central document in the history of alchemy and has been investigated in detail by William R. Newman.

Another important manuscript from this era is the fourteenth-century work by John of Rupescissa (1300–1362), *De Consideratione Quintae Essentiae*. A Latin translation of a recipe for the quintessence of antimony by Hans Sloan (1660–1753) is given by McCallum [5]. The essential points are: (1) Take French stibnite and grind it fine, (2) Extract with rectified vinegar until all the red color is removed, and (3) Distill off the vinegar from the solution and collect the red residuum. It was used orally and was claimed to be very sweet. The sweetness is likely to be due to lead(II) acetate; this substance was used throughout human history to adulterate wine and to provide sweetness. French antimony ores often contained lead, such as Boulangerite ($Pb_3Sb_2S_6$). They also often contained arsenic in the form of Realgar (As_4S_4). This soft, red, molecular solid (formed of As_4S_4 molecules) can be dispersed in solution and is readily powdered. Perhaps it is the source of the red color.

1.2 Antimony and the metals industry

Antimony played an important role in the invention of printing in the fifteenth century. Type metal is an alloy containing metallic antimony, along with lead and tin. The early recipes called for a mixture of stub nails and antimony sulfide, and it was thought that they were producing a special form of iron [8]. But, this recipe produces metallic antimony. The product is then mixed with molten lead, and in some cases tin. The "hardened" lead is the secret to movable type printing. The most famous book on early printing technology is *Mechanick Exercises: Or, the Doctrine of Handy-Works Applied to the Art of Printing (1683, London)* by Joseph Moxon (1627–1691) [9]. Even at this later date, the notion of a "special iron hardener" as the secret was presented. Joseph Moxon was a Fellow of the Royal Society and was an intimate of people like Robert Hooke.

In the sixteenth century, extensive discussions of antimony occur in books like *De Re Metallica* [10]. The mining community in Germany was both very active and quite prosperous. While they extracted metals from many different ores, their usual goal was to separate the gold and silver from the other metals. Antimony played a part in the purification of gold. But, now the metal needed to be obtained in pure form. A new ingredient is added to the classical mix for extracting metals from their ores: saltpeter (KNO_3). The role of the saltpeter was to separate the sulfur from the stibnite as an oxidized and volatile substance. In addition, salt and vitriol were added to provide a liquefied "slag" to dissolve the unwanted salts and allow the antimony to sink to the bottom as a pure regulus. This pathway did not allow impurities like iron or lead to significantly contaminate the metallic antimony.

Another path to pure gold was to treat a gold–silver alloy (electrum) with pure stibnite. In the furnace, the silver was converted to the sulfide, and gold amalgamated with the antimony. The alloy was then roasted in a blast furnace to remove the antimony as the volatile oxide (the process is known as "cupellation") [11].

1.3 Antimony and Basil Valentine

The seventeenth century was characterized by the alchemical use of antimony. One of the key early books was *The Triumphal Chariot of Antimony* (1604) attributed to the pseudonymous Basil Valentine. The recipes contained in the book form the basis for all future work on antimony. The primary goal of the work was to prepare medicines that are beneficial without being poisonous. "When Antimony is properly prepared, it is as harmless as cassia or manna" [12]. Chemistry is all about being able to change matter into a useful form. The theory of matter adopted by Valentine was a form of the alchemical mercury, sulfur, and salt notion as constituents of all metals. In addition, each metal could be prepared in special forms called the Spirit of X, such as the Spirit of Antimony. Another name for these special forms was the Philosophical X, such as the Philosophical Mercury. Once substances had been separated into their philosophical components, the path to the Philosophers Stone was open. While this theory was not correct, it did result in a great deal of experimentation and many new substances were discovered.

One of the characteristics of antimony is the large number of uses to which it can be put. By the time of Basil Valentine, antimony had already been subjected to a systematic study and many of its reactions documented. One of the more interesting phenomena associated with antimony is the large number of different colors exhibited by antimony compounds. As a pure metal, it is a silvery color, and conducts electricity. Valentine calls antimony "one of the seven wonders of the world."

By the seventeenth century, many chemical processes were known and practiced. One class of processes became known as the "dry way," and included heat and subjection to chemical reactions with other solids, liquids, and gases. The other approach was called the "wet way," and included subjection to strong acids and organic solvents. Valentine used them all!

One of the phenomena known to the German miners was that antimony could be prepared as a colored glass (and the German word for antimony is "Spiesglas"). A nice discussion of antimony glasses is contained in the modern book *Secrets of Alchemy* by Larry Principe [13].

Valentine believed that the goal of the alchemist was first to separate the Philosophical mercury, sulfur, and salt that made up antimony. Only then could it become its quintessence, a true Stone (*Lapis*, called by Valentine the Fire Stone). He also compares this antimony substance to potable gold, an alchemical arcana.

The patient reader of this text is finally rewarded with very detailed recipes. Valentine preferred Hungarian Antimony as the starting material (stibnite, Sb_2S_3, is mined at many places in this country [14]). In order to effectively process it, pulverize it and place it in a calcinatory furnace over a coal fire. Watch the smoke and continue until there is no more given off. (The secret ingredient in the recipe is the stirring with an iron spoon. This produces globules of metallic antimony). These brittle globules are pulverized and subjected to calcination until only a white ash is left (antimony trioxide, Sb_2O_3). This ash is then heated in a crucible in a very hot furnace until it fully melts to a clear liquid. (True liquid antimony oxide is a very complex substance, comparable to other glass formers like B_2O_3. Valentine warns that unless great patience is shown, the liquid will merely recrystallize when cooled). The text recommends quenching the liquid on a flat copper dish and producing a yellowish glass. (Principe prefers a marble slab). Valentine also gives a recipe for a mixed glass containing borax.

Now Valentine employs the wet way. He pounds the antimony glass and places it in a Cucurbit (a broad-bottomed glass vessel), with glacial acetic acid (called highly rectified vinegar by Valentine). This is "digested" over a moderate heat for many days, with occasional daily shaking. The liquid abstracts red-colored elements of the glass powder, and is continued until no more color is given off. Valentine then discusses processing of the vinegar solution: drying to a powder, washing with pure rain water until the acidity is removed, and extracting with highly rectified spirit of wine (95% ethyl alcohol). The powder was also recommended for external use on wounds and bruises, and was called Balm of Antimony. Another form for antimony is an "Oil". (An oil is a viscous fluid). The supernatant solution prepared as above is distilled in a recirculating furnace for a month. Valentine calls the oil "the quintessence of antimony." "It purifies the body by means of perspiration, and by means of urinary and salivary excretions."

Valentine developed a series of complicated distillation apparati (Fig. 1.6).

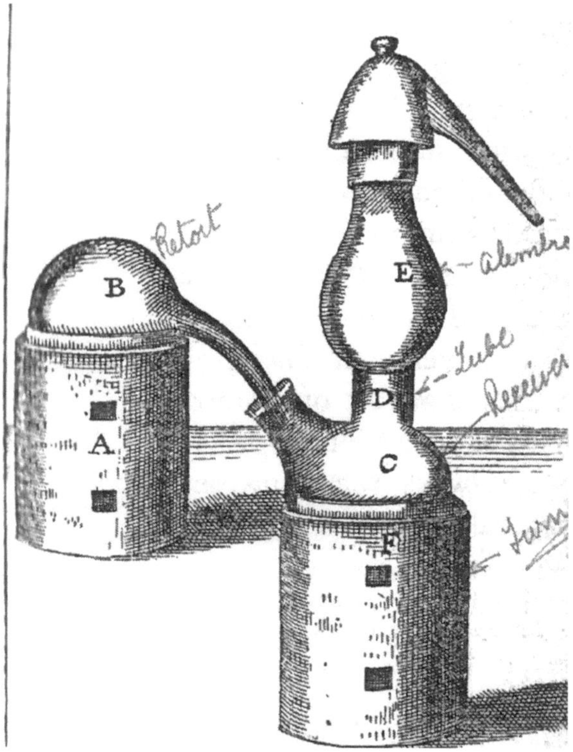

Fig. 1.6: Multiple distillation apparatus from *The Triumphal Chariot of Antimony* [12].

This equipment is used to "distill" stibnite and sal ammoniac (ammonium chloride, NH_4Cl). (The product of this chemical reaction is antimony trichloride ($SbCl_3$). It is usually called "butter of antimony" in the alchemical literature). The liquid is then treated with distilled water and antimony oxychloride (SbClO) precipitates out as a "feathery white" substance. (Usually called "flowers of antimony" in the alchemical literature). The antimony oxychloride is then placed in another type of distillation furnace called a "pelican," which sends the distillate back into the main vessel. Concentrated sulfuric acid is added and the system refluxed until the reaction is complete. At this point, ethyl alcohol (spirit of wine) is added and the process is repeated. The "sediment" is removed and the supernatant is distilled a final time. Then rose water is added to produce the "Arcana of Antimony."

Stibnite can also be reacted with concentrated nitric acid in a sealed cucurbit. The resulting solution is filtered to remove any adventitious impurities, placed in a glass vessel, and distilled to remove the remaining liquid phase. A yellow powder remains. Rain water is added and heated in a water bath; then the supernatant is distilled to dryness and a red powder remains. This substance is suspended in glacial acetic acid and heated for three days. The solution becomes blood red and deposits a red powder. This powder is suspended in spirit of wine, and finally the alcohol is distilled off leaving a "fixed red powder of wondrous efficacy."

Valentine describes another preparation that he calls "Flowers of Antimony," but it does not correspond with the antimony oxychloride feathery crystals discussed above. While it starts with pure white antimony oxide, the crystals are multiply sublimed and produce yellow and red deposits at the higher levels. (These correspond to glassy forms of antimony oxide). The red particles are mixed with colcothar of vitriol (iron(II) sulfate, $FeSO_4$) and sublimed. Once again, Valentine suggests "extracting" the powder with spirit of wine, distilling to dryness, and worshipping the product.

Stibnite can also be reacted with "crude tartar" (potassium carbonate, K_2CO_3) in a "wind furnace" to produce "Kermes mineral", a mixture of antimony oxides and sulfides. It can then be washed with water and treated with lye to become "liver of antimony" (because of its dark red color).

One of the most pyrotechnic pathways to produce the star regulus of antimony (pure metallic antimony) is to mix stibnite with crude tartar (K_2CO_3) and pure saltpeter (KNO_3). The ingredients are pounded together and melted until the reaction ceases. The liquid is then cast onto a copper dish and allowed to crystallize. The "slag" covers the antimony and protects the metal from oxidation and is removed at the end. This process is repeated three times to purify the antimony. A typical star regulus of antimony is shown in Fig. 1.7.

Fig. 1.7: Star regulus of antimony produced and photographed by The Chymistry of Isaac Newton Project (by permission of William R. Newman).

Many compounds of antimony can be prepared from pure antimony regulus. Heating pulverized pure antimony in a round glass vessel over a gentle sand bath leads to sublimed antimony oxide of great purity. But, rather than the white crystalline form, it is the red amorphous form. It is deliquescent and Valentine recommends placing it on a clean smooth stone in a humid cave for six months until a pure red liquid is obtained. Further time in the cave leads to crystallization. Recrystallization eventually produces the pure white oxide.

1.4 Antimony in the seventeenth century

One of the most prolific manufacturers of antimony compounds in the seventeenth century was Rudolph Glauber (1604–1670). He synthesized “butter of antimony” ($SbCl_3$) both from antimony regulus (Sb) plus spirit of salt (HCl) and from stibnite plus corrosive sublimate ($HgCl_2$). It is a white oil.

Two of the most famous alchemists of the seventeenth century were George Starkey (1628–1665) and Robert Boyle (1627–1691) [15]. Starkey was educated at Harvard University in New England (BA 1647, MA 1649), but emigrated to England in 1650 to escape persecution in America. He became an alchemist by interning with an iatrochemical physician in Boston during his undergraduate days (William Palmer) and interacting with the three most famous American alchemists of the time: John Winthrop, Jr., Richard Leader, and Robert Child. He shared some of his knowledge with Robert Boyle in a letter of 1651.

Although Isaac Newton (1642–1727) is more famous for inventing the calculus and classical mechanics, he spent more time studying antimony! An extensive look at this activity has recently become available: *Newton the Alchemist: Science, Enigma and the Quest for Nature’s “Secret Fire”* by William R. Newman [16]. Newton’s goal in his work with antimony was to produce the Philosopher’s Stone. He adopted a stance that viewed stibnite as the “mother of all metals” because it contained both a “mercury” and a “sulfur.” In his attempt to separate stibnite into its philosophical constituents, he often used the “wet way” and dissolved the stibnite in *aqua regia* (a strong mixture of nitric and hydrochloric acid that can dissolve gold itself). After more than 50 years of effort, he failed in his search. But, he became the most accomplished experimental chemist in England and served as the Master of the Mint.

1.5 Antimony and Nicholas Lemery

One of the most celebrated chemists in France at the turn of the eighteenth century was Nicolas Lemery (1645–1715). He published the seminal book on antimony in this era: *Traite de l’antimoine: contenant l’analyse chimique de ce mineral, et un recueil d'un grand nombre d'operations rapportees a l'Academie Royale des Sciences, avec les raisonnemens qu'on a crus necessaires. Ouvrage utile aux physiciens et a tous ceux qui praqtiquent la medicine* (1707) [17] (Fig. 1.8).

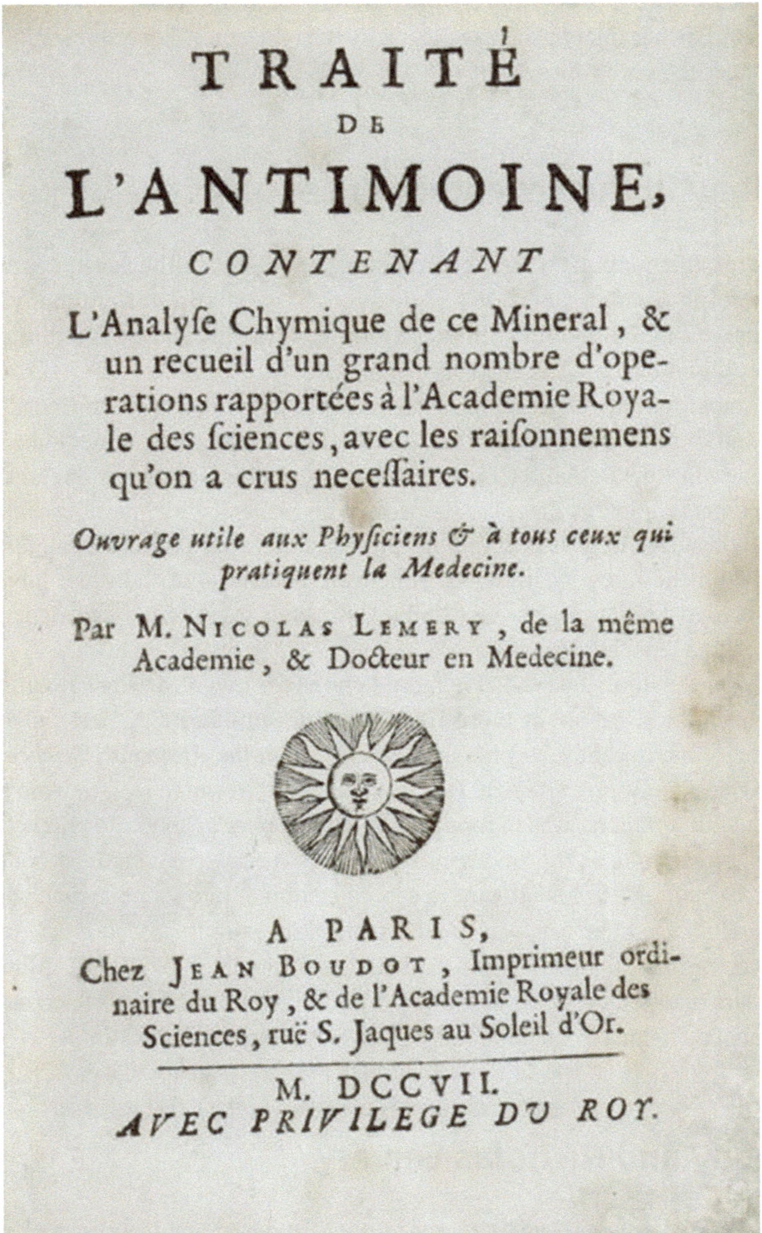

TRAITÉ
DE
L'ANTIMOINE,
CONTENANT
L'Analyſe Chymique de ce Mineral, & un recueil d'un grand nombre d'operations rapportées à l'Academie Royale des ſciences, avec les raiſonnemens qu'on a crus neceſſaires.

Ouvrage utile aux Phyſiciens & à tous ceux qui pratiquent la Medecine.

Par M. NICOLAS LEMERY, de la même Academie, & Docteur en Medecine.

A PARIS,
Chez JEAN BOUDOT, Imprimeur ordinaire du Roy, & de l'Academie Royale des Sciences, ruë S. Jaques au Soleil d'Or.

M. DCCVII.
AVEC PRIVILEGE DU ROY.

Fig. 1.8: *Traite de l'antimoine* by Nicolas Lemery (Science History Institute, by permission).

Most of the preparations given by Lemery are merely clear recipes previously presented by Valentine. Lemery is best known as a Professor in the Academie of Sciences that lectured on the Chemistry of Medicine. He popularized the "cup of antimony" (white wine allowed to stand in a cup of pure antimony for a few days) as an emetic in France (Fig. 1.9). The figure also shows a piece of "Kermes mineral," a mixture of antimony oxides and sulfides obtained by reacting stibnite with salt of tartar (K_2CO_3). It was powdered and used as an iatrochemical medicine.

Fig. 1.9: Antimony cup and Kermes mineral described in Lemery.
https://mariekehendriksen.nl/2013/05/09/alchemy-on-the-amstel-a-visit-to-the-ritman-library/

1.6 Antimony and Herman Boerhaave

The final figure from the alchemical period that published extensively on antimony was Herman Boerhaave (1668–1738). His magnum opus, *Elementa Chemiae*, was published in 1732 and was used either in its original Latin or translated into many languages for another 100 years [18]. Boerhaave was Professor of Medicine, Botany and Chemistry at the University of Leyden. His lectures were attended by aspiring physicians from all over the Latinate world. While he was still thinking in terms of alchemical notions such as the mercury-sulfur theory of metals, he actually carried out all the recipes given in his textbook, and many more. He was the best experimental chemist of his era. He discusses 20 different preparations of antimony and comments on both their chemistry and their use as pharmaceuticals.

Boerhaave employed both the furnace and strong acids in his recipes. He described in detail what was seen during the reactions. He first discusses one of Isaac Newton's favorite reactions: the dissolution of "crude antimony" (stibnite) in *aqua regia*. He warns about the danger of the "red fumes." Boerhaave knew how poisonous antimony compounds could be, and warned his students to be careful during their preparations of their own pharmaceuticals.

One of the simplest recipes consists of merely powdering the stibnite, placing it in a crucible, heating it to fusion for an hour (until the mass stops fuming), letting the mass cool, and separating metallic antimony from the mass. In order to produce a more refined pure antimony, another approach was used: deflagration with tartar (K_2CO_3) and niter (KNO_3). This demonstration always drew applause, since it is highly pyrotechnic. Multiple deflagrations yield the famous star regulus of antimony shown above.

One of the oldest recipes used in the mines employed iron "horse shoe nails" with ground stibnite. They were heated to fusion and then deflagrated with saltpeter (KNO_3). The liquid was poured into melting cones and sold all over the world. This initial product could be refined as above.

One of the most famous antimony medicines was diaphoretic antimony. Ground stibnite was mixed with three parts of niter and deflagrated. The product was collected in a large ladle and heated to fusion for 15 minutes. The resulting substance was a white powder that produced sweating. (The extra saltpeter oxidized the antimony all the way to the oxide, Sb_2O_3). Boerhaave recognized that this substance was a "calx" of antimony. He knew many oxides of metals, including mercury oxide. But, the technology for dealing with the gases given off by heating these compounds had not yet been mastered in the early eighteenth century.

One of the most dangerous antimony compounds of this era was the "butter of antimony" ($SbCl_3$). It was synthesized from crude stibnite and Venetian corrosive sublimate ($HgCl_2$). Even the powder of these materials was deadly. The solids were placed in a large retort and well-luted to a large receiver. The gentle heating in a sand-bath was enough to bring over the white, oily liquid antimony chloride. This substance was used in the eighteenth century to remove external growths, like warts. Boerhaave discusses the general class of metal chlorides and notes that they are often volatile and extremely dangerous.

The theory of alchemy asserted that metals were a compound of mercury and sulfur. One goal of alchemy was to separate these constituents from the crude metal. Boerhaave showed that the purer he made his antimony regulus, the more brittle it became, contrary to the claim that it was the residual sulfur that made metals brittle. This disproved one of the most revered notions of alchemy. He also distilled a sample of common mercury more than 500 times. No amount of purification and heating removed its residual sulfur. It remained common mercury. All claims of the preparation of "sophic" mercury were shown to be mistaken. The "sulfur of antimony" was nothing but the common sulfur contained in the stibnite. It was time for chemistry to move on before an improved understanding of antimony would be obtained. But many compounds of antimony, including colored amorphous antimony oxides, were described accurately.

1.7 Antimony in the nineteenth century

The state of chemistry in the early nineteenth century is well exhibited in the classic book *The Elements of Experimental Chemistry* (1817) by William Henry (1774–1836) [19]. Much had happened since the death of Boerhaave. Perhaps the most important development was the atomic theory of chemistry constructed by John Dalton (1766–1844). Now each substance could be discussed in terms of its atomic composition. Another important development for chemistry was the invention of electrochemistry. Humphry Davy (1778–1829) and Michael Faraday (1791–1867) used this experimental technique to synthesize many substances. Along with the purely empirical parts of this enterprise, chemists like Jacob Berzelius (1779–1848) developed the concept of oxidation state. Atoms could exist in a multitude of possible states. Antimony has an especially rich collection of oxidation states.

In 1817, many errors still haunted experimental chemistry, and even the measured atomic composition of stibnite (Sb_2S_3) was imprecise. Roasting in air produces oxides of antimony, not just one form. This polyvalence continued to confuse chemists throughout the nineteenth century. In addition, antimony oxide can be either crystalline or amorphous. Berzelius identified four distinct oxides, but modern analytical chemistry gives other compositions, from Sb_2O_3 to Sb_2O_5. Henry was aware that antimony oxide was a complex material.

Henry was also aware of the highly complex nature of roasted stibnite. In addition to the true antimony oxide, there are silicate impurities in Hungarian antimony. The fused mass can be quenched to a glass. This glass is highly heterogeneous and consists of mixtures of antimony sulfide and oxide. Henry notes some of the components as equivalent to crocus metallorum and liver of antimony, depending on the ratio of stibnite to antimony oxide.

Another highly complex material is created by reacting stibnite with potash (K_2CO_3). The result is a “hydrosulfureted oxide” of antimony, a mixture of oxides and sulfides called “Kermes mineral” (Fig. 1.9).

The classic antimony oxide formed by deflagration with saltpeter produces a substance that depends on the ratio of stibnite to KNO_3. High enough ratios can even make Sb_2O_5.

By 1817, Davy had perfected the synthesis of gaseous chlorine, Cl_2, and his brother, John Davy (1790–1868), synthesized antimony trichloride directly from the powdered regulus and pure chlorine gas.

William Henry was aware of most of the current research in chemistry in 1817, and, while he had a much deeper understanding of the nature of antimony and its compounds, not a great deal of progress had been made. By the later part of the nineteenth century, a good marker for the study of antimony is provided by the textbook written by Josiah Parsons Cooke (1827–1894), the Erving Professor of Chemistry at Harvard University in Massachusetts: *First Principles of Chemical Philosophy* (1870) [20]. Cooke was one of the developers of the periodic approach to understanding the chemistry of the elements.

Cooke classed antimony with nitrogen, phosphorus, arsenic, and bismuth. One of the immediate rewards for viewing antimony this way was the understanding of stibine (SbH_3). It can be synthesized from antimony trichloride and hydrogen gas:

$$SbCl_3 + 3\,H_2 \rightarrow SbH_3 + 3\,HCl$$

There is a rich collection of stibine-based compounds. Stibine is a classic polyatomic molecule and is a gas at room temperature. Cooke discusses many reactions of the pure metallic antimony in atomic terms and explains the products. Antimony dissolves in *aqua regia* and produces nitrogen dioxide. Pure antimony can also dissolve in hot sulfuric acid to produce sulfur dioxide and antimony trisulfate. One of Cooke's favorite solvents for antimony metal was a mixture of nitric and tartaric acids.

Antimony trichloride can be reacted with chlorine gas to produce antimony pentachloride ($SbCl_5$). The assigned oxidation state for the antimony is then 5+. When this molecule is reacted with pure water vapor, an acid, comparable to phosphoric acid, is produced (H_3SbO_4). But, another antimonic acid is also observed: ($H_4Sb_2O_7$). The multivalent antimony also yields the compound $HSbO_3$. When this substance, metantimonic acid, is gently heated, it eliminates water to become antimonic anhydride, Sb_2O_5. This joins with the other pure antimony oxides: Sb_2O_3 and Sb_2O_4. The tetraoxide is the basis of the mineral Cervantite.

Antimony trichloride can also be reacted with water to produce antimony oxychloride (SbOCl). This is one of the common alchemical substances, but it is not stable when the acid conditions are neutralized. It was called mercurius vitae by the alchemists, and they had no idea what it was.

One of the most famous iatrochemical medicines was tartar emetic. It was formed by dissolving antimony(III) oxide (Sb_2O_3) in a solution of cream of tartar ($KC_4H_5O_6$). Cream of tartar was readily available as a common precipitate on the inside of wine casks. (Modern chemists use it as a good buffer system, pH = 3.557). Cooke clearly identified the crystalline tartar emetic as a salt of potassium, SbO^+, tartrate ($C_4H_2O_6^{2-}$), and one water of crystallization. After heating to 200 °C, the dehydrated product is $KSbC_4H_2O_6$. These kinds of salts are ubiquitous for metals that form compounds like M_2O_3.

Antimony forms both alloys and true metallic compounds with many other elements. Cooke cites the rich phase diagram with zinc. Another group of alloys includes "type metal" and Brittania metal, a common base for silver plated table ware. Antimony will alloy with gold and mercury. Metallic antimony remains a valuable metal in Cooke's world.

The richest chemistry associated with antimony remains its compounds with sulfur. While antimony glance, the common ore of the mines, Sb_2S_3, remains readily available, there are dozens of sulfur minerals with antimony in them. The other constituents include silver, copper, lead, arsenic, and iron.

When hydrogen sulfide, H_2S, is reacted with a solution of antimony pentachloride, an orange-colored precipitate is formed, Sb_2S_5. This compound is the gateway to many sulfoantimoniates, like K_3SbS_4.

1.8 Antimony in the twentieth century

Rather than fading into the alchemical past, antimony proved its value in the modern period of chemistry as well. A major book devoted to antimony appeared in 1919: *Antimony: It's History, Chemistry, Mineralogy, Geology, Metallurgy, Uses, Preparations, Analysis, Production and Valuation; With Complete Bibliographies* by Chung Yu Wang [14].

One of the best stories associated with antimony is the uncertainty of its atomic weight. Wang chose 120 as the best atomic weight, but the current accepted value is 121.76 g mol^{-1}. A brilliant study by Jean-Basptiste Dumas in 1859, based on antimony trichloride, yielded 121.88 g mol^{-1} with silver equal 107.92 g mol^{-1} as a reference. (This would yield 121.82 g mol^{-1} in today's standard). Josiah Parsons Cooke also confirmed values in this range in 1877. But Wang lists more than 30 published values in the late nineteenth century from 119.7 to 122.36 g mol^{-1}. Working with antimony is not easy, and the existence of many different compounds of the same two elements complicates the analysis. Cooke devoted his life to the accurate measurement of atomic weights, and his best student at Harvard, T.W. Richards (1868–1928), was the first American to receive the Nobel Prize in Chemistry for his work on atomic weights.

By 1919 the phase diagram for pure antimony metal was known and other allotropic forms had been studied. Yellow antimony is not a metal and is metastable, like white phosphorus. Gaseous antimony (obtained under vacuum above 300 °C) is green. Black antimony is produced by rapid cooling of the gas and is amorphous.

The existence of stibine (antimony hydride, SbH_3) leads to a whole chemistry of antimony molecules such as trimethyl stibine. It can even be prepared as an alkyl salt, tetramethylstibonium chloride.

Metallic antimony burns in pure fluorine to produce antimony trifluoride, SbF_3, a white crystalline solid. Many mixed solids also form with alkali metals. Antimony chloride also forms mixed solids of this type. Antimony chloride is reactive towards formamide and produces the compound $Sb(NH_2CO)_3$. Similar chemistry is observed for antimony tribromide and tri-iodide. Antimony can also form the pentafluoride, and can react with hydrofluoric acid to form meta-fluoro-antimonic acid: $HSbF_6$.

Antimony sulfide can also be prepared in several amorphous forms with different colors: red, black, and violet. In the liquid state (above 555 °C), pure metallic antimony is soluble in the sulfide, as are many other metallic sulfides. (Only the brave work in this world!) Antimony selenides and tellurides also form.

And antimony phosphide, SbP, a white metallic substance, is now known.

The chemistry of antimony is rich and varied. While new insights into the structure and bonding of antimony compounds needed to wait for the invention of quantum chemistry, the classical chemistry of antimony was complete by 1919.

References

[1] Parr JG. Man, Metals and Modern Magic. Cleveland, American Society for Metals, 1958.

[2] Thomson T. The History of Chemistry. London, Henry Colburn and Richard Bentley, 1835.

[3] Mahmood ZA, Zoha SM, Usmanghani K, Hasan MM, Ali O, Jahan S, Saeed A, Zaihd R, Zubair M. Kohl(Surma): Retrospect and prospect. Pak J Pharm Sci 2009, 22, 107–122.

[4] Partington JR. A History of Chemistry. London, Macmillan & Co., 1970.

[5] McCallum RI. Antimony in Medical History. Edinburgh, The Pentland Press, 1999.

[6] Garland H, Bannister CO. Ancient Egyptian Metallurgy. London, Charles Griffin & Co., 1927.

[7] Newman WR. The "Summa Perfectionis" of Pseudo-Geber: A Critical Edition, Translation and Study. Leiden, E.J. Brill, 1991.

[8] Biringuccio V De La Pirotechnia Venice, 1540.

[9] Moxon J. Mechanick Exercises: Or, the Doctrine of Handy-works. London, Applied to the Art of Printing, 1683.

[10] Agricola G. De Re Metallica. Basil, 1556.

[11] Macquer PJ. Dictionnaire de Chymie. Paris, 1769.

[12] Valentine B. The Triumphal Chariot of Antimony (Currus Triumphalis Antimonii). Erfurt, 1604.

[13] Principe LM. The Secrets of Alchemy. Chicago, University of Chicago Press, 2013.

[14] Wang CY. Antimony: Its History, Chemistry, Mineralogy, Geology, Metallurgy, Uses, Preparations, Analysis, Production, and Valuation; With Complete Bibliographies. London, Charles Griffin and Co., 1919.

[15] Newman WR, Principe LM. Alchemy Tried in the Fire: Starkey, Boyle and the Fate of Helmontian Chymistry. Chicago, University of Chicago Press, 2002.

[16] Newman WR. Newton the Alchemist: Science, Enigma, and the Quest for Nature's Secret Fire. Princeton, Princeton University Press, 2019.

[17] Lemery N. Traite de L'Antimoine. Paris, Chez Jean Boudot, 1707.

[18] Boerhaave H. Elementa Chemiae. Leyden, 1732.

[19] Henry W. The Elements of Experimental Chemistry. Philadelphia, 1817.

[20] Cooke JP. First Principles of Chemical Philosophy. Boston, 1870.

Juraj Majzlan

Chapter 2
Primary and secondary minerals of antimony

2.1 Introduction

Element number 51, antimony, has an average crustal abundance of 0.2 ppm [1], which is roughly comparable to that of silver or cadmium. After comparing its abundance with its mineralogical diversity, Christy [2] found that antimony is much more diverse than expected from its average crustal concentration (Fig. 2.1). He generalized this observation for a number of elements that adopt a particular electronic configuration, multiple oxidation states, and specific coordinations. All of these features are typical for antimony, and are responsible for its rich and variable mineralogy. Antimony can be present in low oxidation states, such as those in the mineral gudmundite

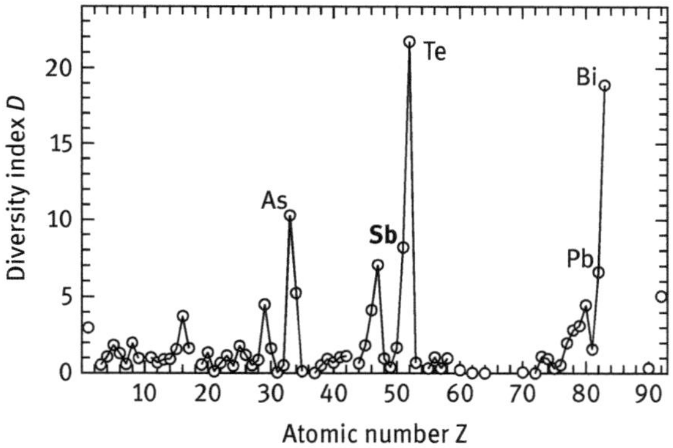

Fig. 2.1: Mineralogical diversity of a number of chemical elements. Mineralogical diversity is calculated as a function of the number of minerals of a particular element relative to its crustal abundance. Lines connect the elements located next to each other in the periodic table. Data from [2].

Acknowledgments: I appreciate greatly the many extraordinary mineral images acquired by Pavel Škácha, Martin Števko, Olga Kaspera (Crystal Classics, with consent of Ian Bruce), and Uwe Haubenreißer used in this publication. I thank Martin Chovan, Tomáš Klimko, Anežka Borčinová Radková, Robert Hovorič, and Jaroslav Pršek for their willingness to provide original images, and their permission to use them here. The sketch in Fig. 2.11 was masterfully drawn by Sarah Bucka. I am thankful for the continuous support and funding from the *Deutsche Forschungsgemeinschaft* and the ANKA synchrotron in Karlsruhe that allowed much of our work on antimony to be done.

Juraj Majzlan, Institute of Geosciences, Friedrich-Schiller University, Burgweg 11, 07749 Jena, Germany, juraj.majzlan@uni-jena.de

https://doi.org/10.1515/9783110668711-002

(FeSbS), or as native antimony (Sb), or even in rare intermetallic compounds (e.g., dyscrasite, Ag_3Sb). Much more common are the minerals of Sb^{3+} (e.g., stibnite, Sb_2S_3, or valentinite, Sb_2O_3), or the minerals of Sb^{5+} (e.g., tripuhyite, $FeSbO_4$).

This rich and variable mineralogy provides many topics for mineralogical and geochemical studies of ore deposits and occurrences. At the same time, it presents a problem for mineral processing as well as for the reduction of environmental damage that can be caused by antimony-bearing waste.

In the 1980s, the amounts of antimony produced in China and the Soviet Union [3] were still comparable, but at this time China predominates, producing an estimated 90% of the global production [4]. Those types of deposits that deliver the most antimony, through their mineralogy, dictate which are the most commercially important antimony minerals. The same minerals, once exposed to the atmosphere, are the key sources of potential environmental problems. Therefore, those minerals will also be given emphasis in this work, even though some of these less common minerals will need to be discussed.

The goal of this work was to describe the main features of the mineralogy of the element antimony. A detailed account would require a deeper crystallographic discussion of the complicated group of sulfosalts that is beyond the scope of this chapter (see [5]). This text is divided into parts that address: i) the carriers of antimony in rock-forming minerals, ii) the primary minerals in ore deposits, iii) the transition from low-temperature hydrothermal overprint to initial weathering stages, and iv) the secondary minerals of antimony. In each part, the main minerals are identified, and for easy orientation their names and formulae are summarized in the tables or figures that accompany each section.

2.2 Antimony in rocks

There is considerable variation in the concentration of antimony among common rock types, but these values rarely exceed 1 ppm [6]. These are the average concentrations, which do not answer the question of antimony reservoirs that may be within these rocks.

The earliest studies on the speciation of antimony within rocks [7] were impaired by their sample preparation, namely, from analyzing bulk separates from rocks. The reservoirs of antimony in common rock types could be determined by extensive analytical work of the minerals present in rocks from different sites around the world. However, such an analytical database by and large is not available. An alternative approach, that of the determination of partition coefficients, is more viable. The partition coefficient D_{Sb} is defined as:

$$D_{Sb}(\text{reservoir A}/\text{reservoir B}) = \text{wt.\% Sb in reservoir A}/\text{wt.\% Sb in reservoir B}.$$

The partition coefficients are functions of the chemical compositions of the reservoirs, including oxygen and sulfur fugacity. Additionally, they depend on temperature and pressure. In the following text, these variables will largely be neglected. Our purpose was to discuss the broad features of antimony's behavior in common crustal rocks, and not the specific variations in the values of D_{Sb} either with the physical or chemical properties of the reservoirs. Partition coefficients between certain melts, rock-forming silicates, and oxides are graphically summarized in Fig. 2.2.

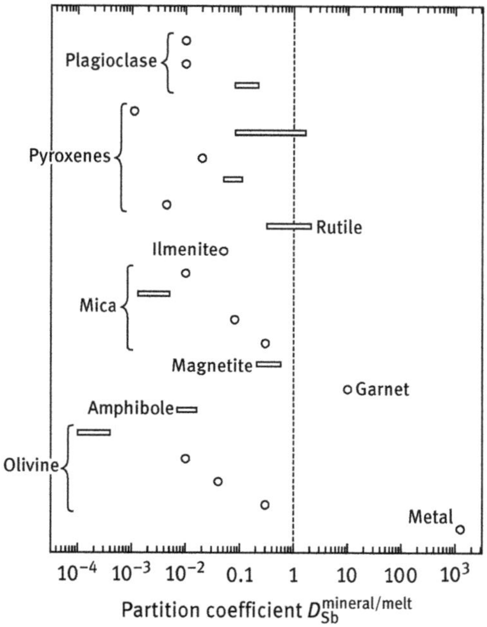

Fig. 2.2: Partition coefficients for antimony between pairs of minerals and melts. Data from [6].

Two types of systems were investigated in some detail with respect to the geochemical behaviors of antimony. The first one consists of mafic and ultramafic rocks that become saturated with respect to a metallic or sulfide liquid at high temperatures. Antimony partitions strongly into metallic, iron-rich melts (D_{Sb}(metal/basalt) = 1,250 [8]). The partition coefficients between the sulfide melt and silicate melt (D_{Sb}(sulfide melt/silicate melt)) are lower, and average ≈4 [9]. This means that antimony preferentially enters the sulfide melt. Upon decreased temperature, this sulfide melt partially solidifies to a phase called the monosulfide solid solution (*mss*, e.g., [10]), a Fe-S phase that rapidly decomposes when the temperature drops further. The partition coefficients show that the *mss* does not take up antimony, neither from the sulfide melt nor from the silicate melt [9, 11]. Antimony is predicted to partition slightly more (more than into the *mss*) in another high-temperature sulfide phase – the intermediate solid solution (*iss*). The known partition coefficients suggest that antimony remains in the sulfide liquid up to a high degree of crystallization. The models predict that accessory Pd-Pt-Sb minerals (e.g., mertieite) could form, if at all, in "the last dregs of crystallization" [11]. These predictions agree well with the observations from natural systems (see below).

Another set of measurements was performed on the partitioning of antimony between both silicate melts and silicate/oxide minerals in the absence of sulfide or metallic melts. The available partition coefficients (Fig. 2.2) indicate that antimony prefers the remaining melt over most rock-forming minerals. For example, micas, olivine, or pyroxenes consistently contain between 10^2 and 10^4 times less antimony than does the melt; however, there are exceptions. Pyroxenes rich in aluminum have partition coefficients (D_{Sb}) slightly higher than 1 [12]. The oxidation state of antimony in these experiments was not controlled, and it appears to be mixed (3+/5+). A phase with a greater affinity for antimony is rutile, with a D_{Sb} up to 2.1 [13]. The highest partition coefficients, up to 10, were measured for garnets [14].

All of these results demonstrate that antimony accumulates in accessory minerals such as rutile or garnets of a certain composition. If such minerals are not present, or they are scarce, antimony remains in the melt or magmatic fluid, and can escape from the solidifying rocks.

There are a number of studies that report partition coefficients among metamorphic minerals; however in these studies, antimony was consistently not analyzed. Nonetheless, it should be assumed that rutile or garnets also accumulate antimony in such rocks. Antimony as Sb^{5+} is considered to be "extremely compatible" with the structure of rutile [15]. Accessory amounts of rutile in eclogitic rocks can carry more than 90% of the total antimony in these rocks [16]. These authors further assume that rutile controls the antimony budget not only in subducted altered oceanic crust, but also in subducted pelagic sediments, even though the antimony concentration in rutile may not exceed 100 ppm. Rutile is considered to be a pathfinder mineral for certain types of ore deposits owing to its elevated antimony concentrations [17]. Occasionally, much higher antimony concentrations in rutile have been reported, for example, in one case up to 2.25 wt.% Sb_2O_5 [18].

The role of garnets was not investigated in such great detail. They are able to take up significant amounts of antimony, to the point where antimony is one of the major and essential elements. Such garnet end-members are dzhuluite ($Ca_3SbSnFe^{3+}_3O_{12}$), bitikleite ($Ca_3Sb^{5+}Sn^{4+}Al_3O_{12}$), and usturite ($Ca_3Sb^{5+}ZrFe^{3+}_3O_{12}$) (e.g., [19]). These are all very rare as minerals, but they document the capabilities of certain garnets to uptake and retain antimony. Interestingly, the high partition coefficients (garnet/hydrous basanitic melt, $D_{Sb} = 10$) were all reported for magnesian, but not calcian, garnets [14]. On the other hand, the magnesian garnets in eclogite-facies metamorphic rocks are not considered as reservoirs of antimony [20]. More work would be needed to discern the role of garnets and their variable chemistry in the retention and release of antimony.

The last mineral that needs to be mentioned as a potential reservoir of antimony in sedimentary and metamorphic rocks is pyrite. Extensive analytical data sets indicate that sedimentary pyrite contains 10–100 ppm (median values) of antimony [21, 22]. Field studies suggest that antimony is liberated during pyrite-to-pyrrhotite transformation under high-grade metamorphic conditions [23]. This antimony is collected by metamorphic fluids, and can be accumulated in ore deposits under favorable conditions.

2.3 Primary minerals

The primary minerals of antimony occur in many types of ore deposits, but most of these minerals are only of mineralogical interest. Practically all of production is based on stibnite ores [3, 4]; but other minerals such as jamesonite or tetrahedrite may occasionally also be used. Pyrite, arsenopyrite, and other arsenic sulfides must be suppressed because they are not acceptable in the product (see [3]). In complex ores, antimony is commonly an unwanted or a penalty element (e.g., [24]). It can either be recovered by special processes or be suppressed [25]. Suppression of antimony and its transfer into the waste stream can improve the quality of the product (e.g., copper); however, this brings another liability, namely, the environmental hazards of antimony in tailings (see below).

In the following sections, antimony minerals in selected types of ore deposit are summarized. Examples of macroscopic specimens of selected primary minerals of antimony are depicted in Fig. 2.3. This text is not intended to provide an exhaustive overview of all ore deposit types and associated antimony minerals. Instead, it focuses on those deposits that deliver commercial antimony, or are unusually rich in terms of antimony minerals. The information on the most important antimony minerals in these specific deposit types is summarized in Tab. 2.1.

2.3.1 Antimony minerals in orogenic, Carlin-type, and amagmatic deposits

Deposits of these types were formed by hydrothermal fluids whose origin is not well defined and understood. The most likely source of the fluids is metamorphic devolatilization; the link to magmatism is often unclear. Because of these unifying features, these deposit types are treated together in textbooks on ore deposits (e.g., [26]), and they are also grouped together here. They may be related by the concepts of crustal continuum [27] or tectonic dispersion [28], which are also instrumental in the explanation of some of their mineralogical features. Mao et al. [29] expressed the opinion that among the large Chinese antimony deposits, there are systems of deeper orogenic and shallower Carlin-type deposits, generated by the same fluid, but which deposited its contents at different crustal levels.

Fig. 2.3: Photographs of selected primary minerals of antimony. Scale bars in the images represent 1 mm unless specified otherwise. Photographs g and h reproduced with permission from Crystal Classics (crystalclassics.co.uk), all other photographs courtesy of Pavel Škácha. (a) chalcostibite, Dúbrava, Slovakia; (b) antimonselite, Příbram, Czech Republic (CZ); (c) schafarzikite, Pernek, Slovakia; (d) polybasite, Jáchymov, CZ; (e) pyrargyrite, Ratibořské Hory, CZ; (f) stephanite, Příbram, CZ; g) stibnite with gold and quartz, Kremnica, Slovakia; (h) stibnite with realgar, Baia Sprie, Romania; (i) chovanite, Livello S. Olga, Miniera M. Arsiccio, Stazzema, Italy.

Tab. 2.1: Summary of primary antimony minerals of different types of ore deposits. More minerals, details about their occurrences, and their formulae can be found in the text and tables of the respective sections in this chapter.

Deposit category – brief notes	
Type	**Minerals of antimony**
Orogenic, Carlin-type and amagmatic deposits – global producers of most antimony. Distinction between these types is sometimes difficult.	
orogenic Sb-Au deposits	dominant: stibnite common to subordinate: Pb-Sb sulfosalts (zinkenite, boulangerite, etc.), berthierite, jamesonite subordinate: tetrahedrite, chalcostibite
Carlin-type deposits	dominant: stibnite rare, locally abundant: getchellite only in sites with Hg enrichment: livingstonite only in sites with Tl enrichment: weissbergite and other Tl-Sb sulfosalts
amagmatic deposits (black-shale hosted)	dominant: stibnite abundant: berthierite subordinate: antimony, gudmundite
Epithermal deposits – rare commercial sources of antimony.	
high-sulfidation deposits	dominant: enargite-stibioenargite and luzonite-famatinite solid solutions common: tetrahedrite-tennantite, in transition to intermediate-sulfidation deposits
epithermal Sn-Ag-Sb deposits	common: stibnite rare: franckeite, cylindrite, ramdohrite rare: Ag-Sb (see below) and Pb-Sb sulfosalts
low- and intermediate-sulfidation deposits	common, usually in small quantities: Ag-Sb sulfosalts (pyrargyrite, polybasite, argentotetrahedrite, miargyrite,diaphorite) rare: stibnite
Orthomagmatic deposits – antimony minerals are accessory. These deposits are mined mostly for Ni, Cu, and platinum-group elements.	
orthomagmatic deposits	rare: mertieite (as a primary magmatic mineral) rare: sudburyite, naldretteite, geversite (products of hydrothermal remobilization)

An important point is that these deposits deliver the vast majority of antimony to the world market. Hence, their mineralogy is also the mineralogy of the most commercially important antimony minerals. Minerals that occur in these deposits and are mentioned in this section are summarized in Tab. 2.2.

Tab. 2.2: Minerals mentioned in the section on Primary minerals in orogenic, Carlin-type, amagmatic, and epithermal deposits. Nominally antimony-free minerals are included in the lower panel of the table.

Minerals of antimony			
andorite	$AgPbSb_3S_6$	livingstonite	$HgSb_4S_8$
antimonselite	Sb_2Se_3	miargyrite	$AgSbS_2$
argentotetrahedrite	$Ag_6Cu_4(Fe,Zn)_2Sb_4S_{13}$	plagionite	$Pb_5Sb_8S_{17}$
berthierite	$FeSb_2S_4$	polybasite	$Ag_{13}Cu_3Sb_2S_{11}$
boulangerite	$Pb_5Sb_4S_{11}$	pyrargyrite	Ag_3SbS_3
bournonite	$CuPbSbS_3$	ramdohrite	$Ag_3Pb_6Sb_{11}S_{24}$
chalcostibite	$CuSbS_2$	robinsonite	$Pb_4Sb_6S_{13}$
cylindrite	$Pb_3Sn_4FeSb_2S_{14}$	schafarzikite	$FeSb_2O_6$
diaphorite	$Ag_3Pb_2Sb_3S_8$	semseyite	$Pb_9Sb_8S_{21}$
dyscrasite	Ag_3Sb	stephanite	Ag_5SbS_4
famatinite	Cu_3SbS_4	stibioenargite	Cu_3SbS_4
franckeite	$Fe(Pb,Sn)_6Sn_2Sb_2S_{14}$	stibnite	Sb_2S_3
getchellite	$AsSbS_3$	tetrahedrite	$Cu_{10}(Fe,Zn)_2Sb_4S_{13}$
gudmundite	$FeSbS$	valentinite	Sb_2O_3
heteromorphite	$Pb_7Sb_8S_{19}$	vrbaite	$Hg_3Tl_4As_8Sb_2S_{20}$
jamesonite	$Pb_4FeSb_6S_{14}$	weissbergite	$TlSbS_2$
kermesite	Sb_2S_2O	zinkenite	$Pb_9Sb_{22}S_{42}$
Nominally antimony-free minerals			
arsenopyrite	$FeAsS$	pyrrhotite	$Fe_{1-x}S$
enargite	Cu_3AsS_4	sphalerite	ZnS
luzonite	Cu_3AsS_4	tennantite	$Cu_{10}(Fe,Zn)_2As_4S_{13}$
pyrite	FeS_2		

The main commodity of the orogenic deposits is gold, but the Sb-Au orogenic deposits represent an important subtype of this category. The principal antimony mineral is always stibnite (Figs. 2.4 and 2.5(a), e.g., [30–33]). It is commonly accompanied by a number of Pb-Fe-Sb phases, such as berthierite and jamesonite, and a wide range of Pb-Sb sulfosalts [34–37]. The most abundant Pb-Sb sulfosalt is zinkenite; although other minerals (such as boulangerite, robinsonite, plagionite, semseyite,

Fig. 2.4: A vein of massive stibnite from an orogenic Sb-Au deposit. The beige rim of the vein consists of quartz with finely dispersed arsenopyrite and pyrite; the country rocks are strongly altered granites. Locality – Dúbrava, Slovakia. Figure courtesy of Martin Chovan.

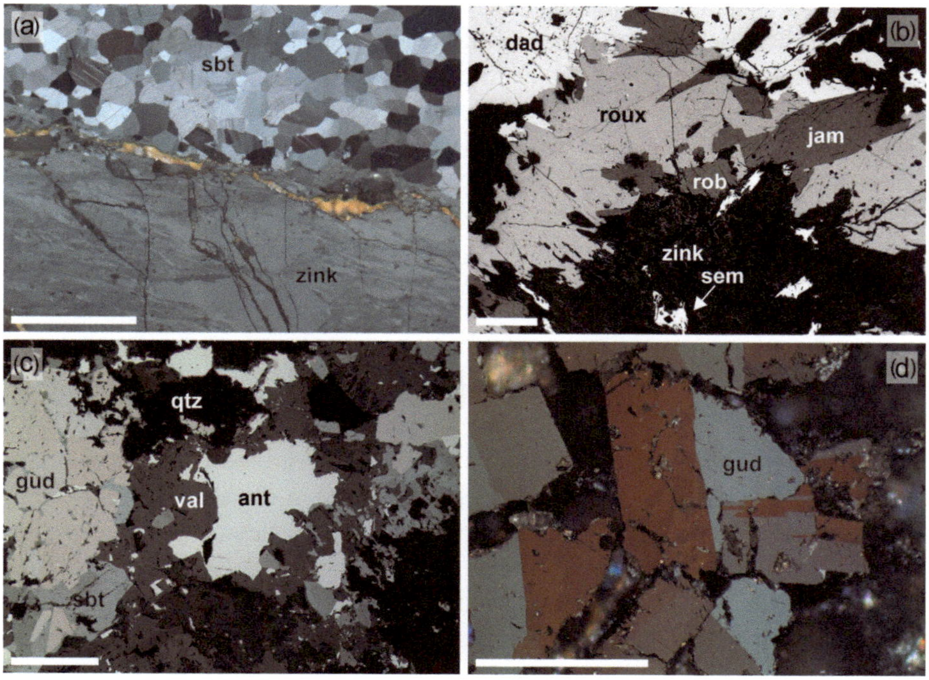

Fig. 2.5: Microscopic images of assemblages of antimony minerals in orogenic and amagmatic deposits. Scale bar in every image represents 200 µm. (a) Massive aggregate of stibnite (sbt) and zinkenite (zink). Locality – Dúbrava, Slovakia. Reflected polarized light, partially crossed nicols. (b) Complex assemblage of Pb-Sb sulfosalts zinkenite (zin), robinsonite (rob), semseyite (s), jamesonite (jam), rouxelite (roux), and dadsonite (dad). Locality – Kl'ačianka, Slovakia. Back-scattered electron (BSE) image. Figure courtesy of Robert Hovorič. (c) Complex assemblage of early antimony (ant) and gudmunite (gud), replaced by stibnite (sbt) and latest valentinite (val) in quartz (qtz). Locality – Pezinok, Slovakia. Reflected polarized light. Figure courtesy of Martin Chovan. (d) Twinned gudmundite (gud) crystals in quartz. Locality – Pezinok, Slovakia. Reflected polarized light, partially crossed nicols. Figure courtesy of Martin Chovan.

heteromorphite, and others) are also present (Fig. 2.5(b)). Cu-Sb minerals, such as the minerals of the tetrahedrite group, chalcostibite (Fig. 2.3(a)), or bournonite, occur in such ores, as well. Arsenopyrite and pyrite are particularly abundant and typical for these deposits; they are nominally free of antimony, but may contain substantial antimony concentrations in some deposits (e.g., up to 4.5 wt.% Sb in arsenopyrite [38]).

The main antimony mineral of the Carlin-type deposits is also stibnite, either in the classical deposits in the western part of the United States [39] or in the large Chinese deposits of somewhat uncertain genesis and affiliation [40]. Locally, arsenic may combine with antimony to form the mineral getchellite. In systems enriched in mercury, livingstonite may be found; in those with elevated thallium concentrations, Tl-Sb sulfosalts, for example, weissbergite or vrbaite, are encountered [41]. Elevated selenium concentrations may lead to an extensive S-Se substitution in stibnite, up to the selenian end-member antimonselite (Fig. 2.3(b) [42]). Overall, the abundance of stibnite dwarfs that of the other antimony minerals throughout these deposits; local geochemical variations in individual deposits provide for significant mineralogical diversity.

The amagmatic deposits of antimony include those hosted by black shales [43, 44] or other types of rocks [45]. They too are dominated by stibnite. In the cases of low fO_2 and fS_2 values in the fluid, these deposits develop an early assemblage of gudmundite, berthierite, and native antimony (Figs. 2.5(c), (d) and 2.6). It is interesting that the evolution of some of these deposits terminates with episodes with unusually high fO_2, and with precipitation of the primary antimony sulfoxide kermesite, and the oxides valentinite or schafarzikite (Fig. 2.3(c) [37, 45]).

2.3.2 Antimony in epithermal deposits

Epithermal deposits are related to volcanic or subvolcanic geologic formations of variable composition and age and were generated by a mixture of magmatic and meteoric fluids [47]. According to a number of features, such as alteration styles or associated volcanic rocks, they are divided into high-, intermediate-, and low-sulfidation types. However, the decisive feature for the distinctions among these different types is the sulfide mineralogy. In that respect, antimony plays an important role. The antimony minerals summarized in this section are listed in Tab. 2.2 together with their formulae.

Epithermal deposits are usually mined for gold, silver, or base metals (Pb, Zn, Cu). Antimony is rarely the main commodity from these deposits [48], but its mineralogical variability in these deposits warrants a separate section.

High-sulfidation Au-Cu deposits are characteristic in their sulfide associations: pyrite, enargite, luzonite, and famatinite [47]. Antimony is one of the major constituents of the sulfides in these deposits, and is harbored by the enargite-stibioenargite and luzonite-famatinite (both Cu_3AsS_4-Cu_3SbS_4) solid solutions. Because these minerals are abundant in high-sulfidation epithermal deposits, in most cases they preclude the formation of other antimony minerals. In gold deposits with lower copper

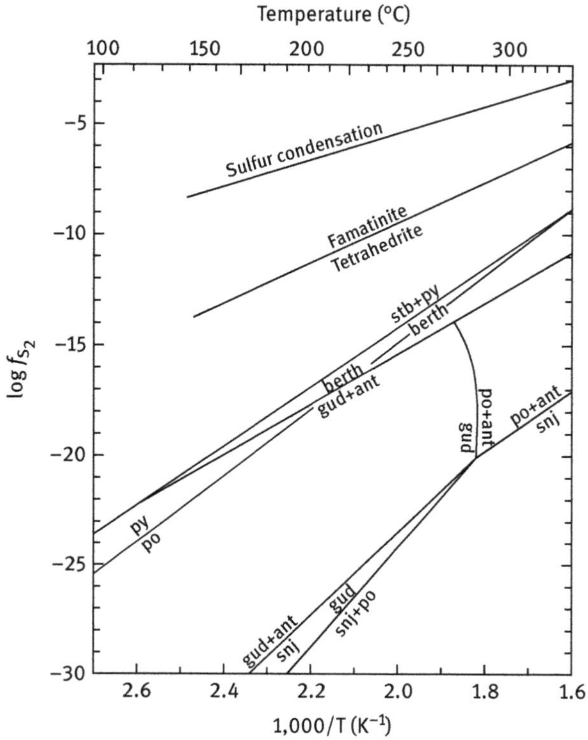

Fig. 2.6: Sulfidation curves for assemblages involving primary antimony minerals. Formulae of most minerals from this figure are given in Tab. 2.2. Abbreviations of minerals: sbt = stibnite, berth = berthierite, gud = gudmundite, py = pyrite, po = pyrrhotite, ant = antimony (Sb), snj = seinäjokite ($FeSb_2$). Simplified and redrawn from [46, 47].

contents, stibnite may occur, but only rarely and in small amounts [49]. In a number of high-sulfidation deposits, minerals of the tetrahedrite group (tetrahedrite, tennantite) occur [50], with variable antimony contents. Tetrahedrite and famatinite are related by the reaction:

$$Cu_{12}Sb_4S_{13}(\text{tetrahedrite}) + (3/2)\,S_2 = 4\,Cu_3SbS_4(\text{famatinite}),$$

showing that under the highest sulfidation states, famatinite alone should be the principal antimony carrier (Fig. 2.6). High-sulfidation assemblages can be overprinted by later events and produce the Ag-Sb sulfosalts pyrargyrite, stephanite, or polybasite [51] that are typical of intermediate- or low-sulfidation deposits.

High-sulfidation Sn-Ag-Sb deposits are less common, although large examples are known, especially in Bolivia [52]. Antimony is more abundant only in the uppermost portions of those deposits, represented by stibnite and a number of rare sulfosalts such as franckeite, cylindrite, and ramdohrite [53], or Ag-Sb and Pb-Sb sulfosalts [54]. These deposits are the rare examples of epithermal deposits whose main commodity may be antimony [48].

Antimony is a common constituent of many intermediate- and low-sulfidation epithermal deposits. It is most commonly harbored by silver sulfosalts such as pyrargyrite or polybasite (Figs. 2.3(d), (e)). They are related to simple sulfides by the reactions:

$$(3/2)\ Ag_2S + (1/2)\ Sb_2S_3 = Ag_3SbS_3(\text{pyrargyrite})$$

$$(13/2)\ Ag_2S + (3/2)\ Cu_2S + Sb_2S_3 = Ag_{13}Cu_3Sb_2S_{11}(\text{polybasite}),$$

showing that these sulfides are not indicators of the sulfidation state of the ore-bearing fluids. In this case, the sulfidation state is indicated by minerals without antimony, such as arsenopyrite, sphalerite, or pyrrhotite.

Antimony and silver are usually deposited in the late stages of ore precipitation. Besides the previously mentioned pyrargyrite and polybasite, they form a rich assemblage of sulfosalts such as argentotetrahedrite, stephanite, miargyrite, diaphorite, andorite, and others (Figs. 2.3(d)–(f)). Pb-Sb sulfosalts, for example boulangerite, also rarely may be found [55]. Under very low sulfide activities, dyscrasite may be formed [56]. Stibnite may occur if the concentrations of base-metal elements and silver are too low [57] (Figs. 2.3(g), (h)); normally, with a few exceptions, stibnite is only an accessory mineral in epithermal deposits [48]. Stibnite precipitates from hot fluids in geothermal power plants that tap into currently active low-sulfidation systems [58].

2.3.3 Antimony minerals in orthomagmatic Ni-Cu-PGE deposits

Orthomagmatic deposits of nickel, copper, and the platinum-group elements (PGE) form by crystallization of an immiscible sulfide melt in magmatic chambers of ultramafic and mafic rocks [59]. Their mineralogy is complex and includes a number of minerals of antimony, with examples of these summarized in Tab. 2.3. They occur in trace amounts, but because of their PGE content, they can be of economic interest.

Tab. 2.3: Examples of PGE-Sb minerals found in orthomagmatic deposits.

anyuiite	$Au(Pb,Sb)_2$	naldretteite	Pd_2Sb
borovskite	Pd_3SbTe_4	stibiopalladinite	Pd_5Sb_2
breithauptite	$NiSb$	stumpflite	$Pt(Sb,Bi)$
geversite	$Pt(Sb,Bi)_2$	sudburyite	$(Pd,Ni)Sb$
isomertieite	$Pd_{11}Sb_2As_2$	ullmannite	$NiSbS$
mertieite-I	$Pd_{11}(Sb,As)_4$	vavrinite	Ni_2SbTe_2
mertieite-II	$Pd_8(Sb,As)_3$	vincentite	$(Pd,Pt)_3(As,Sb,Te)$

Primary magmatic minerals of antimony in these deposits are rare. The most typical mineral from this group is mertieite, thought to crystallize from an evolved sulfidic magma [60, 61], associated with magmatic pyroxene, plagioclase, or spinel. The conclusions from observations of natural assemblages are supported by experimental determinations of partition coefficients for antimony. Antimony (together with tin and arsenic) prefers sulfide melt over either monosulfide solid solution (*mss*) or silicate melt [9]. Hence, sulfide melt is predicted to become enriched in antimony, tin, and arsenic during crystallization, and the formation of antimony minerals in the late magmatic stages is feasible.

The unifying theme of most antimony-mineral occurrences in orthomagmatic deposits is the hydrothermal or metamorphic-hydrothermal overprint of the magmatic assemblages [62]. The minerals commonly found here are sudburyite or naldretteite (Tab. 2.3). Spiridonov et al. [63] described a general sequence of PGE precipitation at Noril'sk (Russia), indicating that antimony predominantly participates in the late mineralization stages. Antimony does not associate with refractory PGE's such as ruthenium, osmium, or iridium; however, this dissociation appears to have a geochemical rationale. Late hydrothermal fluids, often with formidable salinity recorded in the fluid inclusions [64] or by the presence of $PdCl_2$ [65], are efficient at transporting palladium and, to a lesser extent, platinum. Therefore, most of the antimony minerals recorded in these settings contain palladium or platinum (Tab. 2.3). Some authors assume that antimony is transported as aqueous $Sb(OH)_3^0$. Precipitation is triggered by reduction to Sb^0 in the fluid and reaction with Pd(aq) [66].

Antimony minerals related to hydrothermal overprint of orthomagmatic deposits are often associated with metamorphic or alteration minerals such as chlorites, epidote, or amphiboles. The source of antimony is sought in fluids derived from an external source or from the magmatic base-metal minerals such as pyrrhotite. Even though in this regard there are hardly any data for antimony, the latter option is more probable because of the consistent enrichment of late hydrothermal ores in antimony minerals. Apart from the Pd-Pt-Sb phases, Ni-Sb minerals such as breithauptite or ullmannite are also reported occasionally. Breithauptite can also be enriched in Pd [67].

2.4 Transition from low-temperature hydrothermal overprint to weathering

Antimony, unlike many other metals and metalloids, does not only form sulfides, oxides, and oxysalts (e.g., sulfates) in nature. It has the peculiar capability of entering structures with both reduced sulfur and oxygen, and less commonly reduced sulfur and chlorine. This ability adds to the mineralogical diversity of antimony. In addition, it constitutes a bridge between low-temperature hydrothermal assemblages and the early stages of the weathering of primary antimony minerals (Fig. 2.7).

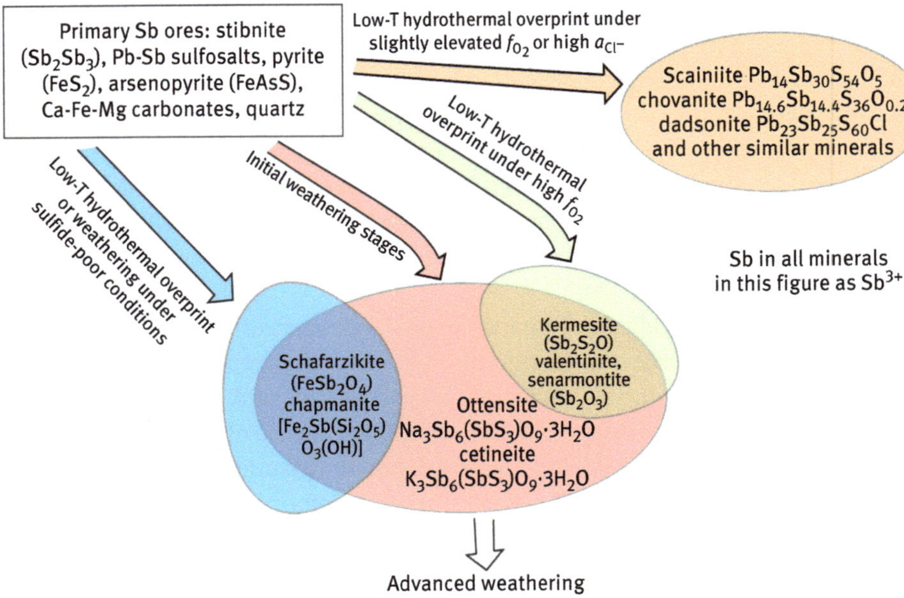

Fig. 2.7: Processes of low-temperature hydrothermal overprint and initial weathering, plus the associated minerals of antimony.

Some of the antimony deposits with common stibnite and Pb-Sb sulfosalts evolve into their terminal stages marked by high fO_2 or fluids with high salinity. These deposits may contain a number of rare Pb-Sb oxosulfosalts, for example scainiite [68] or chovanite (Fig. 2.3(i) [69]), or chlorosulfosalts such as dadsonite [70].

Occasionally, the final hydrothermal overprint leads to the formation of the antimony sulfoxide kermesite, and the antimony oxides senarmontite or valentinite [37, 45, 46]. In these cases, all of these minerals are primary and of hypogene origin.

Exactly these minerals (kermesite, senarmontite, and valentinite) are common secondary, weathering products of stibnite (e.g., [71–74]). They are the bridge between the low-temperature hydrothermal overprint with elevated oxygen fugacity and the weathering conditions under atmospheric oxygen concentration.

The secondary sulfoxides do not only include kermesite. The rare minerals ottensite and cetineite are examples of more complex compositions, with reduced sulfur and oxygen in the same structure [75, 76].

Another bridge between the low-temperature overprint and weathering are antimony minerals which form in sulfide-poor environments. One of these is the sheet silicate chapmanite (Fig. 2.8(a)), considered to be secondary [77, 78], and it is accompanied by the low-temperature hydrothermal kermesite. Another is schafarzikite (Fig. 2.3(c)), known as a primary mineral from a very few localities (e.g., [79]). On the other hand, schafarzikite was labeled as one of the “ultimate sinks of antimony” in the environment [80]. Because schafarzikite is a mineral of Fe^{2+} and Sb^{3+},

Fig. 2.8: Photographs of selected late primary or secondary minerals of antimony. (a) compact aggregate of yellowish-green chapmanite. Locality – Bräunsdorf bei Freiberg, Germany. Photograph courtesy of Uwe Haubenreißer. (b) Pseudomorph of yellow stibiconite after stibnite crystals. Locality – Lojane, Republic of North Macedonia. Reproduced with permission from Crystal Classics (crystalclassics.co.uk); (c) Stibnite crystals partially replaced by yellow stibiconite. Locality – Gyöngyösoroszi, Hungary. Reproduced with permission from Crystal Classics (crystalclassics.co.uk). (d) acicular tripuhyite crystals on tetrahedrite. Locality – Rudňany, Slovakia. Photograph courtesy of Martin Števko. (e) Brandholzite crystals on strongly weathered stibnite, dolomite, and quartz. Locality – Pezinok, Slovakia. Photograph courtesy of Martin Števko. (f) radial aggregates of klebelsbergite crystals. Locality – Kremnica, Slovakia. Photograph courtesy of Martin Števko.

we consider it unlikely that such a mineral can be a common secondary antimony phase. The notion that champanite is a secondary phase also needs to be tested.

2.5 Secondary minerals

There are many secondary minerals of antimony which reflect the chemical composition of primary sources which are being weathered. Representative specimens of the secondary antimony minerals are shown in Figs. 2.8(b)–(f). Minerals mentioned in this section are summarized in Tab. 2.4. Most of the primary, hypogene minerals contain elements in their reduced state, which will react with the oxygen and water in near-surface environments. However, many primary antimony minerals are so rare that their overall contribution to the weathering and release of

Tab. 2.4: Minerals mentioned in the section on Secondary minerals of antimony. The list includes antimony minerals and nominally antimony-free minerals important for the understanding of antimony dispersion in field settings. Primary minerals, mentioned in this section, are not listed. Minerals listed in italics are of uncertain status or discredited. □ = vacancy.

Minerals of antimony			
bindheimite	$(Pb,\square)_2Sb_2(O,OH)_7$	klebelsbergite	$Sb_4O_4(SO_4)(OH)_2$
bottinoite	$NiSb_2(OH)_{12}\cdot 6H_2O$	*partzite*	$(Cu,\square)_2Sb_2(O,OH)_7$
brandholzite	$MgSb_2(OH)_{12}\cdot 6H_2O$	richelsdorfite	$Ca_2Cu_5Sb(AsO_4)_4(OH)_6Cl\cdot 6H_2O$
byströmite	$MgSb_2O_6$	roméite	$(Ca,\square)_2Sb_2(O,OH)_7$
camérolaite	$Cu_6Al_3Sb(OH)_{24}(SO_4)\cdot 2H_2O$	schafarzikite	$FeSb_2O_6$
cervantite	Sb_2O_4	senarmontite	Sb_2O_3
clinocervantite	Sb_2O_4	stibiconite	$Sb_3O_6(OH)$
cualstibite	$Cu_2AlSb(OH)_{12}$	theisite	$Cu_5Zn_5(AsO_4,SbO_4)_2(OH)_{14}$
kermesite	Sb_2S_2O	tripuhyite	$FeSbO_4$
mopungite	$NaSb(OH)_6$	valentinite	Sb_2O_3
Nominally antimony-free minerals			
apatite	group of minerals, $Ca_5(PO_4)_3(OH,F,Cl)$		
copiapite	group of minerals, $(Fe^{2+},Mg,Al,Fe^{3+})Fe(SO_4)_6(OH)_2\cdot 20H_2O$		
dussertite	$BaFe_3(AsO_4)(AsO_3OH)(OH)_6$		
feroxyhyte	δ-FeOOH		
ferrihydrite	≈$FeOOH\cdot nH_2O$		
goethite	α-FeOOH		
jarosite	group of minerals, $(K,Na,H_3O)Fe_3(OH)_6(SO_4)_2$		
lepidocrocite	γ-FeOOH		

antimony into the environment is negligible. Therefore, the main primary minerals must remain in focus when secondary mineralogy is considered. This section is structured according to this principle. The chemical systems are arranged approximately in their order of importance and abundance of the primary minerals (cf. Tab. 2.1). Some rare systems are not considered, even though limited information does exist.

A semiquantitative measure of the abundance of the secondary antimony minerals is the number of worldwide sites where they have been identified (Fig. 2.9). This statistic can be distorted in several ways. First, some of these minerals can also be of primary origin. A survey of hundreds of localities and their descriptions in many different languages would be very tedious. A larger problem is that the mode of origin (primary/ secondary) is often not specified, or is specified yet not backed up by evidence. Second, some of these minerals may occur predominantly in microscopic form, and thus may require advanced techniques (e.g., microdiffraction at a synchrotron source) for their identification, or simply may not be attractive enough to draw the attention of mineral collectors. Therefore, the data in Fig. 2.9 are helpful in assessing the abundance of individual minerals, but should be viewed with some caution.

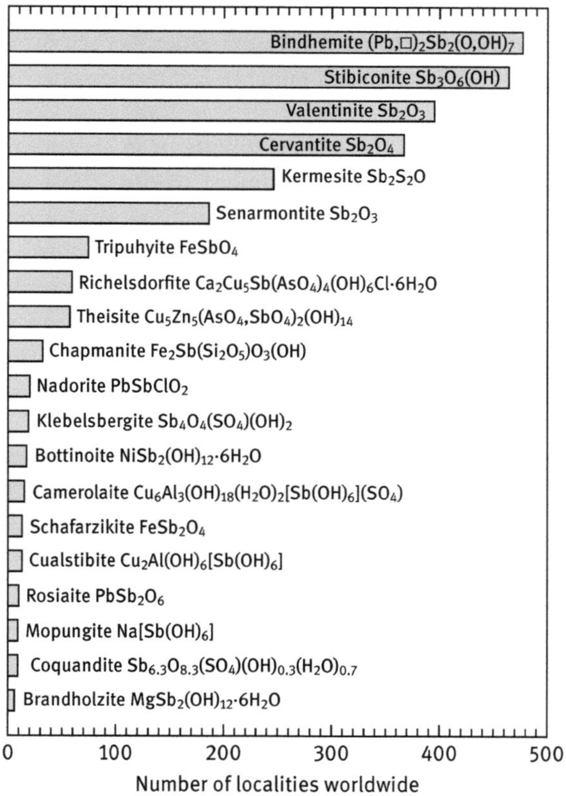

Fig. 2.9: Number of localities for common secondary antimony minerals worldwide. Data from [81]. For details, see text.

2.5.1 Antimony oxides and hydroxides

The Sb^{3+} oxides valentinite and senarmontite (both Sb_2O_3) have already been discussed. They are initially produced during the weathering of stibnite (the most commercially important antimony mineral). The stable polymorph at low temperatures is senarmontite [82].

It is of interest to not only contemplate those minerals which are known, but also about those minerals which might be expected, but have not yet been described. One such of these unknown minerals would possess the formula Sb_2O_5. Its existence has been invoked by misleading phase diagrams of natural assemblages [83]. Early on, it was noted that "antimonic acid cannot be dehydrated by heating in air to give products of constant and reproducible weight without simultaneous reduction of some of the Sb^V to Sb^{III}" [84]. These authors doubted that pure anhydrous Sb_2O_5 exists. It does, but it must be synthesized with excessive oxygen overpressure [85]. Hence, not only is Sb_2O_5 not known as a mineral, but its existence in nature is rather unlikely.

Instead of Sb_2O_5, the partial oxidation of antimony in stibnite or in secondary Sb_2O_3 minerals gives rise to a group of mixed-valence antimony oxides. These include the anhydrous oxides $Sb^{3+}Sb^{5+}O_4$ cervantite and clinocervantite, and a hydrated oxide from the roméite group (referred to in older literature as stibiconite).

All of these minerals form powdery crusts or small crystals on stibnite (Figs. 2.8(b), (c) and 2.10(a)). They are usually spatially associated with stibnite relics. In some cases, they replace stibnite crystals or aggregates completely (i.e., they form pseudomorphs). These minerals are probably very common, with the exception of

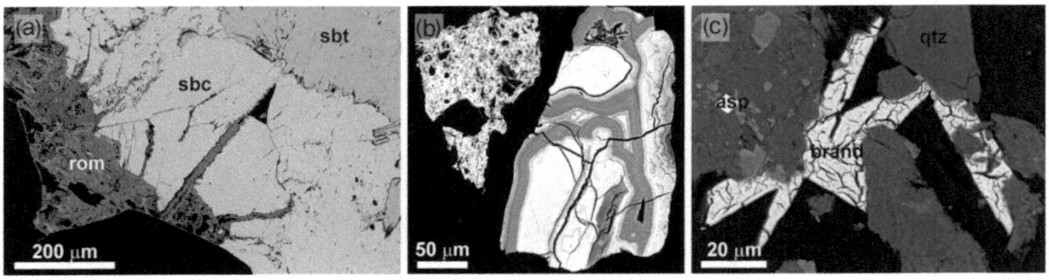

Fig. 2.10: Back-scattered electron (BSE) images of assemblages of secondary antimony minerals. (a) Primary stibnite (sbt) weathering to a thick rim of stibiconite (sbc), replaced from the outer rim by roméite (rom). Locality – Dúbrava, Slovakia. Figure courtesy of Tomáš Klimko. (b) Two grains of antimony weathering products in tailings. The grain on the left-hand side is a porous pseudomorph after stibnite, either valentinite or senarmontite. The other grain is a neoformed aggregate of nanoscopic tripuhyite crystals. Locality – Pezinok, Slovakia. Reproduced from [90] with permission from the Mineralogical Society of America. (c) Brandholzite (brand) crystals from fresh tailings, together with quartz (qtz) grains and fragments of altered rocks, with a small arsenopyrite (asp) crystal in one of them. Locality – Beaver Brook, Newfoundland, Canada. Reproduced from [91], with permission from Elsevier.

clinocervantite [86, 87]. However, their identification usually relies on simple means, such as visual observation. Due to their unspectacular nature, in most cases they were often not given the necessary attention to establish their identity with any certainty (e.g., [88]). Modern confirmation is required for many of the earlier descriptions [89]. The antimony oxides associate with stibnite, kermesite, and rarely with antimony and with each other.

2.5.2 Adsorption of antimony and incorporation of antimony into nominally antimony-free minerals

Upon release into aqueous solutions at many contaminated sites, antimony adsorbs onto the surfaces of preexisting or actively precipitating minerals. The crystallization of stable and refractory antimony minerals is usually hindered kinetically (see below). Hence, nominally antimony-free minerals must be considered if the environmental mobility of antimony is to be understood.

The most potent adsorbers of antimony seem to be iron oxides, because of both their large surface areas and the abundance of reactive sites on their surfaces. At low loadings or short contact times, antimony adsorbs to the most reactive sites; for the surface of goethite, these are the edge-sharing configuration between Fe(III) and Sb(V) octahedra [92, 93]. Longer times or higher loading may induce adsorbtion at the lower-affinity, corner-sharing sites (e.g., [94]). Under conditions of enhanced dissolution-precipitation, antimony is incorporated into the structure of goethite [95], and possibly also other iron oxides. It has been proposed that such incorporation is a transient step toward the transformation to tripuhyite [80], but a thermodynamic analysis suggests that this transformation is viable only at high antimony loadings [95]. At elevated temperatures, Sb(V) retarded the transformation of ferrihydrite to more crystalline phases, but ultimately converted to goethite, feroxyhyte, or tripuhyite [96]. However, tripuhyite formed only at very high Sb/Fe ratios. In field settings, the antimony concentration in goethite reaches 25 wt.% [74]. It is not clear if all of this antimony is in the structure or on the surface. The possibility of nanodomains of other phases cannot be excluded, as these analyses were done with an electron microprobe (beam diameter 5 μm). Similar antimony concentrations can be attained in ferrihydrite that usually also contains substantial arsenic concentrations [74].

Minerals of the jarosite group are also known to pick up antimony, albeit only rarely in substantial amounts [97]. Jarosite was reported to contain 5.9% Sb (average value [87]), and dussertite can contain up to 8% Sb (calculated from the formula given in [98]). Upon reductive dissolution, antimony is released; but much of it is recaptured by lepidocrocite, goethite, or green rust [97]. The similarity of the ionic radii of Fe^{3+} (HS, 0.65 Å) and Sb^{5+} (0.60 Å) indicates the possibility of Fe^{3+}-Sb^{5+} substitution (see also [98]), and this suggests that other secondary ferric minerals may be able to store antimony in their structures. With rare exceptions (e.g., for jarosite and copiapite [99]), such analytical data are not available. A more concerted effort at sites with elevated antimony concentrations would be needed to determine which ferric sulfates or arsenates are important antimony carriers.

Experimental studies have shown that antimony strongly adsorbs onto surfaces of other common minerals, such as apatite [100], clays [101], and others. It is not clear how important these minerals are for the scavenging of antimony in field settings.

2.5.3 System Fe-Sb-O

The association of stibnite, the most commercially important antimony mineral, with pyrite or arsenopyrite, leads to the simultaneous release of iron, antimony, and arsenic at many sites. This elemental combination renders the system Fe-Sb-O the most important one for the secondary mineralogy of antimony.

The two minerals in this system are tripuhyite and schafarzikite. Both have been named as “ultimate sinks” of antimony in the environment [80], because of their low solubility. As mentioned above, schafarzikite with Fe^{2+} and Sb^{3+} in its structure is an unlikely agent of environmental importance. Tripuhyite, on the other hand, seems to be very common, even though an inconspicuous, weathering mineral. Tripuhyite contains Fe^{3+} and Sb^{5+} [102] and is therefore a much more likely candidate for a weathering product of Fe-Sb primary assemblages.

Tripuhyite has consistently been detected in studies of secondary antimony minerals that applied advanced techniques for mineral identification [73, 74, 87, 90, 94, 103–106]. The mineral is easily confused for goethite, it commonly forms nanosized crystals and its Fe/Sb ratio deviates far from 1 [74]. Macroscopic crystals of secondary tripuhyite are very rare, but they can be found (Fig. 2.8(d)). All of these properties make it difficult to detect tripuhyite with the standard and bulk mineralogical techniques. These studies and our experience show that the simultaneous presence of iron and antimony in a weathering environment invariably leads to the precipitation of tripuhyite. This process is slow (see below), but results in an efficient immobilization of antimony as the sparingly soluble tripuhyite [80]. The available evidence indicates that tripuhyite crystallizes from pore solutions in microenvironments rich in antimony (Fig. 2.10(b)). Formation by re-arrangement of the structure of Sb-rich goethite [95] is possible, but not verified.

The nanosized crystals of tripuhyite and its complicated local structure, including defects in the structure, are well known outside of the field of mineralogy. The synthetic $FeSbO_4$ is used in the selective oxidation of propene and other hydrocarbons, accounting for about “one-quarter of the organic precursor chemicals and intermediates used in the manufacture of further industrial products” [107]. The information amassed in the materials science literature (e.g., [108, 109]) can elucidate the field observations of the non-stoichiometry, which makes the identification of this mineral, even with chemical microanalysis, exceedingly difficult. It remains unknown whether this mineral acts as catalyst (or not) for reactions among organic species in nature.

2.5.4 System $CaO–PbO–CuO–Sb_2O_3–Sb_2O_5–H_2O$: Minerals of roméite group

Minerals that belong to this chemical system commonly form as weathering crusts on stibnite (Fig. 2.10(a)), Pb-Sb sulfosalts, or sulfides of the tetrahedrite group. Calcium is usually supplied from omnipresent calcite or dolomite. If iron is available, they may eventually be converted into tripuhyite.

One common link between these minerals is their pyrochlore-like crystal structure. The last mineral classification of the pyrochlore supergroup was so exceedingly complicated that soon after completion it had to be revised [110]. I will adhere only loosely to this classification; it is hardly of any use for field studies. The crystal chemistry of this complex group was addressed by [111]. The possibilities of heterovalent substitution (Ca^{2+} for Sb^{3+}), vacant sites, and substitutions on the anionic position (OH^- for O^{2-}) make understanding of these structures difficult. Natural materials are rarely suitable for detailed crystal chemical investigations because of their heterogeneity, either intrinsic or caused by admixtures of other minerals.

One of the minerals of this group, stibiconite (Figs. 2.8(b), (c)), was already mentioned among the antimony oxides. This mineral contains Sb^{3+} and Sb^{5+}. Another very common phase is roméite (Fig. 2.10(a)), with an ideal composition from $CaSb_2O_5(OH)_2$ to $Ca_2Sb_2O_7$. It has been commonly identified as a weathering product of stibnite or tetrahedrite [73, 74, 87, 104, 112]. Another, similar substance often reported in the literature has the composition $Ca[Sb(OH)_6]_2$ (e.g., [113, 114]), and has been mentioned also in books and compendia (e.g., [115, 116]). The myth of the existence of such a crystalline phase with well-defined properties is stubborn and cannot be easily rooted out. It draws back on the solubility data of Johnson et al. [117], who reported the existence and solubility of such a material, although this material seems to only be an amorphous mixture, with an undefined structure and composition. There is no available evidence whatsoever that such a mineral exists. The natural calcium antimonate is roméite, and, because of its complicated structure and chemistry, may not be easy to identify as such. Other combinations of calcium and antimony such as the recently described mineral smamite [118] may occur, but are very rare.

Lead-antimony sulfosalts may weather to another member of the roméite group, referred to previously as bindheimite. Much of what has been written about roméite here also applies to bindheimite. The variable structure and chemistry complicate the identification. The substance $Pb[Sb(OH)_6]_2$ [117] is unlikely to exist, and should not be used to explain geochemical or mineralogical observations in natural materials.

The last mineral from this group is partzite, which has been recently discredited [119]. Still, it was shown that a Cu-Sb pyrochlore-like phase originates during the weathering of tetrahedrite [105, 120]. A similar phase, without much Cu, was detected among weathering products in realgar-stibnite tailings [112]. In both cases, these pyrochlores contain an appreciable amount of As^{5+}, whose preferred tetrahedral coordination is incompatible with the pyrochlore structure. The nature of these phases remains unknown.

2.5.5 System $MgO–NiO–Na_2O–Sb_2O_5–H_2O$

The antimonates of Mg, Ni, and Na include the minerals brandholzite (Fig. 2.8(e)), mopungite, bottinoite, and the anhydrous byströmite. This structurally incoherent group of minerals is connected by their high solubilities, and therefore a similar mode of occurrence. The antimonate with the highest solubility, $KSb(OH)_6$, is not yet known from nature, although its existence could be assumed.

These minerals appear to be spatially linked to the immediate vicinity of weathering antimony minerals. They seem to be restricted to moist (but not wet) environments, either inside old dumps or in old mines [121–123]. Brandholzite forms by the weathering of stibnite and dissolution of dolomite. Bottinoite could form by the weathering of ullmannite or breithauptite.

The description of brandholzite from fresh tailing deposits (Fig. 2.10(c) [91]) is intriguing. In most studies of secondary antimony mineralogy in mine wastes, only old tailings were investigated. In addition, heavy minerals are separated from the tailings, usually with methods that make extensive use of water, such as panning or settling. In all of these procedures, brandholzite would probably dissolve and remain unnoticed. There are some hints that highly soluble antimony minerals may be common in fresh tailings or mine wastes. For example, the high concentration of exchangeable and water-soluble antimony in tailings [124] could be attributed to these soluble minerals, such as brandholzite. These observations imply that these minerals are common, but only transient, and difficult to detect.

2.5.6 Other secondary minerals of antimony

Antimony readily combines with many elements to form different minerals, most of them only mineralogical rarities, or “freaks” [125].

Sulfates of antimony, for example klebelsbergite (Fig. 2.8(f)), have been described from a few sites [123, 126, 127]. From our observations, they form when stibnite weathers quickly in the presence of pyrite/marcasite and in the absence of carbonates [123]. The acidic, sulfate-rich solutions attack stibnite and quickly crystallize the Sb sulfates. They are spatially restricted to the immediate vicinity of stibnite.

Copper and antimony enter complex structures of the minerals theisite, camérolaite, richelsdorfite, and cualstibite. When they occur, they do so in small amounts

[125]. Interestingly, they are iron-poor (see Fig. 2.12 in [105]). The interpretation of these analytical data is that if iron had been present, tripuhyite would form instead.

2.5.7 General features of secondary mineralogy of antimony

The number of the secondary minerals of antimony seems to be overwhelming. Is there any discernible pattern among these minerals? What have we learned about the dispersion of antimony in the environment? There are several points that need attention in this respect.

1. The most common primary antimony mineral is stibnite, found in association with pyrite, arsenopyrite, quartz, and carbonates. This assemblage defines the geochemical restrictions for the common secondary antimony minerals, namely, the enrichment in antimony, iron, calcium, and magnesium. The abundance of carbonates implies that the weathering of stibnite usually proceeds under near-neutral conditions. Strongly acidic environments with abundant antimony are seldom.
2. Stibnite may undergo rapid transformation to the oxide minerals valentinite, senarmontite, or stibiconite, with no or only partial oxidation of Sb^{3+} to Sb^{5+} (e.g., [128]). These minerals form pseudomorphs after stibnite or other primary antimony minerals.
3. In the presence of carbonates, brandholzite forms rapidly [91, 123]. We found that centimeter-sized brandholzite crystals can grow within months in moist underground spaces. They form from a thin film of aqueous solutions on mine walls (Fig. 2.11). Seasonal addition of water (water trickling into the mine after rain or snow melts) can lead to their dissolution, and an increase of aqueous concentrations of antimony. Such a process can explain the statement "antimony is mobile as an Sb(V) species under common environmental conditions" [83, 128].
4. Pieces of stibnite in outcrops and old dumps are also rapidly covered by cervantite or kermesite, and later by minerals of the roméite group (Fig. 2.11); calcium can be supplied by carbonates, and lead by Pb-Sb sulfosalts. Although reliable data on solubilities are lacking, field observations suggest that roméite and similar minerals are less soluble than brandholzite (or mopungite or bottinoite).
5. Given sufficient time and the presence of iron, antimony will eventually combine with this element to form tripuhyite. Tripuhyite forms in tailings at a slow rate (Fig. 2.11). We found that after decades of weathering, tripuhyite crystallites hardly exceed 10 nm [90]. Antimony-rich goethite may be also present in crystals of similar size. The very low solubility of tripuhyite can explain another statement from the literature, namely that "the element [antimony] is geochemically immobile in solution and in soils" [73].
6. Neither of the conflicting statements on the environmental mobility of antimony is false. They reflect the interplay between the kinetics of formation and thermodynamic stability (Fig. 2.12) [123]. The most soluble phases (e.g., brandholzite)

form rapidly, feed the discharging solutions with antimony, and eventually disappear. Even the less soluble pyrochlores (e.g., roméite) form quickly, but they also eventually dissolve. Tripuhyite forms slowly, but once it exists it locks antimony into an insoluble mineral. The antimony-rich goethite may be a similar sink (Fig. 2.11, cf [95]).

7. Most of the other secondary antimony minerals are rare. They simply reflect the presence of other elements, the absence of iron, or an unusual composition of the aqueous phase (e.g., low pH). In terms of antimony dispersion in the environment, they are hardly of importance.

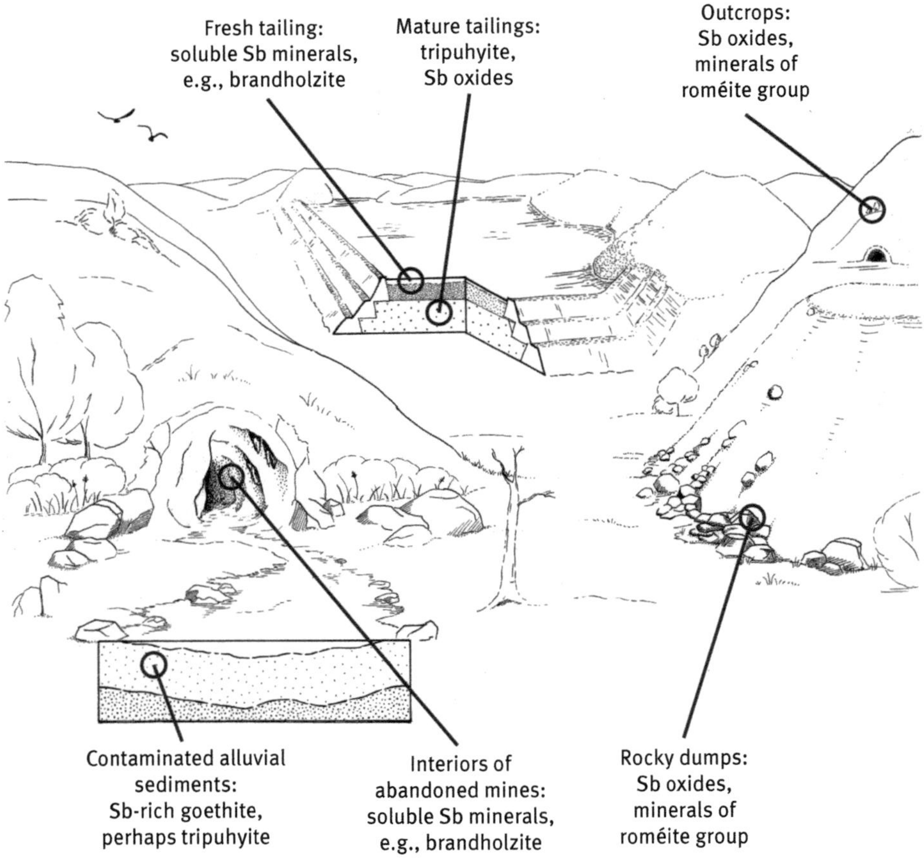

Fig. 2.11: Weathering environments of ores with dominant stibnite, accompanied by pyrite, arsenopyrite, quartz, and carbonates. For each environment, the main weathering products are specified. The variability of the weathering products documents the interplay between the thermodynamic and kinetic constraints on the individual phases. For details, see text. The sketch was drawn by Sarah Bucka.

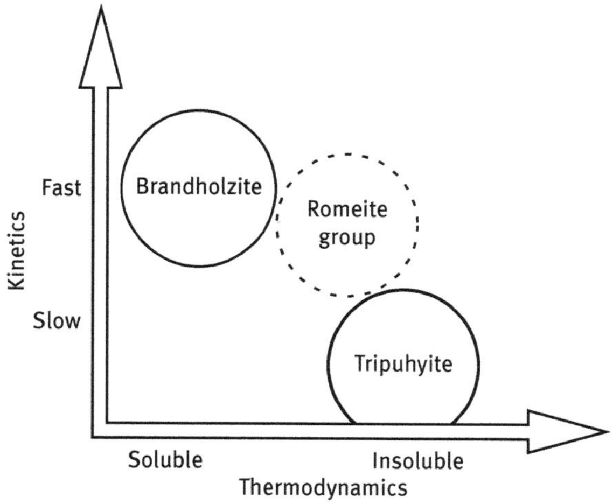

Fig. 2.12: Cartoon representation of the relative solubility and relative rate of crystallization of the main products of weathering of stibnite-pyrite-arsenopyrite assemblage under near-neutral conditions. For details, see text.

References

[1] Taylor SR. Abundance of chemical elements in the continental crust: A new table. Geochim Cosmochim Acta 1964, 28, 1273–1285.

[2] Christy AG. Causes of anomalous mineralogical diversity in the periodic table. Mineral Mag 2015, 79, 33–49.

[3] Lager T, Forssberg KSE. Current processing technology for antimony-bearing ores. A review, part 2. Miner Eng 1989, 2, 543–556.

[4] Anderson CG. The metallurgy of antimony. Chem Erde 2012, 72(S4), 3–8.

[5] Makovicky E. Crystal structure of sulfides and other chalcogenides. Rev Mineral Geochem 2006, 61, 7–125.

[6] GERM Reservoir Database. (Accessed on April 17, 2020, at https://earthref.org/GERM/).

[7] Esson J, Stevens RH, Vincent EA. Aspects of the geochemistry of arsenic and antimony, exemplified by the Skaergaard intrusion. Mineral Mag 1965, 35, 88–107.

[8] Klöck W, Palme H. Partitioning of siderophile and chalcophile elements between sulfide, olivine, and glass in a naturally reduced basalt from Disko Island, Greenland. Proc Lunar Planet Sci Conf 1988, 18, 471–483.

[9] Li Y, Audétat A. Partitioning of V, Mn, Co, Ni, Cu, Zn, As, Mo, Ag, Sn, Sb, W, Au, Pb, and Bi between sulfide phases and hydrous basanite melt at upper mantle conditions. Earth Planet Sci Lett 2012, 355–356, 327–340.

[10] Stone WE, Fleet ME, MacRae ND. Two-phase nickeliferous monosulfide solid solution (mss) in megacrysts from Mount Shasta, California: A natural laboratory for nickel-copper sulfides. Am Mineral 1989, 74, 981–993.

[11] Liu Y, Brenan J. Partitioning of platinum-group elements (PGE) and chalcogens (Se, Te, As, Sb, Bi) between monosulfide-solid solution (MSS), intermediate solid solution (ISS) and sulfide liquid at controlled fO_2–fS_2 conditions. Geochim Cosmochim Acta 2015, 159, 139–161.

[12] Hill E, Wood BJ, Blundy JD. The effect of Ca-Tschermaks component on trace element partitioning between clinopyroxene and silicate melt. Lithos 2000, 53, 203–215.
[13] Klemme S, Prowatke S, Hametner K, Günther D. The partitioning of trace elements between rutile and silicate melts: Implications for subduction zones. Geochim Cosmochim Acta 2005, 69, 2361–2371.
[14] Adam J, Green T. Trace element partitioning between mica- and amphibole-bearing garnet lherzolite and hydrous basanitic melt: 1. Experimental results and the investigation of controls on partitioning behaviour. Contrib Mineral Petr 2006, 152, 1–17.
[15] Meinhold G. Rutile and its applications in earth sciences. Earth-Sci Rev 2010, 102, 1–28.
[16] Zack T, Kronz A, Foley SF, Rivers T. Trace element abundances in rutiles from eclogites and associated garnet mica schists. Chem Geol 2002, 184, 97–122.
[17] Agangi A, Reddy SM, Plavsa D, Fougerouse D, Clark C, Roberts M, Johnson TE. Antimony in rutile as a pathfinder for orogenic gold deposits. Ore Geol Rev 2019, 106, 1–11.
[18] Graham J, Morris RC. Tungsten- and antimony-substituted rutile. Mineral Mag 1973, 39, 470–473.
[19] Galuskina IO, Galuskin EV, Kusz J, Dzierżanowski P, Prusik K, Gazeev VM, Pertsev NN, Dubrovinsky L. Dzhuluite, $Ca_3SbSnFe^{3+}{}_3O_{12}$, a new bitikleite-group garnet from the Upper Chegem Caldera, Northern Caucasus, Kabardino-Balkaria, Russia. Eur J Mineral 2013, 25, 231–239.
[20] Spandler C, Pirard C. Element recycling from subducting slabs to arc crust: A review. Lithos 2013, 170–171, 208–223.
[21] Gregory DD, Large RR, Halpin JA, Baturina EL, Lyons TW, Wu S, Danyushevsky L, Sack PJ, Chappaz A, Maslennikov VV, Bull SW. Trace element content of sedimentary pyrite in black shales. Econ Geol 2015, 110, 1389–1410.
[22] Gregory DD, Cracknell MJ, Large RR, McGoldrick P, Kuhn S, Maslennikov VV, Baker MJ, Fox N, Belousov I, Figueroa MC, Steadman JA, Fabris AJ, Lyons TW. Distinguishing ore deposit type and barren sedimentary pyrite using laser ablation-inductively coupled plasma-mass spectrometry trace element data and statistical analysis of large data sets. Econ Geol 2019, 114, 771–786.
[23] Cave B, Pitcairn I, Craw D, Large R, Thompson J, Johnson S. A metamorphic mineral source for tungsten in the turbidite-hosted orogenic gold deposits of the Otago Schist, New Zealand. Miner Deposita 2017, 52, 515–537.
[24] Minz FE, Bolin N-J, Lamberg P, Bachmann K, Gutzmer J, Wanhainen C. Distribution of Sb minerals in the Cu and Zn flotation of Rockliden massive sulphide ore in north-central Sweden. Miner Eng 2015, 82, 125–135.
[25] Awe SA, Sundkvist J-E, Bolin N-J, Sandström Å. Process flowsheet development for recovering antimony from Sb-bearing copper concentrates. Miner Eng 2013, 49, 45–53.
[26] Pirajno F. Hydrothermal Processes and Mineral Systems. Heidelberg, Springer, 2009.
[27] Groves DI, Goldfarb RJ, Gebre-Mariam H, Hagemann SG, Robert F. Orogenic gold deposits – a proposed classification in the context of their crustal distribution and relationship to other gold deposit type. Ore Geol Rev 1998, 13, 7–27.
[28] Kretschmar U, McBride D. The Metallogeny of Lode Gold Deposits. A Syngenetic Perspective. Amsterdam, Elsevier, 2015.
[29] Mao J, Qiu Y, Goldfarb R, Zhang Z, Garwin S, Ren F. Geology, distribution, and classification of gold deposits in the western Qinling belt, China. Miner Deposita 2002, 37, 352–377.
[30] Mossman DJ, Leblanc ML, Burzynski JF. Antimony-gold deposits of North Atlantic Acadian-Hercynian domain. Trans Inst Min Metall 1991, 100, B227–233.
[31] Jiada W. Antimony vein deposits of China. Ore Geol Rev 1993, 8, 213–232.

[32] Zhai W, Sun X, Yi J, Zhang X, Mo R, Zhou F, Wei H, Zeng Q. Geology, geochemistry, and genesis of orogenic gold–antimony mineralization in the Himalayan Orogen, South Tibet, China. Ore Geol Rev 2014, 58, 68–90.
[33] Hu R, Fu S, Huang Y, Zhou M-F, Fu S, Zhao C, Wang Y, Bi X, Xiao J. The giant South China Mesozoic low-temperature metallogenic domain: Reviews and a new geodynamic model. J Asian Earth Sci 2017, 137, 9–34.
[34] Wagner T, Schneider J. Lead isotope systematics of vein-type antimony mineralization, Rheinisches Schiefergebirge, Germany: A case history of complex reaction and remobilization processes. Miner Deposita 2002, 37, 185–197.
[35] Bellot J-P, Lerouge C, Bailly L, Bouchot V. The Biards Sb-Au–bearing shear zone (Massif Central, France): An indicator of crustal-scale transcurrent tectonics guiding late Variscan collapse. Econ Geol 2003, 98, 1427–1447.
[36] Neiva AMR, Andráš P, Ramos JMF. Antimony quartz and antimony–gold quartz veins from northern Portugal. Ore Geol Rev 2008, 34, 533–546.
[37] Majzlan J, Chovan M, Hurai V, Luptáková J. Hydrothermal mineralisation of the Tatric Superunit (Western Carpathians, Slovakia): I. A review of mineralogical, thermometry and isotope data. Geol Carpath 2020, 71, 85–112.
[38] Zachariáš J, Frýda J, Paterová B, Mihaljevič M. Arsenopyrite and As-bearing pyrite from the Roudný deposit, Bohemian Massif. Mineral Mag 2004, 68, 31–46.
[39] Arehart GB. Characteristics and origin of sediment-hosted disseminated gold deposits: A review. Ore Geol Rev 1996, 11, 383–403.
[40] Hu R-Z, Su W-C, Bi X-W, Tu G-Z, Hofstra AH. Geology and geochemistry of Carlin-type gold deposits in China. Miner Deposita 2002, 37, 378–392.
[41] Boev B, Jovanovski G, Makreski P. Geology and mineralogy of Allchar Sb-As-Ti-Au deposit. Geol Macedon 2012, 3, 215–232.
[42] Liu -J-J, Liu J-M, Liu C-Q, Lu W-Q, Liu S-R, Su W-C. Mineralogy of the stibnite-antimonselite series. Int Geol Rev 1999, 41, 1042–1050.
[43] Ortega L, Vindel E, Beny C. C-O-H-N fluid inclusions associated with gold-stibnite mineralization in low-grade metamorphic rocks, Mari Rosa mine, Caceres, Spain. Mineral Mag 1991, 55, 235–247.
[44] Guillemette N, Williams-Jones AE. Genesis of the Sb-W-Au deposits at Ixtahuacan, Guatemala: Evidence from fluid inclusions and stable isotopes. Miner Deposita 1993, 28, 167–180.
[45] Normand C, Gauthier M, Jébrak M. The Québec antimony deposit: An example of gudmundite-native antimony mineralization in the ophiolitic mélange of the southeastern Québec Appalachians. Econ Geol 1996, 91, 149–163.
[46] Williams-Jones AE, Normand C. Controls of mineral paragenesis in the system Fe-Sb-S-O. Econ Geol 1997, 92, 308–324.
[47] Sillitoe RH, Hedenquist JW. Linkages between volcanotectonic settings, ore-fluid compositions, and epithermal precious metal deposits. Soc Econ Geol Geochem Soc Special Publication 2003, 10, 285–313.
[48] Goldfarb RJ, Hofstra AH, Simmons SF. Critical elements in Carlin, epithermal, and orogenic gold deposits. Rev Econ Geol 2016, 18, 217–244.
[49] White NC, Hedenquist JW. Epithermal gold deposits: Styles, characteristics and exploration. SEG Newsletter 1995, 23, 9–13.
[50] Vikre PG. Ledge formation at the Sandstorm and Kendall gold mines, Goldfield, Nevada. Econ Geol 1989, 84, 2115–2138.
[51] Fanlo I, Subías I, Mateo E. Supergene enrichment of primary Cu-Ag assemblages in Ag at the Pardos deposit, Iberian range, Spain. Can Mineral 2010, 48, 415–430.

[52] Sillitoe RH, Steele GB, Thompson JFH, Lang JR. Advanced argillic lithocaps in the Bolivian tin-silver belt. Miner Deposita 1998, 33, 539–546.
[53] Kelly WC, Turneaure FS. Mineralogy, paragenesis and geothermometry of the tin and tungsten deposits of the Eastern Andes, Bolivia. Econ Geol 1970, 65, 609–680.
[54] Keutsch F, de Brodtkorb MK. Metalliferous paragenesis of the San Jose mine, Oruro, Bolivia. J South Am Earth Sci 2008, 25, 485–491.
[55] Slater ET, McDonald AM, Kontak DJ. Resolving primary and retrograde sulfide and sulfosalt textures in the epithermal Ag-Zn-Pb-Sn-rich Cortaderas zone, Pirquitas mine, Argentina. Can Mineral 2019, 57, 117–143.
[56] Zhang H, Cai Y, Zhang Y, Ni P, Li S, Ding J, Pan Y, Bao T. Mineralogical characteristics of silver minerals from the Dongyang Gold deposit, China: Implications for the evolution of epithermal metallogenesis. J Geochem Explor 2018, 195, 143–156.
[57] Koděra P, Lexa J, Fallick AE, Wälle M, Biroň A. Hydrothermal Fluids in Epithermal and Porphyry Au Deposits in the Central Slovakia Volcanic Field. In: Garofalo PS, Ridley JR, eds, Gold-Transporting Hydrothermal Fluids in the Earth's Crust, Vol. 402. Geol Soc. London, Special Publications, 2014, 177–206.
[58] Wilson NJ, Webster-Brown JG, Brown KL. Controls on stibnite precipitation at two New Zealand geothermal power stations. Geothermics 2007, 36, 330–347.
[59] Naldrett AJ. Magmatic Sulfide Deposits: Geology, Geochemistry and Exploration. Heidelberg, Springer, 2004.
[60] Augé T, Salpeteur I, Bailly L, Mukherjee MM, Patra RN. Magmatic and hydrothermal platinum-group minerals and base-metal sulfides in the Baula complex, India. Can Mineral 2002, 40, 277–309.
[61] Tolstykh ND, Zhitova LM, Shapovalova MO, Chayka IF. The evolution of the ore-forming system in the low sulfide horizon of the Noril'sk 1 intrusion, Russia. Mineral Mag 2019, 83, 673–694.
[62] Wang CY, Prichard HM, Zhou M-F, Fisher PC. Platinum-group minerals from the Jinbaoshan Pd–Pt deposit, SW China: Evidence for magmatic origin and hydrothermal alteration. Miner Deposita 2008, 43, 791–803.
[63] Spiridonov EM, Kulagov EA, Serova AA, Kulikova IM, Korotaeva NN, Sereda EV, Tushentsova IN, Belyakov SN, Zhukov NN. Genetic Pd, Pt, Au, Ag, and Rh mineralogy in Noril'sk sulfide ores. Geol Ore Depos 2015, 57, 402–432.
[64] Molnár F, Watkinson DH, Jones PC. Multiple hydrothermal processes in footwall units of the North Range, Sudbury Igneous complex, Canada, and implications for the genesis of vein-type Cu-Ni-PGE deposits. Econ Geol 2001, 96, 1645–1670.
[65] Li C, Naldrett AJ. Platinum-group minerals from the deep copper zone of the Strathcona deposit, Sudbury, Ontario. Can Mineral 1993, 31, 31–44.
[66] Liu YK, Mungall JE, Ames DE. Hydrothermal redistribution and local enrichment of platinum group elements in the Tootoo and Mequillon magmatic sulfide deposits, South Raglan Trend, Cape Smith Belt, New Quebec Orogen. Econ Geol 2016, 111, 467–485.
[67] Spiridonov EM, Serova AA, Kulikova IM, Korotaeva NN, Zhukov NN. Metamorphic-hydrothermal Ag-Pd-Pt mineralization in the Noril'sk sulfide ore deposit, Siberia. Can Mineral 2016, 54, 429–452.
[68] Orlandi P, Moëlo Y, Meerschaut A, Palvadeau P. Lead-antimony sulfosalts from Tuscany (Italy). I. Scainiite, $Pb_{14}Sb_{30}S_{54}O_5$, the first Pb-Sb oxy-sulfosalt, from Bucadella Vena mine. Eur J Mineral 1999, 11, 949–954.
[69] Topa D, Sejkora J, Makovicky E, Pršek J, Ozdín D, Putz H, Dittrich H, Karup-Møller S. Chovanite, $Pb_{15-2x}Sb_{14+2x}S_{36}O_x$ ($x \approx 0.2$), a new sulphosalt species from the Low Tatra Mountains, Western Carpathians, Slovakia. Eur J Mineral 2012, 24, 727–740.

[70] Makovicky E, Topa D, Mumme WG. The crystal structure of dadsonite. Can Mineral 2006, 44, 1499–1512.
[71] Hill M. Costerfield Au-Sb prospect, Melbourne Trough, VIC. In: Butt CRM, Smith RE, Conceptual Models in Exploration Geochemistry, Vol. 4. Elsevier 1980, 214.
[72] Lazareva EV, Tsimbalist VG, Bortnikova SB, Shuvaeva OV. As and Sb Redistribution within a Cyanide-Bearing Tailings Impoundment. In: Wanty RB, Seal RRII, Water-Rock Interaction. Leiden, Balkema, 2004, 1557–1560.
[73] Diemar GA, Filella M, Leverett P, Williams PA. Dispersion of antimony from oxidizing ore deposits. Pure Appl Chem 2009, 81, 1547–1553.
[74] Lalinská-Voleková B, Majzlan J, Klimko T, Chovan M, Kučerová G, Michňová J, Hovorič R, Göttlicher J, Steininger R. Mineralogy of the Fe-As-Sb weathering products in mine wastes and soils at several Sb deposits in Slovakia. Can Mineral 2012, 50, 481–500.
[75] Sabelli C, Vezzalini G. Cetineite, A New Antimony Oxide-Sulfide Mineral from Cetine Mine, Tuscany, Italy. Monat, Neues Jahrb Miner, 1987, 419–425.
[76] Sejkora J, Hyršl J. Ottensite, a new mineral from Qinglong, Guizhou Province, China. Mineral Record 2007, 38, 77–81.
[77] Polák S. Once again on chapmanite from Pezinok. Mineral Slov 1988, 20, 475–476. in Slovak.
[78] Pauliš P, Dolníček Z, Vrtiška L, Pour O, Žák K, Ackerman L, Veselovský F, Pašava J, Kadlec T, Malíková R. Mineralogy of the stibnite deposit at Chříč near Rakovník (Czech Republic). Bull Mineral Petr 2019, 27, 148–166. in Czech.
[79] Sejkora J, Ozdín D, Vitálos J, Tuček P, Čejka J, Ďuďa R. Schafarzikite from the type locality Pernek (Malé Karpaty Mountains, Slovak Republic) revisited. Eur J Mineral 2007, 19, 419–427.
[80] Leverett P, Reynolds JK, Roper AJ, Williams PA. Tripuhyite and schafarzikite: Two of the ultimate sinks for antimony in the natural environment. Mineral Mag 2012, 76, 891–902.
[81] Mineralogical database of Hudson Institute of Mineralogy. (Accessed on May 10, 2020, at https://www.mindat.org/).
[82] Zotov AV, Shikina ND, Akinfiev NN. Thermodynamic properties of Sb(III) hydroxide complex $Sb(OH)_3(aq)$ at hydrothermal conditions. Geochim Cosmochim Acta 2003, 67, 1821–1836.
[83] Vink BW. Stability relations of antimony and arsenic compounds in the light of revised and extended Eh-pH diagrams. Chem Geol 1996, 130, 21–30.
[84] Stewart DJ, Knop O, Ayasse C, Woodhams FWD, Pyrochlores VII. The oxides of antimony: An X-ray and Mössbauer study. Can J Chem 1972, 50, 690–700.
[85] Schwarzmann E, Rumpel H, Berndt W. Synthese von Diantimonpentoxid Sb_2O_5. Z Naturforsch 1977, B 32, 617–618.
[86] Basso R, Lucchetti G, Zefiro L, Palenzona A. Clinocervantite, β-Sb_2O_4, the natural monoclinic polymorph of cervantite from the Cetine mine, Siena, Italy. Eur J Mineral 1999, 11, 95–100.
[87] Courtin-Nomade A, Rakotoarisoa O, Bril H, Grybos M, Forestier L, Foucher F, Kunz M. Weathering of Sb-rich mining and smelting residues: Insight in solid speciation and soil bacteria toxicity. Chem Erde 2012, 72, 29–39.
[88] Mason B, Vitalian CJ. The mineralogy of the antimony oxides and antimonates. Mineral Mag 1953, 30, 100–112.
[89] Handbook of Mineralogy. (Accessed on June 12, 2020, at http://www.handbookofmineralogy.org/pdfs/stibiconite.pdf).
[90] Majzlan J, Lalinská B, Chovan M, Bläß U, Brecht B, Göttlicher J, Steininger R, Hug K, Ziegler S, Gescher J. A mineralogical, geochemical, and microbiogical assessment of the antimony- and arsenic-rich neutral mine drainage tailings near Pezinok, Slovakia. Am Mineral 2011, 96, 1–13.
[91] Borčinová Radková A, Jamieson HE, Campbell KM. Antimony mobility during the early stages of stibnite weathering in tailings at the Beaver Brook Sb deposit, Newfoundland. Appl Geochem 2020, 115, 104528.

[92] Scheinost AC, Rossberg A, Vantelon D, Xifra IO, Kretzschmar R, Leuz AK, Funke H, Johnson CA. Quantitative antimony speciation in shooting range soils by EXAFS spectroscopy. Geochim Cosmochim Acta 2006, 70, 3299–3312.
[93] Mitsunobu S, Takahashi Y, Terada Y, Sakata M. Antimony(V) incorporation into synthetic ferrihydrite, goethite, and natural iron oxyhydroxides. Environ Sci Technol 2010, 44, 3712–3718.
[94] Ackermann S, Gieré R, Newville M, Majzlan J. Mineral and chemical composition of the crust of weathering bullets from Swiss shooting ranges at Losone and Lucerne. Sci Total Environ 2009, 407, 1669–1682.
[95] Burton ED, Hockmann K, Karimian N. Antimony sorption to goethite: Effects of Fe(II)-catalyzed recrystallization. ACS Earth Space Chem 2020, 4, 476–487.
[96] Bolanz RM, Bläß U, Ackermann S, Ciobota V, Rösch P, Tarcea N, Popp J, Majzlan J. The effect of antimonate, arsenate and phosphate on the transformation of ferrihydrite to goethite, hematite, feroxyhyte, and tripuhyite. Clays Clay Miner 2013, 61, 11–25.
[97] Hudson-Edwards KA. Uptake and release of arsenic and antimony in alunite-jarosite and beudantite group minerals. Am Mineral 2019, 104, 633–640.
[98] Kolitsch U, Slade PG, Tiekink ERT, Pring A. The structure of antimony dussertite and the role of antimony in oxysalt minerals. Mineral Mag 1999, 63, 17–26.
[99] Robinson C The role of jarosite and copiapite in the chemical evolution of acid drainage waters, Richmond mine, Iron Mountain, California. PhD thesis, Queen's University, Kingston, 2000.
[100] Leyva AG, Marrero J, Smichowski P, Cicerone D. Sorption of antimony onto hydroxyapatite. Environ Sci Technol 2001, 35, 3669–3675.
[101] Xi J, He M, Zhang G. Antimony adsorption on kaolinite in the presence of competitive anions. Environ Earth Sci 2014, 71, 2989–2997.
[102] Berlepsch P, Armbruster T, Brugger J, Criddle AJ, Graeser S. Tripuhyite, $FeSbO_4$, revisited. Mineral Mag 2003, 67, 31–46.
[103] Mitsunobu S, Takahashi Y, Utsunomiya S, Marcus MA, Terada Y, Iwamura T, Sakata M. Identification and characterization of nanosized tripuhyite in soil near Sb mine tailings. Am Mineral 2011, 96, 1171–1181.
[104] Borčinová Radková A, Jamieson H, Lalinská-Voleková B, Majzlan J, Števko M, Chovan M. Mineralogical controls on antimony and arsenic mobility during tetrahedrite-tennantite weathering at historic mine sites Špania Dolina-Piesky and Ľubietová-Svätodušná, Slovakia. Am Mineral 2017, 102, 1091–1100.
[105] Majzlan J, Kiefer S, Herrmann J, Števko M, Chovan M, Lánczos T, Sejkora J, Langenhorst F, Lazarov M, Gerdes A, Radková A, Jamieson H, Milovský R. Synergies in elemental mobility during weathering of tetrahedrite $[(Cu,Fe,Zn)_{12}(Sb,As)_4S_{13}]$: Field observations, electron microscopy, isotopes of Cu, C, O, radiometric dating, and water geochemistry. Chem Geol 2018, 488, 1–20.
[106] Keim MF, Staude S, Marquardt K, Bachmann K, Opitz J, Markl G. Weathering of Bi-bearing tennantite. Chem Geol 2018, 499, 1–25.
[107] Grau-Crespo R, Catlow CR, de Leeuw NH. A computer modeling study of redox processes on the $FeSbO_4$ (100) surface. J Catal 2007, 248, 77–88.
[108] Zenkovets GA, Kryukova GN, Tsybulya SV. Structural arrangement of Fe-Sb-O catalysts. Kinet Catal 2002, 43, 731–735.
[109] Huang Y, Ruiz P. The nature of antimony-enriched surface layer of Fe-Sb mixed oxides. Appl Surf Sci 2006, 252, 7849–7855.
[110] Christy AG, Atencio D. Clarification of status of species in the pyrochlore supergroup. Mineral Mag 2013, 77, 13–20.

[111] Bosi F, Christy AG, Hålenius U. Crystal-chemical aspects of the roméite group, $A_2Sb_2O_6Y$, of the pyrochlore supergroup. Mineral Mag 2017, 81, 1287–1302.
[112] Đorđević T, Kolitsch U, Serafimovski T, Tasev G, Tepe N, Stöger-Pollach M, Hofmann T, Boev B. Mineralogy and weathering of realgar-rich tailings at a former As-Sb-Cr mine at Lojane, North Macedonia. Can Mineral 2019, 57, 403–423.
[113] Okkenhaug G, Zhu Y-G, Luo L, Lei M, Li X, Mulder J. Distribution, speciation and availability of antimony (Sb) in soils and terrestrial plants from an active Sb mining area. Environ Poll 2010, 159, 2427–2434.
[114] Cornelis G, van Gerven T, Vandecasteele C. Antimony leaching from MSWI bottom ash: Modelling of the effect of pH and carbonation. Waste Manage 2012, 32, 278–286.
[115] Hockmann K, Schulin R. Leaching of Antimony from Contaminated Soils. In: Selim HM, Competitive Sorption and Transport of Heavy Metals in Soils and Geological Media. Leiden, CRC Press, 2013, 119–146.
[116] Bandara T, Vithanage M. Phytoremediation of Shooting Range Soils. In: Ansari AA, Gill SS, Gill R, Lanza GR, Newman L, Phytoremediation. Management of Environmental Contaminants, Vol. 3. Heidelberg, Springer, 2016, 469–490.
[117] Johnson CA, Moench H, Wersin P, Kugler P, Wenger C. Solubility of antimony and other elements in samples taken from shooting ranges. J Environ Quality 2005, 34, 248–254.
[118] Plášil J, Kampf AR, Meisser N, Lheur C, Brunsperger T, Škoda R. Smamite, $Ca_2Sb(OH)_4[H(AsO_4)_2]\cdot 6H_2O$, a new mineral and a possible sink for Sb during weathering of fahlore. Am Mineral 2020, 105, 555–560.
[119] Mills SJ, Christy AG, Rumsey MS, Spratt J. Discreditation of partzite. Eur J Mineral 2016, 28, 1019–1024.
[120] Ertl A, Brandstätter F. Protopartzite or thrombolite from magnesite quarry Veitsch, Sattlerkogel, Styria, Austria. Joannea-Mineral 2000, 1, 27–30.
[121] Friedrich A, Wildner M, Tillmanns E, Merz PL. Crystal chemistry of the new mineral brandholzite, $Mg(H_2O)_6[Sb(OH)_6]_2$, and of the synthetic analogues $M^{2+}(H_2O)_6[Sb(OH)_6]_2$ (M^{2+} = Mg, Co). Am Mineral 2000, 85, 593–599.
[122] Sejkora J, Ozdín D, Ďuďa R. The supergene mineral association with brandholzite from Pernek, Malé Karpaty Mountains, Slovak Republic. J Geosci 2010, 55, 149–160.
[123] Majzlan J, Števko M, Lánczos T. Soluble secondary minerals of antimony in Pezinok and Kremnica (Slovakia) and the question of mobility or immobility of antimony in mine waters. Environ Chem 2016. doi: 10.1071/EN16013.
[124] Kossoff D, Hudson-Edwards KA, Dubbin WE, Alfredsson M, Geraki T. Cycling of As, P, Pb and Sb during weathering of mine tailings: Implications for fluvial environments. Mineral Mag 2012, 76, 1209.
[125] Roper AJ, Williams PA, Filella M. Secondary antimony minerals: Phases that control the dispersion of antimony in the supergene zone. Chem Erde 2012, 72(S4), 9–14.
[126] Filella M, Philippo S, Belzile N, Chen Y, Quentel F. Natural attenuation processes applying to antimony: A study in the abandoned antimony mine in Goesdorf, Luxembourg. Sci Total Environ 2009, 407, 6205–6216.
[127] Števko M, Ozdín D, Bačík P, Bálintová T. Klebelsbergite $Sb_4O_4(OH)_2SO_4$ from the Kremnica Au-Ag deposit, Slovak Republic. Bull mineral-petrolog Odd Nár Muz 2010, 17, 25–28.
[128] Ashley PM, Craw D, Graham BP, Chappell DA. Environmental mobility of antimony around mesothermal stibnite deposits, New South Wales, Australia and Southern New Zealand. J Geochem Explor 2003, 77, 1.

Robert R. Seal II

Chapter 3 Economic geology and environmental characteristics of antimony deposits

3.1 Introduction

Antimony is commonly listed as a critical mineral, particularly in the United States and European Union [1]. Its criticality, or supply risk, is derived from a combination of economic vulnerability, disruption potential of supply, and trade exposure [2]. Disruption potential relates a country's ability and willingness to supply a commodity. Commodities for which supply is concentrated in the fewest countries have the greatest potential for supply disruption, trade exposure, and economic vulnerability. In 2018, approximately 61% of the world production of antimony was mined from China, followed by Russia (20%) and Tajikistan (10%) [3] (Fig. 3.1). Several reviews of antimony as a critical mineral have been recently published [4, 5].

The uses of antimony can be divided into three main categories: metal products, non-metal uses, and flame retardants [5]. Most metallic antimony use is in lead-acid batteries. Antimony trioxide (Sb_2O_3) combined with halogenated compounds is used as a fire retardant in plastics, fabrics, and other applications. Other non-metallic uses include as a catalyst for plastics, and in the glass industry. Emerging uses include data storage and novel photovoltaic cells. Recycling of batteries represents an important reuse of antimony, but other uses of antimony do not lend themselves to recycling.

This chapter describes mineral deposit types that are primary sources of antimony, and environmental effects related to their mining. Antimony can be recovered as either a primary commodity from some deposits or as a byproduct commodity from some gold or silver deposits. Environmental risks associated with antimony mining include those related to mining in general, such as the acid-generating potential of solid mine waste, and some issues specific to antimony, as described below.

Disclaimer: Any use of trade, firm, or product names is for descriptive purposes only and does not imply endorsement by the U.S. government.

Robert R. Seal II, U.S. Geological Survey, 954 National Center, Reston, VA 20192, USA, rseal@usgs.gov

https://doi.org/10.1515/9783110668711-003

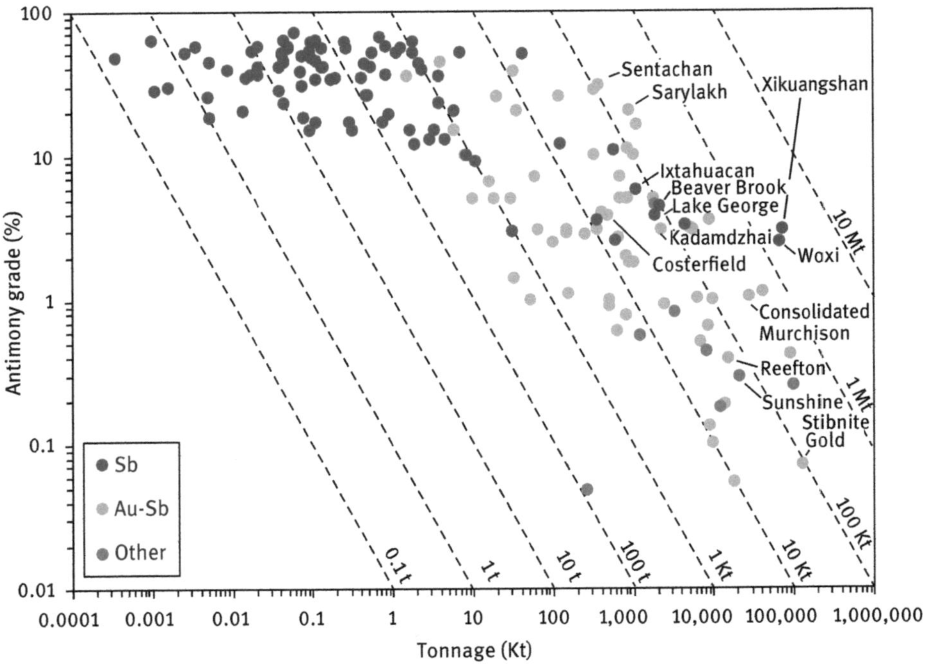

Fig. 3.1: Scatter plot showing grade and tonnage of antimony-bearing deposits [20, 21]. Diagonal dashed lines indicate the amount of contained antimony. Values for the Sunshine Mine were estimated from published data [22, 23].

3.2 Economic geology

3.2.1 Factors influencing ore genesis

Antimony deposits, in general, fail to conform to a simple classification scheme. However, many share a common set of features that relate to both source and depositional processes. These features include regionally extensive sedimentary rocks (or metamorphic equivalents) that include siliciclastic lithologies (sandstone, siltstone, shale), fracture permeability related to regional or local faulting or eruptive activity, and a heat source such as igneous intrusions or regional metamorphism.

Shales, particularly black shales, have potential to serve as fertile source rocks for hydrothermal mobilization of antimony and other trace elements during metamorphism. Fine-grained sedimentary rocks are enriched in antimony relative to other rock types. The continental crust averages 0.2 mg kg^{-1} antimony [6], yet shales average 1.5 mg kg^{-1} [7], and black shales average 5.0 ± 0.5 mg kg^{-1} [8]. Although the source, transport, and concentration mechanisms will likely vary among deposit types, a fertile source region may be essential to the formation of economic deposits. Pitcairn et al. [9]

demonstrated a significant depletion in antimony and other elements (arsenic, gold, mercury, molybdenum, silver, and tungsten) with increased metamorphic grade in the Alpine and Otago schists of New Zealand. These findings suggest that high-grade metamorphism (>400 °C) may mobilize antimony into hydrothermal fluids that may form quartz-antimony veins at lower temperature under favorable conditions. In terms of transport, factors such as temperature, oxidation state, and pH influence the solubility of antimony and its deposition. Antimony occurs predominantly as $Sb(OH)_3^0$, $HSb_2S_4^-$, or $Sb_2S_2(OH)_2^0$ complexes in hydrothermal fluids [10]. Antimony shows a dramatic decrease in solubility by two to three orders of magnitude as temperature decreases from 350 to 150 °C. The influence of decreasing temperature on stibnite solubility is consistent with the common association with quartz, which also exhibits prograde solubility [11]. Naumov et al. [12] found similar ore-forming conditions for antimony deposits in their compilation of fluid inclusion data. Mineralization occurred at a temperature from 50 to 300 °C, with earlier antimony-poor paragenetic stages being at higher temperatures, mostly below 400 °C. Under hydrothermal stibnite-stable conditions, decreasing pH from slightly alkaline conditions to neutral or slightly acidic conditions can result in stibnite deposition. In terms of the role of oxidation state, a decrease in oxidation state from sulfate-stable conditions to sulfide-stable can lead to stibnite precipitation.

Antimony deposits can undergo weathering that affects the character of the ore. The most prominent feature of weathering is the oxidation of sulfide minerals. Oxidation of antimony-bearing minerals yields a variety of oxide and sulfoxide minerals. Secondary antimony oxide and sulfoxide minerals associated with stibnite commonly include senarmonite (Sb_2O_3), valentinite (Sb_2O_3), cervantite ($Sb^{III}Sb^{V}O_4$), stibiconite ($Sb^{III}Sb_2^{V}O_6(OH)$), kermesite (Sb_2S_2O), and tripuhyite ($FeSbO_4$) [13–15]. In polymetallic ores, bindheimite ($Pb_2Sb_2O_7$; aka oxyplumboromeite [16]) and partzite ($Cu_2Sb_2(O,OH)_7$) can be found [14, 15, 17].

3.2.2 Antimony deposit classification

Ore-grade concentrations of antimony are not common, but antimony occurs in a variety of hydrothermal mineral deposit types of all ages, including epithermal veins, pegmatites, and replacement and hot-spring deposits [18]. Antimony deposits can be divided into the following two broad categories: primary antimony producers and byproduct antimony producers. However, the differences between these two groups are somewhat gradational. For example, orogenic gold deposits are commonly classified based on trace element associations: hypozonal Au-As deposits, mesozonal Au-As-Te deposits, and epizonal Au-Sb deposits [19]. Thus, increased demand for antimony due to its emerging criticality has made byproduct antimony from some gold deposits an attractive target. Likewise, the recognition that gold can be closely associated with primary antimony deposits has caused some of these antimony deposits to be reexamined for their gold content.

The distinction between primary and byproduct antimony deposits also corresponds to the empirical differences between simple stibnite deposits and complex polymetallic deposits [4]. Simple quartz-stibnite vein and replacement deposits account for most of the current and recent mine production [5]. They can form in several different types of hydrothermal systems, including the peripheral parts of orogenic gold deposits, intrusion-related gold deposits, porphyry copper and molybdenum deposits, polymetallic mesothermal vein deposits, and sediment-hosted Carlin-type gold deposits. Byproduct production has occurred from some orogenic gold deposits and from polymetallic vein deposits. However, many mines that hold potential for byproduct recovery do not do so because of economic constraints.

The size of antimony-producing deposits being mined has varied over time. Mining in the first half of the twentieth century with demand driven by World Wars I and II mostly exploited small, high-grade deposits, especially in the United States. These deposits typically contained between 0.1 and 1,000 t of total antimony (Fig. 3.1). More modern mines that produce antimony as a primary commodity tend to exploit larger deposits of lower grades. These deposits typically have between 10 Kt and 10 Mt of contained antimony. Deposits mined for their gold content as a primary commodity also contain significant amounts of antimony. Several other deposit types mined for metals such as silver, zinc, or tin can also carry antimony.

3.2.2.1 Simple quartz-stibnite deposits

Simple quartz-stibnite deposits occur either as veins or as carbonate replacements. The most significant deposits include those in China, Canada, and Bolivia. Notable deposits include the Beaver Brook and Lake George (Canada), Xikuangshan (China), Sarylakh and Sentachan (Russia), Kharma (Bolivia), and Consolidated Murchison (South Africa) deposits. The U.S. Antimony Mine (Thompson Falls) in Montana and Yellow Pine deposit in Idaho are the most important deposits that fit within this category in the United States. However, the Yellow Pine deposit is now included within the larger Stibnite Gold prospect, which has demonstrated greater value from gold resources at the site, with antimony intended for byproduct recovery [24].

The two prime examples of simple stibnite deposits in Canada are the Beaver Brook deposit in central Newfoundland [25], and the Lake George deposit in New Brunswick [26, 27]. These two deposits share many geologic similarities. Both occur in fractures of fault systems hosted by Ordovician to Silurian siliciclastic sedimentary rocks, and they are both located near Siluro-Devonian granitic intrusions. The Beaver Brook deposit contains 2.2 million metric tons of ore grading an average of 4.41% antimony [25, 28]. The Lake George deposit consists of two orebodies totaling slightly less than 2 million metric tons at average grades ranging between 3% and 4.2% antimony [27]. The mineralization at both deposits is dominated by quartz-stibnite veins with lesser amounts of carbonate minerals (calcite or dolomite) and minor amounts of pyrite (Fig. 3.2). Arsenopyrite and native antimony are important accessory phases at Lake George [26, 27]. At Lake George, the veins range in thickness from 0.5 to 1.5 m [26, 27]. The predominant alteration assemblages at Lake George are siliceous and phyllic; the siliceous alteration typically extends to less than 5 cm from the edge of the vein, whereas the phyllic alteration can extend to more than 10 m from the vein [27]. Mineralization was inferred to have occurred at temperatures ranging from 300 to 420 °C from fluids ranging in salinity from 1.5 to 18 equivalent wt.% NaCl (8.2% average); quartz deposition occurred at the higher end of the temperature range and stibnite deposition at the lower end [27].

Fig. 3.2: Photography of the Hibbard quartz-stibnite vein at the Lake George antimony mine, New Brunswick, Canada. Photo by R. Seal (1982).

The Xikuangshan deposit in Hunan Province, China, is China's and the world's largest source of antimony; it was discovered in 1540 and modern mining has been conducted for over a century [29–31]. The deposit consists of four large orebodies and several smaller ones (Fig. 3.3). The deposit replaces Devonian limestones and consists predominantly of stibnite, with trace amounts of pyrite, pyrrhotite, and sphalerite. Mineralization was localized near the crests of anticlines (Figs. 3.3 and 3.4). The primary gangue minerals are quartz and calcite along with some secondary barite and fluorite [30, 31]. The ore is mostly in two stratiform, 2.5- to 5-m-thick mineralized beds or lenses that extend for several hundred to thousands of meters within a 45-m-thick carbonate sequence. The deposit covers an area of approximately

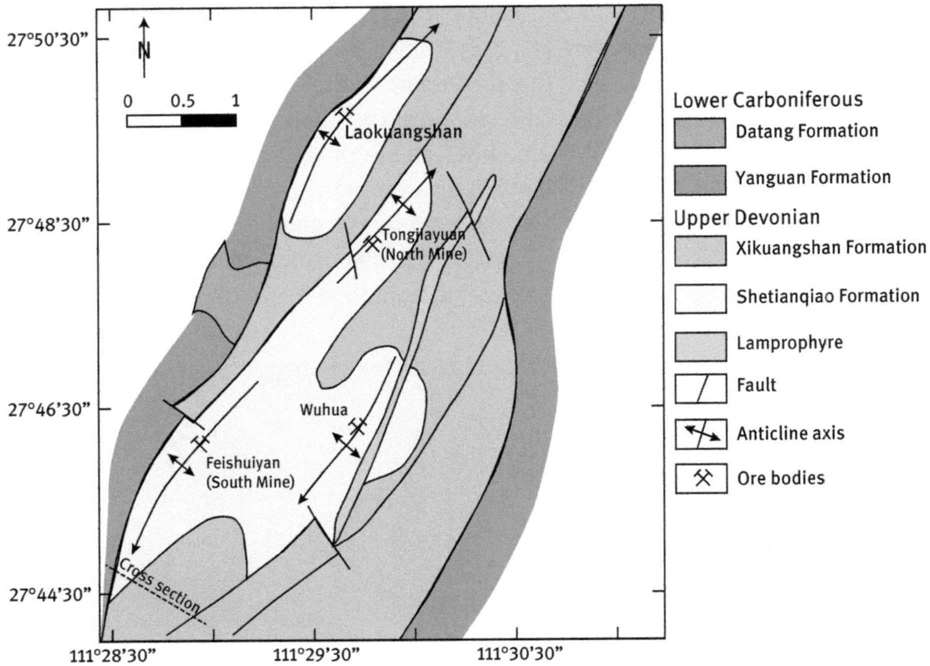

Fig. 3.3: Geologic map of the Xikuangshan district, China. Location of cross section in Fig. 3.4 is shown by dashed line. Modified from A'xiang and Jiantang [33].

16 square kilometers (km^2), has an average grade of 4% antimony, and contains more than 2 Mt of antimony (Fig. 3.3) [32]. The cross section of the district (Fig. 3.4) illustrates the importance of major fault systems to the localization of the ores, and the close association of stibnite mineralization to silicification. Quartz-stibnite and calcite-stibnite ore assemblages account for more than 90% of the ore [31]. Mineralization was divided between an early stage with quartz, stibnite, barite, and fluorite, and a later stage with calcite and lesser quartz and stibnite. The dominant alteration that accompanied mineralization was silicification; there is also some sericitic, argillic, and pyritic alteration [32]. Igneous rocks are notably absent in the vicinity of the Xikuangshan deposit [30]. A'xiang and Jiantang [33] inferred boiling conditions at Xikuangshan and reported fluid inclusion homogenization temperatures from quartz ranging from 124 to 347 °C from fluid with salinity ranging from 0.5 to 3.9 equivalent wt.% NaCl; homogenization temperature from fluid inclusions in stibnite ranged from 112 to 325 °C and salinities ranged from 0.2 to 15.4 equivalent wt.% NaCl.

The U.S. Antimony Mine in western Montana is the second largest simple quartz-stibnite vein deposit in the United States. It had significant production in the past, but was closed in 1983; the production and reserves of the mine are estimated to be nearly 15,400 t of antimony [34]. The deposit is located on the north side of a major east-west-trending wrench fault system (the Lewis and Clark line).

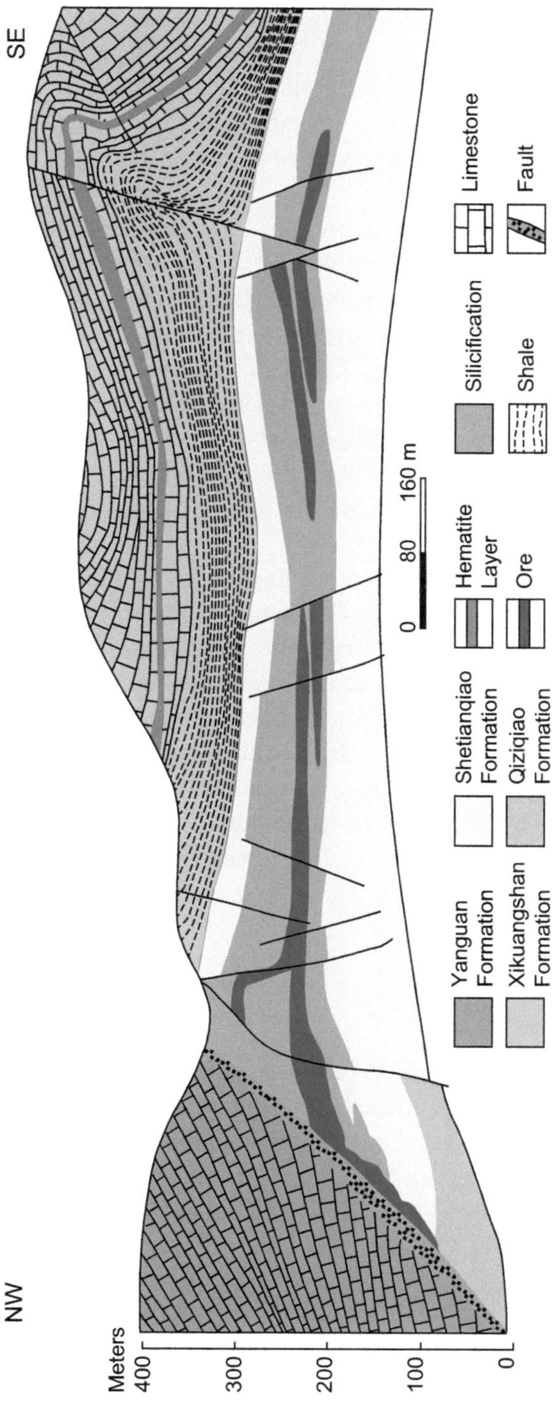

Fig. 3.4: Geologic cross section of the Xikuangshan district, China oriented from NW to SE. Modified from A'xiang and Jiantang [33].

The quartz-stibnite veins occur near the contact between quartzite and phyllite in the uppermost Prichard Formation. The veins can be as large as 1 km long and 1 m wide and contain 60% stibnite; accessory sulfide minerals include arsenopyrite, pyrite, sphalerite, and chalcopyrite [34]. The deposit formed from hydrothermal fluids ranging in temperature from 225 to 265 °C with a salinity of approximately 5 equivalent wt.% NaCl [34].

The Cordillera Oriental in Bolivia contains more than 500 known antimony-gold deposits, which are typified by the Kharma deposit. The Kharma deposit consists of fissure veins dominated by stibnite with subordinate amounts of sulfides (pyrite, arsenopyrite, and sphalerite) and sulfosalts (plagionite, fülöppite, and veenite), and minor amounts of aurostibite [35]. The gold content is generally low (about 1 mg kg^{-1}), and the gold is difficult to recover. Quartz is the main gangue mineral, accounting for up to 90% of the veins, followed by siderite and calcite. The vein system is hosted by a fault zone that cuts Paleozoic fine-grained clastic rocks (argillites), which were deformed and later intruded by granitic rocks from the Permian to Jurassic periods.

Collectively, the features that unite simple antimony deposits as a class involve both geologic setting and ore formation conditions. Key factors include the regional presence of siliciclastic rocks that serve as a potential source of antimony available for remobilization during either metamorphic or magmatic heating, and major fault or fracture zones. Most stibnite deposition occurs at temperatures below 350 °C, where antimony solubility shows a precipitous decrease with decreasing temperature [10, 12], which is consistent with its ubiquitous association with quartz, that was predominantly deposited due to cooling [11]. The predominant modes of hydrothermal alteration for the fracture-filling and carbonate replacement in simple antimony deposits are silicification and sericitization.

3.2.2.2 Byproduct deposits

Byproduct antimony deposits are generally divided between gold-antimony deposits and silver-base metal vein deposits. Significant examples include deposits in Russia, South Africa, Australia, France, and the United States. Important or representative deposits from these countries include the Sarylakh and Sentachan (Russia), Consolidated Murchison (South Africa) deposits, and the Stibnite Gold prospect and the Sunshine Mine (Idaho, USA).

The largest antimony deposits in Russia are the Sarylakh and the Sentachan antimony-gold deposits in the Yakutiya region of eastern Russia, which are located approximately 400 km apart along a major fault zone [36]. Bortnikov et al. [36] estimated that the Sarylakh deposit contains 180,000 t of antimony ore grading 20% antimony along with 40 t of gold ore grading 8 g t^{-1} gold, and that the Sentachan deposit contains 110,000 t of antimony ore grading 30% antimony along with 20 t of gold ore grading 35 g t^{-1} gold (Fig. 3.1). The Sarylakh and Sentachan deposits are hosted by deformed Triassic siliciclastic sedimentary rocks (sandstone, siltstone, and shale) that later were intruded by granitic rocks. The area also experienced volcanic activity. The orebodies are controlled by faults and are dominated by quartz and stibnite. Important, although less abundant, minerals include ankerite, arsenopyrite, berthierite, bournonite, chalcostibite, chalcopyrite, galena, native gold, pyrite, and sphalerite. The antimony mineralization and the gold mineralization were paragenetically distinct from one another at these deposits [35]. At Sentachan, the mineralized zones reach a maximum thickness of 6 m; at Sarylakh, the thickness varies from 0.8 to 17 m. The deposits are inferred to have formed between 190 and 340 °C from hydrothermal fluids containing between 1.6 and 6.8 equivalent wt.% NaCl [36].

The Consolidated Murchison Mine is the most important antimony deposit on South Africa's Antimony Line, and it is representative of mineralization along a significant Archean shear zone [37]. Mineralization was predominantly hosted by quartz-muscovite schists and occurs as quartz-carbonate veins bearing stibnite and berthierite with associated pyrite, arsenopyrite, and gold [38]. Notable deposits in France include La Lucette in the Massif Armoricain and Brioude-Massiac in the Massif Central [39]. Both yielded over 39 Kt of antimony. Stibnite in both areas was deposited at the end of a polymetallic paragenesis. Hydrothermal temperatures initially ranged from 300 to 400 °C, but fell below 300 °C during stibnite precipitation.

The Stibnite Gold prospect (Yellow Pine deposit) in Idaho contains the largest antimony resource in the United States [18] and produced intermittently from 1938 through 1992. The antimony occurs partly in small high-grade quartz-stibnite veins, rarely more than 0.5 m wide, and in large low-grade zones of disseminated stibnite in shear zones in quartz monzonite of the Idaho batholith [40]. Two periods of mineralization are recognized: the first consisted of pyrite, arsenopyrite, and gold; the second consisted of antimony and silver [40]. The Yellow Pine deposit is currently being re-evaluated as part of the Stibnite Gold Project, which has proven significant gold reserves [24].

The Sunshine silver mine in the Coeur d'Alene mining district, Idaho (USA), which exploited a polymetallic silver vein system, historically has been an important byproduct producer of antimony [22, 41, 42]. The ore occurs in steeply dipping vein systems that strike over 700 m and have a vertical extent of approximately 1,400 m. They are composed of tetrahedrite ($[Cu,Ag]_{12}Sb_4S_{13}$), galena, chalcopyrite, sphalerite, pyrite, and arsenopyrite as ore minerals and quartz and siderite as gangue [42]. Antimony was recovered from tetrahedrite [22]. Ore deposition in the Coeur d'Alene district occurred at temperatures ranging from 250 to 350 °C and salinities ranging from 5 to 10 equivalent wt.% NaCl [43].

As with the simple antimony deposits, the gold-antimony deposits similarly have close associations with regional siliciclastic rocks and fault systems. The polymetallic nature of these ores may represent a higher temperature, earlier phase of mineralization. The Cu-, Pb-, Zn-, Ag-rich character of the veins at the Sunshine mine where antimony is recovered as a byproduct is distinct from other byproduct gold-antimony deposits.

3.3 Ore processing

The metallurgy of antimony has been reviewed by Anderson [44], from which this overview is mostly derived unless otherwise noted. Stibnite is typically separated to produce an antimony concentrate by crushing and finely grinding the ore (typically <74 μm diameter), followed by froth flotation or gravity concentration. Antimony recoveries of greater than 90% are common. Flotation separation of antimony oxide minerals is more difficult [45]. Antimony can be recovered by either pyrometallurgical or hydrometallurgical techniques.

The antimony content of the ore has traditionally dictated the pyrometallurgical approach. Low-grade ores (5–25% Sb) are volatilized, medium-grade ores (25–40% Sb) are smelted in a blast furnace, and high-grade ores (40–60% Sb) are treated by liquation or iron precipitation. Volatilization involves roasting the stibnite at 1,000 °C to vaporize antimony trioxide (Sb_2O_3) as described by the reaction:

$$2\ Sb_2S_3 + 9\ O_2 \rightarrow 2\ Sb_2O_3 + 6\ SO_2$$

which is then condensed and captured in flues, condensing pipes, or a baghouse. Antimony oxides can be reduced to melt by heating at 1,200 °C as described by the reaction:

$$Sb_2O_3 + 3\,CO \rightarrow 2\,Sb + 3\,CO_2$$

Liquation melts stibnite in a reducing atmosphere at 550 to 600 °C, collects the molten sulfide, and recrystallizes it as described by the reactions:

$$Sb_2S_3(s) \rightarrow Sb_2S_3(l) \rightarrow Sb_2S_3(s)$$

The antimony sulfide may be an end product; it can be reduced to metallic antimony by iron precipitation, or be oxidized by roasting followed by reduction to metal. Iron precipitation involves heating antimony sulfide under reducing conditions with excess scrap iron to produce antimony metal as described by the reaction:

$$Sb_2S_3(s) + 3\,Fe \rightarrow 2\,Sb + 3\,FeS$$

Blast furnace smelting can be used with antimony sulfide, antimony oxide, or mixed concentrates. Concentrates mixed with flux are heated in blast furnaces at 1,300 to 1,400 °C. Operative reactions include:

$$2\,Sb_2S_3 + 9\,O_2 \rightarrow 2\,Sb_2O_3 + 6\,SO_2$$

$$Sb_2S_3 + 2\,Sb_2O_3 + 9\,O_2 \rightarrow 6\,Sb + 3\,SO_2$$

Antimony melt and slag are produced. The slags generally contain less than 1% antimony.

Hydrometallurgy can be used on both simple and complex antimony-bearing materials. Hydrometallurgical processing of antimony requires a leaching step followed by an electrodeposition step. Two different solvent systems can be used: an alkaline sulfide method and an acidic chloride method. The alkaline sulfide method is the most common. The alkaline sulfide leach solution is a mixture of sodium sulfide and sodium hydroxide; the relevant leaching reactions are:

$$Na_2S + Sb_2S_3 \rightarrow 2\,NaSbS_2$$

$$NaSbS_2 + Na_2S \rightarrow Na_3SbS_3$$

The Na_3SbS_3 is then converted to Na_3SbS_4. Antimony is recovered from solution by electrowinning via the cathodic reaction:

$$SbS_3^{3-} + 3\,e^- \rightarrow Sb^0 + 3\,S^{2-}$$

Antimony recovery from gold-antimony deposits depends on how closely associated the gold is with the antimony. If the stibnite is essentially devoid of gold, then a stibnite concentrate can be produced and processed using the options described above. In cases where gold occurs as microscopic inclusions with the stibnite, then a different approach is required [46]. Because antimony consumes appreciable cyanide during cyanidation of gold ores, the antimony must be extracted before gold can be recovered. Gold and other metals, such as silver and copper, are essentially insoluble in the alkaline sulfide method described above. Antimony leached by the alkaline sulfide solution can be recovered by electrowinning. The residue is then pressure oxidized with nitrogen species, sulfuric acid leaching process, which then renders it amenable to cyanide leaching to recover gold.

For silver-rich vein deposits that contain the mineral tetrahedrite, a similar process can be used to recover antimony, silver, and copper, such as the one developed at the Sunshine Mine in Idaho (USA) [22, 46]. A tetrahedrite concentrate is produced. An alkaline sodium sulfide leach is used to extract antimony from the concentrate, which is then recovered by electrowinning. The resulting high-grade copper-silver residue is then pressure oxidized and dissolved. The silver is recovered as silver chloride, which is then converted to elemental silver. The copper is recovered by traditional solvent extraction and electrowinning.

The metallurgical approach used to extract antimony is influenced by the grade of the deposit, and its status as a primary or byproduct commodity. Pyrometallurgy is commonly used to produce antimony oxides, antimony sulfide, and antimony metal. Hydrometallurgy, considered to be more environmentally acceptable, can be used to retrieve antimony and associated commodities like gold, copper, and silver from complex ores. Antimony from hydrometallurgy can be converted to oxides, metal, and other forms.

3.4 Environmental characteristics

The geogenic environmental risks related to mining arise from the weathering of ores within mine workings, processed ores (tailings), and waste rock. Because most modern primary antimony deposits have ore grades of less than 10% Sb, more than 90% of the material removed by mining will be waste, mostly in the form of fine-grained (<1 mm) mill tailings. The amount of waste rock will depend on whether it is an underground mine or open pit. The majority of simple antimony deposits are mined underground, whereas many of the gold-antimony deposits are open pits.

Key characteristics of tailings and waste rock are their potentials to generate or neutralize acid, their trace element endowments, and the solubility of those trace elements. The acid-generating potential (AP) and the acid-neutralizing potential (NP) of mine waste are commonly evaluated using a technique known as acid-base accounting, where the AP is estimated on the basis of the sulfide-sulfur content of the sample and the NP is estimated either on the carbonate content of the sample or by wet chemical titration [47]. Both AP and NP are expressed in units of kilograms of calcium carbonate per ton of waste (kg $CaCO_3$ t^{-1}). The results are interpreted using the neutralization potential ratio ($NPR = NP\ AP^{-1}$), where values less than 1 are deemed probably acid-generating, those above 2 are deemed non-acid-generating, and those between 1 and 2 have uncertain acid-generating potential. Limited data are available for solid mine waste from simple antimony deposits. Zhou et al. [48] reported data from tailings, waste rock, and slag from Xikuangshan (China), and Seal and Piatak [49] reported data from tailings from Pezinok (Slovakia) and Beaver Brook (Canada). Collectively, the data span the range from NPR values greater than 2 to less than 1 (Fig. 3.5). However, the NP and AP values of these mine waste materials are low (<100 kg $CaCO_3$ t^{-1}) compared to other base- and precious-metal deposit types that can reach AP values as high as 700 kg $CaCO_3$ t^{-1} or higher for pyrite-rich tailings and NP values as high as 900 kg $CaCO_3$ t^{-1} for tailings from limestone-hosted ores [49].

In addition to antimony, arsenic is a common contaminant of concern for antimony deposits. Their geochemical behavior is similar throughout the life cycle of ore genesis and weathering in the environment. However, antimony is recovered as a commodity from these deposits, whereas arsenic is sent to waste streams. Data for the antimony and arsenic content of mill tailings from simple antimony deposits are limited to examples from China, Slovakia, and Canada; collectively, reported antimony concentrations ranged from approximately 1,300 to 10,000 mg kg^{-1} and arsenic from approximately 330 to 11,500 mg kg^{-1} [49–52].

The mine drainage composition of mine sites and smelter sites depends on the weathering of waste materials and the solubility of weathering products. Roper et al. [14] compiled occurrences of secondary antimony minerals from the literature. Weathering products reflect the progressive, oxidative weathering of antimony-bearing sulfide minerals—principally stibnite—and include trivalent, pentavalent,

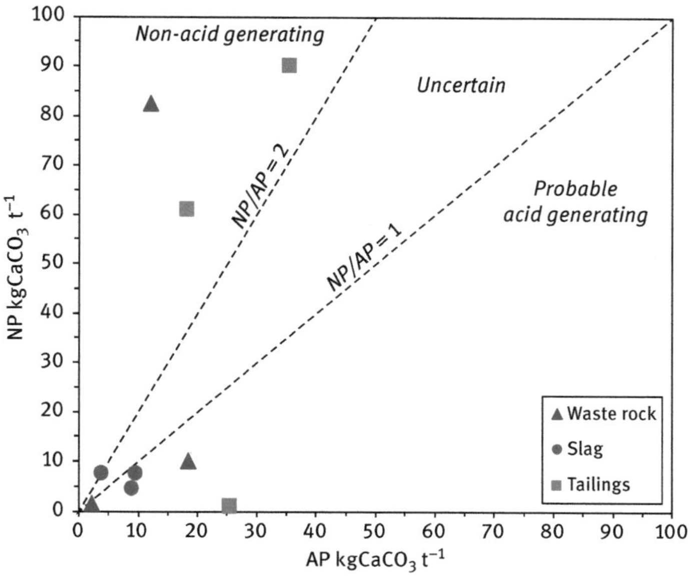

Fig. 3.5: Neutralization potential (NP) and acid-generating potential (AP) characteristics of tailings, waste rock, and slag from primary antimony mines. Data are from Seal and Piatak [49] and Zhou et al. [48].

and mixed valent species. The most common are senarmontite (Sb_2O_3), valentinite (Sb_2O_3), stibiconite ($Sb^{III}Sb^{V}_2O_6(OH)$), kermesite (Sb_2S_2O), cervantite ($Sb^{III}Sb^{V}O_4$), bindheimite ($Pb_2Sb_2O_7$), and tripuhyite ($FeSbO_4$). Note that the validity of stibiconite and bindheimite, among others, as mineral species is open to debate and requires further investigation, which is beyond the scope of this chapter [14, 16]. The common secondary minerals may locally influence antimony concentrations in mine drainage under appropriate conditions. In contrast, the less common minerals may represent short-term storage of antimony during dry periods but are acutely remobilized during rain events. For example, Multani et al. [53] noted the higher solubility of mopungite ($NaSb(OH)_6$) and brandholzite ($Mg[Sb(OH)_6]_2{\cdot}6H_2O$) compared to other secondary antimony minerals. Radková et al. [54] found that brandholzite was the most abundant secondary antimony mineral on the surface of mill tailings at Beaver Brook (Canada) when sampled during a dry period, but it was absent when sampling after a rain event. Thus, these more soluble secondary antimony minerals may act like the better studied, efflorescent metal sulfate salts that form from the oxidative weathering of sulfide minerals in mine waste settings, and store metals and acidity during dry spells, and subsequently release acutely toxic concentrations of metals and acidity by rapid dissolution during rain events [55].

Water chemistry data from simple antimony deposits are available in and around mine sites and smelter sites in Alaska [56, 57], China [48, 58–61], Canada [54],

(Sardinia) Italy [62], and New Zealand [63, 64] (Fig. 3.6). These data include surface water and groundwater data. Waters designated as mine waters refer to water discharging from mine workings or seeping from waste piles. Mine-influenced surface water and groundwater are from sites downgradient from sources and in most cases probably represent mixtures of impaired and unimpaired water. The background surface water data are from sites upstream from known sources of contamination. The drinking water standards from the World Health Organization (WHO) and the U.S. Environmental Protection Agency (USEPA) are shown in Fig. 3.6 for comparison, as are solubility curves for common secondary antimony minerals—senarmontite, cervantite, and tripuhyite. The curves were calculated using The Geochemist's Workbench [65] with the Minteq database. The solubility constant for tripuhyite is from Leverett et al. [66]. For the purpose of illustration, the oxidation-reduction potential for the cervantite and senarmontite curves was fixed assuming an Sb(V)/Sb(III) ratio of 1. The ratio represents stoichiometric dissolution of cervantite, and was used in the calculations for senarmontite merely for consistency. The activity of iron for the tripuhyite calculations was based on the assumption of the presence of goethite, a common ferric hydroxide in mine drainage settings. In actual mine systems, the prevailing redox state is complex and highly dynamic. In iron-rich acid mine drainage settings, ferrous iron dominates the initial stages of sulfide oxidation, but will covert to ferric iron as drainage begins to equilibrate with atmospheric oxygen. Thus, the solubility curves portrayed in Fig. 3.5 are only approximate and subject to underlying assumptions about oxidation state of the water.

One of the most striking features of the water chemistry data is that most pH values fall between 6.5 and 8.5. The narrow, near-neutral pH range reflects the limited amount of pyrite, the primary source of acidity, found in these ores and solid wastes, and the common presence of carbonate minerals in the host rocks and hydrothermal assemblages for these deposits. The highest antimony concentrations tend to be found at higher pH values, which are consistent with the higher solubility of oxyanions, such as antimonate and antimonite, at higher pH. Ashley et al. [17] found similar results for stibnite deposits in Australia and New Zealand. Neither senarmontite nor cervantite—trivalent or mixed valent antimony phases, respectively—appears to influence antimony concentrations, which suggests that pentavalent antimony predominates in these waters. The apparent lack of influence of tripuhyite on water chemistry in these settings may be due to the lack of appreciable iron in these deposits, the limited solubility of iron in this pH range, or the kinetic limitations of heterogeneous reactions involving iron precipitates at neutral pH. The low concentration of iron may also preclude the sorption of antimony and arsenic on to ferric hydroxides as an additional means of lower antimony and arsenic concentrations in mine drainage [53].

As a group, surface water and groundwater data from these areas provide a fairly consistent picture of the geochemistry of mined environments (Fig. 3.6(a)). The background (uncontaminated) water mostly falls below the WHO (0.02 mg L^{-1}) [67],

USEPA (0.006 mg L^{-1}) [68], and Chinese (0.005 mg L^{-1}) [69] drinking water standards. The mine waters, mine-influenced surface waters, and the mine-influenced groundwaters all span a similar range of antimony concentrations that mostly exceed drinking water standards. Only the mine water samples generally range to lower and higher pH values than USEPA secondary standards (pH 6.5–8.5). Speciation studies at Beaver Brook [54] and Xikuangshan [60] documented predominantly pentavalent antimony (>95%) in surface water and groundwater. The surface water and groundwater samples show a crude correlation between antimony and arsenic (Fig. 3.6(b)). Approximately equal numbers of samples exceed both antimony and arsenic drinking water standards. The majority of samples are enriched in antimony relative to arsenic on a mass basis.

The environmental risks at Xikuangshan (China) warrant special attention because of the size of the antimony deposit, the complexity of its mining and ore beneficiation processes, and the long history of operation. The environmental impact in the Xikuangshan district is the culmination of a legacy of mining that spans over a century since the commencement of mining and smelting in the late 1890s [60]. The Xikuangshan case study is based on multiple studies [48, 50, 59, 60, 69, 70]. The Xikuangshan district in Lengshuijiang, Hunan province, covers an area of 70 km^2. Although approximately 200 deposits of various sizes have been discovered in the district, modern, large-scale mining has been focused in the historical North (Tongjiayuan) Mine and more recently in the South (Feishuiyan) Mine areas (Fig. 3.3). The subsurface mining produced a network of stopes [50]. The surface comprises a mining industrial complex including mine and mill buildings, two major smelters with associated slag piles, a tailings storage facility at the southwestern end of the districts, and waste rock. The area experiences a subtropical, monsoonal climate with average annual rainfall of 1,354 mm, with-May to July being the wettest months [69].

Despite the extensive contamination of groundwater, surface water, sediment, and soil, limited characterization of source materials has been done in the district. Zhou et al. [48] reported acid-base accounting data for three samples of waste rock, three samples of metallurgical slag, and one sample of mill tailings from Xikuangshan that indicated low acid-generating and acid-neutralizing potentials for these materials. The NPR values of the samples of waste rock and slag span a range from probable acid-generating to non-acid-generating characteristics, whereas the sole sample of mill tailings has non-acid-generating characteristics (Fig. 3.5). He [50] reported a single analysis of mill tailings that contained 1,291 mg kg^{-1} Sb and 332 mg kg^{-1} As, which is at the low end of the range for antimony mill tailings discussed above.

Surface-water data that include antimony and other trace elements are reported by Fu et al. [68], Liu et al. [60], Wu et al. [70], Wen et al. [59], and Zhou et al. [48]. Groundwater data are reported by Wen et al. [59] and Zhou et al. [48]. The compilation of over 100 groundwater, surface water, and mine water analyses in and around

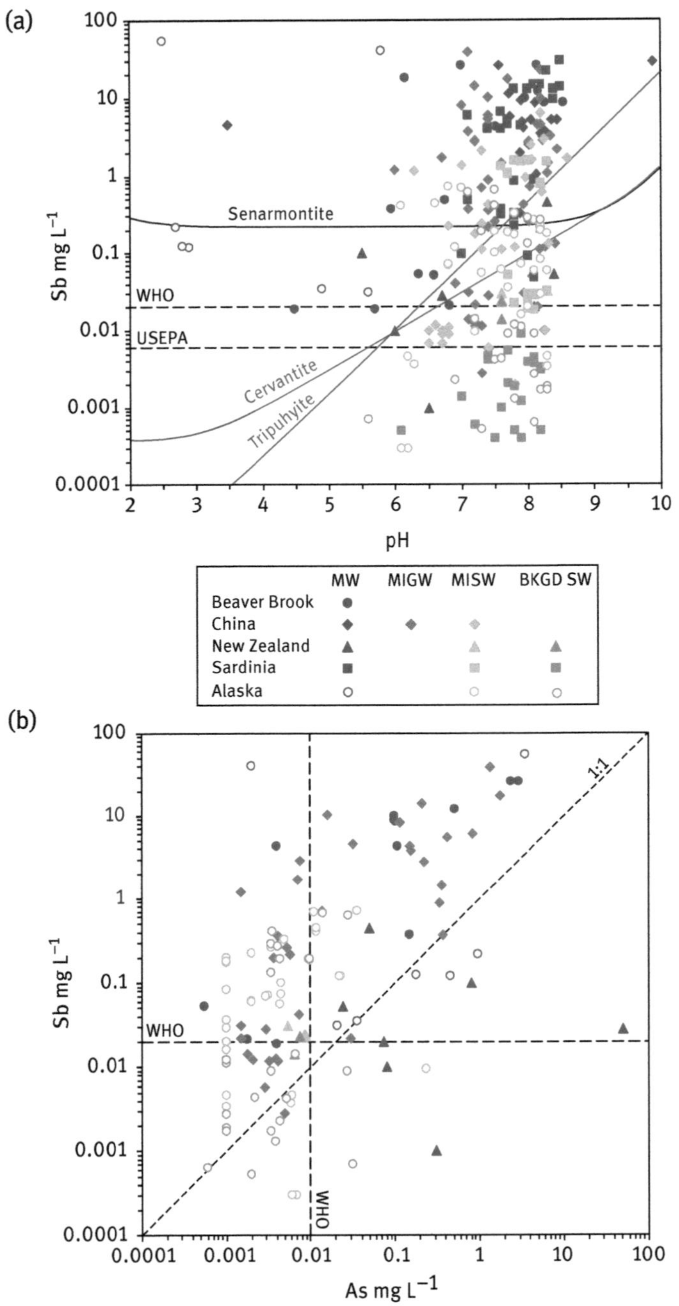

Fig. 3.6: Plots of water chemistry data in the vicinity of antimony mines and smelters from various sites reported in the literature (n = 271). The compilation includes total, dissolved, and unspecified concentrations. Waters are categorized as mine waters (MW), mine-influenced surface water (MISW), mine-influenced groundwater (MIGW), and background (upgradient) surface water (BKGD SW). Only

the Xikuangshan mining district clearly documents the impact that the extensive mining history has had on the environment. All but three of the samples exceed antimony drinking water standards. Surface water and groundwater impacts are most severe in the areas immediately adjacent to the North Mine, South Mine, and Longwachi tailings storage facility (Fig. 3.7). Arsenic concentrations, although more limited, reached a maximum concentration of 1.765 mg L^{-1} [48]. The highest antimony concentrations are found in mine water samples, and in isolated surface water and groundwater samples near the historical North Mine [59]. Stream sediment samples in the district show a steady decrease in antimony concentrations with distance from the mines, starting with concentrations of 7,316 mg kg^{-1} and decreasing to 57 mg kg^{-1} [61]. He [50] investigated soil antimony concentrations within 1 km of the mine area and found elevated concentrations ranging from 97.1 to 4,489.9 mg kg^{-1}.

Dietary health risks in an area encompassing approximately 300 km^2 around the Xikuangshan district were investigated by Wu et al. [70]. The study area included 16 villages with a combined population of over 10,000. They collected 51 drinking water samples and found that all samples exceeded the Chinese drinking water standard for antimony. Their study concluded that drinking water constituted only 19% of the local daily dietary intake of antimony, and that locally grown rice (33%), vegetables (26%), and meat and poultry (19%) comprised most of the balance. In the case of Xikuangshan, the long history of extensive mining, much of it with limited regulation, has left an indelible environmental imprint on the surrounding environment.

The environmental characteristics of gold-antimony deposits are not as well studied as those of simple antimony deposits. Some of the most detailed investigations have been done around the Hillgrove mine, New South Wales, Australia, and the Reefton district, New Zealand [17, 63, 71]. The Hillgrove mine has produced both antimony and gold, whereas the Reefton district produced a sulfide (stibnite, arsenopyrite, pyrite) concentrate for its gold content. At Hillgrove, the mineralized veins are dominated by quartz and stibnite, with lesser amounts of carbonate, arsenopyrite, and gold, among other minor minerals [17].

The environmental features at the Hillgrove mine are summarized from Ashley et al. [17]. At Hillgrove, stibnite is the predominant antimony mineral but arsenopyrite associated with the ores can contain in excess of 5,000 mg kg^{-1} Sb. Surface water and seepage from the tailings storage facility generally conform to the characteristics of antimony mine waters summarized above (Fig. 3.6): pH was near neutral, ranging from 6.2 to 8.5; antimony reached a maximum of 55 mg L^{-1} (among the

Fig. 3.6 (continued)
those categories for each site for which data are available are shown in the legend. Drinking water standards from the World Health Organization (WHO) [67] and the U.S. Environmental Protection Agency (USEPA) [68] are shown as dashed lines. a. pH versus antimony. The solubility limits for various secondary minerals, calculated using The Geochemist's Workbench, are shown by solid lines; b. antimony versus arsenic.

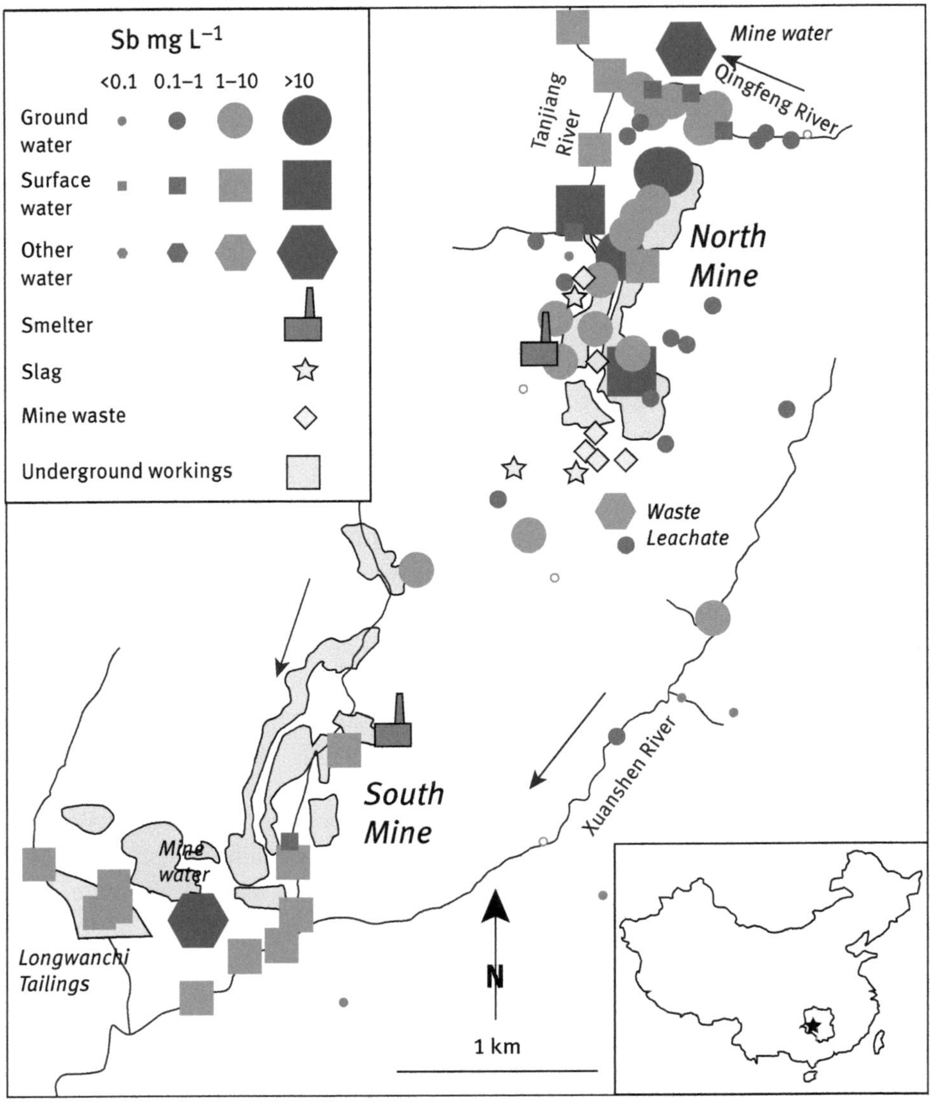

Fig. 3.7: Map showing various mine complex features and antimony concentrations of groundwater, surface water, and other mine waters in the Xikuangshan district. Data are compiled from Liu et al. [60], Wen et al. [59], and Zhou et al. [48]. Open low-concentration samples have antimony concentrations below the WHO drinking standard of 0.02 mg L^{-1}.

highest concentrations reported globally); arsenic reached 7.2 mg L^{-1}, and sulfate ranged 366 to 3,530 mg L^{-1}. Iron flocculants that form as waters emerge from tailings piles effectively scavenge antimony and arsenic from solution; iron precipitates reach concentrations of 11.2% Sb and 10.3% As. The receiving stream had antimony concentrations between 0.47 and 1.80 mg L^{-1} and arsenic concentrations between 0.013 and 0.285 mg L^{-1}. The river that receives the impacted stream rapidly diluted concentrations to near-background levels (0.002 mg L^{-1} Sb; 0.003 mg L^{-1} As). Upstream, natural background concentrations in stream sediments were estimated at 1 mg kg^{-1} Sb and 7.2 mg kg^{-1} As; up to 5 km downstream of the mine, concentrations in sediments ranged between 80 and 4,640 mg kg^{-1} Sb and 102 to 1,490 mg kg^{-1} As. Over the next 20 km, sediment concentrations decreased to 190 to 250 mg kg^{-1} Sb and 90 to 170 mg kg^{-1} As.

The environmental features at the Reefton mine are summarized from Milham and Craw [63] and Ashley et al. [71]. Early mining at Reefton exploited ores with low Sb contents (<100 mg kg^{-1}) but higher As contents (>1,000 mg kg^{-1}). Newer breccia ores commonly contain stibnite and have higher Sb contents. Recent mine waters contain 0.1 to 1 mg L^{-1} Sb. Mine waters issuing from historical adits had concentrations as high as 0.076 mg L^{-1} but got diluted to near-background levels (0.003 mg L^{-1}) approximately 2 km downstream. The mine produced a gold-bearing sulfide (stibnite, arsenopyrite, pyrite) concentrate that was shipped off-site for processing. Historical mining in the district produced ores and mine drainage with elevated As and Sb concentrations, with As being more abundant on a mass basis. More recent mining that exploited stibnite-bearing ores had ore concentrations with subequal amounts of antimony and arsenic. Mine pit waters were slightly enriched in arsenic, whereas plant process waters were slightly enriched in antimony.

3.5 Conclusions

Antimony should continue to be a critical commodity, especially as nascent applications to computer data storage and photovoltaic cells evolve. Because the grades and tonnages of antimony in orogenic gold-antimony deposits rival the grades and tonnages found in simple quartz-antimony deposits, byproduct production of antimony should increasingly satisfy future demand, which will broaden the supply base. The spectrum of metallurgical approaches will allow ore-processing approaches to be customized to various end users. Environmental risks related to antimony deposits are problematic. Exceedances of antimony drinking water standards by several orders of magnitude are nearly ubiquitous in mine-influenced surface water and groundwater in antimony mining districts. Antimony concentrations approaching 40 mg L^{-1} have been reported. Common secondary antimony minerals seem to exert little to no influence on maximum concentrations. Further, the near-neutral pH of antimony mine waters limits dissolved iron concentrations that otherwise could serve as an effective sorbent for both antimony and arsenic. Future sustainable supply of antimony will require both enhanced recovery of byproduct antimony from gold-antimony and polymetallic vein deposits and viable remedies to elevated antimony (and arsenic) concentrations in mine waters.

References

[1] Hayes SM, McCollough EA. Critical minerals: A review of elemental trends in comprehensive criticality studies. Resour Policy 2018, 59, 192–199.

[2] Nassar NT, Brainard J, Gulley A, Manley R, Matos GR, Lederer G, Bird LR, Pineault D, Alonso E, Gambogi J, Fortier SM. Evaluating the mineral commodity supply risk of the U.S. manufacturing sector. Sci Adv 2020, 6, eaay8647.

[3] US Geological Survey (USGS). Mineral commodity summaries 2020. U.S. Geological Survey, 2020, (Accessed June 25, 2020 at https://doi.org/10.3133/mcs2020)

[4] Schwarz-Schampera U. Antimony. In: Gunn G, ed., Critical Metals Handbook. Hoboken, N.J., Wiley and Sons, 2014, 70–98.

[5] Seal RR II, Schulz KJ, DeYoung JH Jr. Antimony, chap. C *of*. In: Schulz KJ, DeYoung JH Jr., Seal RR II, Bradley DC, eds., Critical Mineral Resources of the United States – Economic and Environmental Geology and Prospects for Future Supply. U.S. Geological Survey Professional Paper 1802, 2017, C1–C17.

[6] Taylor SR, McLennan SM. The geochemical evolution of the continental crust. Rev Geophys 1995, 33, 241–265.

[7] Li Y-H, Schoonmaker JE. Chemical Composition and Mineralogy of Marine Sediments. In: McKenzie FT, ed., Sediments, Diagenesis, and Sedimentary Rocks. Oxford, United Kingdom, Elsevier-Pergamon, 2003, 1–35.

[8] Ketris MP, Yudovich YE. Estimations of Clarkes for carbonaceous biolithes—World averages for trace element contents in black shales and coals. Int J Coal Geol 2009, 78, 135–148.

[9] Pitcairn IK, Teagle DAH, Craw D, Olivo GR, Kerrich R, Brewer S. Sources of metals and fluids in orogenic gold deposits – Insights from the Otago and Alpine schists, New Zealand. Econ Geol 2006, 101, 1525–1546.

[10] Williams-Jones AE, Normand C. Controls on mineral parageneses in the system Fe-Sb-S-O. Econ Geol 1997, 92, 308–324.

[11] Rimstidt JD. Gangue Mineral Transport and Deposition. In: Barnes HL, ed., Geochemistry of Hydrothermal Ore Deposits, 3rd Edition. Wiley, New York, 1997, 487–515.

[12] Naumov VB, Dorofeeva VA, Mironova OF. Physicochemical parameters of the origin of hydrothermal mineral deposits: Evidence from fluid inclusion. V. Antimony, arsenic, and mercury deposits. Geochem Int 2018, 56, 901–914.

[13] Aral H. Antimony mineralization in the northern Murat Dagi (Western Turkey). Econ Geol 1989, 84, 780–787.

[14] Szakall S, Papp G, Sajo I, Kovács Á. Antimony oxide minerals from Hungary. Acta Mineralogica-Petrographica 2000, 41, 31–62.

[15] Roper AJ, Williams PA, Filella M. Secondary antimony minerals: Phases that control the dispersion of antimony in the supergene zone. Chemie der Erde 2012, 72, 9–14.

[16] Atencio D, Andrade MB, Christy AG, Gieré R, Kartashov PM. The pyrochlore supergroup of minerals: Nomenclature. Can Mineral 2010, 48, 673–698.

[17] Ashley PM, Craw D, Graham BP, Chappell DA. Environmental mobility of antimony around mesothermal stibnite deposits, New South Wales, Australia and southern New Zealand. J Geochem Explor 2003, 77, 1–14.

[18] Miller MH. Antimony. In: Brobst DA, Pratt WP, eds., United States Mineral Resources. U.S Geological Survey Professional Paper, Vol. 820, 1973, 45–50.

[19] Groves DI, Goldfarb RJ, Robert F, Hart CJR. Gold deposits in metamorphic belts: Overview of current understanding, outstanding problems, future research, and exploration significance. Econ Geol 2003, 98, 1–29.

[20] Bliss J, Orris G. Grade and Tonnage Model of Disseminated Sb Deposits – Model 27e. In: Cox DP, Singer DA, eds., Mineral Deposit Models. U.S. Geological Survey Bulletin 1693, 1986, 187–188.
[21] Berger VI Descriptive, and grade and tonnage model for gold-antimony deposits. U.S. Geological Survey Open-File Report 93–194, 1993.
[22] Anderson CG, Nordwick SM, Krys LE Processing of antimony at the Sunshine Mine. Conference Proceedings, Residues and Effluents: Processing and Environmental Considerations 1992, 349–366.
[23] Long KR Grade and tonnage models for Coeur d'Alene-type polymetallic veins. U.S. Geological Survey Open-File Report 98–583, 1998.
[24] Huss CE, Kirkham GD, Martin CJ, Marek JM, Anderson AR, Kinder RC, Kowalewski PE Stibnite Gold project – Prefeasibility study technical report. Tucson, Arizona, M3 Engineering + Technology Corp. for Midas Gold Corp., December 15, 2014, (Accessed June 25, 2020 at http://www.midasgoldcorp.com/i/pdf/Midas-Gold-Stibnite-Gold-Project-2014-PFS.pdf.)
[25] Lake JWL, Wilton DHC. Structural and Stratigraphic Controls on Mineralization at the Beaver Brook Antimony Deposit, Central Newfoundland. In: Pereira CPG, Walsh DG, eds., Current research: St. John's, Newfoundland and Labrador. Canada, Newfoundland and Labrador Department of Natural Resources, Geological Survey Report, 2006, 135–146.
[26] Scratch RB, Watson GP, Kerrich R, Hutchinson RW. Fracture-controlled antimony-quartz mineralization, Lake George deposit, New Brunswick – Mineralogy, geochemistry, alteration, and hydrothermal regimes. Econ Geol 1984, 79, 1159–1186.
[27] Seal RR II, Clark AH, Morrissy CJ, Lake G. Southwestern New Brunswick—A Silurian, Multi-Stage, Polymetallic (Sb-W-Mo-Au-base metal) Hydrothermal Centre. In: Taylor RP, Strong DF, eds., Recent Advances in the Geology of Granite-Related Mineral Deposits. Canadian Institute of Mining and Metallurgy. Special, Vol. ume 39, 1988, 252–264.
[28] Tallman P, Evans DTW. Geology of stibnite mineralization at the Hunan Line prospects, central Newfoundland: Current Research. Newfoundland Department of Mines and Energy, Geological Survey Branch, Report 94–1, 1994, 263–271.
[29] Panov BS, No L. The Xiguanshan antimony deposit, China. Int Geol Rev 1989, 31, 100–110.
[30] Wu J. Antimony vein deposits of China. Ore Geol Rev 1993, 8, 213–232.
[31] Fan D, Zhang T, Ye J. The Xikuangshan Sb deposit hosted by the Upper Devonian black shale series, Hunan, China. Ore Geol Rev 2004, 24, 121–133.
[32] Yang D-S, Shimizu M, Shimazaki H, Li X-H, Xie Q-L. Sulfur isotope geochemistry of the supergiant Xikuangshan Sb deposit, central Hunan, China—Constraints on sources of ore constituents. Resour Geol 2006, 56, 385–396.
[33] A'xiang H, Jiantang P. Fluid inclusions and ore precipitation mechanism in the giant Xikuangshan mesothermal antimony deposit, South China: Conventional and infrared microthermometric constraints. Ore Geol Rev 2018, 95, 49–64.
[34] Hofstra AH, Marsh EE, Todorov TI, Emsbo P. Fluid inclusion evidence for a genetic link between simple antimony veins and giant silver veins in the Coeur d'Alene mining district, ID and MT, USA. Geofluids 2013, 13, 475–493.
[35] Dill HG, Weiser T, Bernhardt IR, Kilibarda CR. The composite gold-antimony vein deposit at Kharma (Bolivia). Econ Geol 1995, 90, 51–66.
[36] Bortnikov NS, Gamynin GN, Vikent'eva OV, Prokof'ev VY, Prokop'ev AV. The Sarylakh and Sentachan gold-antimony deposits, Sakha-Yakutia – A case study of combined mesothermal gold-quartz and epithermal stibnite ores. Geol Ore Deposits 2010, 52, 339–372.
[37] Jaguin J, Poujol M, Boulvais P, Robb LJ, Paquette JL. Metallogeny of precious and base metal mineralization in the Murchison Greenstone Belt, South Africa – Indications from U-Pb and Pb-Pb geochronology. Mineralium Deposita 2012, 47, 739–747.

[38] Davis DR, Paterson DB, Griffiths DHC. Antimony in South Africa. J S Afr Inst Min Metall 1986, 86, 173–193.
[39] Munoz M, Courjault-Radé P, Tollon F. The massive stibnite veins of the French Palaeozoic basement: A metallogenic marker of Late Variscan brittle extension. Terra Nova 1992, 4, 171–177.
[40] White DE. Antimony deposits of a part of the Yellow Pine district, Valley County, Idaho. U.S. Geol Surv BullU.S 922–I, 1940, 247–279.
[41] Mauk JL, White BG. Stratigraphy of the Proterozoic Revett Formation and its control on Ag-Pb-Zn vein mineralization in the Coeur d'Alene District, Idaho. Econ Geol 2004, 99, 295–312.
[42] Bryan RC, Kunter RS, Prenn NB NI 43–101 Technical Report, Mineral Resource and Preliminary Assessment of the Sunshine Silver Mine Project. Tetra Tech, Golden, Colorado (USA), 2012 (Accessed June 25, 2020 at: https://www.miningdataonline.com/reports/SunshineSilver Mine_PEA_03212014.pdf)
[43] Leach DL, Landis GP, Hofstra AH. Metamorphic origin of the Coeur d'Alene base- and precious-metal veins in the Belt basin, Idaho and Montana. Geology 1988, 16, 122–125.
[44] Anderson CG. The metallurgy of antimony. Chemie der Erde 2012, 72, 3–8.
[45] Lager T, Forssberg KSE. Beneficiation characteristics of antimony minerals: A review – part 1. Miner Eng 1989, 2, 321–336.
[46] Anderson CG, Nordwick SM. Pretreatment using alkaline sulfide leaching and nitrogen species catalyzed pressure oxidation on a refractory gold concentrate. JOM 1996, 23–341.
[47] Hageman PL, Seal RR, Diehl SF, Piatak NM, Lowers HA. Evaluation of selected static methods used to estimate element mobility, acid-generating and acid-neutralizing potentials associated with geologically diverse mining wastes. Appl Geochem 2015, 57, 125–139.
[48] Zhou J, Nyirenda MT, Xie L, Li Y, Zhou B, Zhu Y, Liu H. Mine waste acidic potential and distribution of antimony and arsenic in waters of the Xikuangshan mine, China. Appl Geochem 2017, 77, 52–61.
[49] Seal RR II, Piatak NM, Environmental attributes and resource potential of mill tailings from diverse mineral deposit types. Proceedings, International Mine Water Association Conference, Mine Water and Circular Economy, Wolkersdorfer C, Sartz L, Sillanpää M, Häkkinen A, eds, Lappeenranta, Finland, 2017, 603–609.
[50] He M. Distribution and phytoavailability of antimony at an antimony mining and smelting area, Hunan, China. Environ Geochem Health 2007, 29, 209–219.
[51] Majzlan J, Lalinská B, Chovan M, Bläß U, Brecht B, Götticher J, Steininger R, Hug K, Ziegler S, Gescher J. A mineralogical, geochemical, and microbiological assessment of the antimony- and arsenic-rich neutral mine drainage tailings near Pezinok, Slovakia. Am Mineral 2011, 96, 1–13.
[52] Guo F, Wu FC, Yu F, Bai YC, Fu ZY, Guo WJ. Fate and removal of antimony in response to stringent control activities after a mine tailing spill. Sci Total Environ 2019, 693, article 133604.
[53] Multani RS, Feldmann T, Demopoulos GP. Antimony in the metallurgical industry: A review of its chemistry and environmental stabilization options. Hydrometallurgy 2016, 164, 141–153.
[54] Radková AB, Jamieson HE, Campbell KM. Antimony mobility during the early stages of stibnite weathering in tailings at the Beaver Brook Sb deposit, Newfoundland. Appl Geochem 2020, 115, article 104528.
[55] Chou I-M, Seal RR II, Wang A. The stability of sulfate and hydrated sulfate minerals near ambient conditions and their significance in environmental and planetary sciences. J Asian Earth Sci 2013, 62, 734–758.
[56] Eppinger RG, Briggs PH, Crock JG, Meier AL, Sutley SJ, Theodorakos PM. Environmental-geochemical study of the Slate Creek antimony deposit, Kantishna Hills, Denali National Park and Preserve, Alaska, chap. 8 *of*. In: Wilson FH, Galloway JP, eds., Studies by the U.S. Geological Survey in Alaska, 2000. U.S., Geological Survey Professional Paper 1662, 2002, 123–141.

[57] Ritchie VJ, Ilgen AG, Mueller SH, Trainor TP, Goldfarb RJ. Mobility and chemical fate of antimony and arsenic in historic mining environments of the Kantishna Hills district, Denali National Park and Preserve, Alaska. Chem Geol 2013, 335, 172–188.
[58] Wang X, He M, Xi J, Lu X. Antimony distribution and mobility in rivers around the world's largest antimony mine of Xikuangshan, Hunan Province, China. Microchem J 2011, 97, 4–11.
[59] Wen B, Zhou J, Zhou A, Liu C, Xie L. Sources, migration and transformation of antimony contamination in the water environment of Xikuangshan, China: Evidence from geochemical and stable isotope (S, Sr) signatures. Sci Total Environ 2016, 569–570, 114–122.
[60] Liu F, Le XC, McKnight-Whitford A, Xia Y, Wu F, Elswick E, Johnson CC, Zhu C. Antimony speciation and contamination of waters in the Xikuangshan antimony mining and smelting area, China. Environ Geochem Health 2010, 32, 401–413.
[61] He M, Zhang C, Zhong Q, Shan J, Wang T, Liu H, Jiao Y, Song K, Yin X. Antimony in China. This volume.
[62] Cidu R, Biddau R, Dore E, Vacca A, Marini L. Antimony in the soil-water-plant system at the Su Suergiu abandoned mine (Sardinia, Italy): Strategies to mitigate contamination. Sci Total Environ 2014, 497–498, 319–331.
[63] Milham L, Craw D. Antimony mobilization through two contrasting gold ore processing systems, New Zealand. Mine Water Environ 2009, 28, 136–145.
[64] Wilson NJ, Craw D, Hunter K. Antimony distribution and environmental mobility at an historic antimony smelter site, New Zealand. Environ Pollut 2004, 129, 257–266.
[65] Bethke CM. Geochemical and Biogeochemical Reaction Modeling, 2nd Edition. Cambridge University Press, 2010.
[66] Leverett P, Reynolds JK, Roper AJ, Williams PA. Tripuhyite and schafarzikite: Two of the ultimate sinks for antimony in the environment. Mineral Mag 2012, 76, 891–902.
[67] World Health Organization (WHO). Guidelines for drinking-water quality (4th ed., incorporating the first an addendum): Geneva, Switzerland, World Health Organization, 2017 (Accessed June 25, 2020, at https://www.who.int/water_sanitation_health/publications/drinking-water-quality-guidelines-4-including-1st-addendum/en/.)
[68] US Environmental Protection Agency, 2009, Drinking water contaminants – Standards and regulations – National primary drinking water regulations – Table of contaminants: U.S. Environmental Protection Agency Web page (Accessed June 25, 2020, at https://www.epa.gov/ground-water-and-drinking-water/national-primary-drinking-water-regulations)
[69] Fu Z, Wu F, Amarasiriwardena D, Mo C, Liu B, Zhu J, Deng Q, Liao H. Antimony, arsenic and mercury in the aquatic environment and fish in a large antimony mining area in Hunan, China. Sci Total Environ 2010, 408, 3403–3410.
[70] Wu F, Fu Z, Liu B, Mo C, Chen B, Corns W, Liao H. Health risks associated with dietary co-exposure to high levels of antimony and arsenic in the world's largest antimony mine area. Sci Total Environ 2011, 409, 3344–3351.
[71] Ashley PM, Craw D, Tighe MK, Wilson NJ. Magnitudes, spatial scales and processes of environmental antimony mobility from orogenic gold-antimony mineral deposits, Australasia. Environ Geol 2006, 51, 499–507.

Philip Nuss

Chapter 4 Review of the anthropogenic flows and stocks of antimony

4.1 Introduction

The recent decades have seen a strong increase in the global extraction of raw materials such as fossil fuels, biomass, metals, and non-metallic minerals. Globally, raw materials extraction has increased from around 6 Gt (billion metric tons) in 1900 to more than 90 Gt in 2017 [1]. Future extraction could further double to around 160 to 180 Gt by around 2050 [1–3]. According to estimates of the United Nations Environment Programme International Resource Panel (UNEP IRP), the extraction and processing of raw materials are associated with about half of global greenhouse gas (GHG) emissions and more than 90% of land- and water-related impacts [1, 4]. Therefore, managing raw materials more sustainably is an essential lever for tackling climate change [5] and environmental protection in general.

Metals and metalloids (elements with properties that are intermediate between those of metals and nonmetals) represent an important backbone of modern economies [6] and are important for the transition to a low-carbon and resource-efficient economy [7, 8] as they are widely used, for example, in renewable energy systems and energy storage, electric mobility, and digitalization. Metals and metalloids are also essential for achieving the United Nations Sustainable Development Goals (SDGs) [9, 10].

An increasing number of institutions and activities at global, regional, and national levels exist to support more sustainable materials and resource management [11]. These include, for instance, the UNEP IRP that was formed in 2007 to "build and share the knowledge needed to improve our use of resources worldwide" [12]. The UNEP IRP includes a working group on global metal flows that produces scientific assessments of the global stocks and flows of metals [13], their environmental impacts [14], and more recently also their governance [15]. The Organization for Economic Co-operation and Development (OECD) promotes the sustainable use and management of (raw) materials by fostering resource productivity (a measure of the total amount of materials used by an economy in relation to a benefit such as economic growth or welfare) and

Disclaimer: This chapter does not necessarily reflect the opinion or the policies of the German Federal Environment Agency (UBA).

Philip Nuss, German Environment Agency (UBA), Section I1.1 – Fundamental Aspects, Sustainability Strategies and Scenarios, Sustainable Resource Use, Woerlitzer Platz 1, 06844 Dessau-Rosslau, Germany. philip@nuss.me

https://doi.org/10.1515/9783110668711-004

waste management, as well as through the development of material flows data and methods [3, 16, 17]. The EU has recently published its updated circular economy action plan to increase the circular use or raw materials [18] as part of the EU Green Deal with the overarching goal of making Europe climate neutral by 2050 [19]. At the country-level, different policy programs for more sustainable materials management exist [20]. For example, in Germany the Resource Efficiency Program (ProgRess) aims at a more sustainable extraction and use of natural resources and to preserve the natural foundations of life for the long term [21–23].

The anthropogenic cycles of the elements (largely in focus by the above-mentioned institutions and activities) are embedded in their larger natural (biogeochemical) cycles in which the exchange between the anthroposphere and the environment, and the subsequent transport and fate of elements in nature, are examined. Given that increasing amounts of metals and metalloids are nowadays used, accumulated, and eventually released from the anthroposphere to nature, material flow analysis (MFA) [24, 25] is a useful tool to provide estimates about the sources and magnitude of emissions as well as anticipated future exchanges between the anthroposphere and nature considering material end-use patterns and related in-use stock dynamics. MFA is an analytical method to quantify the flows and stocks of materials in a defined system (e.g., a country or region). In MFA, "materials" is used as an umbrella term for both substances and goods [24]. Understanding the flows, stocks, sources, and sinks of materials in man-made (anthropogenic) and natural systems is crucial for better managing the existing material cycles, understanding their transport and fate, and making educated guesses about future materials use and resource constraints.

Materials flows and stocks information is also essential for policy makers working on resource policies as it sheds light, for example, on the level of circularity of man-made materials cycles or pollution sources – information required when assessing subsequent human and ecosystem risks and when implementing protection measures. Furthermore, considering research on both the natural and anthropogenic cycling of materials helps to more holistically view element cycles and provides a bridge between the research communities of biogeochemists and MFA practitioners/natural resource managers [26].

Carrying out an MFA study requires information from a wide range of scientific disciplines, including environmental science, engineering, economics, chemistry, and earth science. Designing policy solutions for better materials management requires transforming this information subsequently into tangible actions and measures for policy making. However, knowledge gaps about the natural and anthropogenic cycles still exist for many metals, and the available information is of varying completeness and quality [27].

The **focus of this chapter** is on the anthropogenic cycle of antimony, which can be sketched using MFA, and a discussion in the context of sustainable resource and materials management. For this, the chapter provides an overview of published studies on antimony material flows and stocks with details on the different life-cycle stage (from extraction of antimony ores to the processing and manufacture of semi-finished and final products, the use phase, and subsequent waste management including recycling at end-of-life). This includes an overview of available estimates of the environmental losses, which provide a direct link to natural cycles of antimony and which are discussed in greater detail in other chapters of this book.

4.2 Anthropogenic antimony cycle

Antimony is a soft, lustrous, silver-grey metalloid [28] that has been used by humans since the bronze age [29, 30]. It is a naturally occurring element that was used by the Egyptians as black stibnite (Sb_2S_3) for the use of cosmetics as early as 3000 to 4000 BC and later in the form of different antimony compounds, for instance, in medicines [29, 31]. World production of antimony increased significantly during the twentieth century [32] (see Fig. 4.1). Antimony is mainly used in lead-acid batteries (where the lead plate grids are strengthened through the addition of antimony) and in flame retardants (where antimony is used together with halogenated compounds).

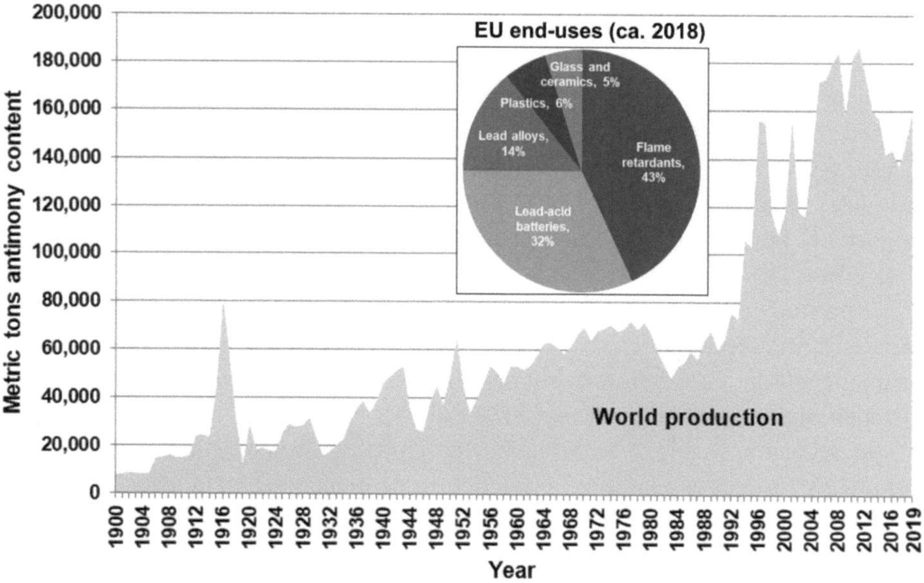

Fig. 4.1: Antimony mine production over time* and current end-use shares [28, 32, 34].
*World mine production from 1900 to 2005 based on United States Geological Survey (USGS) historical statistics [32] and from 2006 to 2019 on USGS Mineral Commodity Summaries [34].

During the past century, China has been the main global provider of antimony accounting for more than 87% of global production associated in large part to mined ore deposits located in the province of Hunan [33]. Current uses of antimony include flame retardants (43%), lead-acid batteries (32%), lead alloys (14%), catalysts and stabilizers in the manufacture of PET (polyethylene terephthalate) (6%), and the use as a decolorizing agent in glass and ceramics (5%) (all percentages represent estimates for the European Union (EU) as reported by the European Commission [28]).

Antimony is considered "critical" by the European Union (EU) because the region is completely reliant on imports [35–38]. This has sparked efforts to promote secondary antimony sources [39] and to better monitor the anthropogenic antimony cycle in the EU [40] including losses to the environment. Antimony was also found to be a critical raw material, for example, by the United States [41].

The main anthropogenic sources of antimony include the combustion of fossil fuels, waste incineration, mine drainage, smelter waste, shooting, and road traffic (e.g., brake abrasion particles) [42, 43]. Global atmospheric antimony emissions from anthropogenic activities in 2010 are estimated at around 1,904 t with fuel combustion found as the major source [44]. Antimony has a toxicity profile similar to that of arsenic [45], is poisonous by inhalation and ingestion, and has been found to be carcinogenic [46]. It was categorized as a priority pollutant by the EU and the United States Environmental Protection Agency (EPA) [47, 48], and is covered in the list of hazardous substances in the Basel convention [49]. The existence of increased levels of antimony in natural systems (e.g., soils and sediments, water bodies, and biological systems) has received increased global attention [33].

However, only a limited number of estimates of the anthropogenic antimony cycle have been published to date (Tab. 4.1). These include MFA studies at national scale for Switzerland [45], China [50], Japan in 2000 [51, 52] and 2003 [53] (in Japanese), the United States [54], at regional level for the EU-28 [40], and at global level [55]. In addition, one study at city scale exists for Stockholm in Sweden [56]. The EU has started with its Raw Materials Information System (RMIS) [57] to integrate MFAs such as for antimony [40] into a database to support a more sustainable and circular use of raw materials under various EU policies [18, 58, 59].

As an example, Fig. 4.2 highlights the global anthropogenic antimony cycle around 2011 from raw materials extraction to waste management visualizing MFA modeling estimates reported by Haarman [55]. In the Sankey diagram, the magnitude of antimony flows is shown by the width of the arrows and possible losses from the anthroposphere to the environment are highlighted with arrows pointing toward the box entitled "environment (natural cycle)" (i.e., losses taking place during the stages of production, fabrication and manufacturing, use, and end-of-life management).

Tab. 4.1: Overview of publicly available MFA studies on antimony (ordered by approximate year for which the MFA estimates were published).

Country	Year	Details	Source
Japan	2000	MFA based on monetary input output tables (MIOTs) showing the flow between different economic sectors and stocks for Sb and a range of other metals (Au, Ag, W, Pt, Pd, Ti, Sn, Pb, Cu, Al, Zn, Fe, Co, Mg, Ni, Mo Mn, Cr).	[51, 52]
USA	2000	Focuses on recycling of Sb in the United States. The recycling rate for Sb was found at around 20%.	[54]
Switzerland	2001	Quantifies imports, exports, inland use and disposal, and releases into the environment (incl. soils). Flow patterns are similar to other industrialized countries.	[45]
Japan	2003	Provides estimates of domestic Sb supply, stocks, and generated waste for the period from 1970 to 2015 and an MFA for 2003. The main emissions source into air was found to be automotive brake abrasion dust.	[53]
Stockholm (Sweden)	2005	Estimates Sb flows in an urban area. Brake linings were found as the major emissions source of Sb, however, other sources such as from flame retardants were also significant.	[56]
Global	2011	Extensive study on the global flows and stock of Sb including a static and dynamic MFA model. Includes a future outlook of Sb uses and a chapter on environmental impacts.	[55]
EU-28	2012	Material system analysis (MSA) carried out for the EU-28. The EU is dependent on the imports of Sb (giving it also a high criticality score especially on the supply risk axis). The study provides estimates of the in-use stock of Sb.	[40]
China	2013	Quantifies Sb flows for whole China. Sb recycling is low and significant amounts of Sb were discharged into the biosphere between 2006 and 2016.	[50]

4.2.1 Production and processing

According to the dynamic MFA model of Haarman [55], around 189 Gg of antimony from ore, 43 Gg from old scrap, and 4 Gg from net metal scrap were globally used as inputs in production in 2011 (Fig. 4.2). The United States Geological Survey (USGS) reports that, globally, the consumption of primary and secondary antimony was around 188 Gg in 2016 [60]. China continues to be the main producer of mined antimony accounting for about 61% of global mine production followed by Russia, and Tajikistan in 2018 (Fig. 4.3).

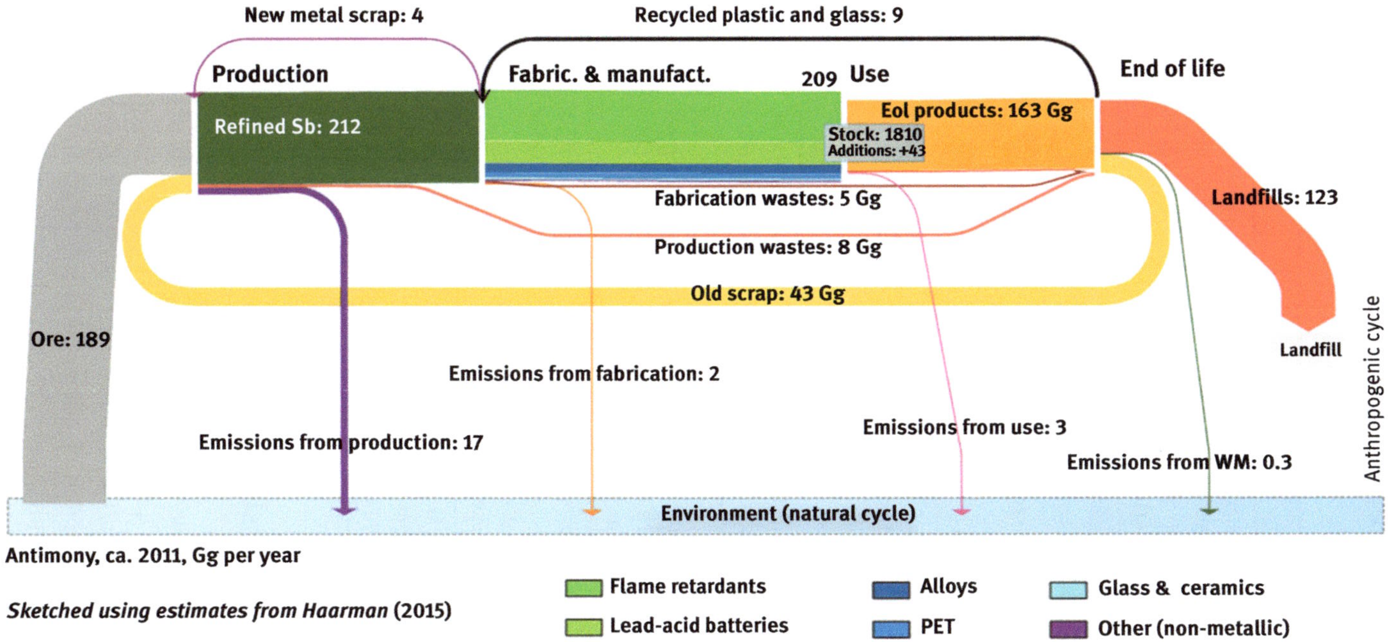

Fig. 4.2: Global anthropogenic antimony cycle around 2011 visualizing the modeling estimates from Haarman [55] based on dynamic MFA[1]. All numbers are in Gigagrams (Gg) antimony.

1 Dynamic material flow analysis (MFA) allows assessing the past, present, and future stocks and flows of materials in the anthroposphere [80].

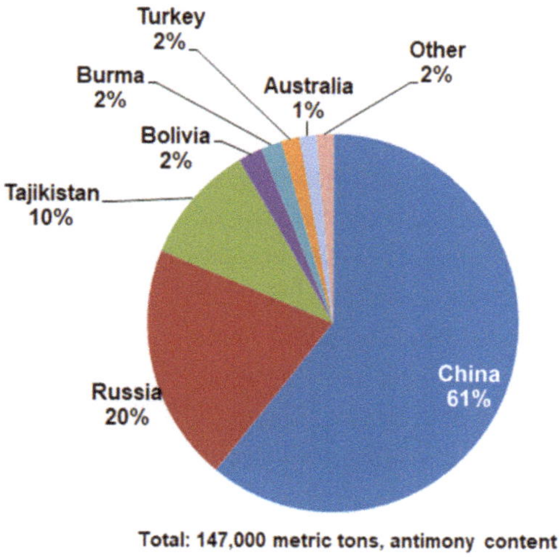

Fig. 4.3: Antimony world mine production by country in 2018 showing data from the USGS [34].

In China, the largest and highest grade deposits are located in the South in Hunan and Yunnan provinces as well as the Zhuang Autonomous Region [60]. The most important antimony ore is stibnite (Sb_2S_3) [30]. In 2019, world antimony reserves are estimated at around 1,500 Gg, with 33% of this located in China, 24% in Russia, and 21% in Bolivia [34]. Typically, antimony metal is obtained from the ore via pyrometallurgical methods [30]. Similar to the mining stage, the production of unwrought antimony metal also largely takes place in China (around 87% on average during 2010–2014), followed by Vietnam (11%), the Kyrgyz Republic (1%), Russia (0.6%), and Bolivia (0.5%) (estimates from [36] based on data from UN Comtrade [61]).

During the mining and beneficiation stage, antimony that is not recovered is usually discharged in tailings ponds and considered by Haarman [55] as emissions to soil. The element might eventually leach into water bodies. During roasting, a fraction of antimony can enter the atmosphere, with an air emission factor of 75 grams antimony per ton of primary antimony used by Haarman in his global MFA [55] based on data from Tian and colleagues [44]. For example, large quantities of antimony are found in the local environments surrounding mining and smelting operations in China [43]. In terms of atmospheric emissions of antimony in China in 2010, a total of 649 t were emitted with fuel combustion being the major source (426 t), followed by non-ferrous metals production (163 t), waste incineration (69 t), brake wear (26 t), and pig iron and steel production (12 t) [44]. However, large uncertainties exist for the actual antimony emissions during the production and processing stage.

4.2.2 Manufacturing and use

Refined antimony metal is further used in the manufacture of semi-finished products (e.g., antimony oxide, alloys, chemicals, flame-retarded plastics and rubber) and final products (e.g., electrical and electronic equipment, cables and wires, textiles, PET containers, lead-acid batteries, articles made with antimony alloys, glass and ceramics, and other products) [40]. Globally, the majority of antimony is currently used in the form of antimony trioxide (ATO) in the production of flame retardants [28, 30, 55]. For this, ATO is combined with halogenated (i.e., brominated or chlorinated) flame-retardant compounds and is used, for example, in plastics, cable coatings, furniture, car seats, and fabrics [28, 40]. ATO also acts as an opacifier in additives to plastics [30]. Furthermore, ATO finds use as a catalyst in the production of polyethylene terephthalate (PET) [62], which is with around 6% the fourth largest end-use for antimony globally [55].

Because antimony is not recovered from plastics (and other uses such as fabrics and coatings), the fate of the metalloid is closely linked to the subsequent flow of these products [63]. Another non-metallurgical use includes hexahydroxyantimonate as a degassing agent in the manufacture of high-quality clear glass. Metallic uses of antimony include lead-acid batteries and lead and other alloys. The main application of antimony as an alloying element for lead-acid batteries has been declining, but still represents the main metallic product with a significant fraction of antimony originating from recycling of used batteries [30]. Estimates of global antimony consumption figures around 2011 are highlighted in Tab. 4.2.

Tab. 4.2: Estimated antimony consumption figures at global level around 2011 (flow values are taken from Table 8 of the study by Haarman [55]).

Application	Details	Flow value / Gg	%
Flame retardants	ATO combined with halogenated flame-retardant compounds used, for example, in in plastics, cable coatings, furniture, car seats, and fabrics.	115	54
Lead-acid batteries	The lead plate grids are strengthened through the addition of antimony.	54	25
Lead and other alloys	Includes, for example, rolled and extruded lead, ammunition, and other alloys (primarily with lead and tin).	24	11
PET	ATO as a catalyst in the production of PET and as a pigment for color.	12	6
Glass & ceramics	Sb is used as a degassing and decolorizing agent.	4	2
Other	For example, organic Sb compounds such as tartar emetic and fuadin, pesticides, ammunition primers, vulcanization of red rubber, mordants, and lubricating agent [30].	2	1
Total		**211**	**100**

Figure 4.4 illustrates end-uses over time for the United States using data from the United States Geological Survey [32].

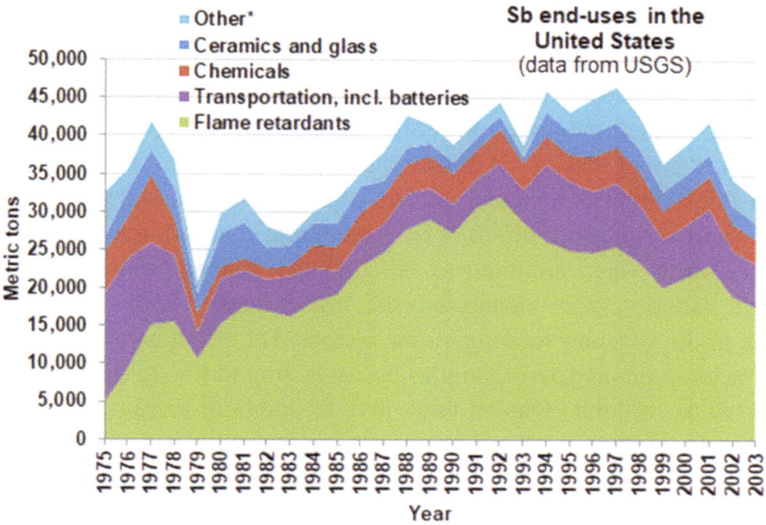

Fig. 4.4: Antimony end-uses from 1975 to 2003 reported by the USGS [32].
*Other uses include, for example, ammunitions, cable coverings, fireworks, metal castings, paper, pigments, rubber products, and sheet, pipe, and type metal [32, 64].

In the United States, the use of antimony in flame retardants has significantly increased since 1975, while the use of antimony in lead alloys used in batteries decreased until 2003. Due to drawbacks[2] in the use of antimony as an alloying element in battery grids, antimony is increasingly being replaced by calcium in grid plates. As a result, the average antimony content in lead alloys used in batteries has been decreasing over time [64]. More recently and at global level, the antimony loading in flame retardants seems to have been decreasing due to higher prices and legislative requirements to make changes to the composition within this application [65]. During the decade until 2030, Roskill expects a 1–2% annual growth for non-metallic applications, but a decline in metallic demand [65]. Similarly, other sources [36, 66] state that future demand for antimony in flame retardants could rise driven by increasingly tight fire regulations, while its use in batteries is less certain and depends on the growth of the automotive sector and substitution with more environmentally friendly options.

Limited information exists on antimony losses during the fabrication and manufacturing process. As an emissions proxy, Haarman [55] largely bases the

2 These include: (1) During discharge, antimony tends to migrate from the positive to the negative plate where it lowers hydrogen overvoltage and speeds up the breakdown of water in the electrolyte. (2) Antimony deposits on the negative plate causing self-discharge of the battery [54].

estimates of global losses on modeling assumptions used for the anthropogenic lead cycle [67, 68]. In Switzerland, the main sources of antimony emissions to the environment were found to be brake linings and shooting ranges [45]. Similarly, in Japan and in Stockholm (Sweden) the main emissions sources included brake linings [53, 56]. A study by Ciacci and colleagues [69] found that about two-thirds of all miscellaneous antimony uses (i.e., pesticides, ammunition primers, fireworks, vulcanizing agent, mordents, cable coverings, metal castings, and brake pads) are either dissipated during the use phase or currently unrecyclable. Limited information is available on the transformation and subsequent fate and transport of antimony in different environmental compartments [48].

In-use stocks occur when antimony is incorporated in long-living products such as lead-acid batteries, other metal products, or as a flame retardant in products that remain in the economy for longer time periods. For example, the average lifetime of lead-acid antimony-based batteries in use is around five to seven years and for bearing metals around eighteen years [45]. In addition, products such as household appliances or textiles, where antimony is used in flame retardants, can have lifetimes of several years as well. On the other hand, applications such as ammunition or brake linings in which antimony is used display shorter lifetimes and generally no significant accumulation within the anthroposphere is found.

The total in-use stock of antimony reported in the literature ranges from 430 t for the city of Stockholm in Sweden in 2005 [56] to around 1,810,000 t at global level in 2011 [55]. The in-use stock of single countries ranges from 10,000 t for Switzerland in 2001 to around 96,000 to 103,100 t for Japan in the early 2000s [51, 53]. The total in-use stock of antimony in the EU-28 is estimated at about 135,000 t in 2012 [40].

4.2.3 Waste management

Antimony products at end-of-life (EoL) are sent to different waste treatment options, including recycling, incineration, and landfilling. Globally, the end-of-life recycling rate (EoL-RR) for antimony in 2011 was estimated at around 32% [55] while for the EU in 2012 an EoL-RR of up to 28% is reported [28, 40]. This is in line, for instance, with the estimate for the United States (EoL-RR ~ 22%) [54]. The total amount of antimony input into the production system from recycling of post-consumer scrap versus overall input (the end-of-life recycling input rate (EoL-RIR) [70]) is reported at around 23% globally [55] and at about 28% for the EU-28 [28, 40], largely due to the recovery from lead-acid batteries. According to Roskill, secondary supply from lead alloys (e.g., batteries) is expected to be sufficient to meet demands for metallic uses within the next decade, so that future demands for primary antimony might be largely triggered by non-metallic uses (mostly flame retardants) in the near future [65].

4.2.4 Comparison of available MFAs from the literature

The socio-metabolic flows and stocks of antimony reported in the literature are divided by population figures to obtain the per capita antimony flows (Fig. 4.5) and in-use stocks (Fig. 4.6) for different countries (the United States, Switzerland, Japan, and China), regions (EU-28 and the World), and the city of Stockholm.

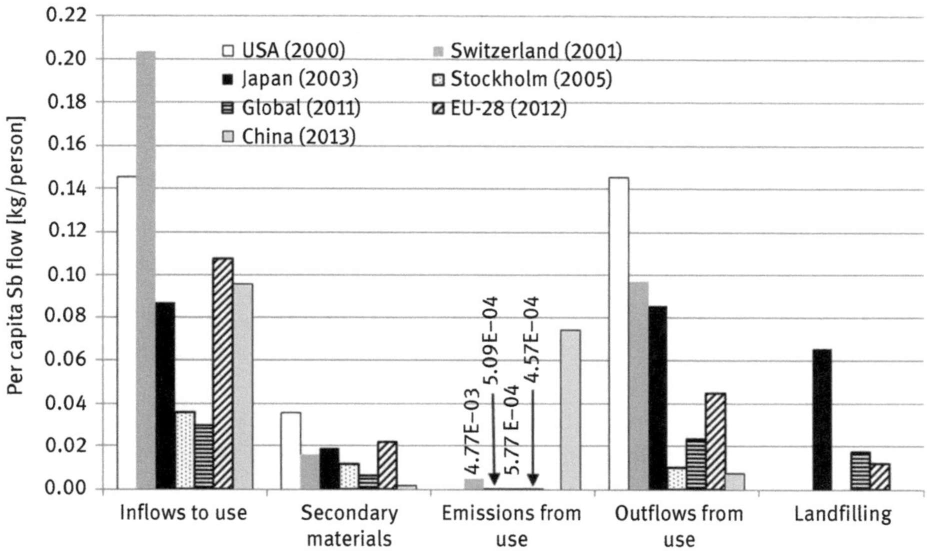

Fig. 4.5: Per capita flows from existing antimony MFA studies[a] (inspired by Haarman [55] and with details on the data used provided in the appendix table).
[a]MFA data: USA (2000) [54], Switzerland (2001) [45], Japan (2003) [52], Stockholm (2005) [56], Global (2011) [55], EU-28 (2012) [40], China (2013) [50]. Population data: countries [71]; Stockholm [72]; World [73].

This comparison shows that per capita inflows to the use phase (consisting of both primary and secondary antimony) range from 0.03 kg Sb per person for the world to 0.20 kg Sb per person for Switzerland (Fig. 4.5). This highlights that materials use is generally higher in rich countries when compared to the global average and lower-income nations [74]. The highest per capita inflows are found for high-income countries or regions such as the United States, Switzerland, Japan, and the EU-28. It is notable that for China, which is classified as an upper-middle-income country [75], per capita inflow to use is found with about 0.10 kg Sb per person, also within the range of other high-income countries. On a per capita basis, secondary materials inputs are highest for the United States (0.035 kg Sb per person), the EU-28 (0.022 kg Sb per person), and Japan (0.019 kg Sb per person) largely due to the recycling of old antimony scrap generated from used lead-acid batteries.

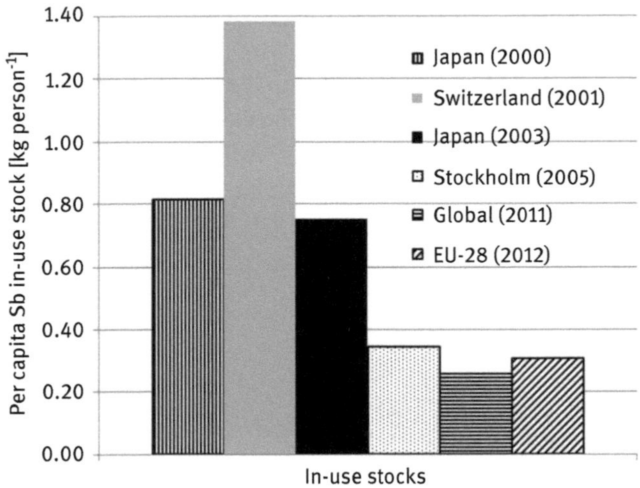

Fig. 4.6: Per capita stocks from existing MFA studies on antimony (inspired by Haarman [55] and with details on the data used provided in the appendix table).
[a]MFA data: Japan (2000) [51], Switzerland (2001) [45], Japan (2003) [52], Stockholm (2005) [56], Global (2011) [55], and EU-28 (2012) [40]. Population data: countries [71]; Stockholm [72]; World [73].

The in-use stocks in high-income countries ranges from 0.75 to 1.38 kg Sb per person and is clearly above the estimate for the world (0.26 kg Sb per person) (Fig. 4.6). The per capita stock in the EU is reported at 0.31 kg Sb per person and only slightly above the global average. The reason for this is unclear as no additional details are provided on the stock estimates provided for the EU [40]. The largest part of the in-use stocks equals long-lived products such as lead-acid batteries, metal products, and flame retardants (e.g., in household appliances).

Antimony emissions from use for China reported by Chu and colleagues were significantly higher than for all other MFAs (Fig. 4.5) [50]. In 2013, the authors report a total of 100.44 Gg of antimony emissions during the use phase, which equal about 0.074 kg Sb per person. According to the authors, the main reason for this is the large use of antimony in compounds which are consumed and then discharged to the biosphere [50]. Globally, emissions to use are found at 4.57×10^4 kg Sb per person in 2011 and other estimates range at around 4.77×10^3 kg Sb for Switzerland and 5.77×10^4 kg Sb for Stockholm (all number are emissions of antimony per person per year). For the latter two studies, the main emissions sources to the environment were brake pads [45, 56] and shooting ranges [45]. Globally, natural antimony flows consisting of riverine flux to ocean, eolian dust, seaspray, net primary production (NPP), extraterrestrial matter, volcano, and soil erosion equal 34 ± 11 Gg Sb per year [76]. Anthropogenic flows, on the other hand, from coal burning, petroleum burning, mining, biomass burning, Human Appropriation of Net Primary Production (HANPP), and construction are estimated at 167 ± 1.34 Gg Sb per year according to the same source. This means that anthropogenic fluxes are greater than respective natural fluxes.

Outflows from use capture the total domestic discharge and range between 0.008 kg Sb per person for China and 0.145 kg Sb per person for the United States. Outflows from use undergo either waste treatment (e.g., landfilling) or recycling. In Japan, the EU-28, and the world, significant amounts of antimony are sent to landfills (Fig. 4.5), with the possibility for antimony to eventually leach back into the environment.

4.3 Conclusions and perspective

Antimony is a metalloid that has been used for thousands of years by humans in various forms. Global demand for antimony has greatly increased in recent decades owing to its use in flame retardants and lead-acid batteries. Antimony is a toxic element that is now found at elevated levels in soils, sediments, and biological systems. The current anthropogenic antimony cycle is characterized by high levels of primary resource use (about 190 Gg y^{-1} at global scale). Losses from human usage into the environment occur at several stages during the life cycle. The largest loss is during waste management to landfills (ca. 123 Gg y^{-1} globally) and only about 20–30% of all antimony exiting the use phase is currently being recycled. Emissions during production occur in the form of mine tailings and during the roasting stage as emissions to the atmosphere. Emissions from use can be significant if large fractions of antimony are used in compound form, for example, in ammunitions or chemicals, or catalysts. The EoL-RIR, as a measure of the total antimony input to the production system from recycling versus overall input, is reported at around 20–30%, largely due to the recovery from lead-acid batteries. However, for most non-metallic uses recycling is currently limited. In the future, the use of primary antimony in lead-acid batteries (which in the EU make up about 32%) is likely to decrease as substitutes are being used and the demand can be increasingly met from recycling of lead-acid batteries. However, a growth area is the application of antimony in glass panels for photovoltaic solar cells (Roskill [77] as cited in Henckens and colleagues [78]).

From a policy perspective, several options exist to further optimize the anthropogenic antimony cycle. These include:

- Phase out the use of antimony in dissipative uses or substitute with other materials. According to criticality studies [28] and the wider literature [30, 55, 78], antimony is relatively easy to substitute in some of its current uses. This includes, for example, its use in the manufacture of pigments and glass (chromium, tin, titanium, zinc, and zirconium are highlighted as possible substitutes by the EU [28]) or the use of calcium in lead alloys [55]. Alumina trihydrate, magnesium hydroxide, and zinc borate are highlighted as potential substitutes for antimony in flame retardants, although performance might be lower [28]. Similarly, substitutes exist for many dissipative applications such as brake pads, pigments and pesticides,

and lubricants [55]. Substitution is among the most promising options for reducing global antimony demand [78].
- Increasing recycling rates for antimony-containing products in order to lower demands for primary antimony and reduce environmental pressures associated with primary production. The recycling rate of lead-acid batteries is already high [54] and expected to be sufficient to meet demands for antimony in batteries in the near future [65]. However, antimony recycling from other metallic applications (e.g., lead sheets, tubes, pipes) is low and for most non-metallic uses (with the exception of PET bottles) nonexistent [78]. This can be partly explained by the fact that recycling from non-metallic uses is currently not economic, even though it can be technically feasible, and metallic applications such as lead alloys used in construction have long lifetimes, so materials available for recycling are currently low compared to primary materials input [78]. Therefore, increasing recycling seems to be limited at this point in time.
- Increasing the material efficiency of mining and the subsequent production and fabrication and manufacturing processes. As the majority of emissions during the anthropogenic cycle of antimony occur during these stages, lowering the losses can help to lower the overall antimony demand.
- Reduce overall consumption of high-income countries by fostering lifestyle changes complementing technological advancements. To date, worldwide growth in affluence is closely coupled to increases in resource consumption and environmental pollution [79]. Therefore, in addition to the above-mentioned technological or engineering-type approaches toward a more sustainable use of materials (antimony) use, also the overall materials demand has to be lowered. This can be triggered by a myriad of policy measures aiming, for example, to formulate sustainable consumption corridors, increasing the lifespan of products, fostering sharing approaches, education, and finding alternative measures of wealth to gross domestic product (GDP).

Similar to other materials, the demand for antimony has significantly increased in recent decades and elevated levels in the environment are now observed. MFA can help to elucidate the anthropogenic counterparts of the natural material cycles usually in focus by biogeochemists. This helps in understanding the current and anticipated magnitude of anthropogenic antimony flows to the environment. Future research should continue to combine information on both natural and anthropogenic antimony cycles and provide concrete recommendations to policy- and decision-makers on how to more sustainably manage this scarce element in the future.

Tab. 4.3: Material flow analysis parameters available in the literature. Flows and stocks in metric tons of antimony and population data in number of people.

Geographical area	Year	Inflows to use	Secondary materials	In-use stocks	Emissions from use	Outflows from use	Landfilling	Details on data source	Reference	Population[a]
Japan	2000	–	–	103,100	–	–		– Table 1 in publication	[51, 52]	126,843,000
USA	2000	41,000	10,000	–	–	41,000		– Figure 1 in publication. – Recycling flow equals old and new scrap inputs. – EoL flow equals the sum of dissipative used and old scrap generated.	[54]	282,162,411
Switzerland	2001	1,470	115	10,000	35	700		– Figure 1 in publication. – Consumptive losses are the sum of flows to the environment from consumption and waste management. – Recycling equals the flow from waste management to consumption.	[45]	7,229,854
Japan	2003	11,100	2,400	96,000	65	10,900	8,400	– Figure 6 in publication. – Recycling equals material recovery. – Consumptive losses equals emissions to air. – EoL flows equals the total of domestic discharge.	[53]	127,718,000

(continued)

Tab. 4.3 (continued)

Geographical area	Year	Inflows to use	Secondary materials	In-use stocks	Emissions from use	Outflows from use	Landfilling	Details on data source	Reference	Population[a]
Stockholm (Sweden)	2005	45	15	430	0.72	15.86		– Tables 10 and 11 in publication. – EoL flows equal the sum of waste and recycling flows.	[56]	1,247,968
Global	2011	209,000	47,260	1,810,000	3,220	163,000	123,000	– Based on dynamic model in appendix C of the publication	[55]	7,041,194,000
EU-28	2012	47,572	9,710	135,000	–	19,960	5,480	– Based on Figure 13 in publication. – Inflows to use equals the sum of primary material, secondary material, processed material, product, and waste. – EoL flows equal the sum of domestic waste, non-functional recycling and secondary material. – Landfilling equals additions to landfill and tailings.	[40]	441,395,932
China	2013	130,000	2,400	–	100,440	10,560		– Figure 2 in publication. – Recycling equals the sum of old and new scrap.	[50]	1,357,380,000

[a]Population data for the year of the MFA were taken from the following sources: individual countries [71]; Stockholm [72]; world [73].

References

[1] UNEP. Global Resources Outlook. 2019: Natural Resources for the Future We Want. 2019.
[2] Hatfield-Dodds S, Schandl H, Newth D, Obersteiner M, Cai Y, Baynes T, et al. Assessing global resource use and greenhouse emissions to 2050, with ambitious resource efficiency and climate mitigation policies. J Clean Prod 2017, 144, 403–414.
[3] OECD. Global Material Resources Outlook to 2060: Economic Drivers and Environmental Consequences. Paris: OECD Publishing; 2019. Available: https://doi.org/10.1787/9789264307452-en
[4] Cabernard L, Pfister S, Hellweg S. A new method for analyzing sustainability performance of global supply chains and its application to material resources. Sci Total Environ 2019, 684, 164–177.
[5] UNFCCC. Paris Agreement. 2015.
[6] Greenfield A, Graedel TE. The omnivorous diet of modern technology. Resour Conserv Recycl 2013, 74, 1–7.
[7] EC. EU Raw Materials Scoreboard 2018. Brussels, Belgium; 2018. Available: https://op.europa.eu/en/publication-detail/-/publication/117c8d9b-e3d3-11e8-b690-01aa75ed71a1
[8] Sovacool BK, Ali SH, Bazilian M, Radley B, Nemery B, Okatz J, et al. Sustainable minerals and metals for a low-carbon future. Science 2020, 367, 30–33.
[9] Mancini L, Vidal Legaz B, Vizzarri M, Wittmer D, Grassi G, Pennington D. Mapping the Role of Raw Materials in Sustainable Development Goals. A preliminary analysis of links, monitoring indicators, and related policy initiatives. Luxemburg, Publications Office of the European Union, 2019.
[10] UN. Sustainable Development Goals. 2019 [cited 7 Nov 2019]. Available: https://sustainabledevelopment.un.org/sdgs
[11] Mancini L, Nuss P. Responsible materials management for a resource-efficient and low-carbon society. Resources 2020, 9, 68.
[12] UNEP IRP. UNEP International Resource Panel: About us. 2020 [cited 23 Aug 2020]. Available: https://www.resourcepanel.org/about-us
[13] UNEP. Metal Stocks in Society: A Scientific Synthesis. Paris: International Panel for Sustainable Resource Management Working Group on the Global Metal Flows; 2010. Available: http://www.resourcepanel.org/reports/metal-stocks-society
[14] van der Voet E, Salminen R, Eckelman M, Mudd G, Norgate T, Hischier R. Environmental Risks and Challenges of Anthropogenic Metals Flows and Cycles. United Nations Environment Programme (UNEP), 2013.
[15] Ayuk ET, Pedro AM, Ekins P, Gatune J, Milligan B, Oberle B, et al. Mineral Resource Governance in the twenty-first century: Gearing extractive industries towards sustainable development. Nairobi, Kenya., A Report by the International Resource Panel: United Nations Environment Programme, 2020.
[16] OECD. Measuring Material Flows and Resource Productivity: Volume I. The OECD Guide. Paris: The Organisation for Economic Co-operation and Development (OECD); 2008. Available: https://www.oecd.org/environment/indicators-modelling-outlooks/MFA-Guide.pdf
[17] OECD. Measuring Material Flows and Resource Productivity: Volume II. The Accounting Framework. Paris: The Organisation for Economic Co-operation and Development (OECD); 2008. Available: https://www.oecd.org/env/indicators-modelling-outlooks/MFA-Accounting-Framework.pdf
[18] EC. Circular Economy Action Plan: For a cleaner and more competitive Europe. European Commission (EC); 2020. Available: https://ec.europa.eu/environment/circular-economy/pdf/new_circular_economy_action_plan.pdf

[19] EC. The European green deal. 2019. Report No.: COM/2019/640 final. Available: https://eur-lex.europa.eu/legal-content/EN/TXT/?qid=1576150542719&uri=COM%3A2019%3A640%3AFIN
[20] EEA. More from Less – Material Resource Efficiency in Europe: 2015 Overview of Policies, Instruments and Targets in 32 Countries. Copenhagen: European Environment Agency (EEA); 2016. Report No.: EEA Report No. 10/2016.
[21] BMU. German Resource Efficiency Programme (ProgRess), Berlin, 2012.
[22] BMUB. German Resource Efficiency Programme II – Programme for the sustainable use and conservation of natural resources. Berlin: Federal Ministry for the Environment, Nature Conservation, Building and Nuclear Safety (BMUB); 2016. Available: https://www.bmu.de/publikation/german-resource-efficiency-programme-ii-programme-for-the-sustainable-use-and-conservation-of-natu/
[23] BMU. Deutsches Ressourceneffizienzprogramm ProgRess III. Berlin: Federal Ministry for the Environment, Nature Conservation, and Nuclear Safety (BMU); 2020. Available: https://www.bmu.de/download/deutsches-ressourceneffizienzprogramm-progress-iii/
[24] Brunner PH, Rechberger H. Handbook of Material Flow Analysis: For Environmental, Resource, and Waste Engineers, 2nd Edition. CRC Press, 2016.
[25] Graedel TE. Material flow analysis from origin to evolution. Environ Sci Technol 2019, 53, 12188–12196.
[26] Nuss P, Blengini GA. Towards better monitoring of technology critical elements in Europe: Coupling of natural and anthropogenic cycles. Sci Total Environ 2018, 613–614, 569–578.
[27] Chen W-Q, Graedel TE. Anthropogenic cycles of the elements: A critical review. Environ Sci Technol 2012, 46, 8574–8586.
[28] EC. Study on the EU's list of Critical Raw Materials (2020): Critical Raw Materials Factsheets (Final). Brussels: European Commission (EC); 2020. Available: https://ec.europa.eu/docsroom/documents/42883/attachments/2/translations/en/renditions/native
[29] Smichowski P. Antimony in the environment as a global pollutant: A review on analytical methodologies for its determination in atmospheric aerosols. Talanta 2008, 75, 2–14.
[30] Grund SC, Hanusch K, Breunig HJ, Wolf HU Antimony and antimony compounds. Ullmann's Encyclopedia of Industrial Chemistry. Wiley-VCH Verlag GmbH & Co. KGaA; 2000. Available: http://onlinelibrary.wiley.com/doi/10.1002/14356007.a03_055.pub2/abstract
[31] Anderson CG. The metallurgy of antimony. Geochemistry 2012, 72, 3–8.
[32] Kelly T, Matos G Historical statistics for mineral and material commodities in the United States: Supply-demand statistics / end-use statistics for antimony. U.S. Geological Survey (USGS); 2015. Report No.: Data Series 140. Available: Historical Statistics for Mineral and Material Commodities in the United States.
[33] Herath I, Vithanage M, Bundschuh J. Antimony as a global dilemma: Geochemistry, mobility, fate and transport. Environ Pollut 2017, 223, 545–559.
[34] USGS. Mineral commodity summary: Antimony. U.S. Geological Survey (USGS); 2020. Available: https://www.usgs.gov/centers/nmic/antimony-statistics-and-information
[35] EC. Critical raw materials resilience: Charting a path towards greater security and sustainability. European Commission (EC); 2020. Report No.: COM/2020/474 final. Available: https://eur-lex.europa.eu/legal-content/EN/TXT/?uri=CELEX:52020DC0474
[36] EC. Study on the review of the list of Critical Raw Materials: Critical Raw Materials Factsheets. Brussels, Belgium: DG GROW; 2017. Available: https://publications.europa.eu/en/publication-detail/-/publication/7345e3e8-98fc-11e7-b92d-01aa75ed71a1/language-en
[37] EC. Critical raw materials for the EU. Brussels, Belgium: European Commission (EC); 2010. Available: http://ec.europa.eu/enterprise/policies/rawmaterials/documents/index_en.htm

[38] EC. Report on critical raw materials for the EU. Brussels, Belgium: European Commission (EC); 2014. Available: http://ec.europa.eu/enterprise/policies/raw-materials/files/docs/crm-report-on-critical-raw-materials_en.pdf
[39] Dupont D, Arnout S, Jones PT, Binnemans K. Antimony recovery from end-of-life products and industrial process residues: A critical review. J Sustain Metall 2016, 2, 79–103.
[40] BIO by Deloitte. Study on data for a raw material system analysis: Roadmap and test of the fully operational MSA for raw materials. Prepared for the European Commission, DG GROW; 2015. Available: https://ec.europa.eu/growth/tools-databases/msa/sites/default/files/deliverables/MSA_Final_Report.pdf
[41] USDOI. Final list of critical minerals 2018: A notice by the interior department on 05/18/2018. United States Department of the Interior, Office of the Secretary; 2018. Report No.: 2018–10667. Available: https://www.federalregister.gov/documents/2018/05/18/2018-10667/final-list-of-critical-minerals–2018
[42] Álvarez-Ayuso E, Otones V, Murciego A, García-Sánchez A, Regina IS. Mobility and phytoavailability of antimony in an area impacted by a former stibnite mine exploitation. Sci Total Environ 2013, 449, 260–268.
[43] He M, Wang X, Wu F, Fu Z. Antimony pollution in China. Sci Total Environ 2012, 421–422, 41–50.
[44] Tian H, Zhou J, Zhu C, Zhao D, Gao J, Hao J, et al. A comprehensive global inventory of atmospheric antimony emissions from anthropogenic activities, 1995–2010. Environ Sci Technol 2014, 48, 10235–10241.
[45] Mathys R, Dittmar J, Johnson A Antimony in Switzerland: A substance flow analysis. Bern: Swiss Federal Office for the Environment FOEN; 2007. Report No.: Environmental studies no. 0724. Available: https://www.bafu.admin.ch/bafu/en/home/topics/chemicals/publications-studies/publications/antimony.html
[46] Beyersmann D, Hartwig A. Carcinogenic metal compounds: Recent insight into molecular and cellular mechanisms. Arch Toxicol 2008, 82, 493. doi: 10.1007/s00204-008-0313-y.
[47] EPA. Priority Pollutant List. United States Environmental Protection Agency (EPA); 2014. Available: https://www.epa.gov/sites/production/files/2015-09/documents/priority-pollutant-list-epa.pdf
[48] Filella M, Belzile N, Chen Y-W. Antimony in the environment: a review focused on natural waters: I. Occurrence. Earth-Sci Rev 2002, 57, 125–176.
[49] UN. Basel Convention on the control of transboundary movements of hazardous wastes and their disposal. United Nations; 2019. Available: www.basel.int/Portals/4/download.aspx?d=UNEP-CHW-IMPL-CONVTEXT.English.pdf
[50] Chu J, Mao J, He M. Anthropogenic antimony flow analysis and evaluation in China. Sci Total Environ 2019, 683, 659–667.
[51] Murakami S. Material Flows and Stocks of Metals Surrounding Japan. Sendai, Japan, 2006.
[52] Murakami S, Yamanoi M, Adachi T, Mogi G, Yamatomi J. Material flow accounting for metals in Japan. Mater Trans 2004, 45, 3184–3193.
[53] Tsunemi K, Wada H. Substance flow analysis of antimony for risk assessment of antimony and antimony compounds in Japan. J Jpn Inst Met Mater 2008, 72, 91–98.
[54] Carlin J Antimony recycling in the United States in 2000. U.S. Geological Survey (USGS); 2006. Report No.: U.S. Geological Survey Circular 1196–Q. Available: http://pubs.er.usgs.gov/publication/cir1196Q/
[55] Haarman A. The Anthropogenic Antimony Cycle: Dynamic Analysis of Global Flows and Stocks of Antimony and Associated Environmental Impacts. Masters Thesis, TU Delft, Leiden University. 2015. Available: http://publicationslist.org/data/arthur.haarman/ref-2/2015_Arthur%20Haarman%20-%20The%20Anthropogenic%20Antimony%20Cycle.pdf

[56] Månsson NS, Hjortenkrans DST, Bergbäck BG, Sörme L, Häggerud AV. Sources of antimony in an urban area. Environ Chem 2009, 6, 160–169.
[57] EC. EU Raw Materials Information System (RMIS). 2020 [cited 29 Aug 2020]. Available: https://rmis.jrc.ec.europa.eu/
[58] EC. Monitoring framework for the circular economy. Brussels: European Commission (EC); 2018. Report No.: SWD(2018) 17 final. Available: https://eur-lex.europa.eu/legal-content/EN/TXT/?qid=1516265440535&uri=COM:2018:29:FIN
[59] EC. The raw materials initiative – meeting our critical needs for growth and jobs in Europe. European Commission (EC); 2008. Available: https://ec.europa.eu/growth/sectors/raw-materials/policy-strategy_en
[60] USGS. 2016 Minerals yearbook: Antimony. U.S. Geological Survey (USGS); 2016. Available: https://www.usgs.gov/centers/nmic/antimony-statistics-and-information
[61] United Nations. UN Comtrade Database. In: United Nations Commodity Trade Statistics Database [Internet]. 2014 [cited 14 May 2014]. Available: http://comtrade.un.org/
[62] James E, Turner A. Mobilisation of antimony from microplastics added to coastal sediment. Environ Pollut 2020, 264, 114696.
[63] Filella M, Hennebert P, Okkenhaug G, Turner A. Occurrence and fate of antimony in plastics. J Hazard Mater 2020, 390, 121764.
[64] Linden D, Reddy T. Handbook of Batteries. McGraw-Hill Education Ltd; 3rd edition (see also https://www.amazon.de/-/en/David-Linden/dp/0071359788), 2001.
[65] Roskill. Antimony: Outlook to 2030 (Website Summary). Roskill Information Services; 2020. Report No.: 14th edition. Available: https://roskill.com/market-report/antimony/
[66] Schwarz-Schampera U. Antimony. In: Gunn G, editor, Critical Metals Handbook. John Wiley & Sons, 2014, 70–98. Available: http://onlinelibrary.wiley.com/doi/10.1002/9781118755341.ch4/summary.
[67] Mao JS, Dong J, Graedel TE. The multilevel cycle of anthropogenic lead: I. Methodology. Resour Conserv Recycl 2008, 52, 1058–1064.
[68] Mao JS, Dong J, Graedel TE. The multilevel cycle of anthropogenic lead: II. Results and discussion. Resour Conserv Recycl 2008, 52, 1050–1057.
[69] Ciacci L, Reck BK, Nassar NT, Graedel TE. Lost by design. Environ Sci Technol 2015, 49, 9443–9451.
[70] Talens Peirò L, Nuss P, Mathieux F, Blengini GA Recycling Indicators based on EU flows and raw materials system analysis data. Luxembourg: Publications office of the European Union; 2018. Report No.: EUR 29435 EN. Available: https://ec.europa.eu/jrc/en/publication/towards-recycling-indicators-based-eu-flows-and-raw-materials-system-analysis-data
[71] World Bank. Population total. 2020 [cited 25 Sep 2020]. Available: https://data.worldbank.org/indicator/SP.POP.TOTL
[72] World Population Review. Stockholm Population. 2020 [cited 25 Sep 2020]. Available: https://worldpopulationreview.com/world-cities/stockholm-population
[73] UN. Total Population – Both Sexes. 2020 [cited 9 Sep 2020]. Available: https://population.un.org/wpp/Download/Standard/Population/
[74] UNEP. Global Material Flows and Resource Productivity. An Assessment Study of the UNEP International Resource Panel. Paris, UNEP International Resource Panel, 2016.
[75] World Bank. World Bank Country and Lending Groups: Country Classification. 2020 [cited 26 Sep 2020]. Available: https://datahelpdesk.worldbank.org/knowledgebase/articles/906519-world-bank-country-and-lending-groups
[76] Sen IS, Peucker-Ehrenbrink B. Anthropogenic disturbance of element cycles at the Earth's surface. Environ Sci Technol 2012, 46, 8601–8609.

[77] Roskill. Study of the Antimony Market. London, United Kingdom, Roskill Consulting Group Limited, 2011.
[78] Henckens MLCM, Driessen PPJ, Worrell E. How can we adapt to geological scarcity of antimony? Investigation of antimony's substitutability and of other measures to achieve a sustainable use. Resour Conserv Recycl 2016, 108, 54–62.
[79] Wiedmann T, Lenzen M, Keyßer LT, Steinberger JK. Scientists' warning on affluence. Nat Commun 2020, 11, 3107.
[80] Müller E, Hilty LM, Widmer R, Schluep M, Faulstich M. Modeling metal stocks and flows: A Review of dynamic material flow analysis methods. Environ Sci Technol 2014, 48, 2102–2113.

Mengchang He, Chengjun Zhang, Qianyun Zhong, Jun Shan, Tianning Wang, Huiji Liu, Yonghong Jiao, Kaiyue Song and Xiaobo Yin

Chapter 5 Antimony in China

5.1 Introduction

Antimony is considered one of the valuable non-ferrous elements in the world, and plays a vital role in economic construction. However, antimony and its compounds also pose a severe threat to the global environment, because of bioaccumulation and long-distance migration. China has more abundant antimony resources and reserves than any other country, and annual production accounts for 80% of the world's total antimony production [1]. There are 117 known antimony reserves in China, which are locate in 18 provinces (or districts). The antimony reserve was documented as 3.279 Mt in 2018 [2]. The ample storage of antimony and rapid economic development place China in an increasingly important position in global anthropogenic antimony emissions. As a precondition to environmental contamination, remediation, and management of the antimony industry, essential information is a requisite. In this chapter, we review and classify previous studies, monitoring data, and national policy to establish the basics of antimony distribution and industrial background in China.

Acknowledgments: This work was supported by the National Natural Science Foundation of China (Grant No. 42030706, 21677014).

Mengchang He, State Key Laboratory of Water Environment Simulation, School of Environment, Beijing Normal University, Beijing 100875, China. hemc@bnu.edu.cn
Chengjun Zhang, State Key Laboratory of Water Environment Simulation, School of Environment, Beijing Normal University, Beijing 100875, China. zhangcj@mail.bnu.edu.cn
Qianyun Zhong, State Key Laboratory of Water Environment Simulation, School of Environment, Beijing Normal University, Beijing 100875, China. zhqy16@mail.bnu.edu.cn
Jun Shan, State Key Laboratory of Water Environment Simulation, School of Environment, Beijing Normal University, Beijing 100875, China. shanjun@mail.bnu.edu.cn
Tianning Wang, State Key Laboratory of Water Environment Simulation, School of Environment, Beijing Normal University, Beijing 100875, China. wangtn@mail.bnu.edu.cn
Huiji Liu, State Key Laboratory of Water Environment Simulation, School of Environment, Beijing Normal University, Beijing 100875, China. liuhuiji@mail.bnu.edu.cn
Yonghong Jiao, State Key Laboratory of Water Environment Simulation, School of Environment, Beijing Normal University, Beijing 100875, China. Jiaoyh@mail.bnu.edu.cn
Kaiyue Song, State Key Laboratory of Water Environment Simulation, School of Environment, Beijing Normal University, Beijing 100875, China. songky3311@mail.bnu.edu.cn
Xiaobo Yin, State Key Laboratory of Water Environment Simulation, School of Environment, Beijing Normal University, Beijing 100875, China. yxb@mail.bnu.edu.cn

https://doi.org/10.1515/9783110668711-005

5.2 Ore resources and production

China has the most abundant antimony vein-type ore deposits in the world, which account for 55% of the world's reserve of antimony. Most of the deposits are in the provinces of Guangxi, Hunan, Guizhou, East Yunnan, and Guangzhou within the South China-Circum-Pacific antimony ore belt [3]. There are four antimony ore belts in China (Tabs. 5.1 and 5.2): the South China-Circum-Pacific antimony ore belt, West Yunnan-Mediterranean antimony ore belt, and Qinling-Central Asia antimony ore belt are all situated south of Yellow River, and the Jilin-Daxinganling-Tianshan-Outer Baikal antimony ore belt lies in the north of China.

Tab. 5.1: Antimony ore belts in China.

Ore belt	Range (provinces)	Sb reserves / Mt	Percentage of total Sb reserves of China	Typical deposits	Number of deposits	Reference
South China-Circum-Pacific Sb ore belt	Guangxi	0.685	20.9	Dachang	51	[3, 4, 7, 119, 120]
	Hunan	0.826	25.2	Xikuangshan Woxi Longshan Zhazixi	117	
	Jiangxi Zhejiang Anhui			Muli		
	Guizhou	0.413	12.6	Qinglong Banpo	44	
	Guangdong	0.079	2.4	Lejiawan		
	Yunnan	0.308	9.4	Baoshan	108	
West Yunnan-Mediterranean Sb ore belt	West Yunnan			Shimeduo		
Qinling-Central Asia Sb ore belt	Gansu	0.190	5.8	Yawan		
	Shanxi			Gongguan Qingtonggou Caiao		
	Henan			Dahegou		
	Xizang	0.485	14.8	Meiduo	83	
	Hubei				38	
	Qinghai				36	

Tab. 5.1 (continued)

Ore belt	Range (provinces)	Sb reserves / Mt	Percentage of total Sb reserves of China	Typical deposits	Number of deposits	Reference
Jilin-Daxinganling-Tianshan-Outer Baikal Sb ore belt	Jilin			Sanhe		
	Heilongjiang					
	Inner Mongolia					

Antimony ore mining, separation, and smelting sites are usually close to the antimony ore deposits. Therefore, most are in the provinces of Hunan, Guangxi, Yunnan, Guizhou, Jiangxi, Shanxi, and Gansu. The classes of antimony concentrates are decided by the mineral composition of the antimony ore deposit (Tab. 5.3). Due to the increasing demand and the improvement of antimony smelting skills, the antimony production of China increases year by year. China accounts for 80–90% of the world's antimony production; the antimony production of China was around 150,000 t y^{-1} from 2005 to 2015 [4]. However, in recent years, antimony production has been reduced because of stricter environmental standards since 2008. But after the eco-upgrade, the production has been maintained at a steady level, which is shown by the annual antimony productions of 97,683, 89,600, and 100,000 t in 2017, 2018, and 2019, respectively [5, 6].

Most of the antimony mining in China is underground and the conventional methods are cemented, mortar filling for column roof, full mining for retaining irregular pillars and segmental open space shrinkage method. There are three common ways of antimony mineral separation: hand selection, gravity separation, and flotation [4]. These three methods are usually used in combination to achieve better separation. Concerning antimony smelting, thermometallurgy is the commonly used method in antimony smelting plants in China [7]. The main antimony products are: antimony concentrate, refined antimony, antimonous oxide, antimonous sulfide. Furthermore, since the beginning of the twenty-first century, China has paid more attention to the further processing of antimony, and the main refined products are: high-purity antimonous oxide, fine grain antimonous oxide, colloidal antimony pentoxide, antimony oxide as catalyst, dust-free Sb(III) oxide, sodium antimonite, antimony acetate, antimony alloy, high-purity antimony, organic antimony compounds, Sb-containing flame retardant base material, etc. [4].

Tab. 5.2: Typical antimony deposits in China.

Deposit name	Province	Mineral composition	Element association	Tonnage / Mt Sb	Grade (%Sb)	Ore belt	Reference
Xikuangshan	Hunan	Stibnite	Sb	0.859	3.5–4.5	South China-Circum-Pacific Sb ore belt	[3, 121]
Dachang	Guangxi	Pyrite, pyrrhotite, cassiterite, sphalerite, chalcopyrite, stibnite, scheelite	Sn-Cu-Zn-Sb-W	0.317	3.69	South China-Circum-Pacific Sb ore belt	[3, 121–125]
Woxi	Hunan	Stibnite, scheelite, native gold	Sb-Au-W	0.167	1.34–4.26	South China-Circum-Pacific Sb ore belt	[3, 121, 126]
Banxi	Hunan	Stibnite, pyrite, arsenopyrite	Sb	0.062	0.4–60.5 (average: 21.03)	South China-Circum-Pacific Sb ore belt	[121]
Banpo	Guizhou	Stibnite, pyrite	Sb	0.143	4.83	South China-Circum-Pacific Sb ore belt	[3, 127, 128]
Gongguan, Qingtonggou	Shanxi	Stibnite, cinnabar, pyrite	Hg-Sb	0.017 0.026	1.51 0.98	Qinling-Central Asis Sb ore belt	[3, 129]
Qinglong	Guizhou	Stibnite, pyrite	Sb	0.3	3.94	South China-Circum-Pacific Sb ore belt	[3, 130]
Longshan	Hunan	Stibnite, native gold, pyrite, arsenopyrite	Sb-Au	0.043	3.86	South China-Circum-Pacific Sb ore belt	[3, 131]
Sanhe	Jilin	Stibnite	Sb-Hg	–	–	Jnlin-Daxinganling Tianshan-Outer Baikal Sb ore belt	[3, 121]

Tab. 5.3: Classification of antimony deposits in China.

Elemental composition of the ore minerals	Mineral composition of the ore minerals	Percentage of total Sb reserve in China	Sb concentrate class	Typical deposits	Reference
Sb	Stibnite	67	Antimony concentrate	Xikuangshan Banxi Muli Yawan	[4, 7, 121]
Sb-Au-W	Stibnite, native gold, wolframite, scheelite	21	Sb-Au concentrate	Woxi Longshan Zhazixi	
Sb-W (Sn-Cu-Pb-Zn-Ag-Hg)	Stibnite, scheelite, berthierite, jamesonite, stannite, boulangerite	12	Sb-Hg/W/Pb/Zn concentrate	Dachang Gongguan Banpo	

5.3 Emission sources

Emissions of antimony occur during the production, manufacture, fabrication, and usage stages. Antimony emissions discharged into the biosphere reached 1.916 Mt in the period from 2006 to 2016 [8].

5.3.1 Antimony emissions during antimony production processes

Solid and liquid wastes that contain antimony are produced during the antimony ore mining and separation processes. Uncontrolled emission and release of these wastes may lead to severe antimony pollution. Although some differences in emission nodes exist among different mining and separation factories due to different mining technologies and mineral separation processes utilized, common emission nodes and possible fate are summarized in [4]. As for dust, although it can be produced during the crushing and sieving processes, almost all may be inhibited by watering or collected by cloth bags [4].

Wastes are also produced during the smelting process. During the compounding of raw material, accessories, fuel, and flux, exhaust gas containing dust is produced. Although these dust particles can be effectively collected at the emission node with cloth bags, part of the exhaust gas containing antimony is still discharged into the atmosphere. It is reported that in 2009, 26.7% of the total antimony atmospheric emission in China originated from nonferrous metal smelting. Since thermometallurgy is the most widely used antimony smelting method in China, the most significant water consumption nodes are in the smelting furnace, and in the dust collecting system, and slag water. Usually, the water is recycled water and will not be discharged outside the smelting plant. The primary solid wastes produced during the smelting process are volatile smelting slag from the blast furnace, gypsum residue after flue gas desulfurization, and arsenic alkali residue. Arsenic alkali residue has the highest content of antimony among these three solids (20–40%), and the speciation is mainly sodium antimonite with minor antimony metal [7, 9].

5.3.2 Antimony emissions during coal combustion

Coal combustion significantly contributes to antimony atmospheric emissions [10]. A high concentration of antimony is common in Chinese coals, with the average content estimated to be 2.27 $\mu g\ g^{-1}$ [11]. In particular, coals from Guizhou and Inner Mongolia are extremely enriched in antimony. The concentration of antimony in a Late Permian anthracite sampled from Xingren, Guizhou can be as high as 3,860 $\mu g\ g^{-1}$ [12]. During coal combustion, exhaust gas containing antimony is produced and discharged into the atmosphere. Also, part of the antimony is partitioned into solid residues. The total atmospheric antimony emission in China caused by coal combustion was about 530.86 t in 2005. Emissions are mainly concentrated in the central and eastern regions of China and primarily originate from the industrial and power sectors [13]. In 2009, the national atmospheric emissions of antimony were estimated to be 818 t, 61.8% of which came from coal combustion [14]. Furthermore, the total antimony emissions discharged from coal combustion in China increased from 133.19 t in 1980 to 546.67 t in 2007, with an annual average growth rate of 5.4% [15].

5.3.3 Antimony emissions during antimony manufacture and fabrication

Antimony emissions from manufacture and fabrication are a major contribution to the total anthropogenic antimony emissions. It is estimated that 0.106 Mt of antimony were discharged into the biosphere in 2013, in China, accounting for 41.1% of the total emissions [8].

5.3.4 Antimony emissions during usage of antimony-containing products

Antimony emissions at the usage stage represent a large part of the total antimony emission in China. In 2013, antimony emissions originating from the usage of antimony-containing products was estimated to be 0.100 Mt, accounting for 39.0% of the total antimony emission that year [8]. Common products that contain antimony are PET bottles, brakes, thermoelectric devices, flame retardants. Antimony emission from brake wear is a significant mobile source. It is estimated that, in 2009, there were 37.6 million vehicles with Sb-containing brake pads in China, and 22.2 t of antimony were emitted to the ambient air as brake abrasion dust. Incineration is one of the disposal methods for municipal waste, and atmospheric antimony emission are produced. In 2009, 12.9% of total municipal wastes were disposed of by incineration, producing 60.7 t atmospheric antimony emission [14]. Furthermore, surrounding soils are also influenced by municipal waste incineration [16].

5.4 Distribution in the environment

5.4.1 Atmosphere

The morphological characteristics and spatial distribution of atmospheric antimony in China have been studied. Generally, atmospheric antimony concentration ranges from a few ng m^{-3} to dozens of ng m^{-3} [17]. The level of antimony in the atmosphere varies significantly in different regions and seasons (Tab. 5.4). The seasonal variation of antimony is dependent on the variation of particulate matter (PM). High temperature and humidity in summer, heating, and boiler combustion in winter are all causes of the variation [18–20]. The spatial distribution of atmospheric antimony pollution is closely related to its pollution source [21]. Ao et al. [22] investigated the atmospheric antimony deposition in the Xunyang Hg-Sb mining area, Shaanxi Province [22]. It was found that

Tab. 5.4: Concentration of antimony in atmospheric particulate matter (PM) of China.

City	PM	Atmospheric concentration / ng m^{-3}	Seasonal averaged concentration / ng m^{-3}				References
			Winter	Spring	Summer	Autumn	
Beijing	PM2.5	16.00	10.00	7.50	4.90	8.40	[18]
Tianjin		10.10		7.70	11.20	11.50	
Langfang		8.90	12.90	7.10	5.80	9.70	
Baoding		17.30	34.10	7.60	5.20	22.30	
Nanjing		2.80	5.71	1.84	0.82	2.61	[20]
Chendu	PM10	7.90	9.20	9.10	5.50	7.80	[19]
Guangxi		1.31					[132]
Pearl River Delta		1.80					[133]

the antimony concentration in the precipitation (3.1–7.1 μg L^{-1}) and pot experiment soil (4.9–29 mg kg^{-1}) in the mining area was much higher than the background value (0.49 μg L^{-1} and 3.9 mg kg^{-1}, respectively). Furthermore, the high population density, extensive motor vehicle use and intensive industrial manufacturing in the developed urban areas also result in high atmospheric antimony emissions [18].

After being emitted to the atmosphere, antimony has shown typical long-distance migration on both local and regional scales [14]. Beaudon et al. [23] studied the aerosol deposition of heavy metals in ice cores on the Tibetan Plateau and found that China had the highest concentration of antimony emissions and sediment flux in 1965, mainly originating from coal combustion [23].

The topography, climate, and economic development of China's provinces and cities vary greatly, which significantly affects the atmospheric distribution and migration behavior of antimony. At present, existing studies on the distribution of antimony in the atmosphere are limited and further exploration is needed.

5.4.2 Soils

Antimony contamination in soil has attracted nationwide attention in recent decades. The distribution of antimony in soils is determined by many interacting factors, specifically the geological setting, the presence of volcanic ash, groundwater, antimony mobility, anthropogenic activities, etc. [24].

5.4.2.1 Antimony distribution in soils from mining areas

China has the most abundant antimony resources in the world and owns the world's top three antimony mines [25]. A significant amount of antimony wastes have been released from mining and smelting processes, causing serious antimony contamination in the mine or nearby areas [24].

Generally speaking, higher antimony concentrations tend to be present near the mining and smelting areas. Antimony concentrations range from a dozen to several thousand mg kg^{-1} near the mining and smelting areas (Tab. 5.5), while the geometric mean of antimony concentration in Chinese soil is only 1.06 mg kg^{-1} [26]. Sequential extraction procedures show that antimony in the soil near mining areas is mainly present in the residual fraction followed by antimony bound to organics, iron and manganese oxides, carbonate and in exchangeable form [27].

Antimony concentrations are quite skewed with large standard deviations due to the high heterogeneity of the soils near Xikuangshan [28]. Also, the spatial variation of antimony is significant [29]. The variation of antimony concentrations in surface soils is caused by factors including the distance from the antimony smelter and plant [30] and soil types [31]. As antimony smelters are usually located in mountain valleys, the exhaust

Tab. 5.5: Concentrations of antimony in the soils of mining areas and some cities in China.

Region	System details	Sb / mg kg^{-1}			Reference
		Min	Max	Average	
Xikuangshan, Hunan	Mining areas	100	5,045	659	[82]
	Central smelters		1,214		[32]
	Lianxi River and tde Qifeng River areas		1,547		[72]
	Foot of Xikuangshan	141	8,733	1,315	[30]
	South/north of the Sb mine			145/720	[134]
Guangxi	Sb smelting area of Hechi city	155	30,439	3,171	[31]
	Typical slag heap of Nandan County	49	4,999		[73]
De'an county, Jiangxi	Minging	133	593	362	[117]
Southwestern Guizhou	Smelter plant			35.6	[135]
	Puqing mining areas	0.49	1,431	14.2	[136]
	QingLong mining areas	18.52	124.14	71.33	[137]
	Banpo mining/smelting areas	267	5,633	1,732	[76]
	Banpo smelting factory	4.47	124.91		[93]

fumes do not diffuse easily. Soil antimony contents show a decreasing trend as the distance from the smelter increases [31]. A large number of metal elements will likely leach into the surrounding environment because of surface runoff, and accumulate in the local settlement, resulting in potential serious pollution.

In addition to the variation in the concentration in the surface layer, the levels of antimony and its migration patterns in the soil profile are also quite different among different sampling points [32, 33]. In the profiles of the Hechi smelting area, the highest concentration of antimony appears in the surface soil of 0–2 cm, and decreases rapidly in the range of 0–20 cm, while the level below 20 cm does not change much [34]. The antimony concentration tends to increase below 60 cm in the profile and may pose a potential threat to groundwater quality. Moreover, the migration characteristics of antimony in the soil profile are affected by many factors, such as soil utilization type [34, 35], metal oxide contents (Fe and Mn), Ca and Mg ions, and physical and chemical properties like Eh, pH, and soil water content [33].

5.4.2.2 Antimony distribution in soils in the non-antimony mining areas

Cities can be grouped in three categories according to their average antimony concentration in soils [36–39]. Most parts of China have a low average level of antimony in soils (0.500–1.50 mg kg^{-1}); this antimony level is much lower than those from Southwestern China. Tibet, Hubei, Hunan, and Zhejiang have a moderate average level of antimony (1.50–2.00 mg kg^{-1}), corresponding to the South China-Circum-Pacific antimony ore belt. Yunnan, Guizhou, Guangxi, and Shanghai have high average antimony contents (>2.00 mg kg^{-1}), corresponding to the Western Yunnan Tibet belt [24]. The antimony concentrations range from 0.45 to 2.51 mg kg^{-1}, with an average value of 1.13 mg kg^{-1} in soils from non-antimony mining areas [36, 37, 40, 41]. These concentrations are lower than the maximum permissible pollutant concentration for antimony [42]. However, the antimony concentrations from coal/gold mine sites (Wuda Coalfield, Inner Mongolia/Guizhou) are about 40 times higher than the average antimony concentration mentioned above [36, 38].

5.4.3 Waters

As the largest river in China and the third largest in the world, the Yangtze River is 6.30×10^3 km long, with a drainage area of 1.8 million km^2. As an extensive water system, many mining areas exist in the Yangtze River basin. Typical levels of dissolved antimony reported in the literature for unpolluted waters are less than 1 μg L^{-1} [24]. The Council of the European Communities and China have established the maximum permissible antimony concentrations in drinking water as 10 μg L^{-1} and 5 μg L^{-1} [11], respectively.

5.4.3.1 Abundance of antimony in mining area waters

Recent studies reported that total concentrations of antimony in water samples from Xikuangshan ranged from 16 to 38,288 μg L^{-1}. The highest concentration was found in the water of the tailings dam. Such severely contaminated water was adequately treated and the ecological risk to the surrounding environment is limited. The antimony concentrations in rivers directly contaminated by mine wastewater were tenfold higher than those in rivers that were far from the wastewater outlet. The total antimony concentrations in well reservoir and upstream waters in the study area were 17 μg L^{-1}, 344 μg L^{-1} and 415 μg L^{-1}, respectively. The antimony concentrations in these waters all exceeded the drinking water guideline of antimony in China (5 μg L^{-1}) [43].

In the Dachang multi-metalliferous mine, waters of various types such as adit waters, flotation drainage, leaching waters, and river waters all contained high concentrations of dissolved antimony, reaching up to 5,475 μg L^{-1}. The pH values ranged from 6.51 to 9.82 because the geological background of Karst is rich in carbonates, which increase the mobility of antimony, resulting in approximately 80% of the antimony in the water samples being soluble and mobilised long distances with the water. The downstream sediments contained up to 2,591 mg kg^{-1}, while the downstream waters contained 211 μg L^{-1} [44]. Simulation calculations suggest that the impact of antimony on the surface water can reach a distance of 120 km [45].

In the Banpo antimony mining area, although most of the antimony deposits are closed and abandoned, waste drainage and tailings produced by mining activities are still contributing to the antimony pollution of the Duliujiang River. The adit waters and mine wastes from the old mine dump show up to 13,350 μg L^{-1} and 8,792 μg L^{-1} antimony, respectively. Antimony released from mine wastes has higher mobility than arsenic under oxidizing conditions, especially in neutral and basic conditions [46], which means the dissolved antimony has a strong migration ability in the Duliujiang mainstream and tributary waters (pH of 6.85–9.12). After the dilution by tributary water with a high flow rate, however, an average antimony concentration of only 10 μg L^{-1} remains in the final downstream section of the Duliujiang River basin [47].

5.4.3.2 Antimony concentration in waters from different regions

The long-distance migration and atmospheric deposition of antimony will lead to antimony pollution in inaccessible glacier snow plains. The Tibetan Plateau and surrounding mountains are one of the most concentrated glacier regions on earth. In the Southeastern Tibetan Plateau, the average concentration of antimony in snow/ice and snow pit samples was approximately 2.58 pg g^{-1} with a range of 1.64–9.20 pg g^{-1}, which was comparable to antimony concentrations determined in the Mt. Everest ice core (2.5 pg g^{-1}). Shrinkage of the glaciers affects the water source of large rivers (e.g., the Indus, Ganges, Brahmaputra, Yangtze, and Yellow River). It was estimated that about 10 kg of antimony were released each year through the glacier meltwater [48].

The average concentration of antimony was 65.3 μg L^{-1} with a range of 49.5–86.9 μg L^{-1} in the surface water of the Yangtze River in the Nanjing section [49]. The Lake Taihu, which is surrounded by numerous plain wetlands, is exposed to constant-release antimony point-sources from local wastewater treatment facilities. The mean antimony concentration in the wetland's water was 7.03 μg L^{-1}, including dissolved antimony and antimony attached to suspended solids [50]. The antimony geochemistry in the Yangtze River estuary is mainly controlled by riverine input, atmospheric deposition, the incursion of Taiwan Warm Current, and release processes from the particulate phase. In surface water of this area, the average concentrations for Sb(III) and Sb(V) were 0.152 (0.029–0.736) μg L^{-1} and 0.592 (0.121–2.567) μg L^{-1}, respectively. In the bottom layer, the concentrations were much lower, ranging from 0.023 to 0.116 μg L^{-1} for Sb(III) (with an average of 0.050 μg L^{-1}) and from 0.047 to 0.441 μg L^{-1} for Sb(V) (with an average of 0.194 μg L^{-1}) [51].

The East China Sea (ECS) is one of the most critical marginal seas for the Western Pacific. The horizontal distribution of the total dissolved inorganic antimony concentrations varied from 0.086 to 0.607 μg L^{-1}, with an average of 0.224 μg L^{-1}. The levels were high in the coastal area of the Yangtze River Estuary and Hangzhou Bay and decreased rapidly off the coast in both the surface and bottom layers. The concentrations of dissolved inorganic antimony in the ECS remained at natural levels compared with other areas in the world [51]. Dissolved antimony concentrations in the surface waters of Bohai Bay ranged spatially from 0.386 to 1.075 μg L^{-1}, with a mean value of 0.562 μg L^{-1}, of which Sb(V) was the main chemical species (89% of the total dissolved antimony) [52].

5.4.4 Sediments

5.4.4.1 Antimony distribution in sediments of the mining area

Antimony in aquatic systems is mainly accumulated in bed sediments [24]. Studies on antimony in sediments (shown in Tab. 5.6) mainly focus on surface sediments, especially in rivers. Sediment concentrations in rivers are affected by the flow rate of rivers and the rate of sediment accumulation [53]. From a regional perspective, high antimony concentrations in sediments are mostly located in Southwestern China, where most antimony ore resources are located. Concentrations of antimony in the sediments of the Zijiang River and the Duliujiang River are high with average values of 36.64 mg kg^{-1} and 468.66 mg kg^{-1} [54, 55], respectively. These values are much higher than the background values in the river sediments of China (0.800–3.00 mg kg^{-1}). Higher concentrations of antimony in sediments were mainly detected within or near mining and smelting areas. The maximum antimony concentration found in sediments was 2.77×10^4 mg kg^{-1} at the Dushan mining area. There was a huge difference in antimony concentration between river sediments in rivers that flow through the mine areas compared with those flowing through non mine areas [56]. A study conducted in the Xikuangshan Sb mine area showed a significant decrease in antimony concentrations in sediments with increasing distance from the mine (from 7,316 to 57 mg kg^{-1}) [57].

5.4.4.2 Spatial distribution of antimony in river/lake sediments

The antimony concentrations in sediments from alpine highlands and northern China were much lower due to the relatively lower background antimony concentration. The average antimony concentrations in sediments of the Caohai Lake, Yellow River's lower reaches, Qinghai Lake, Qarhan Salt Lake, Jiaozhou Bay, Nansi Lake, and Daliao

Tab. 5.6: Concentrations of antimony in surface (0–10 cm) sediments of China.

	Sampling site	Sb / mg kg^{-1}			Reference
		Min	Max	Average	
River	Yangtze River estuary and its adjacent sea area	0.44	4.41	1.15	[111]
		0.32	0.968	0.577	[53]
		0.36	1.48		[138]
	Three gorges reservoir area	0.67	1.59	1.14	[139]
	Zijiang River	1.88	184.83	36.64	[55]
	DuliuJiang River	1.21	7,080	468.66	[54]
	Estuarine sediment of the Daliao River	0.27	1.12	0.68	[140]
	Shanghai suburban area			1.4	[141]
Lake	Meiliangwan of Taihu Lake			3.5	[142]
	East Taihu Lake			3.21	
	The whole of Taihu Lake	4.11	95.09	9.56	[143]
	Chenshidang Lake	0.58	5.99	1.72	[144]
	Poyang Lake (Summer)	1.04	7.07	3.27	[145]
	Poyang Lake (Winter)	0.39	1.25	1.13	
	The inlets of Dongting Lake	0.40	16.49	3.28	[112]
		0.79	3.93	1.76	[146]
	East Dongting Lake	0.39	4.04	2.40	[112]
	South Dongting Lake	1.17	5.15	3.26	
	West Dongting Lake	1.16	3.21	2.05	
	The whole of Dongting Lake			2.57	
	Caohai Lake (Yunnan Guizhou Plateau)	1.14	1.69	1.51	[59]
	Qarhan Salt Lake in (Qaidam Basin, Xingjiang Province)	0.14	0.73	0.45	[62]
	Qinghai Lake	2.76	9.25	4.54	[61]
	Jiaozhou Bay	0.45	1.21	0.7	[58]
	Nansi Lake (Shandong Peninsula)	0.92	1.77	1.23	[63]

Tab. 5.6 (continued)

	Sampling site	Sb / mg kg^{-1}			Reference
		Min	Max	Average	
Bays and oceans	Jiaozhou Bay	0.45	1.21	0.7	[58]
	Ledong sea area in South China sea	0.34	1.07		[147]
Mining and smelting areas	The upstream of Qinglong Sb mines (Guizhou Province)	77.48	195.68	131.11	[148]
	The downstream of Qinglong Sb mines (Guizhou Province)	140.28	266.22	207.07	
	Dachang Sb mines (Guangxi Province)	5.06	20.90	12.9	[149]
	Diaojiang River (The downstream of Dachang Sb mines)			347	[150]
	Xikuangshan Sb mines (natural water sediment)	57.1	7,315	1,163	[87]
	Xikuangshan Sb mines (along Lianxi River to the Main Stem of Zijiang River)	57	7,316		[57]
	Zuoxiguo Sb mine (Kaiyuan of Yunnan Province)			462	[56]
	Danzhai Sb mine (Kaiyuan of Yunnan Province)			376	
	Kaiyuan, not directly impacted by mining activities (Yunnan Province)			248	
	Daye Tieshan iron mine (Hubei Province)	0.51	202.3	43.8	[40]
	Yata Au Mine (Guizhou Province)	111	223	149	[151]
	Dushan Sb mine (Guizhou Province)	5,511	27,684		[77]

River were only 1.51, 1.44, 4.54, 0.45, 0.70, 1.23, and 0.90 mg kg^{-1}, respectively [58–63]. In surface sediments and sediment cores from the Songhua River in Northeast China, antimony concentrations ranged between 0.18–0.57 and 0.34–0.81 mg kg^{-1}, respectively [60].

Lake sediments are often used for the historical reconstruction of heavy metal pollution. Lake sediments (60 cm deep) from Gonghai which is located in Ningwu, Shanxi Province record evidence of the past 200 years: The antimony concentration range in the lake sediments is 0.99 to 1.29 mg kg^{-1}, with a mean of 1.08 mg kg^{-1}. During the period 1980–2014, the flux of anthropogenic antimony was 0.11–0.61 mg m^{-1} y^{-1}, with a mean value of 0.31 mg m^{-1} y^{-1} [64]. Antimony is also a typical pollutant in Lugu Lake sediments. Antimony concentrations of 30 cm depth sediments ranged from 0.61 to 3.20 mg kg^{-1}, and the Sb pollution peaked in historical periods (i.e., the 1880s to the 1940s and around the 1920s). Since the mid-1980s, antimony pollution has also increased [65].

5.4.5 Plants

5.4.5.1 Antimony distribution in non-food plants growing in mining areas

Many surveys have been carried out in antimony mining areas in order to investigate the distribution of antimony in plants and to screen hyperaccumulator species for remediation. Antimony concentrations in plants in mining areas range widely, reflecting regional differences and plant species.

Most of the reported data of antimony in plants was collected from the mining area of Xikuangshan. Fu et al. [66] sampled nine native wild plants in the forest subarea. They found that *P. vittata* and *P. aquilinum* were potential plants for phytoremediation, with high TFs (Tab. 5.7). They belong to the root compartment plant type [66]. Wei et al. [67] collected 65 plant samples at ten sites of Xikuangshan; the highest concentration of antimony was 90.98 mg kg^{-1} accumulated by *Hippochcaete ramosissima* [67]. The level of antimony in 34 plants collected in the survey of the Xikuangshan antimony deposit area ranged from 3.92 to 143.69 mg kg^{-1}, and the antimony level in roots was much higher than that in the above ground part. *P. vittata* species of Pteridaceae family showed the advantage of accumulating antimony from the contaminated environment [68]. As a particular species of plant in China, *Boehmeria nivea* is very common in Xikuangshan. *Boehmeria nivea* showed tolerance and enrichment capacity of antimony, and could be a potential candidate plant for phytoremediation of combined pollution [69, 70]. Some colonized plant species were collected in a field survey of Xikuangshan, relatively high levels of antimony were reported in the leaves of *Aster subulatus* (477 mg kg^{-1}), *B. nivea* (323 mg kg^{-1}), and *H. bipinnatae* (235 mg kg^{-1}) *H. bipinnatae* was regarded as the most suitable species for phytostabilization [71]. Some further studies on the speciation of antimony in plants were conducted by Wei et al. [72]; results showed that the prevalent species in the soil and plant samples were inorganic antimony: both Sb(III) and Sb(V) were found in the plant tissues, and the proportion of Sb(III) was higher than that of Sb(V), especially in aboveground tissues [72].

Tab. 5.7: Antimony concentrations, bioaccumulation factors (BAF = organ/soil), and transformation factors (TF = aboveground/underground) in plant samples from some antimony mine areas.

Organism	Area	Site	Sb in soil / mg kg^{-1}	Organ / mg kg^{-1} dw			BAF			TF	Reference
				shoot/stem	root	leave	shoot	root	leaves		
Zea mays L.	Xikuangshan	Mining area, Hunan Province, Central South China	947	6.38	14.4		0.0067	0.0152		0.44	[78]
			116	6.57	4.99		0.0567	0.043		1.32	
			3016	14.5	48.9		0.0048	0.0162		0.3	
			786	11.1	29.8		0.0142	0.0379		0.37	
			325	6.4	4.93		0.0194	0.015		1.3	
			23.8	4.21	4.51		0.1737	0.1859		0.93	
			370	8.51	26.9		0.023	0.0727		0.32	
Dryopteris erytbrosora	Xikuangshan	Mining area, Hunan Province, Central South China	299	27.53	11.86		0.0921	0.0397		2.32	[66]
Pteris vittata				18.79	54.23		0.0628	0.1814		0.35	
Imperata cylindrica			294	162 (24.98–416)	192 (7.26–376)		0.5510	0.6531		0.84	
Erigeron annuus				73.68	18.00		0.2506	0.0612		4.09	
Oplismenus undulatifolius (A) Bea				84.86	220		0.2886	0.7483		0.39	

Radish	Xikuangshan	Mining area, Hunan Province, Central South China	101		4.4	1.48		0.0436	0.0147	0.34	[82]
			5045		6.82	121.40		0.0014	0.0241	17.80	
			852		10.49	92.34		0.0123	0.1084	8.80	
			513		3.96	34.11		0.0077	0.0665	8.61	
Phytolacca acinosa Roxb	Xikuangshan	Mining area, Hunan Province, Central South China	1,062	>190			0.18				[28]
Boehmeria nivea (L.) Gaudich.			11,798	4,029			0.34				
Miscanthus floridulus (Labill.) Warb.			11,230	683			0.06				
Eremochloa ophiuroides (Munro) Hack.			7,487	128			0.02				
Lactuca tatarica (L.) C.A.Mey			9,433	357			0.04				
Leersia hexandra Swartz			1,938	292			0.15				
Hygroryza aristate (Retz.) Nees			1,575	109			0.07				
Arachis hypogaea L.			1,837	314			0.17				
Comnyza canadensis (L.) Cronq.			527	461			0.87				

(continued)

Tab. 5.7 (continued)

Organism	Area	Site	Sb in soil / mg kg^{-1}	Organ / mg kg^{-1} dw			BAF			TF	Reference
				shoot/stem	root	leave	shoot	root	leaves		
Dicranopteris linearis (Burm.) Underw.	Qinglong Sb mining area	Guizhou Province, Southwest China	19.93	0.4	1.59	0.59	0.0201	0.0798	0.0296	0.62	[25]
Pteridium aquilinum (Linn.) Kuhn var.			72.61	0.36	3.66	0.49	0.0050	0.0504	0.0067	0.23	
Pteridium aquilinum (Linn.) Kuhn var.			44.33	0.55	4.81	0.5	0.0124	0.1085	0.0113	0.22	
Saccharum arundinaceum Retz				0.91	2.98		0.0205	0.0672		0.31	
Dicranopteris linearis (Burm.) Underw			39.02	5.55	8.14	1.36	0.1422	0.2086	0.0349	0.85	
Miscanthus sinensis Anderss				0.61	0.43		0.0156	0.0110		1.42	
Pteridium aquilinum (Linn.) Kuhn var.				0.74	3.64	0.83	0.0190	0.0933	0.0213	0.43	
Artemisia dubia Wall. Ex Bess				0.4	4.07	2.33	0.0103	0.1043	0.0597	0.67	
Cattails	Mine tailing of a Sb mining site	Dushan County of South Guizhou Province	3,216.49	97.63	617.64	183.63	0.0304	0.1920	0.0571	0.46	[75]

Other two typical mining areas are Guangxi and Guizhou provinces in Southwest China, which hold the second and third largest antimony mines. Four species of plants (*Pteris vittata*, *Saccharum arundinaceum*, *Miscanthus floridulus*, and *Thysanolaena maxima*) were collected from three tailing dumps of the Dachang multi-metalliferous mine area. Tolerance and accumulation were observed, and antimony enrichment by *Pteris vittata* was confirmed [73]. Mosses were collected from the areas around the antimony historical smelter located in Hechi, Guangxi Province; the antimony concentration of the mosses ranged from 41.0 to 5.77×10^3 mg kg^{-1}, which was 1.5–213.6 times higher than the background value [74]. Chen et al [75] collected cattail growing in the mine tailing of an antimony mining site in Dushan County of South Guizhou Province; the results showed that the estimated annual turnover of antimony in cattail was 1,278 mg m^{-2}, and it represented a target plant for ecological remediation in combined polluted areas [75]. Some other types of dominating wild plant and rhizospheric soil were also sampled at nine sites from the local mining, milling, and smelting sites in Dushan, The uptake of antimony by local plants varied greatly (21–21,148 mg kg^{-1}), and some species (such as *Chenopodium album Linn.*, *Sedum emarginatum Migo*, and *Sedum lineare Thunb*) accumulated more than 1,000 mg kg^{-1} Sb [76]. Eight plants were sampled from the Qinglong antimony mines. Results showed that *Artemisia dubia* Wall. Ex Bess and *Miscanthus sinensis* Anderss showed higher accumulation capacity for antimony and arsenic [25]. The concentration of heavy metals in *Polygonum capitatum* sampled in Dushan antimony deposit was analyzed. The content distribution of each part was root > leaf > stem > flower, and strong tolerance of *Polygonum capitatum* in contaminated soil with antimony deposits was observed. Antimony in Polygonum cephalum in this mine area posed certain health risks to human health [77].

Transfer coefficients of antimony was also quite different between different tissues of the plants. The enrichment of antimony in the aboveground of plants was less than that in the underground parts [68, 78–80]. However, the bioaccumulation factors (BAFs) varied greatly (ranged from 0.005 to 0.87), which might result in abnormally high antimony levels in plants growing at contaminated sites with different soil types [67, 76]. For most plants, the concentration of antimony in tissues followed: root > leaf > stem [76]. *Pteris vittata* enrichment order was: root > leaf > stem, while antimony in the *Miscanthus floridulus*, and Thysanolaena maxima followed the order of leaf > root > stem [73]. For the rice plants collected in Xikuangshan, antimony contents in tissues decreased in the order of root > shoot > husk > grain, and husk, bran, and aleurone layer tended to concentrate more antimony than other tissues [81]. The concentrations and enrichment coefficients of soil and plants in some mining areas are shown in Tab. 5.7.

5.4.5.2 Antimony distribution in vegetables and crops in mining areas

Food crops and vegetables could be important intake sources of antimony for humans especially for local inhabitants in the mining areas. Various crops and vegetables have been sampled around Xikuangshan. High levels of antimony were found in root and leaves of radish growing in Xikuangshan (the average concentration of antimony in roots and leaves were 5.6 and 54.0 mg kg^{-1}, respectively). However, no phytotoxicity was observed [82]. The mean contents of antimony in 60 vegetable samples ranged from 6.65 to 29.31 mg kg^{-1}. Furthermore, the antimony concentration varied in species; *Coriandrum sativum* L. was the highest, and *Brassica pekinensis* L. the lowest [83]. A range of daily food samples were collected in Xikuangshan; the mean concentration of antimony in rice, corn, and vegetables were 470 mg kg^{-1} (160–930 mg kg^{-1} dw), 300 mg kg^{-1} (90–720 mg kg^{-1} dw) and 379 mg kg^{-1} ww (30–1,660 mg kg^{-1} ww), respectively. The results showed that rice and vegetables (especially leafy vegetables) were two of the dominant oral intake sources of antimony for the residents; the dietary intake of antimony for people living in the vicinity of Xikuangshan (554 mg d^{-1}) is 1.5-fold higher than the tolerable daily intake (TDI, 360 mg d^{-1}) [84]. Asymptomatic accumulation of antimony in plants, especially for agricultural products cultivated in contaminated areas, should be of concern.

A few studies on the contents and risk of antimony in vegetables and grains in the market have been reported. Tan et al. (2016a) [85] randomly collected 703 vegetable samples from supermarkets, farmers' markets, and stalls in 14 cities/prefectures in Hunan Province to investigate the antimony content in food and assess human exposure. The results showed that the total Sb content was at a normal level in all vegetable samples, and antimony concentrations in leaf vegetables were higher than that in root, stem, flower, and fruit vegetables ($p < 0.05$) [85]. Vegetables are the primary source of dietary antimony exposure, with a contribution rate of 83.1% [86].

5.4.6 Animals and microorganisms

5.4.6.1 Antimony distribution in animals

Some animals accumulate antimony from the environment. The ability of antimony accumulation is closely related to antimony bioavailability [87]. Several works about aquatic, amphibian, and terrestrial organisms have been carried out by Chinese researchers. The concentration of antimony in fish in mining areas was much lower than that of arsenic and mercury [88]. In Gaotang Lake, the antimony content in fish was low (<0.04 mg kg^{-1} ww), and did not exceed the Chinese maximum allowable concentrations [89]. As for the antimony distribution inside the fish, it is most enriched in the kidney and liver, followed by muscle, bladder, and skin, and fewest in gill [88].

Comparing amphibian and terrestrial organisms, the antimony enrichment capacity of earthworms is significantly higher than those of algae, crab, frog, fish, and river snail [87]. Grasshopper can be regarded as a biological indicator of heavy metal pollution [90]. Grasshoppers and earthworms are both invertebrates [87], and invertebrates are more likely to accumulate antimony [91].

Also, methylated antimony has been detected in the saliva and urine of humans, which may be related to the metabolism of mammals [92].

5.4.6.2 Antimony distribution in microorganisms

Studying the bacterial community and metabolic potential of antimony pollution in various environments is of great significance for understanding the bioremediation of antimony pollution. Research on antimony and microorganisms involves community structure, metabolism, functional bacteria, metabolism, pathways.

With the help of high-throughput sequencing methods, the community diversity and structural composition of antimony-contaminated areas can be obtained [93, 94], and functional microorganisms with antimony metabolism genes identified [93, 95]. By biogeochemical processes, microbes can reduce soil antimony [93], especially in mining areas [96]. Microorganisms such as Cupriavidus, Moraxella, Bacillus [97], and *Agrobacterium tumefaciens* GW4 [98], can oxidize the more toxic Sb(III) into less toxic Sb(V) [96].

5.5 Environmental and health risk of antimony

Exposure of organisms (plants, animals, microorganisms) and the human body to antimony might cause acute and chronic toxicity, posing potential health risks to humans, through direct contact and intake, and ecological risk to ecosystems.

5.5.1 Human health risk of antimony

Antimony ore mining and smelting, occupational exposure, and other kinds of exposure might significantly affect human health. E-waste recyclers and some related workers, as well as some longtime residents living close to recycling sites, show high levels of antimony in hair [99]. A survey of the population in the Xikuangshan and Qinglong mining areas and Guiyang (non-mining area) showed that antimony concentration in the hair of people living in Xikuangshan (0.250–82.4, mean 15.9 $\mu g\ g^{-1}$) and Qinglong (0.060–45.9, mean 5.15 $\mu g\ g^{-1}$) mining areas were significantly higher than the control group (Guiyang, 0.065–2.87, mean 0.532 $\mu g\ g^{-1}$), and the risk for children was higher than that for adults [100, 101]. Industrial antimony exposure provided clear correlation with human immunity markers of workers [102]. Elevated antimony in the soil near the roadside and the urban regions has also been investigated; possible potential adverse effects on human health and the ecological environment due to antimony pollution were of concern [103].

Nearly all of the studies about the health risk of antimony exposure in China have been performed in Xikuangshan and other typical antimony mining areas. Dietary exposure to antimony was the dominant health risk to residents in proximity to antimony mines in China. The high level of antimony intake was due to the consumption of food, indicating that long-term consumption of vegetables from mining areas might cause a health risk for the people living in the vicinity [83, 84, 86]. Furthermore, according to some investigations in and around Xikuangshan, antimony in drinking water posed a major health risk; drinking water contributed 85–100% of the average daily dose (ADD) of antimony with also concern for the potential health risk for children [88, 104, 105]. Environmental accidents may also cause antimony contamination in edible tissues of vegetables collected from contaminated fields such as those affected by smelting wastewater spill from a Pb/Sb smelter located in Guangxi Province that exceeded the Chinese government and FAO/WHO guidelines [106]. This contamination is generally restricted to mining areas because surveys of antimony levels in rice, vegetables, meat, strawberry, and well water samples collected over much of Hunan Province, show that antimony concentrations were at a reasonable levels [85, 107, 108].

5.5.2 Environmental and ecological risk

Large amounts of antimony discharged into the biosphere might cause potential health and ecological risk. Some antimony mining and smelting vicinities have been severely contaminated and this should be of concern [24, 109]. Long et al. [110] reported that antimony was the first predominant pollutant in Xikuangshan with a concentration that was 1,120 times the corresponding background values. This might result in ecological and food security risk [110].

Antimony is a global toxic element that can be transported over long distances; the ecological risk of antimony downstream of mining areas and some typical reservoir areas have reached the level of moderate pollution and with high potential ecological risk, especially in the south branch of the Yangtze River estuary and Hangzhou Bay [111]. According to recent research in Dongting Lake and the Zijiang River basin located in Hunan Province, antimony is one of the primary contaminants, and the risk level for Dongting Lake was "serious" ecological risk (with an ecological risk index of 689.3) [112, 113]. The ecological risk of antimony in the surface sediments of the Changjiang estuary and the adjacent sea, however, showed low risk or nearly no risk because of low antimony concentrations [53].

There is still a lack of reasonable and integrated ecological risk assessment methods for antimony pollution in the environment. Much work is still needed, such as the improvement of the ecotoxicology database and environmental standards.

5.6 Policy and management

Severe environmental contamination caused by anthropogenic antimony emissions in China has attracted widespread attention during the past years [114]. Considering its threat to ecological and individual health, the Chinese government has taken an active policy approach to prevent antimony pollution.

5.6.1 Emission standards

To mitigate human exposure, China has established strict emission standards for antimony. The recommended level in China for drinking water is 5 $\mu g\ L^{-1}$ [115]. China issued the "Emission standards of pollutants for stannum, antimony and mercury industries" (GB 30770–2012). Emission limits for antimony in emissions to water are set at 1.0 $mg\ L^{-1}$ and 0.3 $mg\ L^{-1}$ for existing and new facilities; but for specially protected areas, the acceptable level should be controlled to under 0.3 $mg\ L^{-1}$. The concentration limits for antimony in air emissions are set at 8.0 $mg\ m^{-3}$ and 4.0 $mg\ m^{-3}$ for existing and new facilities; and for specially protected areas, the acceptable level should be strictly controlled to under 4.0 $mg\ m^{-3}$ [116].

5.6.2 Pollutant emission permits

The government formulates technical standards for pollutant discharge permits, which guide and regulate antimony smelting enterprises, and also put forward feasible technical requirements for pollution prevention and control. The allowable emission limits include the allowable pollutant emission concentration and allowable total emission amount. For air pollutants, the permissible emissions are determined by units of production facilities or outlets, according to GB 13271 and GB 30770. Specifically, the allowable pollutant discharge concentration is the hourly average concentration. For water pollutants, the permitted discharge concentration and permitted discharge amount are determined based on the discharge outlet of the production facility or the total discharge outlet of the enterprise. The allowable discharge concentration is also determined by GB 13271 and GB 30770. The permissible pollutant discharge concentration is a daily mean concentration.

5.6.3 Pollution control

(1) Air pollution control
Nowadays, although China's energy structure is changing from single source to energy diversification, coal will still be the core energy source in coming years [117]. The average concentration of antimony is 7.06 mg kg^{-1} from 1,058 representative coal samples [11]. During the coal combustion process, antimony will inevitably be released into the atmosphere. To improve the quality of the atmospheric environment, the government has adjusted and optimized the layout of antimony industries, strengthening joint prevention and control measures among different regions during recent years. Meanwhile, efforts will be focussed on reducing the production capacity in steel, foundry, coking, and electrolytic industries, and imposing strict limits on the discharge of air pollutants including antimony [118]. Also, the government will strengthen joint prevention, control, and governance in key regions, by unifying early warning grading standards and implementing emergency emission reduction measures in advance.

(2) Water pollution control
As a counter strategy to the current situation of antimony water pollution, the government will fully implement the action plan for preventing and controlling water pollution by regulating industrial, agricultural, and domestic pollution sources [118]. To extend the prevention and control of groundwater pollution, the government defines a centralized drinking water source protection zone and promotes standardization, especially at the source areas and along the routes of the south-to-north water diversion project.

(3) Soil pollution control
In general, the remedial measures of antimony pollution promoted by the government can be listed as follows: classifying management of cultivated land, controlling heavily polluted farmland, and prohibiting the planting of edible agricultural products in contaminated farmland. Projects will be undertaken to improve and protect agricultural land. All polluted enterprises are intended for relocation and reconstruction (or demolition when necessary) under strict environmental supervision.

5.6.4 Representative facility

Nowadays, a positive response to the environmental protection policy has been carried out nationwide. Among them, a water treatment facility constructed by Hsikwang Shan Twinking Star Co., Ltd. is representative. The treatment capacity of the sewage treatment station is about 1×10^5 t per day, and its investment costs reach as high as 15.8 million yuan.

The wastewater first flows into the mine tailings. After treatment by precipitation, purification, water quality, and quantity homogenization process, it enters the distribution pool, which consists of three consecutive reaction tanks. In the primary reaction tank, biological preparations are added to react with antimony and arsenic, forming heavy metal complexes. Then, the wastewater enters the secondary reaction tank, where an hydrolysis reaction takes place by adding lime milk to adjust the pH of the system. Next, a small amount of PAM is added to the third reaction tank to achieve removal of antimony, arsenic, and suspended solids by flocculation reaction. After the former reaction, the wastewater flows into an inclined pipe sedimentation tank for solid–liquid separation. The supernatant, can be emitted after regulating the pH in the clean water reservoir. The sludge finally flows into the wastewater facility for recycling through the collecting tank and thickening tank in turn.

5.7 Conclusion and perspective

As the world leader in antimony reserves and production, China has been facing significant challenges in the treatment of pollution originating from antimony production, manufacture, fabrication, and use. Ambient environments (soil, water, sediment, etc.) near the mining areas are enriched with antimony. The atmosphere is also contaminated because of coal combustion or vehicle brake wear. Serious threats to the ecological environment and human health exist. Actions taken by the Chinese government include stricter standards, scientific policies, and more effective supervision. However, a lot more work is needed concerning the migration and transformation of antimony in the environment and its risk to the environment and human health. Furthermore, policy and management can also be optimized with the improvement in antimony production techniques and a better understanding of the fate of antimony.

References

[1] Zhou S, Hursthouse A, Chen T. Pollution Characteristics of Sb, As, Hg, Pb, Cd, and Zn in Soils from Different Zones of Xikuangshan Antimony Mine. J Anal Methods Chem 2019, 2019.
[2] China Mineral Resources 2019. Ministry of Natural Resources. PRC, 2019.
[3] Wu JD. Antimony vein deposits of China. Ore Geol Rev 1993, 8, 213–232.
[4] Zhao ZL, He MC, Wang JJ, Lin H, Wang F. Pollution and Prevention of Heavy Metals in Nickel Cobalt Tin Antimony Mining and Processing Industry (in Chinese). Beijing, China, Tsinghua University Press, 2015.
[5] The Yearbook of Nonferrous Metals Industry of China 2018 (in Chinese). Beijing, China, China Nonferrous Metals Industry Association, 2018.
[6] Mineral Commodity Summaries 2020. Reston, VA, USA, U.S. Geological Survey, 2020.
[7] Yang XS. Control and Management of Heavy Metal Pollution in Key Nonferrous Metal Smelting Industries (in Chinese). Beijing, China, China Environment Press, 2014.

[8] Chu J, Mao J, He M. Anthropogenic antimony flow analysis and evaluation in China. Sci Total Environ 2019, 683, 659–667.

[9] Guo X, Wang K, He M, Liu Z, Yang H, Li S. Antimony smelting process generating solid wastes and dust: Characterization and leaching behaviors. J Environ Sci 2014, 26, 1549–1556.

[10] Chen J, Liu G, Kang Y, et al. Atmospheric emissions of F, As, Se, Hg, and Sb from coal-fired power and heat generation in China. Chemosphere 2013, 90, 1925–1932.

[11] Qi CC, Liu GJ, Chou CL, Zheng LQ. Environmental geochemistry of antimony in Chinese coals. Sci Total Environ 2008, 389, 225–234.

[12] Dai SF, Zeng RS, Sun YZ. Enrichment of arsenic, antimony, mercury, and thallium in a Late Permian anthracite from Xingren, Guizhou, Southwest China. Int J Coal Geol 2006, 66, 217–226.

[13] Tian H, Zhao D, He M, Wang Y, Cheng K, Qu Y. Atmospheric antimony emission inventories from coal combustion in China in 2005 (in Chinese). China Environ Sci 2010, 30, 1550–1557.

[14] Tian HZ, Zhao D, Cheng K, Lu L, He MC, Hao JM. Anthropogenic Atmospheric Emissions of Antimony and Its Spatial Distribution Characteristics in China. Environ Sci Technol 2012, 46, 3973–3980.

[15] Tian HZ, Zhao D, He MC, Wang Y, Cheng K. Temporal and spatial distribution of atmospheric antimony emission inventories from coal combustion in China. Environ Pollut 2011, 159, 1613–1619.

[16] Li Y, Zhang H, Shao L, Zhou X, He P. Impact of municipal solid waste incineration on heavy metals in the surrounding soils by multivariate analysis and lead isotope analysis. J Environ Sci 2019, 82, 47–56.

[17] Zhu J, Guo J, Wang L, et al. Overview on research on environmentally geochemical characteristics of antimony (in Chinese). Earth Environ 2010, 38, 109–116.

[18] Zhu CY, Tian HZ, Hao Y, et al. A high-resolution emission inventory of anthropogenic trace elements in Beijing-Tianjin-Hebei (BTH) region of China. Atmos Environ 2018, 191, 452–462.

[19] Cheng X, Huang Y, Zhang SP, Ni SJ, Long ZJ. Characteristics, Sources, and Health Risk Assessment of Trace Elements in PM10 at an Urban Site in Chengdu, Southwest China. Aerosol Air Qual Res 2018, 18, 357–370.

[20] Qi L, Chen M, Ge X, Zhang Y, Guo B. Seasonal Variations and Sources of 17 Aerosol Metal Elements in Suburban Nanjing, China. Atmosphere 2016, 7.

[21] He MC, Wang NN, Long XJ, et al. Antimony speciation in the environment: Recent advances in understanding the biogeochemical processes and ecological effects. J Environ Sci 2019, 75, 14–39.

[22] Ao M, Qiu GL, Zhang C, et al. Atmospheric deposition of antimony in a typical mercury-antimony mining area, Shaanxi Province, Southwest China. Environ Pollut 2019, 245, 173–182.

[23] Beaudon E, Gabrielli P, Sierra-Hernandez MR, Wegner A, Thompson LG. Central Tibetan Plateau atmospheric trace metals contamination: A 500-year record from the Puruogangri ice core. Sci Total Environ 2017, 601, 1349–1363.

[24] He M, Wang X, Wu F, Fu Z. Antimony pollution in China. Sci Total Environ 2012, 421, 41–50.

[25] Wen J, Xu S, Chen J, Wu Z. Study of Sb and as pollution characteristics in soils and wild plants of antimony mine (in Chinese). Environ Sci Technol 2013, 36, 18–22.

[26] Wei FS, Chen JS, Wu YY, zhen CJ. Study on soil environmental background value in China (in Chinese). Environ Sci 1991, 12–19.

[27] Chen Q, Xu S, An Y, Chen J, Wu Z. Distribution,speciation and bio-availability of arsenic (As) and antimony (Sb) in soils of antimony mine (in Chinese). Environ Chem 2014, 33, 1301–1306.

[28] Okkenhaug G, Zhu Y-G, Luo L, Lei M, Li X, Mulder J. Distribution, speciation and availability of antimony (Sb) in soils and terrestrial plants from an active Sb mining area. Environ Pollut 2011, 159, 2427–2434.
[29] Wang XQ, Li FB, Yuan CL, et al. The translocation of antimony in soil-rice system with comparisons to arsenic: Alleviation of their accumulation in rice by simultaneous use of Fe(II) and NO3. Sci Total Environ 2019, 650, 633–641.
[30] Fu Z, Wu F, Mo C, Deng Q, Meng W, Giesy JP. Comparison of arsenic and antimony biogeochemical behavior in water, soil and tailings from Xikuangshan, China. Sci Total Environ 2016, 539, 97–104.
[31] Xiang M, Zhang G, Li L, Wei X, Li H. The Characteristics of Heavy Metals in Soil around the Hechi Antimony-Lead Smelter, Guangxi, China (in Chinese). Earth Environ 2010, 38, 495–500.
[32] Yang HL, He MC, Wang XQ. Concentration and speciation of antimony and arsenic in soil profiles around the world's largest antimony metallurgical area in China. Environ Geochem Health 2015, 37, 21–33.
[33] Shangguan YX, Zhao L, Qin Y, Hou H, Zhang N. Antimony release from contaminated mine soils and its migration in four typical soils using lysimeter experiments. Ecotoxicol Environ Saf 2016, 133, 1–9.
[34] Xiang M, Zhang GP, Li L, Wei XF, Cai YB. Characteristics of heavy metals in soil profile and pore water around Hechi antimony-lead smelter, Guangxi, China (in Chinese). Chin J Environ Sci 2012, 33, 266–272.
[35] Yang H, He M. Distribution and Speciation of Selenium, Antimony, and Arsenic in Soils and Sediments Around the Area of Xikuangshan (China). CLEAN – Soil, Air, Water 2016, 44, 1538–1546.
[36] He S, Chen M, YZ L, et al. Trace elements and their environmental problems within the surface loose sediments in the major cities of Guizhou province (in Chinese). Guizhou Geology 2005, 3, 147–155.
[37] MEMSC. The Background Values of Soil Elements in China. Beijing, Meterological Press, 1990, 364.
[38] Wang G, Cao F, Shan B, Meng M, Wang W, Sun R. Sources and assessment of mercury and other heavy metal contamination in soils surrounding the wuda underground coal fire area, inner Mongolia, China. Bull Environ Contam Toxicol 2019, 103, 828–833.
[39] Yan G, Mao LC, Jiang BY, et al. The source apportionment, pollution characteristic and mobility of Sb in roadside soils affected by traffic and industrial activities. J Hazard Mater 2020, 384.
[40] Zhang XJ, Hu MA, Zhao YH, Zhao BS. A Survey of heavy metals pollution in Daye Tieshan area (in Chinese). Enuivonl sci technol 2005, 28, 40–43.
[41] Dai F, Li X. A preliminary study on background levels of thirteen priority metals in soil of Shanghai (in Chinese). Shanghai Environ Sci 2009, 28, 271–274.
[42] Chang APG, Page AL, Asano T. Developing Human Health-Related Chemical Guidelines for Reclaimed Water and Sewage Sludge Applications in Agriculture. World Health Organization: Division of Environmental Health, 2002, 94.
[43] Guo WJ, Fu ZY, Wang H, Song FH, Wu FC, Giesy JP. Environmental geochemical and spatial/temporal behavior of total and speciation of antimony in typical contaminated aquatic environment from Xikuangshan, China. Microchem J 2018, 137, 181–189.
[44] Li L, Liu H, Li H. Distribution and migration of antimony and other trace elements in a Karstic river system, Southwest China. Environ Sci Pollut Res 2018, 25, 28061–28074.
[45] Li L, Zhang GP, Liu H, Xiang M, Wei XF. Antimony and arsenic migration and environmental impacts on river draining in the dachang multi-metalliferous mine area in Guangxi, China (in Chinese). Res Environ Sci 2009, 22, 682–687.

[46] Ning ZP, Xiao TF, Yang F, Jia YL, Sun JL, He LB. Antimony Pollution and Sulfur Isotope Study in the Waters of an Antimony Mine area (in Chinese). Bull Mineral Petrol Geochem 2011, 30, 135–141.
[47] Li L, Tu H, Zhang S, et al. Geochemical behaviors of antimony in mining-affected water environment (Southwest China). Environ Geochem Health 2019, 41, 2397–2411.
[48] Zhang Y, Kang S, Chen P, et al. Records of anthropogenic antimony in the glacial snow from the southeastern Tibetan Plateau. J Asian Earth Sci 2016, 131, 62–71.
[49] Wu B, Zhao DY, Jia HY, Zhang Y, Zhang XX, Cheng SP. Preliminary risk assessment of trace metal pollution in surface water from Yangtze River in Nanjing section, China. Bull Environ Contam Toxicol 2009, 82, 405–409.
[50] Yao C, Jiang X, Che FF, Wang K, Zhao L. Antimony speciation and potential ecological risk of metal(loid)s in plain wetlands in the lower Yangtze River valley, China. Chemosphere 2019, 218, 1114–1121.
[51] Wu XD, Song JM, Li XG, Yuan HM, Li N. Behaviors of dissolved antimony in the Yangtze River Estuary and its adjacent waters. J Environ Monit 2011, 13, 2292–2303.
[52] Duan L, Song J, Li X, Yuan H. The behaviors and sources of dissolved arsenic and antimony in Bohai Bay. Cont Shelf Res 2010, 30, 1522–1534.
[53] Zhuang W, Lai X, Wang Q, Liu Y, Chen Q, Liu C. Distribution characteristics, sources and ecological risk of antimony in the surface sediments of Changjiang Estuary and the adjacent sea, East China. Mar Pollut Bull 2018, 137, 474–480.
[54] Ning Z, Xiao Q, Lan X, et al. Spatial distribution characteristics and potential ecological risk of antimony and selected heavy metals in sediments of Duliujiang River (in Chinese). Environ Sci 2017, 38, 2784–2792.
[55] Zhang ZX, Lu Y, Li HP, Tu Y, Liu BY, Yang ZG. Assessment of heavy metal contamination, distribution and source identification in the sediments from the Zijiang River, China. Sci Total Environ 2018, 645, 235–243.
[56] Sun XX, Li BQ, Han F, Xiao EZ, Xiao TF, Sun WM. Impacts of arsenic and antimony co-contamination on sedimentary microbial communities in Rivers with different pollution gradients. Microb Ecol 2019, 78, 589–602.
[57] Wang XQ, He MC, Xi JH, Lu XF. Antimony distribution and mobility in rivers around the world's largest antimony mine of Xikuangshan, Hunan Province, China. Microchem J 2011, 97, 4–11.
[58] Liang X, Song J, Duan L, et al. Source identification and risk assessment based on fractionation of heavy metals in surface sediments of Jiaozhou Bay, China. Mar Pollut Bull 2018, 128, 548–556.
[59] Ma L, Xia PH, Lin T, Huang XF, Tang XC. Speciation and ecological risks of cobalt and antimony in black-necked crane (Grus nigricollis) habitat sediments in China's Caohai Lake. Polish J of Environ Stud 2020, 29, 1725–1736.
[60] He MC, Lin CY, Zhou YX, Fan YH. Distribution of arsenic and antimony in the sediments from Daliao River watersystem (Northeast of China). In: Zhu YG, Lepp N, Naidu R, editors, Biogeochemistry of Trace Elements: Environmental Protection, Remediation and Human Health Tsinghua University Press. 2007, 780–781.
[61] Xu L, Wu F, Zheng J, et al. Sediment records of Sb and Pb stable isotopic ratios in Lake Qinghai. Microchem J 2011, 97, 25–29.
[62] Yan MJ. Determination and Distribution of Trace Antimony in The Global Core Samples in Qaidam Basin (in Chinese). Chengdu university of technology, 2011.
[63] Wang Q, Liu Y, Zhuang W, Chen Q. Research on geochemical characteristics and environmental risk of Be,Sb and Tl in surface sediments of the Nansihu Lake (in Chinese). Acta Scientiae Circumstantiae 2018, 38, 1968–1982.

[64] Wan DJ, Song L, Yang JS, et al. Increasing heavy metals in the background atmosphere of central North China since the 1980s: Evidence from a 200-year lake sediment record. Atmos Environ 2016, 138, 183–190.
[65] Lin Q, Liu EF, Zhang EL, et al. Reconstruction of atmospheric trace metals pollution in Southwest China using sediments from a large and deep alpine lake: Historical trends, sources and sediment focusing. Sci Total Environ 2018, 613, 331–341.
[66] Fu S, Wei CY, Xiao Y, Li LH, Wu DS. Heavy metals uptake and transport by native wild plants: Implications for phytoremediation and restoration. Environ Earth Sci 2019, 78.
[67] Wei C, Deng Q, Wu F, Fu Z, Xu L. Arsenic, antimony, and bismuth uptake and accumulation by plants in an old antimony mine, China. Biol Trace Elem Res 2011, 144, 1150–1158.
[68] Qi C, Wu F, Deng Q, et al. Distribution and accumulation of antimony in plants in the super-large Sb deposit areas, China. Microchem J 2011, 97, 44–51.
[69] She W, Jie YC, Xing HC, et al. Uptake and accumulation of heavy metal by Ramie(Boehmeria nivea) growing on antimony mining area in Lengshuijiang city of Hunan Province (in Chinese). J Agro-Environ Sci 2010, 29, 91–96.
[70] Ku W, Zhao Y, Lei C, Dong M, Dai M, Yi H. Heavy metal pollution in soils and characteristics of heavy metal accumulation of dominant plants in antimony mining area (in Chinese). Chin J Environ Eng 2012, 6, 3774–3780.
[71] Long J, Tan D, Deng S, Lei M. Uptake and accumulation of potentially toxic elements in colonized plant species around the world's largest antimony mine area, China. Environ Geochem Health 2018, 40, 2383–2394.
[72] Wei C, Ge Z, Chu W, Feng R. Speciation of antimony and arsenic in the soils and plants in an old antimony mine. Environ Exp Bot 2015, 109, 31–39.
[73] Li L, Zhang G, Liu H, Xiang M, Wei X, Li H. Distribution and mobility of Sb and As in topsoils and plants in the Dachang multi-metalliferous mine area, Guangxi, China (in Chinese). Acta Scientiae Circumstantiae 2010, 30, 2305–2313.
[74] Li H, Zhang G, Li L, Wei X, Cai Y, Fu Z. Heavy Metal Deposition Recorded by Mosses around a Hechi Antimony Smelter in Guangxi,China (in Chinese). Res Environ Sci 2012, 25, 968–973.
[75] Chen J, Wu P, Zhang M, Zhang C, Han Z. Accumulation characteristics of Sb and As in cattail growing in antimony mine tailings (in Chinese). Chin J Ecol 2015, 34, 2645–2649.
[76] Ning ZP, Xiao TF, Xiao EZ. Antimony in the Soil-Plant System in an Sb Mining/Smelting Area of Southwest China. Int J Phytoremediat 2015, 17, 1081–1089.
[77] Wang L, Yang AJ, Deng QJ, et al. Distribution characteristics of heavy metals in soil-Polygonum capitatum system at Dushan antimony deposit,Guizhou Province (in Chinese). Chin J Ecol 2017, 36, 3545–3552.
[78] Zhang S, Wang Y, Pervaiz A, Kong L, He M. Comparison of diffusive gradients in thin-films (DGT) and chemical extraction methods for predicting bioavailability of antimony and arsenic to maize. Geoderma 2018, 332, 1–9.
[79] Shtangeeva I, Bali R, Harris A. Bioavailability and toxicity of antimony. J Geochem Explor 2011, 110, 40–45.
[80] Zhong Q, Ma C, Chu J, et al. Toxicity and bioavailability of antimony in edible amaranth (Amaranthus tricolor Linn.) cultivated in two agricultural soil types. Environ Pollut 2020, 257, 113642.
[81] Wu TL, Cui XD, Cui PX, et al. Speciation and location of arsenic and antimony in rice samples around antimony mining area. Environ Pollut 2019, 252, 1439–1447.
[82] He MC. Distribution and phytoavailability of antimony at an antimony mining and smelting area, Hunan, China. Environ Geochem Health 2007, 29, 209–219.

[83] Zeng D, Zhou S, Ren B, Chen T. Bioaccumulation of Antimony and Arsenic in Vegetables and Health Risk Assessment in the Superlarge Antimony-Mining Area, China. J Anal Methods Chem 2015, 2015, 909724.
[84] Wu F, Fu Z, Liu B, et al. Health risk associated with dietary co-exposure to high levels of antimony and arsenic in the world's largest antimony mine area. Sci Total Environ 2011, 409, 3344–3351.
[85] Tan XW, Ma JH, Xiao FY. Evaluation on the Content and Pollution of Total Antimony in Vegetables from Hunan Area (in Chinese). J Anhui Agric Univ Sci 2016a.
[86] Tan XW, Ma JH, Xiao FY, Peng W, Tan YM, Xiao SL. Dietary exposure assessment of antimony in Hunan (in Chinese). Chin J Food Hyg 2016b.
[87] Fu Z, Wu F, Mo C, et al. Bioaccumulation of antimony, arsenic, and mercury in the vicinities of a large antimony mine, China. Microchem J 2011, 97, 12–19.
[88] Fu Z, Wu F, Amarasiriwardena D, et al. Antimony, arsenic and mercury in the aquatic environment and fish in a large antimony mining area in Hunan, China. Sci Total Environ 2010, 408, 3403–3410.
[89] Cheng J, Zhang X, Ren S, Wang T, Tang Z. Metals in wild fish from Gaotang Lake in the area of coal mining, China: Assessment of the risk to human health. Environ Sci Pollut Res 2019, 26, 23754–23762.
[90] Devkota B, Schmidt GH. Accumulation of heavy metals in food plants and grasshoppers from the Taigetos Mountains, Greece. Agric Ecosyst Environ 2000, 78, 85–91.
[91] Ainsworth N, Cooke JA, Johnson MS. Distribution of antimony in contaminated grassland 2. Small mammals and invertebrates. Environ Pollut 1990, 65, 79–87.
[92] Ye L, Jing CY. Antimony exposure and speciation in human biomarkers near an active mining area in Hunan, China. Abstr Pap Am Chem Soc 2018, 256, 1.
[93] Huang BC, Long J, Liao HK, et al. Characteristics of Bacterial Community and Function in Paddy Soil Profile around Antimony Mine and Its Response to Antimony and Arsenic Contamination. Int J Environ Res Public Health 2019, 16.
[94] Huang S, Wang L, Zhao Y. Ecological risk assessment from the perspective of soil heavy metal accumulations in Xiamen city, China. Int J Sustainable Dev World Ecol 2018, 25, 411–419.
[95] Li J, Wang Q, Oremland RS, Kulp TR, Rensing C, Wang G. Microbial Antimony Biogeochemistry: Enzymes, Regulation, and Related Metabolic Pathways. Appl Environ Microbiol 2016, 82, 5482–5495.
[96] Li JK, Yu H, Wu XL, et al. Novel Hyper Antimony-Oxidizing Bacteria Isolated from Contaminated Mine Soils in China. Geomicrobiol J 2018, 35, 713–720.
[97] Li JX, Zhang YX, Zheng SL, Liu FH, Wang GJ. Anaerobic Bacterial Immobilization and Removal of Toxic Sb(III) Coupled With Fe(II)/Sb(III) Oxidation and Denitrification. Front Microbiol 2019, 10.
[98] Li J, Yang B, Shi M, et al. Effects upon metabolic pathways and energy production by Sb(III) and As(III)/Sb(III)-oxidase gene aioA in Agrobacterium tumefaciens GW4. Plos One 2017, 12.
[99] Huang Y, Ni WQ, Chen YW, Wang XL, Zhang JW, Wu KS. Levels and risk factors of antimony contamination in human hair from an electronic waste recycling area, Guiyu, China. Environ Sci Pollut Res 2015, 22, 7112–7119.
[100] Liu BJ, Wu FC, Deng QJ, et al. Pollution characteristics of antimony, arsenic and mercury in human hair at Xikuangshan antimony mining area and Guiyang City, China (in Chinese). Environ Sci 2009, 30, 907–912.
[101] Liu B, Wu F, Li X, et al. Arsenic, antimony and bismuth in human hair from potentially exposed individuals in the vicinity of antimony mines in Southwest China. Microchem J 2011, 97, 20–24.

[102] Wu CC, Chen YC. Assessment of Industrial Antimony Exposure and Immunologic Function for Workers in Taiwan. Int J Environ Res Public Health 2017, 14.
[103] Yan G, Mao LC, Liu SX, et al. Enrichment and sources of trace metals in roadside soils in Shanghai, China: A case study of two urban/rural roads. Sci Total Environ 2018, 631–632, 942–950.
[104] Ye L, Qiu S, Li X, Jiang Y, Jing C. Antimony exposure and speciation in human biomarkers near an active mining area in Hunan, China. Sci Total Environ 2018, 640–641, 1–8.
[105] Fei JC, Min XB, Wang ZX, Pang Z, Liang YJ, Ke Y. Health and ecological risk assessment of heavy metals pollution in an antimony mining region: A case study from South China. Environ Sci Pollut Res 2017, 24, 27573–27586.
[106] Yuan Y, Xiang M, Liu C, Theng BKG. Chronic impact of an accidental wastewater spill from a smelter, China: A study of health risk of heavy metal(loid)s via vegetable intake. Ecotoxicol Environ Saf 2019, 182, 109401.
[107] Tan XW, Xiao SL, Ma JH, et al. Content analysis of total antimony and pollution evaluation of fresh strawberries in Hunan Province (in Chinese). China Fruit Veg 2016c.
[108] Tan XW, Xiao FY, Ma JH, Peng W, Huang Y, Huang HX. Analysis and pollution evaluation of total antimony in well water in Xiangtan (in Chinese). J Dis Monit Control 2017.
[109] Chen QP, Xu SQ, Sun HR, Ren HY, Xu SL, Zhang YX. Speciation distribution and ecological risk assessment of antimony and arsenic in soils of the antimony mining area (in Chinese). Earth Environ 2014, 42, 773–778.
[110] Long JM, Tan D, Deng SH, Lei M. Pollution and ecological risk assessment of antimony and other heavy metals in soils from the world's largest antimony mine area, China. Hum Ecol Risk Assess 2018, 24, 679–690.
[111] Du W, Zhang XY, Ge B. Geochemical characteristics and environmental risk assessment of antimony in surface sediments from the Yangtze River estuary and its adjacent sea area (in Chinese). Bull Sci Technol 2019, 35, 193–200.
[112] Zhang ZX, Zhang N, Li HP, Lu Y, Wang Q, Yang ZG. Risk assessment, spatial distribution, and source identification of heavy metal(loid)s in paddy soils along the Zijiang River basin, in Hunan Province, China. J Soils Sediments 2019, 19, 4042–4051.
[113] Xie YN, Ouyang MF, Huang DZ, et al. Pollution characteristics,sources and ecological risk of heavy metals in sediments from dongting lake and its lake inlets (in Chinese). Environ Chem 2017, 36, 2253–2264.
[114] Zhang XY, Tang LS, Zhang G, Wu HD. Heavy Metal Contamination in a Typical Mining Town of a Minority and Mountain Area, South China. Bull Environ Contam Toxicol 2009, 82, 31–38.
[115] Zheng J, Ohata M, Furuta N. Studies on the speciation of inorganic and organic antimony compounds in airborne particulate matter by HPLC-ICP-MS. Analyst 2000, 125, 1025–1028.
[116] China TMoEPo. Emission standards of pollutants for stannum, antimony and mercury industries. 2014.
[117] Liu C, Liu F, Xie Z, Liang H, Huang D. The speciation and bioavailability of antimony in the soil near antimony mine area of De'an (in Chinese). J East China Inst Technol 2007, 30, 358–361.
[118] You XY, Tan AH, Su YR, Yang J. Current Situation and Countermeasures of Antimony Metallurgical Industry Pollution Control (in Chinese). Hunan Nonferrous Metals 2015, 31, 69–73.
[119] Metallic mineral resources. China Mining Association, Available at http://wwwchinaminin gorg/Facts/2006-10-10/1160461445d1479html 2006.
[120] Wang YL, Chen YC, Wang DH, Xu J, Chen ZH, Liang T. The principal antimony concentration areas in China and their resource potentials (in Chinese). Geol China 2013, 40, 1366–1378.
[121] Hu X, Murao S, Shi M, Li B. Classification and distribution of antimony deposits in China. Resour Geol 1996, 46, 287–297.

[122] Fu M, Changkakoti A, Krouse HR, Gray J, Kwak TAP. An oxygen, hydrogen, sulfur, and carbon isotope study of carbonate-replacement (skarn) tin deposits of the Dachang tin field, China. Econ Geol Bull Soc Econ Geol 1991, 86, 1683–1703.
[123] Fu M, Kwak TAP, Mernagh TP. Fluid inclusion studies of zoning in the Dachang tin-polymetallic ore field, People's-Republic-of-China. Econ Geol Bull Soc Econ Geol 1993, 88, 283–300.
[124] Peng Z, Watanabe M, Hoshino K. Overview of Tin-bearing polymetallic mineralization in the Dachang ore field, Guangxi, China (in Japanese). Resour Geol 1997, 47, 331–340.
[125] Gu LX, Zaw K, Hu WX, et al. Distinctive features of Late Palaeozoic massive sulphide deposits in South China. Ore Geol Rev 2007, 31, 107–138.
[126] Gu XX, Schulz O, Vavtar F, Liu JM, Zheng MH, Fu SH. Rare earth element geochemistry of the Woxi W-Sb-Au deposit, Hunan Province, South China. Ore Geol Rev 2007, 31, 319–336.
[127] Jin ZG, Dai TG, Jiang H, Chen XL. Geochemical characteristics and depth prospecting prediction of Banpo antimony deposit in Dushan County, Guizhou Province (in Chinese). Geol Prospect 2004, 40, 24–27.
[128] Wang YL, Jin SC. Geochemistry Characteristics Comparison of Inclusion in Banpo and Banian Antimony Deposit of Guizhou (in Chinese). Nonferrous Met 2010, 62, 123–128.
[129] Zhang Y, Tang HS, Chen YJ, Leng CB, Zhao CH. Ore geology, fluid inclusion and isotope geochemistry of the Xunyang Hg-Sb orefield, Qinling Orogen, Central China. Geol J 2014, 49, 463–481.
[130] Chen J, Yang RD, Du LJ, et al. Mineralogy, geochemistry and fluid inclusions of the Qinglong Sb-(Au) deposit, Youjiang basin (Guizhou, SW China). Ore Geol Rev 2018, 92, 1–18.
[131] Wang JJ, Hu YZ, Han RS. Geochemical characteristics and its implications of trace elements in Qinglong antimony deposit, Guizhou Province,China (in Chinese). Acta Mineralogica Sinica 2011, 31, 571–577.
[132] Guo GH, Song B, Xia DS, Yang ZJ, Wang FP. Metals and metalloids in PM10 in Nandan County, Guangxi, China, and the health risks posed. Environ Geochem Health 2018, 40, 2071–2086.
[133] Wu YJ, Li GY, Yang Y, An TC. Pollution evaluation and health risk assessment of airborne toxic metals in both indoors and outdoors of the Pearl River Delta, China. Environ Res 2019, 179, 9.
[134] Tan D, Long J, Li B, Ding D, Du H, Lei M. Fraction and mobility of antimony and arsenic in three polluted soils: A comparison of single extraction and sequential extraction. Chemosphere 2018, 213, 533–540.
[135] Zhang JY, Zhong Q. Investigation and analysis on environmental pollution due to a small antimony smeltery in Southwest Guizhou province. J Environ Health 2001, 18, 95–96.
[136] L. D. Geochemical Survey for the Ravine Soils of the Puqing Antimony-Gold Deposit of Guizhou and Prospecting Prediction. Kunming, Kunming University of Science and Technology, 2005, 105.
[137] Sun XX, Li BQ, Han F, et al. Vegetation type impacts microbial interaction with antimony contaminants in a mining-contaminated soil environment. Environ Pollut 2019, 252, 1872–1881.
[138] Duan LQ, Song JM, Yuan HM, Li XG, Li N, Ma JK. Distribution, chemical speciation and source of trace elements in surface sediments of the Changjiang Estuary. Environ Earth Sci 2014, 72, 3193–3204.
[139] Li XH. Ecological Risk Assessment on Sediments in Sb-mining Area Anddevelopment of Restoration Technology (in Chinese). Beijing, Beijing forestry university, 2013.
[140] Lin CY, He MC, Li YX, Liu SQ. Content, enrichment, and regional geochemical baseline of antimony in the estuarine sediment of the Daliao river system in China. Chemie Der Erde-Geochem 2012, 72, 23–28.

[141] Gao YJ, Geng CN, Cao Y. Three assessment methods on the heavy metals contamination in river sediments of Shanghai suburban area (in Chinese). Environ Eng 2015, 33, 121–125.
[142] Ren MY, Wang D, Ding SM, et al Seasonal mobility of antimony in sediment-water systems in algae- and macrophyte-dominated zones of Lake Taihu (China). Chemosphere 2019, 223, 108–116.
[143] Yu T, Zhang Y, Meng W, Hu XN. Characterization of heavy metals in water and sediments in Taihu Lake, China. Environ Monit Assess 2012, 184, 4367–4382.
[144] Mao LC, Ye H, Li FP, Yang M, Tao H, Wen HF. Enrichment assessment of Sb and trace metals in sediments with significant variability of background concentration in detailed scale. Environ Sci Pollut Res 2019, 26, 2794–2805.
[145] Liao MX. Environmental Geochemistry on Arsenic and Antimony In Sediment-Water Interface of Poyang Lake (in Chinese). Chengdu, Chengdu University of Technology, 2005.
[146] Li FF, Li LQ, Fu Z, Yin YQ, Liu Y. Pollution characteristics and ecological risk of heavy metals in the surface sediments of the inlets of Dongting Lake, China (in Chinese). Geochimica 2017, 46, 580–589.
[147] Wang XY, Lang CY, Fang C, Wang JF, Wu XR, Zhou L. Distribution of species of arsenic and antimony in surface sediments of ledong sea area in South China sea (in Chinese). Contemp Chem Ind 2015, 44, 1821–1824.
[148] Wu S. Pollution Characteristics and Ecological Risks of Water Environment in an Antimony Mining Area in Guizhou (in Chinese). Hengyang, University of south China, 2017.
[149] Li XH. Ecological Risk Assessment on Sediments in Sb-mining Area Anddevelopment of Restoration Technology (in Chinese). Beijing, Beijing forestry university, 2013.
[150] Song SQ, Liang LF, Zhou YZ, Wu H, Zhou X. The situation and remedial measures of the cropland polluted by heavy metals from mining along the Diaojiang River (in Chinese). Bull Mineral Petrol Geochem 2003, 22, 152–155.
[151] Zhang GP. Case studies of the water environmental geochemistry of mines in Guizhou Province (in Chinese). Guiyang: State key laboratory of environmental geochemistry, institute of geochemistry. Chin Acad Sci 2005.

Clemens Reimann and Karl Fabian
Chapter 6 Diffuse soil contamination by antimony

6.1 Introduction

For antimony, eight regional- to continental-scale geochemical datasets establish reliable concentration statistics in top- and subsoil from different parts of the world: Australia, the United States, Europe. Comparing the statistical distributions of antimony in these two soil layers by means of cumulative probability (CP) plots provides several quantitative constraints for the relative importance of natural versus anthropogenic influences – with a focus on Europe. It is relatively easy to detect major localized contamination sources like metal smelters by geochemical mapping in their surroundings at the local to regional scale (e.g., [1, 2]). To estimate the input of elements to soil from non-local or diffuse contamination, for example, via long-range atmospheric transport, is much more challenging. Because the European Commission has identified *diffuse pollution* as one of the eight major threats to soil quality [3] it is important to estimate this input. Only in a second step can it then be decided if this diffuse contamination fits a generally accepted definition of diffuse pollution, a term that is often used imprecisely in environmental sciences. Pollution implies that the contaminant has a biological impact. The distinction between contamination and pollution is essential for elements where an anthropogenic contamination overlays a high natural background variation and has no biological effect, such that there is no *diffuse pollution* of soil, even if contamination can be detected.

Many authors have tried to establish the impact of anthropogenic contamination of an element in soil by (1) comparing measured values to upper crust concentrations (e.g., from [4]), and by (2) using ratios of element concentrations in top- and subsoil (e.g., [5, 6]), or 3) by defining more elaborate ratios like geoaccumulation indices, contamination factors, or enrichment factors (EFs – e.g., [7]). However, none of these ratios or factors can work in other than extreme cases of soil contamination, because the assumptions underlying their calculation neglect basic principles of soil formation and natural differentiation. This has been demonstrated repeatedly for a variety of data sets from different regions and projects [8–10].

During the last 10–20 years several high-quality geochemical mapping projects with sample sizes in the vicinity of $n \sim 10^3$ have been carried out by geological

Clemens Reimann, Aspangweg 49, 3433, Königstetten, Austria, clemensreimann@yahoo.co.uk
Karl Fabian, Department of Geoscience and Petroleum, Norwegian University of Science and Technology (NTNU), S.P. Andersens veg 15a, N-7031, Trondheim, karl.fabian@ntnu.no

https://doi.org/10.1515/9783110668711-006

survey organizations. The studied areas cover large regions up to whole continents and, despite their generally low sample densities, result in consistent and reproducible element distribution maps (e.g., [1, 11–17]). In the case of antimony, the distribution patterns resulting from these projects indicate a strong relation of antimony to geological units and natural processes such as weathering. Large-scale human impact would obliterate geological boundaries and overprint natural patterns with anthropogenic patterns like population density or industrial activity. Such effects of anthropogenic antimony input are not distinguishable at the continental scale. In contrast, a strong anthropogenic impact is clearly visible at local scales, for example, when mapping antimony in the surroundings of metal smelters, power plants, or in urban geochemical mapping projects (see [18] for examples).

To obtain a statistically based quantitative estimate of low levels of diffuse contamination we recently developed a method based on the comparison of the cumulative density functions (CDF) of top- and subsoil concentrations of an element on the regional to continental scale [19]. So far, results for lead [19], cadmium [20], and copper and zinc [21] have been presented. Modeled modifications of subsoil distributions in [19] demonstrate that the method can distinguish between influences of dilution or up-concentration, and the effects of diffuse, scattered, or local contamination, given a large enough data set of about 1,000 samples or more. Here the focus will be on studying the CDFs of antimony at the regional to continental scale for minerogenic (five examples from around the world with a focus on Europe) and organic topsoil (three examples from northern Europe). Furthermore, the GEMAS (geochemical mapping of agricultural soils) dataset [12, 22] will be used to visualize the regional distribution of antimony in agricultural (top)soil at the European scale.

6.2 Methods

6.2.1 Datasets

Table 6.1 summarizes information about and references to the datasets used. They cover continental to regional scales each using individually documented sampling and analytical procedures, and quality control results as referenced. For more details the reader is referred to the references provided. Results for antimony concentrations in three transects are included in Tabs. 6.1 and 6.2 to substantiate that a significant enrichment of antimony in organogenic topsoil is a general feature in northern Europe, and to provide a reference to data sources for antimony in plants, which are not discussed here.

Tab. 6.1: Overview of surveys and transects used. N: number of samples, size: area covered, DL: detection limit (mg kg^{-1}).

Survey scale and name	**Material**	**N**	**Size / km^2**	**Extraction**	**DL / mg kg^{-1}**	**Reference**
Continental scale (>1,000,000 km^2)						
Geochemical Survey of Australia	catchment outlet sediments, top- and sublayer (0–10 and 60–80 cm)	1300	6.8 mill.	1:3 M HNO_3:HCl, ICP-MS	0.02	[15, 16]
North American Geochemical Landscapes Project	soil A and C horizon	4813/4780	c. 8 mill.	4-acid (total), ICP-MS	0.05	[17]
European FOREGS Project	top- and subsoil (0–25 and 0–75 cm)	783/840	4.4 mill.	4-acid (total), ICP-MS	0.02	[14]
GEMAS Project	Ap (0–20 cm) and Gr (0–10 cm) soil	2108/2024	5.6 mill.	*aqua regia*, ICP-MS	0.02	[12, 22]
The Baltic Soil Survey (BSS Project)	top- and subsoil (Ap and C horizon – 0–25 and 50–75 cm)	745	1.8 mill.	4-acid (total), ICP-MS	0.1	[11]
Barents Project	soil O and C horizon	1337/1319	1.1 mill.	4-acid (total), ICP-MS	0.02/0.01	[13]
Regional scale (>10,000 but <1,000,000 km^2)						
Kola Project	soil O and C horizon	617/605	200.000	O: conc. HNO_3, ICP-MS; C: *aqua regia*, GFAAS	O: 0.01 C: 0.01	[1]
Nord Trøndelag Project (NTR)	soil O and C horizon	752	25.000	*aqua regia*, ICP-MS	0.02	[32]
Tellus Project, Ireland	top- and subsoil (5–20 and 35–50 cm)	6.862	14.000	*aqua regia*, ICP-MS	0.05	[33]

(continued)

Tab. 6.1 (continued)

Survey scale and name	Material	N	Size / km^2	Extraction	DL / $mg\ kg^{-1}$	Reference
Local scale (transects)						
Oslo	bedrock, soil O and C horizon, 9 different plant materials	40	120	*aqua regia*, ICP-MS	0.02	[34]
South Norway	soil O and C horizon, 4 different plants	44–46	200	*aqua regia*, ICP-MS	0.02	[35]
		(36)–41	100	*aqua regia*, ICP-MS	0.005	[36–38]

Tab. 6.2: Analytical results (median values - *: estimated) for antimony from the projects named in the Methods section. The ratio between top- and subsoil is also provided. Information on sampling, analytical methods and detection limits is provided in Tab. 6.1.

Project	OHO / mg kg^{-1}	Topsoil / mg kg^{-1}	C (Subsoil) / mg kg^{-1}	O (Top) / C (Sub)
ANTIMONY IN MINEROGENIC TOPSOIL				
AUSTRALIA		0.12	0.13	0.92
USA		0.61	0.57	1.1
FOREGS Europe		0.6	0.47	1.3
FOREGS N-Europe		0.22	0.19	1.2
FOREGS S-Europe		0.88	0.83	1.1
GEMAS Europe		0.23		
GEMAS N-Europe		0.11		
GEMAS S-Europe		0.35		
BSS N-Europe		0.24	0.2	1.2
Tellus		0.32	0.2	1.6
ANTIMONY IN ORGANOGENIC TOPSOIL				
Project	**OHO / mg kg^{-1}**	**Topsoil / mg kg^{-1}**	**C (Subsoil) / mg kg^{-1}**	**O (Top) / C (Sub)**
E-Barents	0.17		0.0024	7.2
Kola	0.183		0.009*	20
N-Trøndelag	0.33		0.04	8.2
Transects:				
Gjøvik	0.9		0.17	5.4
Oslo	1.235		0.165	7.5
S-Norway	1.38		0.07	21.2

6.2.2 Mapping

The obvious and central method for studying contamination is to create maps of the element concentration over the sampling area to detect anomalous patterns. Even though this appears to be a straightforward process, it requires some care to represent the relevant information. A most important imperative for an unbiased data analysis is to initially only map *measured* data and not interpretations, interpolations, or predictions. A second rule is to choose the color scale in dependence of the data distribution, for example by using percentiles as color boundaries. The resulting maps can then be compared to geological, climatic, population, or other thematic maps to interpret conspicuous element distribution patterns.

6.2.3 CDF analysis

To distinguish background concentration variations from diffuse, scattered, and local contamination and to account for dilution or concentration processes that change the element concentration between different soil layers, it is helpful to represent the CDF of a data distribution in a cumulative probability (CP) plot. If top- and subsoil are chemically and mineralogically comparable, and neither weathering, nor biosphere processes, nor contamination considerably changes their chemical composition, then their CDFs should be almost identical. Accordingly, deviations between top- and subsoil CDFs can be used to infer information about the above-mentioned processes.

Successful CDF analysis requires a large dataset (> 500–1,000 samples) and low detection limits coupled with high analytical quality over the whole concentration range. Until quite recently, soil data sets from more than 500 samples for antimony have been scarce. Sufficiently low detection limits that resolve at least the full data distribution above the 5% percentile are still rare (for an example see Fig. 6.1). Another problem for elements with low concentrations in the sample material is that their CDFs are often obscured at the low-concentration end by discretization steps due to rounding to accuracy instead of precision. This is a routine procedure in many laboratories that unnecessarily renders valuable data practically useless.

In most environmental studies, like in mineral exploration, the interpretation is at present focused completely on the extremely high concentrations above the 95–98% percentiles of the measured CDF. If the survey area is not too large in comparison to the locus of contamination or mineralization, high element concentrations emitted from a defined source, like a metal smelter or a mineral deposit, can indeed distort the topsoil CDF in these upper concentration ranges. Examples of such massive metal contamination at the local to regional scale are the distributions of copper, cobalt, or nickel in the Kola region [1]. In exploration geochemistry, the upper percentiles of CP plots have been used since the early 1970s to identify unusual metal anomalies related to mineralization (e.g., [23, 24]). In contrast to massive local contamination, diffuse contamination is *per se* not expected to add exceptionally high element concentrations. Rather, it results in the initially low element concentrations at the lower percentiles of the CDF being raised by the added contamination that may be an insignificant fraction at high concentrations where it would not even be detectable.

Quite often a linear concentration shift (LCS) on the *logarithmic* scale between top- and subsoil is observed over the whole concentration range in these diagrams. A linear concentration shift by a factor a means that concentrations less than c in

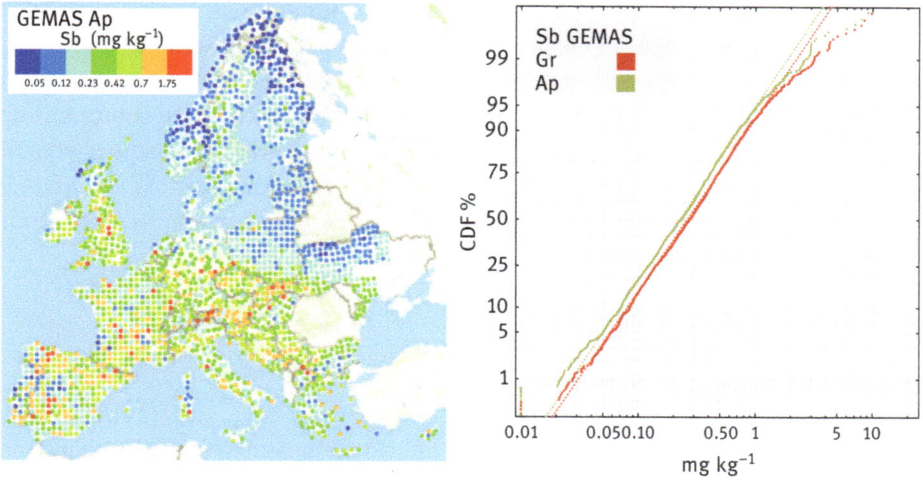

Fig. 6.1: Left: Antimony distribution in European agricultural soil (A_p-layer, 0–20 cm). Right: CP plots of the antimony CDFs for agricultural (Ap, 0–20 cm) and grazing land (Gr, 0–10 cm) soil from the GEMAS project. Data from [12, 22].

the subsoil are equally abundant as concentrations less than $a{\cdot}c$ in the topsoil samples. This cannot be caused by contamination, but indicates a natural dilution or concentration process between the sample materials [19]. For some elements the LCS can be substantial when minerogenic subsoils are compared to organogenic (e.g., forest) topsoils. For some elements the LCS can be positive on the logarithmic scale, with $a > 1$, (e.g., silver, cadmium, mercury, and lead) or negative, with $0 < a < 1$, (e.g., Al), depending on the change in binding characteristics of the elements between the two matrix materials. The strong natural enrichment in the concentration of a number of elements in organic soils and coals has been recognized and described as early as 1937 [25].

Scattered contamination from spatially distributed localized emitters or sources – like smelters, power plants, or traffic – that raises element concentrations only at nearby sample locations is difficult to separate from *diffuse* contamination providing a blanket addition of element concentration from large-scale atmospheric transport and mixing processes that deposit an almost constant amount of contaminants at all sample sites. From the viewpoint of CDF analysis, the amount of ubiquitous diffuse contamination in a study area cannot be substantially higher than the minimum concentration found in the topsoil after correcting for the subsoil CDF through its LCS. Areas with unusual high concentrations in a survey area do not reflect diffuse contamination but outline local or scattered contamination, mineralization, or other element enrichment processes.

Unfortunately, diffuse contamination is only one process that shifts the lower end of the topsoil CDF toward "too high" concentrations. This shift can therefore be used only to estimate the *maximal* amount of diffuse contamination, because other processes, like plants regulating the element concentrations in their organs, also can systematically enrich an element in the topsoil above a minimum level (e.g., cadmium [20]). *Overabundant* concentrations in topsoil become visible by overlaying the shifted subsoil and the topsoil CDFs [19]. They can occur at all concentration levels and may indicate *scattered* local contamination sources in the survey area or a strong climatic influence on element concentrations. The analysis of CDFs and CP plots is explained in more detail using example diagrams demonstrating the various effects, and applied to various elements, in [19–21].

6.3 Results

A percentile-based map of more than 2,000 individual measurements of antimony in European agricultural soil in Fig. 6.1 (data from GEMAS [12, 22]) dominantly features a clearly delineated step between concentration levels below the median in northern and correspondingly higher levels in southern Europe. This step coincides with the southern boundary of the last glaciation [12, 22]. The median antimony concentration in an *aqua regia* extraction is 0.35 mg kg^{-1} in southern Europe in contrast to 0.11 mg kg^{-1} in soils from northern Europe. The same trend is visible in Tab. 6.2: the projects from the far north of Europe (East Barents, Kola, and Nord Trøndelag project) report the lowest antimony median values in the C horizon. The fact that the GEMAS antimony map dominantly displays a geological boundary implies that the majority of antimony in European soils is of natural origin and demonstrates that the antimony background concentration range varies by a factor of more than 3 between northern Europe with a median of 0.11 mg kg^{-1} and southern Europe with a median of 0.35 mg kg^{-1}. The CP plot in Fig. 6.1 (right) demonstrates that it has become possible to produce high-quality antimony datasets with a sufficiently low detection limit to present practically the whole concentration range. Unfortunately, no subsoil was analyzed within the GEMAS project, as the focus was on agricultural (regularly ploughed) soil (0–20 cm, Ap samples) and grazing land soil (Gr).

FOREGS and GEMAS are two independent continental-scale geochemical mapping projects that collected independent soil samples at a low sample density (FOREGS: one site per 5,000 km^2, GEMAS: one site per 2,500 km^2) across the European continent. Because HF extraction was used in FOREGS, the median antimony concentration of 0.6 mg kg^{-1} for the FOREGS project is higher than the value of 0.23 mg kg^{-1} for the *aqua regia* extraction of the GEMAS project. Yet, the FOREGS and GEMAS projects report very similar antimony distribution patterns in European topsoil (see [14] or [18] for the FOREGS maps). This supports the validity and robustness of the low-density mapping approach [26]. It indicates that the antimony concentrations recovered in the two different extractions essentially reflect the same source material. This demonstrates that the low sample density is well suited for outlining the main processes that determine the antimony distribution in soil at the continental scale, considering that the aim of a continental-scale geochemical mapping project is to achieve a general overview of the characteristic background variation in antimony, and not to detect new mineral deposits or to identify all local contamination sources, which are anyway usually known. By assuming that the soil surface expression of a local contamination source or a natural mineralization may cover only a very small area (for examples see [18]), the latter would require a sample density of around ten samples per km^2. Each of the approximately 2,500 cells with a size of 50 km × 50 km – which in GEMAS are represented by a single sample – otherwise would have to be covered by 25,000 samples. A corresponding European map would require consistent sampling and measurement of more than 60 million samples. These considerations emphasize that the similarity of independent low-density antimony maps is by no means self-evident but demonstrates the large-scale nature of variations in antimony background concentrations. This is also true for the internal comparison of the independent GEMAS agricultural soil (Ap) vs. GEMAS grazing land soil (Gr) samples.

Table 6.2 highlights several general features of antimony values as observed in soil:

1) There are substantial differences between the results obtained for *aqua regia* and near-total 4-acid (including HF) extraction. The median value for *aqua regia* soluble antimony in European topsoil is by almost a factor 3 lower in the GEMAS project than in the FOREGS project that used the stronger extraction.
2) For minerogenic soils antimony is generally enriched by a factor of 1.1 to 1.6 in topsoil. This appears to be independent of the extraction and the soil type sampled. The only exception are the Australian soils, where the antimony median value is somewhat lower in topsoil that may be related to winnowing of fines.
3) Antimony median concentrations are quite comparable for minerogenic and organogenic topsoil – this is surprising given the large difference between the two sample materials; the forest soil O horizon consists of decaying plant material. Plants do in general not tend to substantially take up or enrich antimony [18, 27, 28].
4) The ratio between top- and bottom soil is much larger in the projects that used an organic topsoil (forest soil O horizon). When organic topsoil was collected this ratio is between 5 and 20. Unfortunately all projects using organogenic topsoil were carried out in northern Europe, where the antimony concentrations in the C horizon are in general much lower than in southern Europe (Tab. 6.2).
5) Antimony concentrations are much higher in the older and more weathered soil of southern Europe, than in the young (<10,000 years) and much less weathered soil of northern Europe. The difference is almost a factor 3.

The median values for northern Europe (BSS project and FOREGS northern Europe – Tab. 6.2) both using a strong extraction, including HF (Tab. 6.1), are almost identical, although these are two completely independent surveys with independent sample locations.

When studying the CP diagrams for the different projects with a sufficiently large sample number (Figs. 6.2 and 6.3) it becomes clear that there still exist many problems with regard to reliable antimony analyses.

In many projects the detection limit is too high to get a reliable expression of the statistical distribution in the low concentration range. For the Australian data set about 7% of the samples returned values below detection in both layers. In addition, the effects of unnecessary data rounding by the laboratory are clearly visible and affect the data distribution from about the 80th percentile downward. Artificial steps due to rounding are also visible for the FOREGS dataset from the 30th percentile downward. For the Baltic Soil Survey (BSS) more than 10% of all top- and subsoil samples returned values below detection. The Tellus samples are again affected by excessive rounding down from about the 75th percentile. For the East Barents Project

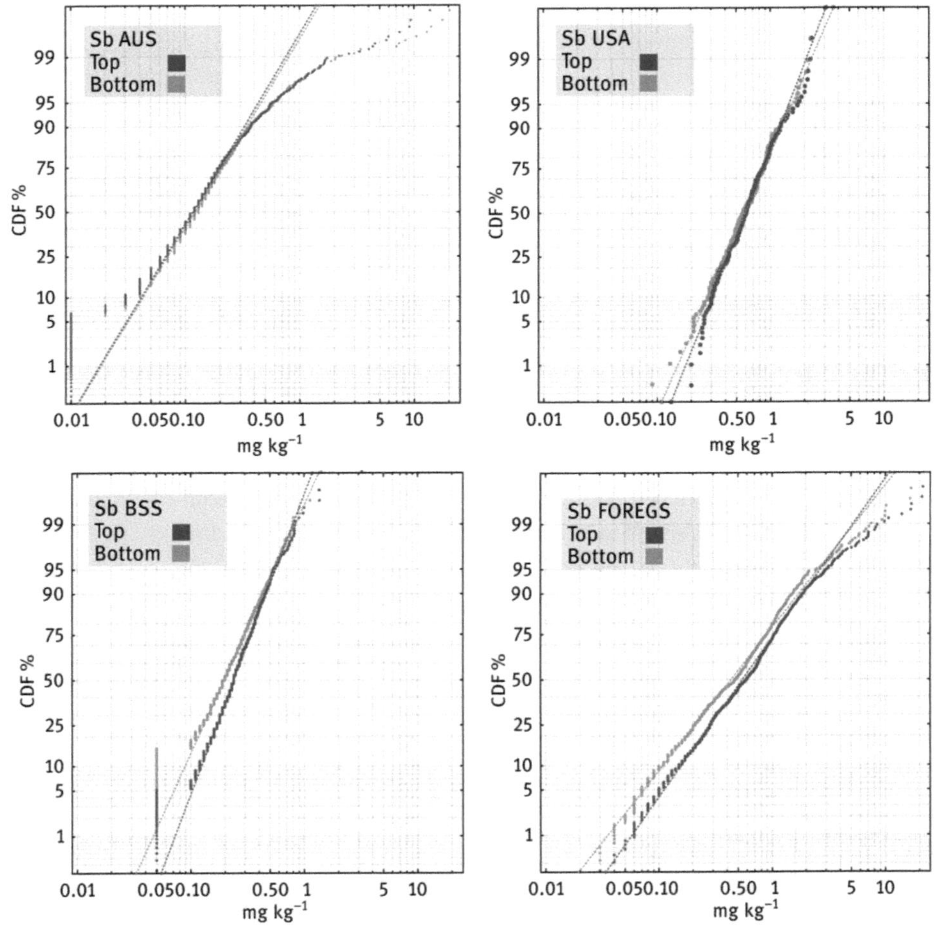

Fig. 6.2: Four CP diagrams for projects using minerogenic topsoil (Australia, USA, FOREGS and BSS projects).

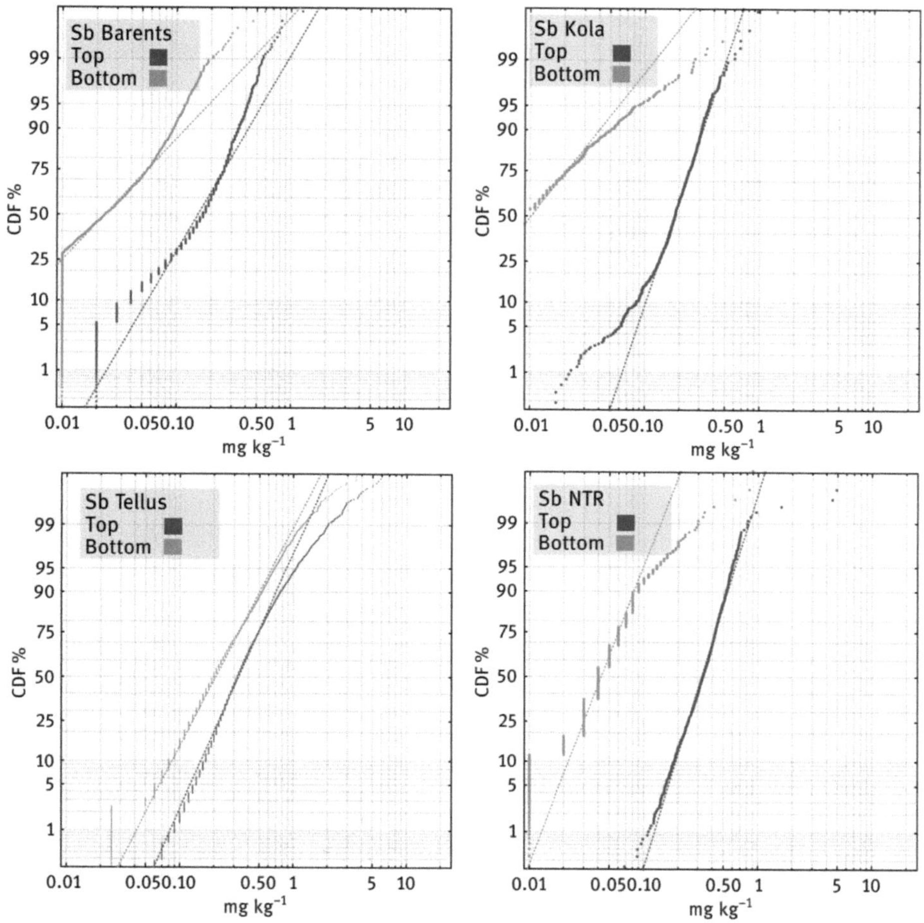

Fig. 6.3: Four additional CP diagrams: Tellus used minerogenic topsoil, and E-Barents, Kola and Nord Trøndelag (NTR) projects used organogenic topsoil.

25% of the C horizon and about 6% of the O horizon samples are below detection and the O horizon samples are severely affected by excessive rounding. For the Kola project about 50% of the C horizon samples are below the detection limit. For the Nord Trøndelag Project, again the C horizon samples are severely affected by a high detection limit and by excessive rounding of over 90% of the C horizon results. All three projects using organogenic topsoil show unusually low antimony concentrations in the subsoil (C horizon). This is probably due to weathering and the history of soil formation (see difference between northern and southern Europe in Fig. 6.1). The bedrock may also play a role as large areas of northern Europe are covered by Archaean to Proterozoic crystalline rocks. Unfortunately, a good lithogeochemistry database providing total antimony concentrations for different bedrock types and ages is still missing.

To better estimate diffuse antimony contamination it would be especially important to obtain reliable antimony concentration data with low detection limit and better resolution. This precision at low values is needed because samples with low natural antimony concentration are most affected by diffuse input across Europe [19]. In so far, it is surprising that the estimates for excess antimony, and LCS values still lead to quite coherent values for all data sets.

The CDFs for the BSS and Tellus projects exemplarily demonstrate the effect that one would expect to see for diffuse contamination [19]: a slightly increased concentration steepens the CDF of the top layer at the lower concentration range, but due to the logarithmic scale almost disappears toward higher concentrations. For the surveys using minerogenic topsoil only a small linear concentration shift is visible (up to 1.39 – Tellus). Excess antimony ranges between 0 (Australia) and 0.054 mg kg^{-1} (BSS, northern Europe). The excess antimony for Europe (FOREGS) is 0.03 mg kg^{-1}. This on a first glance low value – which corresponds to the median excess antimony value for minerogenic soil – still represents about 10% of the median antimony concentration in European minerogenic topsoil, which would actually be a quite substantial fraction if it originates from anthropogenic input (Tabs. 6.2 and 6.3). The highest values of excess antimony in minerogenic soil are observed for the Tellus and BSS projects (Tab. 6.3). These are exactly those projects where at least some of the topsoil samples contain more organic material than usual due to wet and cold climatic conditions that favor the accumulation of organic material in soil. Thus the observed "excess" is probably not exclusively due to diffuse contamination, but will also reflect a constant addition of antimony due to a constant amount of more antimony-rich organic material in the soil samples from these surveys.

The three projects using organogenic topsoil (the forest O horizon) show quite different results: linear concentration shifts of 2.59 (E-Barents) to 4.52 (Nord Trøndelag) reflect the much lower antimony concentrations observed in the C horizons in northern Europe (Tab. 6.2). The median value for excess antimony is now 0.13 mg kg^{-1} compared to the 0.03 mg kg^{-1} reported for minerogenic topsoil. The three projects that use organogenic topsoil (see Tab. 6.2) deliver estimates of excess antimony in a narrow range from 0.1 mg kg^{-1} (E-Barents) to 0.18 mg kg^{-1} (Kola). In this case the excess antimony, besides diffuse contamination, also may reflect biologically adjusted antimony levels through plant interactions, strong organic binding, or climatic effects. Unfortunately, the statistical interpretation based on CDF analysis of the available data cannot differentiate between these processes. Thus the estimate of 0.03 mg kg^{-1} from the projects with minerogenic topsoil provides the most reliable estimate of maximum input of antimony via diffuse anthropogenic contamination at the European scale. The value obtained for the United States, 0.024 mg kg^{-1}, is quite comparable (see Tab. 6.3). In Australia no excess antimony is visible.

The CDF of the O horizon for the Kola Project indicates that there are too many high antimony values in the middle concentration range in the O horizon. This appears to be a climatic effect. Increased bioproductivity from the coast of the Barents

Tab. 6.3: Excess antimony (in mg kg^{-1}) and linear concentration shift (LCS) for eight regional to continental-scale projects. CvM refers to the *p*-value of the Cramer-von Mises test [31]. CvM-values larger 0.05 indicate that the hypothesis that the adapted subsoil antimony distribution agrees with the topsoil distribution cannot be rejected on a 5% level (see [19]).

	MINEROGENIC TOPSOIL		
Project	**excess Sb**	**LCS**	**CvM**
AUSTRALIA	0.0	1.0	0.07
USA	0.024	1.0	0.27
FOREGS (Europe)	0.03	1.19	0.84
BSS (N-Europe)	0.054	0.934	0.11
Tellus	0.048	1.39	0.008
	ORGANOGENIC TOPSOIL		
Project	**excess Sb**	**LCS**	**CvM**
E-Barents	0.10	2.59	0
Kola	0.18	2.88	0
N-Trøndelag	0.13	4.52	0.0003

Sea toward the south leads to a strong north-south gradient in antimony concentrations in the O horizon [18, 29]. While in the O horizon the lowest antimony concentrations occur along the coast, the highest antimony concentrations in the C horizon mark the Caledonian sediments occurring along the coast [1].

A constant diffuse input of antimony to European soil is thus in the range of 0.01 mg kg^{-1} to 0.05 mg kg^{-1} with a most likely value of <0.03 mg kg^{-1}. The EU risk assessment report for diantimony trioxide [30] provides a predicted no effect concentration (PNEC) of 37 mg kg^{-1} dw for soil. This value is by three orders of magnitude higher than the observed diffuse input at the continental scale. Because no adverse effect is expected from these low concentrations, it appears justified to conclude that there is no “diffuse pollution” by antimony in soil, although up to 10% of the soil median antimony content may be due to diffuse *contamination*. True antimony pollution will be observed – and needs to be abated – at the local scale, for example, near metal mines or smelters, and in some extreme cases at urban pollution hotspots (see [18]), which all require mapping at the local scale.

6.4 Conclusions

Large-scale regional to continental-scale surveys of antimony concentration in soils provide a consistent and important view on antimony distribution and indicate possible origin and transport processes. Statistical analysis of topsoil versus subsoil concentrations permits estimation of a maximal amount of diffuse antimony contamination in topsoil. The main results can be summarized as follows:

(a) Reported antimony concentrations strongly depend on the chemical extraction method employed. For GEMAS, using *aqua regia*, the median of 0.23 mg kg^{-1} is by a factor 3 smaller than for a near total 4-acid extraction, including HF as used in the FOREGS project (median 0.6 mg kg^{-1}).

(b) Improvements in analytical quality would considerably help to achieve a more detailed overview of the distribution of antimony in soil, which at present is still hampered by too high detection limits and excessive and usually unnecessarily strict rounding of the reported results by many laboratories.

(c) In Europe a strong difference in antimony concentration is observed between southern (high Sb) and northern Europe (low Sb). The FOREGS project shows a factor 4 difference in the median concentration for southern Europe (0.88 mg kg^{-1}) when compared to northern Europe (0.22 mg kg^{-1}). For the GEMAS project the median for southern Europe is 0.35 mg kg^{-1} and that for northern Europe is 0.11 mg kg^{-1}, a factor of 3 lower. In a European map of antimony in agricultural soil, the concentration break between southern and northern Europe follows the southern border of the last glaciation. This may relate to lithological differences between older crystalline rocks in the north and their weathering products moved south by glaciation, and younger sedimentary rocks and volcanics, in the South. More likely, this antimony concentration difference reflects the huge difference in soil age and thus in weathering. A larger fraction of the total antimony in the older and more weathered soils in southern Europe has been released over time and is accessible to acid extraction.

(d) Median antimony concentrations in topsoils are quite comparable for both minerogenic and organogenic topsoil (the forest soil O horizon). Within the surveys covering northern Europe (0.20 vs. 0.18 mg kg^{-1}) the projects based on organogenic topsoil show a much larger difference between topsoil and subsoil. The ratio of topsoil/subsoil antimony concentrations is close to 1 when minerogenic topsoil is collected and reaches values of 5 to 20 for the projects where organogenic topsoil was collected. This is also reflected in a substantial linear concentration shift visible for these projects. Excess antimony in organic topsoil is in consequence also by almost an order of magnitude higher than observed for the projects collecting minerogenic topsoil.

(e) Notwithstanding the many uncertainties and the limited data quality for antimony concentrations in most projects, it is possible to estimate the maximum amount of diffuse antimony contamination in the European soil samples. Not more than 0.03 mg kg^{-1} antimony can be due to diffuse contamination. In some respect, this is a rather high value, as it corresponds to about 10% of the median antimony concentration. On the other hand, it is small enough to explain why the patterns observed in continental-scale maps of antimony in soil from three continents (BSS, FOREGS, GEMAS, USGS geochemical landscapes, and geochemical atlas of Australia), as well as the maps from more regional scale projects (Kola, Nord Trøndelag, Tellus), are clearly dominated by natural processes. Local contamination or the occurrence of mines is sometimes marked by local anomalies.

References

[1] Reimann C, Äyräs M, Chekushin V, Bogatyrev I, Boyd R, Caritat P de, Dutter R, Finne TE, Halleraker JH, Jæger Ø, Kashulina G, Lehto O, Niskavaara H, Pavlov V, Räisänen ML, Strand T, Volden T. Environmental Geochemical Atlas of the Central Barents Region. Trondheim, Norway, NGU-GTK-CKE Special Publication, Geological Survey of Norway, 1998, 745 pp.

[2] Bonham-Carter GF, Henderson PJ, Kliza DA, Kettles IM. Comparison of metal distributions in snow, peat, lakes and humus around a copper smelter in western Quebec, Canada. Geochem Explor Environ Anal 2006, 6, 215–228.

[3] European Commission. Thematic strategy for soil protection. Report COM 2006, 231, 2006.

[4] Wedepohl KH. The composition of the continental crust. Geochim Cosmochim Acta 1995, 59, 1217–1232.

[5] Facchinelli A, Sacchi E, Mallen L. Multivariate statistical and GIS-based approach to identify heavy metal sources in soils. Environ Poll 2001, 114, 313–324.

[6] Massas I, Ehliotis C, Gerontidis S, Sarris E. Elevated heavy metal concentrations in top soils of an Aegean island town (Greece): Total and available forms, origin and distribution. Environ Monit Assess 2009, 151, 105–116.

[7] Loska K, Cebula J, Pelczar J, Wiechula D, Kwapilinski J. Use of enrichment and contamination factors together with geoaccumulation indexes to evaluate the content of Cd, Cu, and Ni in the Rybnik water reservoir in Poland. Water Air Soil Pollut 1997, 93, 347–365.

[8] Reimann C, Caritat P de. Intrinsic flaws of element enrichment factors (EFs) in environmental geochemistry. Environ Sci Technol 2000, 34, 5084–5091.

[9] Reimann C, Caritat P de. Distinguishing between natural and anthropogenic sources for elements in the environment: Regional geochemical surveys versus enrichment factors. Sci Total Environ 2005, 337, 91–107.

[10] Sucharova J, Suchara I, Hola M, Marikova S, Reimann C, Boyd R, Filzmoser P, Englmaier P. Top-/bottom-soil ratios and enrichment factors: What do they really show?. Appl Geochem 2012, 27, 138–145.

[11] Reimann C, Siewers U, Tarvainen T, Bityukova L, Eriksson J, Gilucis A, Gregorauskiene V, Lukashev VK, Matinian NN, Pasieczna A. Agricultural Soils in Northern Europe: A Geochemical Atlas. Stuttgart, Geologisches Jahrbuch, Sonderhefte, Reihe D, Heft SD 5, Schweizerbart'sche Verlagsbuchhandlung, 2003, 279 pp.

[12] Reimann C, Birke M, Demetriades A, Filzmoser P, O'Connor P. Chemistry of Europe's Agricultural Soils, Part a: Methodology and Interpretation of the GEMAS Data Set, Geologisches Jahrbuch Reihe B, Band B 102. Stuttgart, Schweizerbart Science Publishers, 2014, 523 pp.

[13] Salminen R, Chekushin V, Tenhola M, Bogatyrev I, Glavatskikh SP, Fedotova E, Gregorauskiene V, Kashulina G, Niskavaara H, Polischuok A, Rissanen K, Selenok L, Tomilina O, Zhdanova L. Geochemical Atlas of Eastern Barents Region. Amsterdam, Elsevier, 2004, 548 pp.

[14] Salminen R (Chief-Ed), Batista MJ, Bidovec M, Demetriades A, De Vivo B, De Vos W, Duris M, Gilucis A, Gregorauskiene V, Halamic J, Heitzmann P, Lima A, Jordan G, Klaver G, Klein P, Lis J, Locutura J, Marsina K, Mazreku A, O'Connor PJ, Olsson SÅ, Ottesen R-T, Petersell V, Plant JA, Reeder S, Salpeteur I, Sandström H, Siewers U, Steenfelt A, Tarvainen T. Geochemical Atlas of Europe. Part 1 – Background Information, Methodology and Maps. Espoo, Finland, Geological Survey of Finland, 2005, 526.

[15] Caritat P de, Cooper M. National Geochemical Survey of Australia: The Geochemical Atlas of Australia, Geoscience Australia, Record 2011/20, GeoCat#71973, Vol. 1. Canberra, Australian Government, 2011, 1–268.

[16] Caritat P de, Cooper M. National Geochemical Survey of Australia: The Geochemical Atlas of Australia, Geoscience Australia, Record 2011/20, GeoCat#71973, Vol. 2. Canberra, Australian Government, 2011, 269–557.

[17] Smith DB, Cannon WF, Woodruff LG, Solano F, Ellefsen KJ. Geochemical and Mineralogical Maps for Soils of the Conterminous United States. Denver, U.S, Geological Survey, 2014, 386 pp.

[18] Reimann C, Matschullat J, Birke M, Salminen R. Antimony in the environment: Lessons from geochemical mapping. Appl Geochem 2010, 25, 175–198.

[19] Fabian K, Reimann C, Caritat P de. Quantifying diffuse contamination: Method and application to Pb in soil. Environ Sci Technol 2017, 51, 6719–6726.

[20] Reimann C, Fabian K, Flem B. Cadmium enrichment in topsoil: Separating diffuse contamination from biosphere-circulation signals. Sci Total Environ 2019, 651, 1344–1355.

[21] Reimann C, Fabian K, Flem B, Englmaier P. The large-scale distribution of Cu and Zn in sub and topsoil: Separating topsoil bioaccumulation and natural matrix effects from diffuse and regional contamination. Sci Total Environ 2019, 655, 730–740.

[22] Reimann C, Birke M, Demetriades A, Filzmoser P, O'Connor P. Chemistry of Europe's Agricultural Soils, Part B: General Background Information and Further Analysis of the GEMAS Data Set, Geologisches Jahrbuch Reihe B, Band B 103. Stuttgart, Schweizerbart Science Publishers, 2014, 352.

[23] Sinclair AJ. Selection of threshold values in geochemical data using probability graphs. J Geochem Explor 1974, 3, 129–149.

[24] Sinclair AJ. Application of Probability Graphs in Mineral Exploration, Vol. Special 4. Toronto, Association of Exploration Geochemists, 1976, 95 pp.

[25] Goldschmidt VM. The principles of distribution of chemical elements in minerals and rocks. J Chem Soc, London 1937, 655–673. https://pubs.rsc.org/en/content/articlelanding/1937/jr/jr9370000655#!divAbstract.

[26] Smith DB, Reimann C. Low-density geochemical mapping and the robustness of geochemical patterns. Geochem Explor Environ Anal 2008, 8, 219–227.

[27] Dunn CE. Biogeochemistry in Mineral Exploration. In: Hale M, ed., Handbook of Exploration and Environmental Geochemistry, Vol. 9. Amsterdam, Elsevier, 2007, 460 pp.

[28] He M, Wang N, Long X, Zhang C, Ma C, Zhong Q, Wang A, Wang Y, Pervaiz A, Shan J. Antimony speciation in the environment: Recent advances in understanding the biogeochemical processes and ecological effects. J Environ Sci 2018, 75, 14–39.

[29] Reimann C, Banks D, Kashulina G. Processes influencing the chemical composition of the O-horizon of podzols along a 500 km north-south profile from the coast of the Barents Sea to the Arctic Circle. Geoderma 2000, 95, 113–139.

[30] EU. European Union Risk Assessment Report Diantimony Trioxide, CAS No: 1309-64-4, Einecs No: 215-175-0. Rapporteur Sweden. Luxembourg, Office for official publications of the European Communities, 2008.
[31] von Mises R. Wahrscheinlichkeit, Statistik und Wahrheit. Wien, Julius Springer Verlag, 1928, 192.
[32] Reimann C, Fabian K, Schilling J, Roberts D, Englmaier P. A strong enrichment of potentially toxic elements (PTEs) in Nord Trøndelag (central Norway) forest soil. Sci Total Environ 2015, 536, 130–141.
[33] Young M, Donald A Eds. A Guide to the Tellus data. Belfast, UK, Geological Survey of Northern Ireland, 2013, 233pp.
[34] Reimann C, Arnoldussen A, Englmaier P, Filzmoser P, Finne TE, Garrett RG, Koller F, Nordgulen Ø. Element concentrations and variations along a 120 km long transect in south Norway – anthropogenic vs. geogenic vs. biogenic element sources and cycles. Appl Geochem 2007, 22, 851–871.
[35] Reimann C, Englmaier P, Gough L, Lamothe P, Nordgulen Ø, Smith D. Geochemical gradients in O-horizon soils of southern Norway: Natural or anthropogenic?. Appl Geochem 2009, 24, 62–76.
[36] Flem B, Torgersen E, Englmaier P, Andersson M, Finne TE, Eggen OA, Reimann C. Response of soil C and O horizon and terrestrial moss to various lithological units and mineralization in southern Norway. Geochem Explor Environ Anal 2018, 18, 252–262.
[37] Reimann C, Fabian K, Flem B, Andersson M, Filzmoser P, Englmaier P. Geosphere-biosphere circulation of chemical elements in soil and plant systems from a 100 km transect from southern-central Norway. Sci Total Environ 2018, 639, 129–145.
[38] Andersson M, Reimann C, Flem B, Englmaier P, Fabian K. Element distribution in *Lactarius rufus* in comparison to the underlying substrate along a transect in southern Norway. Appl Geochem 2018, 97, 61–70.

Susan C. Wilson, Lakmini P. Egodawatta and Susan Tandy

Chapter 7
Antimony soil–plant transfer

7.1 Introduction

Plants growing in antimony-contaminated soils provide a potentially important exposure route for humans and animals to antimony via food contamination and grazing feed [1–3]. Antimony can accumulate in plant tissues [4, 5] and may be toxic to plants [6, 7]. Although background antimony concentrations in soil are typically <10 mg kg^{-1} [8, 9], much higher concentrations in soils contaminated by sources such as mining and smelting can result in extremely high concentrations detected in plants growing in these soils [10–12]. For example, up to 11,800 mg kg^{-1} antimony has been reported in soils contaminated by waste rock surrounding one of the world's largest antimony mines, the Xikuangshan (XKS) mine in China [13], and numerous studies report on uptake and high concentrations in the plants (up to 4,029 mg kg^{-1}), including crops, growing in the contaminated soils [12, 14, 15]. Nevertheless, plant concentrations and toxicity depend on a range of environmental factors that control antimony speciation, solution concentrations, soil to root transfer, plant uptake, and translocation. The soil mineralogy, organic matter, pH, redox, and the plant species are important in this regard. Antimony can exist as free ionic forms in soils, precipitated as solids, adsorbed onto soil organic and inorganic constituents, and as exchangeable or structural constituents of primary and secondary minerals, all representing different phytoavailability [9]. Both the environmentally important antimony oxidation states, Sb(III) and Sb(V), as inorganic and organic forms can transfer to plants but with different phytoavailability [16, 17, 18]. Inorganic and Sb(III) compounds are usually considered more toxic to plants [19], but this is not always the case, and varies depending on plant and soil conditions [20, 21], and different antimony species show different toxicity to different plants [22]. Research on factors controlling soil to plant transfer, together with antimony plant concentrations, uptake and translocation mechanisms, toxicity, and detoxification has proliferated over the last 10 years. This is advancing the understanding of ecotoxic effects and providing the science needed to manage risk and mitigate food-chain contamination

Susan C. Wilson, School of Environmental and Rural Science, University of New England, NSW 2351, Australia. swilso24@une.edu.au
Lakmini P. Egodawatta, Department of Planning, Industry and Environment, Sydney, NSW, Australia. Lakmini.Egodawatta@environment.nsw.gov.au
Susan Tandy, Sustainable Agriculture Sciences, Rothamsted Research, North Wyke, Devon, EX20 2SB, UK. susan.tandy@rothamsted.ac.uk

https://doi.org/10.1515/9783110668711-007

with human and ecosystem exposure. In this chapter we summarize the current knowledge and consider the gaps still remaining.

7.2 Soil factors influencing antimony solution concentrations and speciation for plant uptake

Plant accumulation increases with antimony soil concentration over a wide range of concentrations. Positive and significant correlation is evident between plant concentrations and both total and soluble soil antimony concentrations, but the relationship is closer for soluble soil antimony [23]. Considerable scatter in the relationship occurs, however, mainly as a consequence of variation in the soluble antimony and the antimony speciation in different soils. The soluble antimony soil concentration is influenced by its complex biogeochemical cycle and the differential processes for Sb(V) and Sb(III). Sorption-desorption reactions [9] with phyllosilicate clay minerals, (hydr)oxides of iron, manganese, and aluminum, and natural organic matter (NOM) are important, but the extent of adsorption or desorption and the resulting solution concentration available for plant uptake depends strongly on sorbent, competitive ligands, pH, and redox (Fig. 7.1) as discussed in the following sections.

7.2.1 Mineral surfaces

Iron and manganese, and also aluminum (hydr)oxides, are responsible for significant antimony retention in soils, but sorption is highly pH dependent [24–26]. Sorption can occur by outer sphere or inner sphere mono-dentate, bidentate-binuclear, and bidentate-mononuclear complexes which influences the strength of sorption, but the mechanism depends on the antimony species and sorption surface [9]. Sorption of the Sb(V) oxyanion increases with decreasing soil pH as positive surface charge increases, whereas adsorption of the neutral Sb(III) aqueous species is constant over a wide pH range [27, 28] via strong inner sphere bidentate mononuclear complexes [29, 30]. Both Sb(III) and Sb(V) bind weakly to clay minerals compared to Fe/Mn/Al (hydr)oxides, mainly by surface ligand exchange mechanisms at broken clay edges [27], but Sb(III) can show a stronger adsorption affinity than anionic Sb(V), which is influenced by pH [31, 32].

7.2.2 Solid and dissolved soil natural organic matter

Solid and dissolved soil NOM strongly influence antimony soil solution concentrations, and NOM also plays a key role in antimony redox cycling as described in Section

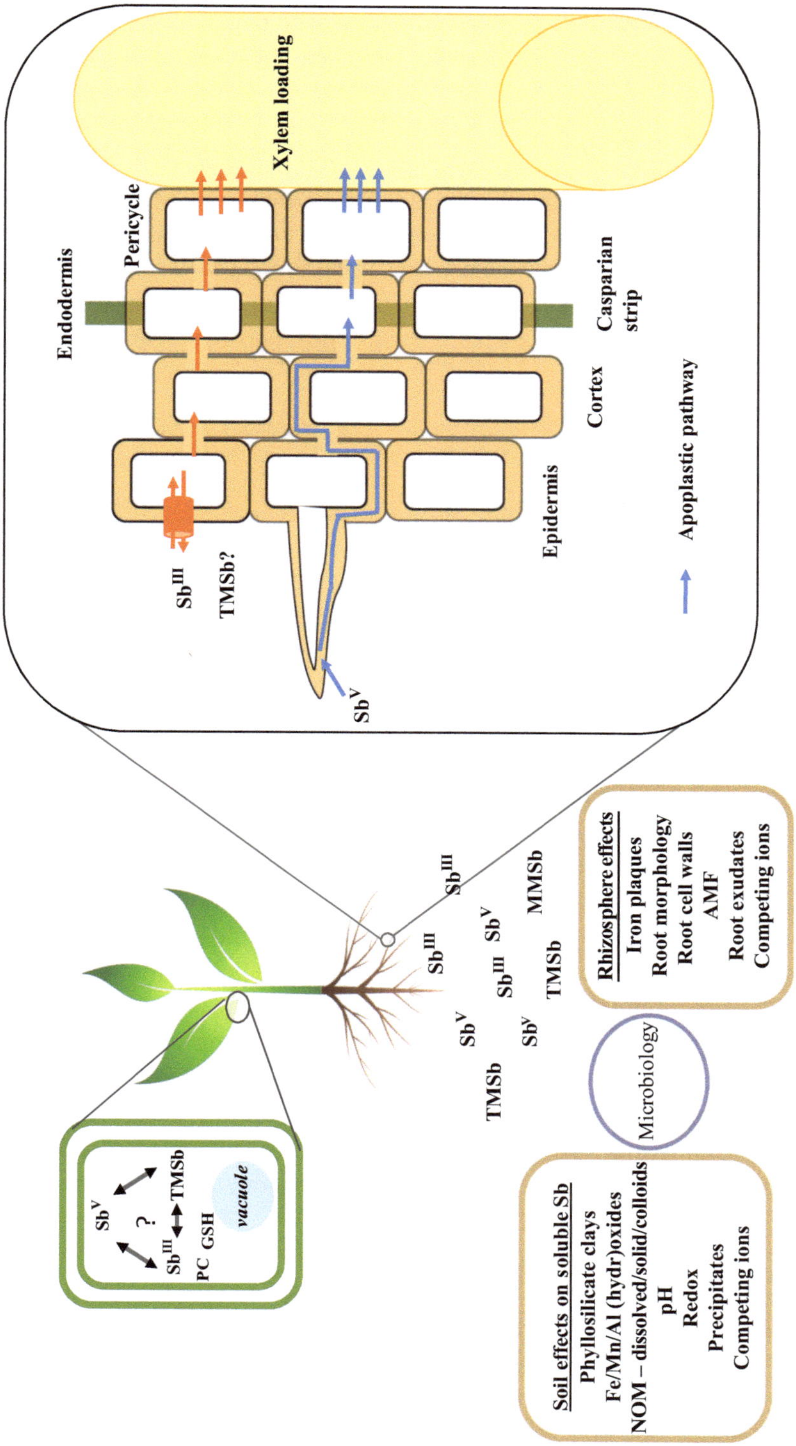

Fig. 7.1: Antimony soil to plant transfer.

7.2.5. Antimony mobility can increase by complexation with both dissolved NOM and colloidal organic ligands [18, 33]. Complexation with both Sb(V) and Sb(III) is pH dependent, decreasing with increasing soil pH. The maximum adsorption for Sb(V) with humic acid is reported at pH < 7 [34, 35] and pH 6 for Sb(III) [36]. Dissolved NOM can also compete with antimony for adsorption sites on iron and aluminum hydroxides and increase solution concentrations [37, 38]. Sorption to solid NOM has been considered to play a minor role in antimony retention in soils, but recent studies show that Sb(V) can preferentially complex with surface active sites in solid NOM in the presence of Fe (hydr)oxides, and sequestration of Sb(III) on thiol and carboxyl/phenol functional groups of solid NOM is important in environments with high NOM concentrations, such as wetlands [39, 40].

7.2.3 Competitive ions

Inorganic co-occurring ions can compete with antimony on soil surfaces to increase solution concentrations. These include the anions NO_3^-, CO_3^{2-}, SO_4^{2-}, and PO_4^{3-} that have been shown to partially decrease Sb(V) adsorption on minerals surfaces [41–44]. The efficiency of Sb(V) desorption by anionic ligands on Fe/Al/Mn (hydr)oxides is reported to increase in the order $NO_3^- < SO_4^{2-} < CO_3^{2-} < PO_4^{3-}$ [43], and Xi et al. [45] demonstrated competitive desorption by phosphorus on bentonite. However, competition with co-occurring ions in soil and the increase in soil solution concentration may not always result in an increased antimony accumulation in plants. For example, many studies report little influence of PO_4^{3-} on plant concentrations due to the differences in molecular configuration of the tetrahedral PO_4^{3-} anion with that of the octahedral Sb(V) anion and the different plant uptake pathways used [6, 46]. Much work has considered the co-competing effects of arsenic on antimony plant concentrations because both metalloids frequently co-occur in soils. In water spinach (*Ipomoea aquatica*) antimony accumulation was enhanced by increasing arsenic in soil [47], and this was also observed in *Pteris cretica* L. (Cretan brake fern) in hydroponic cultures [48]. In contrast, selenium suppressed antimony accumulation in all tissues of paddy rice in a hydroponic experiment [49]. Therefore, the influence of co-competing ions on plant concentrations depends not only on competitive desorption from soils but also on the antimony uptake pathway in plants.

7.2.4 Co-precipitation and precipitation

Antimony is immobilized in co-precipitation and precipitation reactions, that decrease soluble concentrations [50]. Iron, manganese, and aluminum (hydr)oxides are important scavengers in co-precipitation mechanisms during reformation with redox changes in soil [51, 52]. In calcareous soils calcium can reduce antimony in solution and also antimony in plants by forming insoluble Ca-antimonates ($Ca_2Sb_2O_7$) [50, 53, 54], but the influence of calcium on antimony in solution is complex. In other work, the increase in soil pH with increased Ca promoted antimony release, and increased plant antimony concentrations [55]. Further, Sb(III) will precipitate with sulfur in anoxic conditions [56], and the role of the highly insoluble secondary mineral tripuhyite ($FeSbO_4$) and its control of low-level antimony solution concentrations has been the subject of much interest, although its formation is likely to be slow due to kinetic constraints [57].

7.2.5 Redox

Soil redox strongly influences antimony speciation. In natural aerobic conditions, antimony predominantly exists in solution as the +V octahedral antimonate oxyanion ($Sb(OH)_6^-$), which is stable over a wide pH range; but in anaerobic soil conditions such as flooded rice-paddies, the +III tetrahedral neutral antimonite, ($Sb(OH)_3$), becomes thermodynamically stable at environmentally relevant pH values. Plants show differential uptake of the two inorganic antimony species [58] (see Section 7.5 in this chapter).

Soil redox influences plant antimony concentrations via reductive dissolution of the important redox-sensitive Fe/Mn (hydr)oxide antimony sorption surfaces releasing antimony to solution. Dissolved NOM carbon acts as an electron donor in aerobic biotic respiration [59] and, with high concentrations, a rapid decrease in redox occurs as oxygen is depleted, especially in flooded or waterlogged environments. The NOM can also act as an electron shuttle reducing the redox-sensitive elements by coupling the antimony and Fe/Mn redox cycles with organic functional groups such as phenolics and sulphydryl groups [60, 61]. The manganese surfaces are reduced before the iron surfaces according to the geochemical redox ladder [42, 62, 63]. In these conditions, antimony concentrations and speciation in solution show significant variation. Evidence suggests that Sb(V) strongly sorbed on Mn (hydr)oxides under aerobic conditions is released by reductive dissolution [64] and Sb(III) becomes thermodynamically stable as redox is reduced. The Fe(II) also acts as an electron donor in the reduction of Sb(V) to Sb(III) in the soil solution [42]. The Sb(III) is strongly sorbed to Fe (hydr)oxides at a pH ~7, but with a further decrease in redox, may be released into solution [42, 64, 65]. Soil colloids and dissolved NOM are also thought to play a role because solution antimony cannot always be explained by relationships with the soluble ions [18]. This variation in antimony in solution has consequences for plant concentrations that are reported to increase or decrease with changing redox conditions but this is strongly influenced by plant species uptake of the different antimony species [58].

7.2.6 Biomethylation

Antimony undergoes biomethylation by bacteria and fungi in soils under oxic and anoxic conditions to form mono-, di-, and tri-methyl antimony compounds (MMSb, DMSb, and TMSb, respectively) [66, 67]. These reactions predominate in anaerobic conditions and are strongly associated with organic matter [68]. Mono-, di-, and tri-methylated antimony species have been reported in agricultural and garden soils of the Ruhr Basin in Germany with MMSb the most abundant [69], and, in China, TMSb has been detected in rhizosphere soils at an old mining area [70]. TMSb has also been quantified at up to 1.35 mg kg^{-1} in Swiss shooting range soils [18]. Although methylated compounds typically occur in soils at lower concentrations than inorganic forms [18], they are detected in soil solution, and TMSb is reported as relatively mobile [71]. Biomethylation alters antimony toxicity and so the transfer of methylated species from soils to plants under low redox conditions, as recently confirmed for TMSb [18, 72], has consequences for risk assessment. Chapter 11 discusses antimony biomethylation processes.

7.3 Antimony accumulation and species in plants

Antimony is considered nonessential to plants and background antimony concentrations in terrestrial vascular plants typically range from 0.2 to 50 mg kg^{-1} antimony in plant tissue [73]. Much higher plant concentrations are often reported for antimony contaminated soils, however (Tab. 7.1). For example, up to 4,029 mg kg^{-1} antimony was reported in the leaves of *Boehmeria nivea* (L.) Gaudich grown in an antimony mining area in China with total soil antimony >5,000 mg kg^{-1} [13], and some hyperaccumulating ferns (*Pteris cretica*) can also show over 1,000 mg kg^{-1} in fronds [48, 74]. Accumulation, however, varies significantly and, in comparison, much lower antimony concentrations were found in some vegetables (up to a mean of 37 mg kg^{-1} in *Raphanus sativus* L) grown in the same mine area of China, in soils with antimony concentrations up to 3,950 mg kg^{-1} [15]. Zeng et al. [15] reported significant differences in accumulation even across the range of vegetables sampled at the mine area in China, with a maximum aboveground antimony concentration in *Raphanus sativus* L (37 mg kg^{-1}) and the minimum in *Brassica pekinensis* L (5.9 mg kg^{-1}). Typically, total antimony accumulation in plant parts occurs in the order of roots > shoots > leaves > fruits, with high root concentrations and low translocation coefficients, often ≪1 [3, 11, 54, 75–77], although incomplete removal of soil particles from roots may confound root concentrations reported in field soils [78].

Different antimony species show different accumulation patterns in different plants. For example, Tschan et al. [79], growing sunflower and maize in Sb(III) and Sb(V) spiked soil, showed that sunflowers had a higher affinity for Sb(V) whereas antimony speciation had no influence on antimony uptake in maize;

Tab. 7.1: Plant accumulation and speciation studies (plants with >100 mg kg^{-1} Sb in tissues are reported for field and pot studies) (adapted from Feng et al. [74]).

Plant species	Sb exposure	Sb speciation in growth medium	Sb concentration / mg kg^{-1}	Sb species and concentration / mg kg^{-1} (unless shown otherwise)	Experiment details	Reference
Pteris vittata	5 mg kg^{-1}	$KSb(OH)_6$	230[a] (roots)	shoot upto 77% Sb(V), upto 29% Sb(III); root upto 90% Sb(V), TMSb(V) detected	Pot experiment	[87]
Phytolacca acinosa Roxb.	590 mg kg^{-1}	Main as Sb(V)	190[a] (leaves)	Sb(III) 4, Sb(V) 51	Field survey	[13]
Boehmeria nivea (L.) Gaudich	5579 mg kg^{-1}	Main as Sb(V)	4029[a] (leaves)	Sb(III) 27, Sb(V) 507	Field survey	[13]
Miscanthus floridulus (Labill.) Warb.	4284 mg kg^{-1}	Main as Sb(V)	683[a] (leaves)	Sb(III) 37, Sb(V) 97	Field survey	[13]
Eremochloa ophiuroides (Munro) Hack.	1319 mg kg^{-1}	Main as Sb(V)	128[a] (leaves)	Sb(III) 1, Sb(V) 20	Field survey	[13]
Lactuca tatarica (L.) C.A. Mey	4333 mg kg^{-1}	Main as Sb(V)	357[a] (leaves)	Sb(III) 17, Sb(V) 188	Field survey	[13]
Leersia hexandra Swartz	1302 mg kg^{-1}	Main as Sb(V)	292 [a] (leaves)	Sb(III) 15, Sb(V) 73	Field survey	[13]
Hygroryza aristata (Retz.) Nees	930 mg kg^{-1}	Main as Sb(V)	109[a] (leaves)	Sb(III) 4, Sb(V) 55	Field survey	[13]
Arachis hypogaea L.	1364 mg kg^{-1}	Main as Sb(V)	314[a] (leaves)	Sb(III) 1, Sb(V) 15	Field survey	[13]
Comnyza canadensis (L.) Cronq.	478 mg kg^{-1}	Main as Sb(V)	461[a] (leaves)	Sb(III) 0.1, Sb(V) 2.5	Field survey	[13]
Oryza sativa L. cv. Jiahua	1562 mg kg^{-1}	Main as Sb(V)	545 (roots) 14 (shoots)	Sb(III) 81, Sb(V) 384 Sb(III) 0.5, Sb(V) 8	Pot experiment	[54]

(continued)

Tab. 7.1 (continued)

Plant species	**Sb exposure**	**Sb speciation in growth medium**	**Sb concentration / mg kg^{-1}**	**Sb species and concentration / mg kg^{-1} (unless shown otherwise)**	**Experiment details**	**Reference**
Pteris cretica	NA	Sb(V)		roots Sb(III) 19%, Sb(V) 81% shoots Sb(III) 83%, Sb(V) 17%	Hydroponic culture	[88]
Oryza sativa L. cv. Nanjing 45	5 mg L^{-1}	 Sb(V) Sb(III)		% Sb(V) of total Sb roots 97, leaves 98 roots 96, leaves 92	Hydroponic culture	[89]
Pteris vittata	8 mg L^{-1}	 Sb(V) Sb(III)		% Sb(V):Sb(III) roots 98:2 roots 19:81 little transfer to fronds	Hydroponic culture	[90]
Rumex patientia	1472 mg kg^{-1}	Inorganic Sb(V) in rhizosphere soils	438 roots 48 leaves	For all plants - Sb(III) dominant in leaves, TMSb detected in some leaves but not roots, unknown Sb(V) compound dominant in roots	Field survey	[70]
Dryopteris erythrosora	648 mg kg^{-1}	Inorganic Sb(V) in rhizosphere soils	1421 roots 107 leaves		Field survey	[70]
Oplismenus undulatifolius (Ard.) Bea.	469 mg kg^{-1}	Inorganic Sb(V) in rhizosphere soils	141 roots 37 leaves		Field survey	[70]
Fagopyrum dibotrys (D. Dom.) Hara	479 mg kg^{-1}	Inorganic Sb(V) in rhizosphere soils	1072 roots 59 leaves		Field survey	[70]

Pteris vittata	441 mg kg^{-1}	Inorganic Sb(V) in rhizosphere soils	107 roots 39 leaves		Field survey	[70]
Lolium perenne L. cv. Calibre	1 mg L^{-1}	 Sb(V) Sb(III) TMSb		% Sb(V):Sb(III):TMSb roots 61:41:nd[b]; shoot 95:nd:3 roots 53:46:0.5; shoot 72:27:nd roots 26:nd:59; shoot 16:nd:76	Hydroponic culture	[82]
Pteris vittata	1 mg L^{-1}	Sb(V) Sb(III)		roots Sb(V) and Sb(III) roots only Sb(III)	Hydroponic culture	[85]
Lolium perenne L. Calibre	1 mg L^{-1}	 Sb(V) Sb(III) TMSb		% Sb(V):Sb(III) roots 100:0; shoots 49:51 roots 76:24; shoots 60:40 roots - all methylated Sb	Hydroponic culture	[16]
Lolium perenne L. Calibre	1 mg L^{-1}	 Sb(V) Sb(III) TMSb		% Sb(V):Sb(III):TMSb roots 88:3:1; shoots 67:nd:6 roots 5:76:nd; shoots 44:31:4 roots 6:3:11; shoot 8:nd:62	Hydroponic culture	[72]
Secale cereale L. cv. Palazzo	1 mg L^{-1}	 Sb(V) Sb(III) TMSb		% Sb(V):Sb(III):TMSb roots 95:4:1; shoots 84:3:3 roots 5:72:nd; shoots 34:48:2 roots 8:4:47; shoots 1:nd:70	Hydroponic culture	[72]
Triticum aestivum L. cv. Sella	1 mg L^{-1}	 Sb(V) Sb(III) TMSb		% Sb(V):Sb(III):TMSb roots 91:3:1; shoots 61:nd:5 roots 6:76:nd; shoots 47:36:2 roots 14:22:36; shoots 2:3:87	Hydroponic culture	[72]

(continued)

Tab. 7.1 (continued)

Plant species	Sb exposure	Sb speciation in growth medium	Sb concentration / mg kg^{-1}	Sb species and concentration / mg kg^{-1} (unless shown otherwise)	Experiment details	Reference
Festuca pratensis Huds. cv. Preval	1 mg L^{-1}			% Sb(V):Sb(III): TMSb	Hydroponic culture	[72]
		Sb(V)		roots 87:1:1; shoots 79:nd:5		
		Sb(III)		roots 4:62:nd; shoots 57:17:4		
		TMSb		roots 5:3:26; shoots 1:nd:75		
Indica rice (Yexiangyou-No. 3)	1–500 mg L^{-1}	Sb(III) as $KSbC_4H_4O_7$		shoots only Sb(III) at 100 mg L^{-1}, Sb(III) and Sb(V) at higher exposure roots Sb(III) and Sb(V)	Hydroponic culture	[77]
Indica rice Yexiangyou-No. 3	10, 20 mg L^{-1}	Sb(V)		Mainly Sb(V) in roots and shoots but Sb(III) present	Hydroponic culture	[19]
		Sb(III)		Mainly Sb(III) in roots and shoots but Sb(V) present		

[a] Maximum concentration; [b] nd = not detected

Shtangeeva et al. [22] showed that wheat had a higher affinity for Sb(III) whereas rye had a higher affinity for antimony(V). Leafy vegetables often show a larger accumulation factor compared to tuber fruit and leguminous vegetables [12, 80], and this is of concern for human exposure. Rice, another staple in the diets of populations around the world, also represents a cause of concern for human exposure [3]. For example, the daily dietary intake of antimony was 554 µg d^{-1} (for adults with body weight of 60 kg) in the residents living near the XKS mine in China, exceeding the tolerable daily intake value (TDI) of 360 µg d^{-1} set by the World Health Organization [81], and rice and vegetables (including tuber and leafy vegetables) contributed 33% and 26%, respectively, of the total daily antimony dietary intake [2]. Human exposure to antimony through food is described in Chapter 15.

Much has been collated on total concentrations measured in plants growing in Sb-contaminated soils [3, 23, 74, 75], but only chemical speciation analysis of plant parts can provide a complete understanding of uptake mechanisms and toxicity. This has been restricted by the complexity in speciation preservation during extractions [82], absence of appropriate standards and certified reference material, and interference in analyses [83, 84]. Few studies, until recent years, have reported both the antimony speciation in soils and also plant parts (Tab. 7.1). Antimony speciation in these studies has been achieved typically using mild extractants (citric, oxalic, ascorbic acids) and analysis with hydride generation or HPLC-ICP-MS [13, 72]. Recently bulk X-ray absorption near edge spectroscopy (XANES) has shown promise for identification of inorganic and organic antimony species in plant parts but requires rigour in quality control using appropriate standards and sample preservation [16, 85].

It is clear from the studies in Tab. 7.1 that different antimony species are detected in different plants even when grown in the same type of growth medium with the same antimony species. For example, in the study of Okkenhaug et al. [13] between 0.6 and 57.5% of total acid extracted antimony in plants growing in the highly contaminated XKS mine site area in China was citric acid–extractable, assumed to extract Sb(III) and Sb(V). The speciation analysis showed that Sb(III) in plant extracts accounted for 1.8–28% of total antimony and this was higher than the Sb(III) fraction in the water extracts of the soils. Possible explanations included facilitated uptake of trivalent antimony species by the plant, or that a reduction of Sb(V) to Sb(III) occurred after uptake, as also proposed by others in later work [16, 83]. In this study a significant proportion of total plant antimony was not citric acid–extractable, which supports the co-occurrence of an unknown antimony species or methylated antimony species as reported by others [72, 85]. For example, Muller et al. [86] carried out speciation analysis on aqueous extracts of Chinese brake fern *Pteris vittata* growing in Sb(V) spiked composites, and found that in addition to Sb(V), a considerable concentration of Sb(III) and TMSb(V) was present in plant parts. More recent work has confirmed TMSb transfer to ryegrass roots [16, 82]. Therefore, in addition to Sb(III) and Sb(V), it is clear that other antimony species may also transfer from soil to plants, including TMSb, and some transformation within the plant may occur.

7.4 Rhizosphere effects

Antimony in soils is taken up by plants mainly via root tissues. The soil-root boundary represents a unique interface for antimony entry into plants. The growing knowledge on plant root and rhizosphere influences on antimony soil to plant transfers includes the following (Fig. 7.1) [7]:

i) Different plants and genotypes show different root morphology and anatomy, including cell wall characteristics and maturity of apoplastic uptake pathways. These characteristics are known to influence plant metal uptake [91]. Because the roots are the primary entry point for antimony to plants in soils, and cell walls are important for antimony storage [89], root morophology and anatomy would be expected to influence antimony soil to plant transfers, but few studies exist. Liu et al. [92] demonstrated that the pectin, hemicellulose (I and II), and lignin content of cell walls in rice altered as a result of antimony stress and other factors such as selenium in soil, and this regulated antimony uptake in roots due to a cell aging effect at higher concentrations. Work on root morphology to date however, examines the effect of antimony on root structure due to toxic response [7, 93], in contrast to morphological influences on root antimony uptake.
ii) Root exudates, the chemicals excreted into the soil by plants, are commonly grouped into low (organic acids, phenolics, sugars) and high (mucilage as proteins, polysaccharides) molecular weight compounds and play a major role in root soil contact [94]. The root exudates can alter pH in the rhizosphere, compete for sorption sites in soil, and form chelates with metals in soil [7]. In hydroponic experiments using rice, Sb(III) and Sb(V) stimulated the production of oxalic acid and modified concentrations of other organic acids, and the authors speculated that antimony plant uptake would be affected by competition for sorption sites, and by moderation of iron plaque formation on roots [19].
iii) Iron plaque formation on plant root surfaces is critical for modifying antimony uptake and accumulation and has been most studied for rice [95]. These plaques are usually formed via root oxygen production and oxidation of soluble Fe(II) [96]. The influence of iron plaques on antimony concentrations in roots and translocation is complex but studies report that antimony promotes iron plaque formation facilitated by Sb-resistant bacteria [96], iron plaque sequesters antimony and limits uptake of Sb(III) and Sb(V) [89, 97] but also, in contrast, that sequestration can increase Sb(V) concentrations in roots [98]. Iron plaques can also facilitate the oxidation of adsorbed Sb(III) [89]. Effects depend on antimony species and concentration, rice cultivar, and soil water environment, but iron plaque moderation of soil–plant antimony transfers is clearly important for plants growing in wetland environments.
iv) Microorganisms play a key role in the biogeochemical cycling of antimony in soils as already described, but the arbuscular mycorrhizal fungi (AMF) show importance for altering antimony bioavailability and accumulation in plants. This group of

fungi, which forms symbiotic relationships with plant roots, can reduce or increase metal uptake depending on plant and element [99], and the few studies available for antimony show enhanced availability and potential to promote accumulation in plants by increasing exchangeable antimony. Plants studied include Bermunda grass (*Cynodon dactylon*), carrot (*Daucus carota*), lettuce (*Lactuca sativa*), alfalfa (*Medicago sativa*), and maize (*Zea mays*) [78]. The effect is dependent on plant species [78, 100, 101] but the mechanism is not yet fully explained.

The rhizosphere represents a unique zone of intimate soil–root contact and clearly has a critical role in determining antimony plant concentrations. Further advances in analytical capability are still needed to allow a complete appreciation of the rhizosphere influence on antimony soil to plant transfers.

7.5 Plant uptake, translocation and storage

7.5.1 Uptake

Chemical speciation is clearly one of the most important influences on antimony uptake and accumulation in plants (Fig. 7.1) [58]. The molecular configurations and charge of the aqueous antimony inorganic Sb(III) and Sb(V) species (tetrahedral neutral $Sb(OH)_3$; octahedral, negatively charged $Sb(OH)_6^-$) direct the specific uptake pathways and transporters used. Plants usually show a greater capacity for Sb(III) uptake than for Sb(V) [16, 22, 72, 82], which can partially be explained by the negative charge of the Sb(V) anion requiring greater energy for transfer across the negatively charged root surface.

Aquaporin water channels are membrane proteins in cell walls that primarily mediate the transport of water along an osmotic gradient. A form of aquaporin, an aquaglyceroporin, can also conduct small uncharged molecules such as glycerol, but aquaporins are impermeable to charged molecules. Thus, uptake of Sb(III) is considered a passive process through aquaglyceroporins across the root membrane of plants [17, 102], similar to trivalent arsenic that has a similar configuration (Fig. 7.1). Kamiya and Fujiwara [103] first identified the nodulin 26-like intrinsic protein (NIP1:1) Sb(III) transporting aquaglyceroporin in plants [72, 104] (refer to Chapter 12 on detoxification). However, aquaglyceroporin channels show different affinity for different elements and, for example, Meharg and Jardine [105] showed that aquaglyceroporin channels in rice plants had a greater affinity for As(III) than Sb(III) and the uptake of Sb(III) was reduced by 50% in the presence of As(III) in rice growing for 20 min in water containing 0.1 mM As(III). Therefore, the presence of certain other molecules will influence Sb(III) uptake as well as the soil water regime, and different aquaglyceroporin proteins in different plants will influence differential antimony uptake [106].

The uptake mechanism for the Sb(V) anion is different. Uptake is not thought to occur by the same phosphate transporter pathway used by As(V), which has a different molecular configuration, and is tetrahedral in aqueous solution [23]. Tschan et al. [23] first proposed that Sb(V) could enter plants by either i) the root symplast via anionic transporters of low selectivity, or ii) the apoplastic pathway (flowing through cell wall to cell wall) that depends on the concentration gradient of ions between plant root and the soil, and access through a damaged, leaky, or immature Casparian strip (Fig. 7.1). Most evidence suggests the apoplastic route is dominant [16], although use of the low-affinity phosphate uptake system or the symplastic route cannot be discounted because phosphorus has been shown to inhibit Sb(V) uptake in rice and *Pteris vittata* [17, 102].

Understanding plant uptake of methylated antimony is still evolving (refer to Chapter 11 on biomethylation) and has focused on TMSb(V). Spiked hydroponic solutions of ryegrass, fescue, rye, and wheat have demonstrated TMSb(V) uptake in plant roots, but to a smaller extent than Sb(III) and Sb(V), and also translocation to shoots [16, 72, 82]. Uptake without speciation change and via aquaporins was hypothesized by the authors [16]. Little research is available on the solution chemistry of methylated species, but TMSb is thought to occur in the form of the neutral species $(CH_3)_3Sb(OH)_2$ or as the monovalent cation $[(CH_3)_3SbOH]^+$ in aqueous solutions under neutral conditions [107]. Thus, TMSb plant uptake would be expected to be more easily facilitated compared to the Sb(V) anion. Methylated antimony species are considered more mobile in plants than inorganic antimony species based on the limited knowledge available [72].

7.5.2 Translocation, subcellular distribution, and transformations

Antimony can translocate to above ground plant parts but concentrations, speciation, and compartmentalization depend on plant species. Translocation factors are usually low ($\ll$1), but some hyperaccumulating plants can show much greater antimony translocation to plant fronds (see Section 7.3). Inorganic Sb(V), Sb(III) and methylated antimony species have been reported in above-ground plant parts sampled in the field (although analytical capability may have confounded antimony species reported in older studies) [108] and may simultaneously occur in plant tissues even when growing in a medium contaminated with only one of the species (Tab. 7.1). Hydroponic experiments in recent years have helped to elucidate mechanisms and transformations using a range of different plants (ryegrass, wheat, fescue, rye, rice). The work on understanding translocation and subcellular distribution is still evolving (Fig. 7.1), but evidence for the following proposed transformations and distribution exists:

- Sb(III) once taken up is found as Sb(III)-thiol compounds in roots, probably complexed by compounds such as glutathione (GSH) and phytochelatins (PCs), which is proposed as an important sequestration and detoxification mechanism [16].

- Immobilization and vacuole storage of the Sb(III)-thiol complexes may explain the relatively low transfer of antimony to shoots observed in plants grown in Sb(III)-treated nutrient solution.
- Sb(III) may be oxidized to Sb(V) in the plant, with an increase in relative abundance of Sb(V) from roots to shoots reported [77, 82, 85, 89]. A range of mechanisms have been proposed including Sb(III)-induced stress oxidation [109].
- Sb(V) may be reduced to Sb(III) in the plant and evidence exists that this may occur in roots and shoots, and that the Sb(III) is then detoxified by complexation with thiol groups as described above [88].
- Recent evidence suggests that antimony methylation may occur in plants, the methylated compounds are stored in cells, and antimony demethylation may also occur within the plant [70, 72, 82].

Thus, antimony movement in plants is governed by sequestration and transport processes specific to the plant and antimony species. Advances in analytical methodology, including XANES, provide scope for progressing the understanding of mechanisms currently proposed.

7.5.3 Hyperaccumulation

Only a few antimony hyperaccumulating plants had been identified, mainly at field mine sites [4, 74], with above ground plant concentrations reported at >1,000 mg kg^{-1} and root-to-shoot translocation factors of >1 [110]. A series of studies on the well-known arsenic hyperaccumulator Pteris family [74] suggest that antimony uptake and translocation mechanisms in these species may be different to the mechanisms described in Section 7.5.1 because Sb(III) uptake in *P. vittata* was not inhibited by aquaglyceroporin inhibitors [17]. Uptake was also facilitated when antimony co-occurred with arsenic due to As(V) membrane stress modification [87]. These plants exhibit strong cell wall and cytosol compartmentalization [48] and possess the ability to efflux Sb(III) from the roots [90] for antimony detoxification.

7.6 Toxicity and detoxification

Antimony toxicity to plants depends on uptake, concentration, and speciation, but antimony can induce a range of toxic effects (Tab. 7.2, See Chapter 14 for specific ecotoxicological data). Plants studied include ferns, rice, wheat, sunflowers, rye, and maize (Tab. 7.2), and 5–10 mg kg^{-1} is the often cited toxic concentration in plants [111]. Some plants, however, tolerate much higher concentrations, and wheat, as an example, can tolerate >100 mg kg^{-1} in roots [77]. Different antimony species show different toxicity to different plants. Most of the studies reported have focused on Sb(III), which has traditionally been considered more toxic than Sb(V) [109] (Tab. 7.2). Sb(V), however, was reported more toxic for rye [112], and TMSb(V) was reported more toxic than Sb(III) and Sb(V) to aquatic plants [113]. Little work, however, is reported, on the toxicity of methylated species (Tab. 7.2) (refer to Chapter 11 on methylation). Effects of antimony exposure include reduced growth and biomass, and inhibition of the photosynthetic capacity [20, 114]. Increased reactive oxygen species (ROS) production as indicated by high malondialdehyde content (an indicator of lipid peroxidation) has also been observed in ferns *Microlepia hancei* and *Cyclosorus dentatus* (Tab. 7.2) [74]. Plants subjected to stress show a primary response consisting of the rapid production of ROS and also reactive nitrogen species (RNS). The plant can respond with enzymatic and non-enzymatic antioxidant systems that control the concentrations of reactive species to maintain redox homeostasis and functionality of the biological system. With antimony exposures, antioxidant enzyme responses have been reported, including peroxidase (POD), catalase (CAT), and superoxide dismutase (SOD), and also non-enzymatic antioxidant responses, including ascorbate and glutathione, as well as phenolic-type compounds such as flavonoids, phenylpropanoids, glycosides, and carotenoids (Tab. 7.2). Exposure to antimony can also simply disturb the uptake of essential nutrients from the soil and thereby affect distribution in the plant [115]. For example, Shtangeeva et al. [112] demonstrated a significant decrease in the uptake of Ca^{2+} and Na^{+} in wheat with antimony exposure. In addition, the presence of co-contaminants in soils can alter a plant's toxic response due to competitive effects in soils or on plant uptake. For example, selenium co-occurring with antimony reduced the toxic response in rice [115] but arsenic as a co-contaminant increased toxic response in water spinach (*Ipomoea aquatica*) in a synergistic response [47]. Antimony often exists in soil with other elements, thus understanding the synergistic, antagonistic, or additive influences of co-occurring elements warrants further understanding.

Plants possess a number of antimony detoxification mechanisms (see Chapter 12 on detoxification) that include the activation of the antioxidant systems, both enzymatic and non-enzymatic [114, 116] in response to excessive ROS species in the plant tissues. The PCs are also considered to contribute to antimony detoxification by facilitating the redistribution in plants via thiol complexes and sequestration [16, 74]. The role of PCs and antimony effects on PCs are the subject of much current interest [74].

Tab. 7.2: Toxic effects reported in plants.

Plant system or function affected	Plant	Symptom	Experiment	Sb species concentration	Reference
Plant growth and biomass	Ferns - *Microlepia hancei*, *Cyrtomium fortune*, *Cyclosorus dentatus*	Reduced biomass (13–35% decrease)	Hydroponic culture	Sb(III) as $KSbC_4H_4O_7$ 20 mg L^{-1}	[117]
	Oryza sativa L. cv. Weiyou 402	Reduced biomass (14%)	Hydroponic culture	Sb(III) as $KSbC_4H_4O_7$ 5 mg L^{-1}	[49]
	Zea mays	Reduced seedling growth at 50 mg kg^{-1} soil Sb	Pot experiment	Sb(III) 10–1000 mg kg^{-1} in soil	[118]
	Wheat – *Triticum aestivum*	– Root length reduced – Germination inhibited at >50 mg L^{-1} – Less effect on shoots	Hydroponic culture	Sb(III) as $KSbC_4H_4O_7$ 1–500 mg L^{-1}	[77]

(continued)

Tab. 7.2 (continued)

Plant system or function affected	Plant	Symptom	Experiment	Sb species concentration	Reference
Oxidative stress and antioxidant systems	Ferns - *Microlepia hancei*, *Cyclosorus dentatus*	Increased malondialdehyde concentration	Hydroponic culture	Sb(III) as $KSbC_4H_4O_7$ 20 mg L^{-1}	[117]
	Zea mays	Activities of peroxidase superoxide dismutase and catalase altered at >50 mg kg^{-1} soil Sb	Pot experiment	Sb(III) 10–1000 mg kg^{-1} in soil	[118]
	Oryza sativa L. cv. Weiyou 402	Increased malondialdehyde concentration	Hydroponic culture	Sb(III) as $KSbC_4H_4O_7$ 5 mg L^{-1}	[49]
	Helianthus annuus L. cv. Safira	Strong oxidative response for phenolic and the ROS/RNS metabolic processes	Hydroponic culture	Sb(V) upto 1mM	[114]
	Wheat – *Triticum aestivum*	Glutathione, malondialdehyde, superoxide dismutase and peroxidase changes, and increased at higher concentrations	Hydroponic culture	Sb(III) as $KSbC_4H_4O_7$ 1–500 mg L^{-1}	[77]
	Rice (Yangdao No.6)	Sb(III) and Sb(V) both induced oxidative stress	Hydroponic culture	Sb(III) and Sb(V) upto 20 mg L^{-1}	[109]
	Indica rice (Yexiangyou No.3)	Sb(III) increased oxidative stress compared to Sb(V) based on malondialdehyde	Hydroponic culture	Sb(III) and Sb(V) upto 20 mg L^{-1}	[19]

Photosynthesis	*Synechocystis* sp.	– Inhibited photosystem II activity – Disrupted electron transport – Decreased photochemical efficiency	Hydroponic culture	Sb(III) 1.0–10 mg L^{-1}	[119]
	Zea mays	– Chlorophyll synthesis inhibition – Decreasing photochemical efficiency	Pot experiment	Sb(III)	[118]
	Helianthus annuus L. cv. Safira	– Chlorophyll content declined – Decreased photosynthesis efficiency	Hydroponic culture	Sb(V) upto 1mM	[114]
	Acorus calamus	Both Sb species unfavourably influenced photosynthesis from 250 mg kg^{-1} Sb(III). 500 mg kg^{-1} Sb(V)	Pot experiment	Sb(III) and Sb(V) upto 2000 mg kg^{-1}	[20]
Uptake of essential elements	Indica rice (Fengmeizhan)	Decreased concentration of Ca, Mg, K, Fe Mn, Cu and Zn	Hydroponic culture	Sb(III) as $KSbC_4H_4O_7{\cdot}1/2H_2O$ 5 mgL^{-1}	[115]
	Wheat – *Triticum astevium*	– Variations in macro and trace elements – Reduced Ca, K, Cu in roots	Hydroponic culture	Sb(V) as $Sb(NO_3)_5$	[112]
	Rye *Secale cereale* L.	– Variation in Mg, Ti, K, Co, Mn for both Sb forms	Hydroponic culture	Sb(III) as $SbCl_3$, Sb(V) as $SbCl_5$ Both at 75 mg L^{-1}	[22]
	Rice (Yangdao No.6)	– Both Sb species induced changes and some negative effects on shoot and root concentrations of some elements (Ca, K, Cu, K, Mg, Cu, Mn, Zn) – Sb(III) more toxic than Sb(V)	Hydroponic culture	Sb(III) and Sb(V) upto 20 mg L^{-1}	[109]
Protein synthesis and function	Wheat – *Triticum astevium*	Gene expression of transporter proteins influenced at low Sb concentrations	Hydroponic culture	Sb(III) as $KSbC_4H_4O_7$ 1–500 mg L^{-1}	[77]
Hormone synthesis	Indica rice (Yexiangyou No.3)	Sb(III) increased endogenous hormone abscisic acid (ABA) but not Sb(V)	Hydroponic culture	Sb(III) and Sb(V) upto 20 mg L^{-1}	[19]

7.7 Conclusion

Iron, manganese, and aluminium hydr(oxides), NOM, redox, and pH are the key soil factors that influence antimony speciation and concentrations in soil solution for soil to plant transfer. While much is understood about antimony transformations and sorption/desorption mechanisms in soils and effects on soil solution concentrations, some experimental observations in changing redox environments still cannot be explained. In particular, the role of colloidal particles and methylated antimony species requires better understanding. The rhizosphere represents a zone of unique microscale Sb-root-soil interface processes, including root iron plaques and AMF, that importantly determine plant antimony uptake. Rhizosphere research for antimony is still in its infancy, however. Antimony strongly sequesters in plant roots with limited translocation to upper plant parts in many plant species, excepting a few known hyperaccumulating species. Uptake of Sb(III) via aquaporins is typically greater than that of Sb(V), but uptake pathways for methylated species are not yet clear. Inorganic and organic antimony species are detected in plant tissues and recent evidence suggests species transformation occurs in plant tissues, but speciation preservation during plant part extractions and analytical capability still limits detection and validation. The antimony toxicity to plants and success of detoxification is also dependent on antimony speciation. Thus, advancing analytical capability is the fundamental requirement for improved understanding and risk quantification in Sb soil to plant transfer.

References

[1] Ngo LK, Pinch BM, Bennett WW, Teasdale PR, Jolley DF. Assessing the uptake of arsenic and antimony from contaminated soil by radish (*Raphanus sativus*) using DGT and selective extractions. Environ Pollut 2016, 216, 104–114.

[2] Wu F, Fu Z, Liu B, Mo C, Chen B, Corn W, Liao H. Health risk associated with dietary co-exposure to high levels of antimony and arsenic in the world's largest antimony mine area. Sci Total Environ 2011, 409, 3344–3351.

[3] Pierart A, Shahid M, Séjalon-Delmas N, Dumat C. Antimony bioavailability: knowledge and research perspectives for sustainable agricultures. J Hazard Mater 2015, 289, 219–234.

[4] Baroni F, Boscagli A, Protano G, Riccobono F. Antimony accumulation in *Achillea ageratum, Plantago lanceolata* and *Silene vulgaris* growing in an old Sb mining area. Environ Pollut 2000, 109, 347–352.

[5] Obiakor MO, Tighe M, Pereg L, Wilson SC. Bioaccumulation, trophodynamics and ecotoxicity of antimony in environmental freshwater food webs. Crit Rev Environ Sci Technol 2017, 47, 2208–2258.

[6] Tschan M, Robinson B, Schulin R. Antimony uptake by *Zea mays (L.)* and *Helianthus annus (L.)* from nutrient solution. Environ Geochem Health 2008, 30, 187–191.

[7] Zhu YM, Yang JG, Wang LZ, Lin ZT, Dai JX, Wang RJ, Yu YS, Liu H, Rensing C, Feng RW. Factors influencing the uptake and speciation transformation of antimony in the soil-plant system, and the redistribution and toxicity of antimony in plants. Sci Total Environ 2020, 140232.
[8] Hammel W, Debus R, Steubing L. Mobility of antimony in soil and its availability to plants. Chemosphere 2000, 41, 1791–1798.
[9] Wilson SC, Lockwood PV, Ashley PM, Tighe M. The chemistry and behaviour of antimony in the soil environment with comparisons to arsenic: A critical review. Environ Pollut 2010, 158, 1169–1181.
[10] Flynn HC, Meharg AA, Bowyer PK, Paton GI. Antimony bioavailability in mine soils. Environ Pollut 2003, 124, 93–100.
[11] Telford K, Maher W, Krikowa F, Foster S, Ellwood MJ, Ashley PM, Lockwood PV, Wilson SC. Bioaccumulation of antimony and arsenic in a highly contaminated stream adjacent to the Hillgrove Mone, NSW, Australia. Environ Chem 2009, 6, 133–143.
[12] Fu Z, Wu F, Mo C, Deng Q, Meng W, Giesy JP. Comparison of arsenic and antimony biogeochemical behavior in water, soil and tailings from Xikuangshan, China. Sci Total Environ 2016, 539, 97–104.
[13] Okkenhaug G, Zhu Y-G, Luo L, Lei M, Li X, Mulder J. Distribution, speciation and availability of antimony (Sb) in soils and terrestrial plants from an active Sb mining area. Environ Pollut 2011, 159, 2427–2434.
[14] Qi C, Wu F, Deng Q, Liu G, Mo C, Liu B, Zhu J. Distribution and accumulation of antimony in plants in the super-large Sb deposit areas, China. Microchem J 2011, 97, 44–51.
[15] Zeng D, Zhou S, Ren B, Chen T. Bioaccumulation of antimony and arsenic in vegetables and health risk assessment in the superlarge antimony-mining area, China. J Anal Methods Chem 2015, 2015.
[16] Ji Y, Sarret G, Schulin R, Tandy S. Fate and chemical speciation of antimony (Sb) during uptake, translocation and storage by rye grass using XANES spectroscopy. Environ Pollut 2017, 231, 1322–1329.
[17] Tisarum R, Chen Y, Dong X, Lessl JT, Ma LQ. Uptake of antimonite and antimonate by arsenic hyperaccumulator *Pteris vittata*: Effects of chemical analogs and transporter inhibitor. Environ Pollut 2015, 206, 49–55.
[18] Grob M, Wilcke W, Mestrot A. Release and biomethylation of antimony in shooting range soils upon flooding. Soil Systems 2018, 2, 34.
[19] Feng RW, Lei L, Su JM, Zhang RR, Zhu YM, Chen WX, Wang LZ, Wang RJ, Dai JX, Lin ZT. Toxicity of different forms of antimony to rice plant: Effects on root exudates, cell wall components, endogenous hormones and antioxidant system. Sci Total Environ 2020, 711, 134589.
[20] Zhou X, Sun C, Zhu P, Liu F. Effects of antimony stress on photosynthesis and growth of *Acorus calamus*. Front Plant Sci 2018, 9, 579.
[21] Duester L, Diaz-Bone RA, Kösters J, Hirner AV. Methylated arsenic, antimony and tin species in soils. J Environ Monit 2005, 7, 1186–1193.
[22] Shtangeeva I, Steinnes E, Lierhagen S. Uptake of different forms of antimony by wheat and rye seedlings. Environ Sci Pollut Res 2012, 19, 502–509.
[23] Tschan M, Robinson BH, Schulin R. Antimony in the soil–plant system–a review. Environ Chem 2009, 6, 106–115.
[24] Tighe M, Lockwood PV, Ashley PM, Murison RD, Wilson SC. The availability and mobility of arsenic and antimony in an acid sulfate soil pasture system. Sci. Total Environ 2013, 463, 151–160.
[25] Casiot C, Ujevic M, Munoz M, Seidel JL, Elbaz-Poulichet F. Antinony and arsenic mobility in a creek draining an antimony mine abandoned 85 years ago (upper Orb basin, France). Appl Geochem 2007, 22, 788–798.

[26] Rakshit S, Sarkar D, Punamiya P, Datta R. Antimony sorption at gibbsite–water interface. Chemosphere 2011, 84, 480–483.
[27] Leuz A-K, Monch H, Johnson CA. Sorption of Sb(III) and Sb(V) to goethite: Influence on Sb(III) oxidation and mobilization. Environ Sci Technol 2006, 40, 7277–7282.
[28] Guo X, Wu Z, He M, Meng X, Jin X, Qiu N, Zhang J Adsorption of antimony onto iron oxyhydroxides: Adsorption behavior and surface structure. J Hazard Mater 2014, 276, 339–345.
[29] Hockmann K, Lenz M, Tandy S, Nachtegaal M, Janousch M, Schulin R. Release of antimony from contaminated soil induced by redox changes. J Hazard Mater 2014, 275, 215–221.
[30] Scheinost AC, Rossberg A, Vantelon D, Xifra I, Kretzschmar R, Leuz A-K, Funke H, Johnson CA. Quantitative antimony speciation in shooting-range soils by EXAFS spectroscopy. Geochim Cosmochim Acta 2006, 70, 3299–3312.
[31] Xi J, He M, Lin C. Adsorption of antimony (V) on kaolinite as a function of pH, ionic strength and humic acid. Environ Earth Sci 2010, 60, 715–722.
[32] Xi J, He M, Kong L. Adsorption of antimony on kaolinite as a function of time, pH, HA and competitive anions. Environ Earth Sci 2016, 75, 136.
[33] Hockmann K, Schulin R. Leaching of antimony from contaminated soils. In: Magdi Selim, H, ed., Competitive Sorption and Transport of Heavy Metals in Soils and Geological Media. CRC Press, Taylor and Francis Group, 2012, 119–145.
[34] Tighe M, Lockwood P, Wilson S. Adsorption of antimony (V) by floodplain soils, amorphous iron (III) hydroxide and humic acid. J Environ Monit 2005, 7, 1177–1185.
[35] Fan Y, Zheng C, Huo A, Wang Q, Shen Z, Xue Z, He C. Investigating the binding properties between antimony (V) and dissolved organic matter (DOM) under different pH conditions during the soil sorption process using fluorescence and FTIR spectroscopy. Ecotoxicol Environ Saf 2019, 181, 34–42.
[36] Buschmann J, Sigg L. Antimony (III) binding to humic substances: Influence of pH and type of humic acid. Environ Sci Technol 2004, 38, 4535–4541.
[37] Verbeeck M, Warrinnier R, Gustafsson JP, Thiry Y, Smolders E. Soil organic matter increases antimonate mobility in soil: An $Sb(OH)_6$ sorption and modelling study. Appl Geochem 2019, 104, 33–41.
[38] Karimian N, Burton ED, Johnston SG, Hockmann K, Choppala G. Humic acid impacts antimony partitioning and speciation during iron (II)-induced ferrihydrite transformation. Sci Total Environ 2019, 683, 399–410.
[39] Bennett WW, Hockmann K, Johnston SG, Burton ED. Synchrotron X-ray absorption spectroscopy reveals antimony sequestration by reduced sulfur in a freshwater wetland sediment. Environ Chem 2017, 14, 345–349.
[40] Besold J, Kumar N, Scheinost AC, Lezama Pacheco J, Fendorf S, Planer-Friedrich B. Antimonite complexation with thiol and carboxyl/phenol groups of peat organic matter. Environ Sci Technol 2019, 53, 5005–5015.
[41] Xi J, He M, Wang K, Zhang G. Adsorption of antimony (III) on goethite in the presence of competitive anions. J Geochem Explor 2013, 132, 201–208.
[42] Rouwane A, Rabiet M, Grybos M, Bernard G, Guibaud G. Effects of NO_3^- and PO_4^{3-} on the release of geogenic arsenic and antimony in agricultural wetland soil: A field and laboratory approach. Environ Sci Pollut Res 2016, 23, 4714–4728.
[43] Biver M, Krachler M, Shotyk W. The desorption of antimony (V) from sediments, hydrous oxides, and clay minerals by carbonate, phosphate, sulfate, nitrate, and chloride. J Environ Qual 2011, 40, 1143–1152.

[44] Kolbe F, Weiss H, Morgenstern P, Wennrich R, Lorenz W, Schurk K, Stanjek H, Daus B. Sorption of aqueous antimony and arsenic species onto akaganeite. J Colloid Interface Sci 2011, 357, 460–465.
[45] Xi J, He M, Lin C. Adsorption of antimony (III) and antimony (V) on bentonite: Kinetics, thermodynamics and anion competition. Microchem J 2011, 97, 85–91.
[46] Egodawatta LP, Holland A, Koppel D, Jolley DF. Influence of soil phosphate on the accumulation and toxicity of arsenic and antimony in choy sum cultivated in individually and co-contaminated soils. Environ Toxicol Chem 2020, 39, 1233–1243.
[47] Egodawatta LP, Holland A, Koppel D, Jolley DF. Interactive effects of arsenic and antimony on *Ipomoea* aquatica growth and bioaccumulation in co-contaminated soil. Environ Pollut 2020, 259, 113830.
[48] Feng R, Wei C, Tu S, Tang S, Wu F. Simultaneous hyperaccumulation of arsenic and antimony in Cretan brake fern: Evidence of plant uptake and subcellular distributions. Microchem J 2011, 97, 38–43.
[49] Feng R, Wei C, Tu S, Tang S, Wu F. Detoxification of antimony by selenium and their interaction in paddy rice under hydroponic conditions. Microchem J 2011, 97, 57–61.
[50] Oorts K, Smolders E, Degryse F, Buekers J, Gasco G, Cornelis G, Mertens J. Solubility and toxicity of antimony trioxide (Sb_2O_3) in soil. Environ Sci Technol 2008, 42. 4378–4383.
[51] Mitsunobu S, Takahashi Y, Terada Y, Sakata M. Antimony (V) incorporation into synthetic ferrihydrite, goethite, and natural iron oxyhydroxides. Environ Sci Technol 2010, 44, 3712–3718.
[52] Burton ED, Hockmann K, Karimian N, Johnston SG. Antimony mobility in reducing environments: The effect of microbial iron (III)-reduction and associated secondary mineralization. Geochim Cosmochim Acta 2019, 245, 278–289.
[53] Johnson CA, Moench H, Wersin P, Kugler P, Wenger C. Solubility of antimony and other elements in samples taken from shooting ranges. J Environ Qual 2005, 34, 248–254.
[54] Okkenhaug G, Zhu Y-G, He J, Li X, Luo L, Mulder J. Antimony (Sb) and arsenic (As) in Sb mining impacted paddy soil from Xikuangshan, China: Differences in mechanisms controlling soil sequestration and uptake in rice. Environ Sci Technol 2012, 46, 3155–3162.
[55] Evangelou MW, Hockmann K, Pokharel R, Jakob A, Schulin R. Accumulation of Sb, Pb, Cu, Zn and Cd by various plants species on two different relocated military shooting range soils. J Environ Manage 2012, 108, 102–107.
[56] Kulp TR, Miller LG, Braiotta F, Webb SM, Kocar BD, Blum JS, Oremland RS. Microbiological reduction of Sb (V) in anoxic freshwater sediments. Environ Sci Technol 2014, 48, 218–226.
[57] Leverett P, Reynolds JK, Roper AJ, Williams PA. Tripuhyite and schafarzikite: Two of the ultimate sinks for antimony in the natural environment. Mineral Mag 2012, 76, 891–902.
[58] Wan X-M, Tandy S, Hockmann K, Schulin R. Changes in Sb speciation with waterlogging of shooting range soils and impacts on plant uptake. Environ Pollut 2013, 172, 53–60.
[59] Verbeeck M, Thiry Y, Smolders E. Soil organic matter affects arsenic and antimony sorption in anaerobic soils. Environ Pollut 2020, 257, 113566.
[60] Lovley DR. Microbial Fe (III) reduction in subsurface environments. FEMS Microbiol Rev 1997, 20, 305–313.
[61] Lovley D, Fraga J, Blunt-Harris E, Hayes L, Phillips E, Coates J. Humic substances as a mediator for microbially catalyzed metal reduction. Acta Hydroch Hydrob 1998, 26, 152–157.
[62] Borch T, Kretzschmar R, Kappler A, Van Cappellen P, Ginder-Vogel M, Voegelin A, Campbell K Biogeochemical redox processes and their impact on contaminant dynamics. Environ Sci Technol 2010, 44, 15–23.
[63] Mitsunobu S, Harada T, Takahashi Y. Comparison of antimony behaviour with that of arsenic under various soil redox conditions. Environ Sci Technol 2006, 40, 7270–7276.

[64] Hockmann K, Tandy S, Lenz M, Schulin R. Antimony leaching from contaminated soil under manganese-and iron-reducing conditions: Column experiments. Environ Chem 2014, 11, 624–631.
[65] Hockmann K, Tandy S, Lenz M, Reiser R, Conesa HM, Keller M, Studer B, Schulin R. Antimony retention and release from drained and waterlogged shooting range soil under field conditions. Chemosphere 2015, 134, 536–543.
[66] Filella M, Williams PA, Belzile N. Antimony in the environment: Knowns and unknowns. Environ Chem 2009, 6, 95–105.
[67] Filella M, Belzile N, Lett M-C. Antimony in the environment: A review focused on natural waters. III. Microbiota relevant interactions. Earth Sci Rev 2007, 80, 195–217.
[68] Frohne T, Rinklebe J, Diaz-Bone RA, Du Laing G. Controlled variation of redox conditions in a floodplain soil: Impact on metal mobilization and biomethylation of arsenic and antimony. Geoderma 2011, 160, 414–424.
[69] Duester L, Diaz-Bone RA, Kosters J, Hirner AV. Methylated arsenic, antimony and tin species in soils. J Environ Monit 2005, 7, 1186–1193.
[70] Wei C, Ge Z, Chu W, Feng R. Speciation of antimony and arsenic in the soils and plants in an old antimony mine. Environ Exp Bot 2015, 109, 31–39.
[71] Yang H, He M. Adsorption of methylantimony and methylarsenic on soils, sediments, and mine tailings from antimony mine area. Microchem J 2015, 123, 158–163.
[72] Ji Y, Mestrot A, Schulin R, Tandy S. Uptake and transformation of methylated and inorganic antimony in plants. Front Plant Sci 2018, 9, 140.
[73] Bowen HJM. Environmental Chemistry of the Elements. London, Academic Press, 1979.
[74] Feng R, Wei C, Tu S, Ding Y, Wang R, Guo J. The uptake and detoxification of antimony by plants: A review. Environ Exp Bot 2013, 96, 28–34.
[75] Shahid M, Khalid S, Dumat C, Pierart A, Niazi NK. Biogeochemistry of antimony in soil-plant system: Ecotoxicology and human health. Appl Geochem 2019, 106, 45–59.
[76] Cidu R, Biddau R, Dore E, Vacca A, Marini L. Antimony in the soil–water–plant system at the Su Suergiu abandoned mine (Sardinia, Italy): Strategies to mitigate contamination. Sci. Total Environ 2014, 497, 319–331.
[77] Ma C, He M, Zhong Q, Ouyang W, Lin C, Liu X. Uptake, translocation and phytotoxicity of antimonite in wheat (*Triticum aestivum*). Sci Total Environ 2019, 669, 421–430.
[78] Pierart A, Dumat C, Maes AQ, Sejalon-Delmas N. Influence of arbuscular mycorrhizal fungi on antimony phyto-uptake and compartmentation in vegetables cultivated in urban gardens. Chemosphere 2018, 191, 272–279.
[79] Tschan M, Robinson B, Johnson CA, Bürgi A, Schulin R. Antimony uptake and toxicity in sunflower and maize growing in Sb III and Sb V contaminated soil. Plant Soil 2010, 334, 235–245.
[80] Wu T-L, Cui X-D, Cui P-X, Ata-Ul-Karim ST, Sun Q, Liu C, Fan T-T, Gong H, Zhou D-M, Wang Y-J. Speciation and location of arsenic and antimony in rice samples around antimony mining area. Environ Pollut 2019, 252, 1439–1447.
[81] Antimony in drinking water, background document for development of WHO guidelines for drinking water quality. World Health Organization, 2003. (Accessed November 20 2020, 2020, at https://www.who.int/water_sanitation_health/water-quality/guidelines/chemicals/antimony/en/.)
[82] Mestrot A, Ji Y, Tandy S, Wilcke W. A novel method to determine trimethylantimony concentrations in plant tissue. Environ Chem 2016, 13, 919–926.
[83] Miravet R, Bonilla E, Lopez-Sanchez J, Rubio R. Antimony speciation in terrestrial plants. Comparative studies on extraction methods. J Environ Monit 2005, 7, 1207–1213.

[84] Maher WA, Krikowa F, Foster SD, Ellwood MJ, Bennett WW. Antimony measurements in environmental matrices: Seven considerations. J Anal At Spectrom 2018, 33, 706–712.
[85] Wan X, Lei M, Chen T. Interaction of As and Sb in the hyperaccumulator *Pteris vittata* L.: Changes in As and Sb speciation by XANES. Environ Sci Pollut Res 2016, 23, 19173–19181.
[86] Müller K, Daus B, Mattusch J, Stärk H-J, Wennrich R. Simultaneous determination of inorganic and organic antimony species by using anion exchange phases for HPLC–ICP-MS and their application to plant extracts of *Pteris vittata*. Talanta 2009, 78, 820–826.
[87] Müller K, Daus B, Mattusch J, Vetterlein D, Merbach I, Wennrich R. Impact of arsenic on uptake and bio-accumulation of antimony by arsenic hyperaccumulator *Pteris vittata*. Environ Pollut 2013, 174, 128–133.
[88] Wang X, Wei Z, Feng R, Tu S. The accumulation and transformation of antimony characteristics in *Pteris cretica* and *Brassica Juncea*. Acta Sci Circumstantiae 2014, 12, 720–727.
[89] Ren JH, Ma LQ, Sun HJ, Cai F, Luo J. Antimony uptake, translocation and speciation in rice plants exposed to antimonite and antimonate. Sci Total Environ 2014, 475, 83–89.
[90] Tisarum R, Lessl JT, Dong X, de Oliveira LM, Rathinasabapathi B, Ma LQ. Antimony uptake, efflux and speciation in arsenic hyperaccumulator *Pteris vittata*. Environ Pollut 2014, 186, 110–114.
[91] Huang L, Li WC, Tam NFY, Ye Z. Effects of root morphology and anatomy on cadmium uptake and translocation in rice *(Oryza sativa* L.). J Environ Sci 2019, 75, 296–306.
[92] Liu Y, Lv H, Yang N, Li Y, Liu B, Rensing C, Dai J, Fekih IB, Wang L, Mazhar, SH. Roles of root cell wall components and root plaques in regulating elemental uptake in rice subjected to selenite and different speciation of antimony. Environ Exp Bot 2019, 163, 36–44.
[93] Wu Q, Feng R, Guo J, Wang R, Xu Y, Fan Z, Mo L. Interactions between selenite and different forms of antimony and their effects on root morphology of paddy rice. Plant Soil 2017, 413, 231–242.
[94] Walker TS, Bais HP, Grotewold E, Vivanco JM. Root exudation and rhizosphere biology. Plant Physiol 2003, 132, 44–51.
[95] Cui X-D, Wang Y-J, Hockmann K, Zhou D-M. Effect of iron plaque on antimony uptake by rice (*Oryza sativa* L.). Environ Pollut 2015, 204, 133–140.
[96] Long J, Zhou D, Li B, Zhou Y, Li Y, Lei M. The effect of an antimony resistant bacterium on the iron plaque fraction and antimony uptake by rice seedlings. Environ Pollut 2020, 258, 113670.
[97] Cai F, Ren J, Tao S, Wang X. Uptake, translocation and transformation of antimony in rice (*Oryza sativa* L.) seedlings. Environ Pollut 2016, 209, 169–176.
[98] Huang Y, Chen Z, Liu W. Influence of iron plaque and cultivars on antimony uptake by and translocation in rice (*Oryza sativa* L.) seedlings exposed to Sb (III) or Sb (V). Plant Soil 2012, 352, 41–49.
[99] Ferrol N, Tamayo E, Vargas P. The heavy metal paradox in arbuscular mycorrhizas: From mechanisms to biotechnological applications. J Exp Bot 2016, erw403.
[100] Wei Y, Chen Z, Wu F, Hou H, Li J, Shangguan Y, Zhang J, Li F, Zeng Q. Molecular diversity of arbuscular mycorrhizal fungi at a large-scale antimony mining area in southern China. J Environ Sci 2015, 29, 18–26.
[101] Wei Y, Su Q, Sun Z, Shen Y, Li J, Zhu X, Hou H, Chen Z, Wu FC. The role of arbuscular mycorrhizal fungi in plant uptake, fractions, and speciation of antimony. Appl. Soil Ecol 2016, 107, 244–250.
[102] Feng R, Lei L, Liu B, Chen W, Zhang R, Wang L, Li Y, Su J, Dai J, Wang R. Effects of different inhibitors such as malonic acid, Na_3PO_4 and $HgCl_2$ on uptake of different forms of antimony in rice plant. Plant Soil 2019, 445, 259–271.

[103] Kamiya T, Fujiwara T. Arabidopsis NIP1; 1 transports antimonite and determines antimonite sensitivity. Plant Cell Physiol 2009, 50, 1977–1981.
[104] Mukhopadhyay R, Bhattacharjee H, Rosen BP. Aquaglyceroporins: Generalized metalloid channels. Biochimica et Biophysica Acta 2014, 1840, 1583–1591.
[105] Meharg AA, Jardine L. Arsenite transport into paddy rice (*Oryza sativa*) roots. New phytologist 2003, 157, 39–44.
[106] Bhattacharjee H, Mukhopadhyay R, Thiyagarajan S, Rosen BP. Aquaglyceroporins: Ancient channels for metalloids. J Biol 2008, 7, 1–6.
[107] Zheng J, Takeda A, Furuta N. Investigating the electrospray mass spectra of inorganic and organic antimony compounds. J Anal At Spectrom 2001, 16, 62–67.
[108] Craig P, Jenkins R, Dewick R, Miller D. Trimethylantimony generation by *Scopulariopsis brevicaulis* during aerobic growth. Sci Total Environ 1999, 229, 83–88.
[109] Zhu Y, Wu Q, Lv H, Chen W, Wang L, Shi S, Yang J, Zhao P, Li Y, Christopher R. Toxicity of different forms of antimony to rice plants: Effects on reactive oxidative species production, antioxidative systems, and uptake of essential elements. Environ Pollut 2020, 114544.
[110] McGrath SP, Zhao F-J. Phytoextraction of metals and metalloids from contaminated soils. Curr Opin Biotechnol 2003, 14, 277–282.
[111] Kabata-Pendias A. Trace Elements in Soils and Plants, 34th Edition. Boca Raton, CRC Press, 2010.
[112] Shtangeeva I, Bali R, Harris A. Bioavailability and toxicity of antimony. J Geochem Explor 2011, 110, 40–45.
[113] Duester L, Van Der Geest H, Moelleken S, Hirner A, Kueppers K. Comparative phytotoxicity of methylated and inorganic arsenic-and antimony species to *Lemna minor, Wolffia arrhiza* and *Selenastrum capricornutum*. Microchem J 2011, 97, 30–37.
[114] Ortega A, Garrido I, Casimiro I, Espinosa F. Effects of antimony on redox activities and antioxidant defence systems in sunflower (*Helianthus annuus* L.) plants. PloS One 2017, 12, e0183991.
[115] Feng R, Wei C, Tu S, Liu Z. Interactive effects of selenium and antimony on the uptake of selenium, antimony and essential elements in paddy-rice. Plant Soil 2013, 365, 375–386.
[116] Chai L-Y, Mubarak H, Yang Z-H, Yong W, Tang C-J MN. Growth, photosynthesis, and defense mechanism of antimony (Sb)-contaminated *Boehmeria* nivea L. Environ Sci Pollut Res 2016, 23, 7470–7481.
[117] Feng R, Wei C, Tu S, Wu F, Yang L. Antimony accumulation and antioxidative responses in four fern plants. Plant Soil 2009, 317, 93–101.
[118] Pan X, Zhang D, Chen X, Bao A, Li L. Antimony accumulation, growth performance, antioxidant defense system and photosynthesis of *Zea mays* in response to antimony pollution in soil. Water Air Soil Pollut 2011, 215, 517–523.
[119] Zhang D, Pan X, Mu G, Wang J. Toxic effects of antimony on photosystem II of *Synechocystis* sp. as probed by in vivo chlorophyll fluorescence. J Appl Phycol 2010, 22, 479–488.

Gudny Okkenhaug and Andreas Botnen Smebye

Chapter 8 Shooting range contamination: antimony occurrence, transport, and fate

8.1 Introduction

The European Union (EU) has more than 650,000 contaminated sites registered [1], possibly up to 2.8 million sites including dark numbers, which puts pressure on the environment. Military activities represent one of the main contaminant sources (3.4%, on average of 22 European Environment Agency (EEA) countries/regions) [2], covering, for example, small arms training activities, artillery/mortar ranges, battle runs, rocket ranges, and grenade courts. Antimony pollution at these sites mainly stems from bullets from small arms with a lead and antimony alloy core covered with an alloy of copper and zinc [3]. Lead bullets exploit antimony as a hardening agent, making up 2–8% of their composition [4] (Fig. 8.1). Although there is a relatively low amount of antimony in bullets, the United States' 3,000 military and 9,000 civilian shooting ranges result in an estimated annual antimony deposition of 1,900 t and 60,000 t lead [5]. In smaller countries such as Switzerland with more than 2,000 shooting ranges, up to 25 t of antimony and 500 t of lead enter the pedosphere every year [6].

Depending on climate and soil conditions, bullets are oxidized, mobilizing metals and antimony. This mobilization makes shooting ranges a significant source for pollution and environmental concern [8, 9] over a long period of time [10]. This has led to increased attention on antimony at small arms shooting ranges in recent years [11] and will thus be the focus of this chapter.

Gudny Okkenhaug, Norwegian Geotechnical Institute, (NGI), Department of Geotechnics and Environment, Oslo, Norway; Norwegian University of Life Sciences (NMBU), Faculty of Environmental Sciences and Natural Resource Management, Ås, Norway, gudny.okkenhaug@ngi.no
Andreas Botnen Smebye, Norwegian Geotechnical Institute, (NGI), Department of Geotechnics and Environment, Oslo, Norway, andreas.botnen.smebye@ngi.no

https://doi.org/10.1515/9783110668711-008

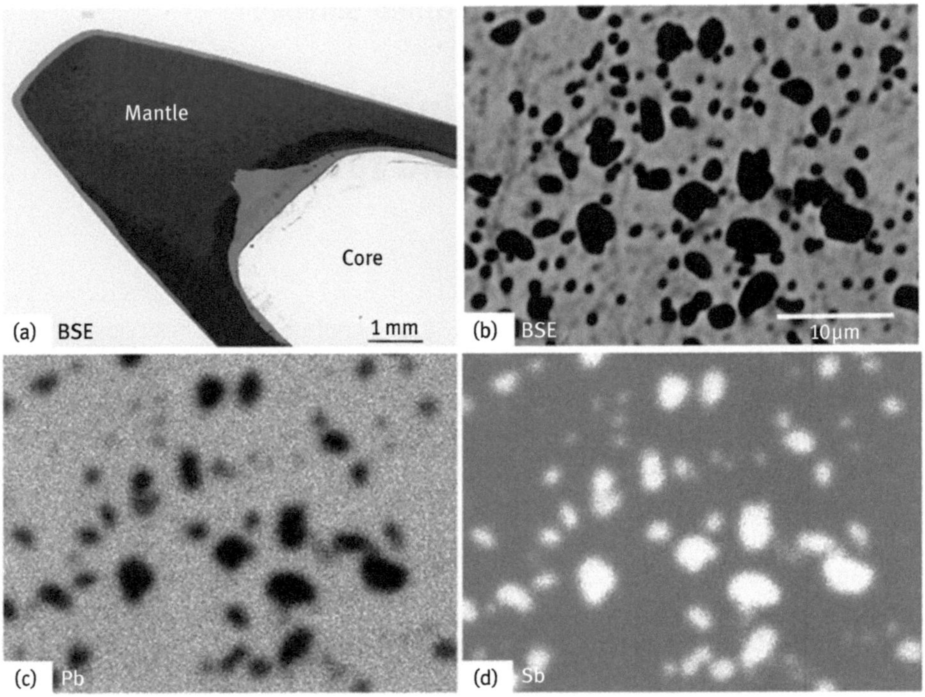

Fig. 8.1: Back-scattered electron (BSE) images of lead bullet: (a) front part in longitudinal section; (b) pristine bullet core (lead matrix in gray, antimony inclusions in black); (c) lead distribution map area shown in (b); antimony distribution map of area shown in (b) [7]. Reprinted with permission from Ackermann et al. (2009).

8.2 Occurrence of antimony at small arms shooting ranges

8.2.1 Distribution of antimony in soil

Contaminant distribution and spreading at small arms shooting ranges are closely related to their design, which can be divided into three areas: (1) firing point; (2) target area and backstop berm; and (3) fire line, the area between the firing point and the target area (Fig. 8.2). A *basis shooting range* has fixed firing points and target area with varying fire line length (typically between 25 and 300 m), whereas *field shooting ranges* have mobile firing points and target areas within the natural terrain with or without backstop berm. High metal and antimony concentrations are normally found around the target areas and backstop berm, while the remaining vicinity is usually more diffusely polluted.

High total concentrations of antimony in shooting range soils, up to 2,200 mg kg^{-1}, or on a microscale even higher, were reported for instances from Switzerland, Norway,

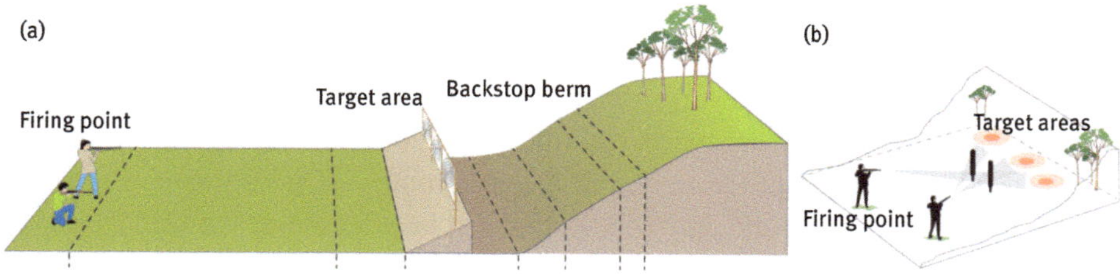

Fig. 8.2: Illustration of small arms shooting ranges: (a) Basis shooting range with firing point, target area, and backstop berm. (b) Field shooting range with firing point and target areas [12]. Reprinted with permission from Norwegian Defence Research Establishment; FFI-report 2010/00116, 2010.

Australia, South Korea, and the United States (Tab. 8.1). Heterogeneous distribution of spent bullets and fragments over large areas explains some of the observed variation between these sites [9, 13]. Bullets usually penetrate only the upper soil (0–10 cm) [14], with an exception for peatlands where high concentrations of antimony were observed down to 20–30 cm under the surface as well [15, 16].

Tab. 8.1: Concentration of antimony and lead in soil and soil porewater found at military shooting ranges.

Conditions, soil type	**Soil concentrations / mg kg^{-1}**		**Soil porewater / µg L^{-1}**		**Reference**
	Sb	**Pb**	**Sb**	**Pb**	
Mineral soil, Poland	<0.01–89.6	2.4–29,000	>80–995[a]	< -654[a]	[17]
Topsoil, acidic mires, Norway	0.02–830	260–13,000	2.9–150[b]	3.0–2,500[b]	[16]
Mineral soil, backstop berm, Norway	40–123	356–1,112	19–349[b]	7–1,495[b]	[13]
Mineral soil, backstop berm, Switzerland	35–17,500	1,400–515,800	–	–	[18]
Mineral soil, Korea	164	17,468	–	–	[19]
Backstop berm, Australia	6.57–252	399–10,403	–	–	[14]
Backstop berm, USA	1–517	11–4,675	–	–	[20]
Mineral soil, backstop berm, Canada	10–1,413	423–66,972	7.2–12.2[b]	6.6–15[b]	[21]

[a] laboratory incubation; [b] in situ.

8.2.2 Weathering of bullets and solid phase speciation

Spent bullets are subjected to corrosion and weathering, depending on climate (temperature, precipitation), soil chemical conditions (e.g., pH, Eh, organic matter content, cation exchange capacity), and mechanical soil treatment [22]. Furthermore, the extent of bullet fragmenting and surface area that can be attacked by corrosion strongly affects the weathering rate [23]. Coarser materials such as sand contain larger particles, with rough surfaces where bullets collide and fragment, than in, for example, a finer silty loam [24]. The exposure of bullet material to weathering processes causes zerovalent metal(loid)s to oxidize, typically forming a weathering crust comprising secondary mineral phases encapsulating surfaces and acting as a control on metal (loid) release to solution [25].

Characterization of polluted shooting range soils has shown that antimony predominantly occurs as either metallic Sb(0) or Sb(V). Using antimony K-edge X-ray absorption spectroscopy (XAS) on soils from Swiss shooting ranges, Scheinost et al. [26] identified only these two antimony species, including unweathered metallic Sb(0) and oxidized Sb(V). They found that oxidized Sb(V) likely consisted of inner-sphere sorption complexes on Fe-oxides, with edge and corner-sharing configuration occurring simultaneously. The presence of Sb(V) in all samples suggests that this species predominates after weathering of metallic Sb(0) in oxic soils, and no occurrence of Sb(III) was found. The association of antimony with Fe-oxides adjacent to weathering bullets was also suggested by Ackermann et al. [7]. They detected a cemented zone around the bullet consisting predominately of Fe-oxides, containing 0.04–0.63 wt.% Sb_2O_5.

Extraction experiments with shooting range soils have also shown antimony binding to Fe-oxides as well as to organic matter. Oxalate extractions on contaminated shooting range soils have indicated that antimony is associated with amorphous Fe-oxides [13]. This high association of antimony to iron (hydr)oxides is well known from other soil systems [27]. In soils with high levels of organic matter such as peatlands, a higher extraction of antimony from the oxidizable organic matter using hydrogen peroxide than oxalate implies that organic matter also forms inner-sphere complexes with antimony [15, 16]. Steely et al. [28] used chromatographic methods (size exclusion chromatography-inductively coupled plasma-mass spectrometry, SEC-ICP-MS) to also find that antimony binds to organic matter.

Studies on bullets from small arms shooting ranges have indicated weathering rates between 0.25 and 1% annually; complete bullet dissolution would then occur within 30–300 years [22]. Okkenhaug et al. [10] carried out a column study on soil from a shooting range backstop berm (123 mg Sb kg^{-1}) where the soil was flushed with water for 64 days representing approximately 140 years of infiltration. The dissolution rate of antimony under the prevailing conditions was calculated to 0.7 mg Sb (kg soil $d)^{-1}$. Thus, 34% of the total amount of antimony leached out during the experiment [10]. Corroding of bullets at shooting ranges represents a long-term source of antimony mobilization in the environment.

8.3 Fate of antimony in soil and soil water at contaminated shooting ranges

8.3.1 Redox speciation of antimony in shooting range soils

Oxidation of metallic Sb(0) in bullets to Sb(III) proceeds relatively fast (days–weeks), as determined by Ilgen et al. [29] in a laboratory system. Exclusive occurrence of Sb(V) in solid phase analysis of oxic shooting range soil confirms this finding [7, 26]. Correspondingly, antimony occurs predominantly as Sb(V) in porewater from oxic shooting range soils [6, 13]. Equivalent findings were found by Barker et al. [24] who carried out a four-year field study on the early-stage weathering of bullets in various soils (sand, sandy loam, loamy sand, silt loam). Sb(V) was the dominant aqueous product in the berm runoff. The presence of aqueous Sb(III) was observed only in a silty loam with low permeability and low redox potential. Antimony binding to organic matter has been shown to facilitate the oxidation of Sb(III) to Sb(V) [28] supporting the described field observations. Similar results of predominance of Sb(V) were found by Lewinska et al. [30] incubating shooting range soils under oxic and anoxic conditions and monitoring content and composition of antimony in porewater. Whereas Sb(V) dominated at concentrations around 80–900 µg L^{-1} under oxic conditions, Sb(III) increased constantly upon prolonged incubation under anoxic conditions.

The redox conditions also affect antimony transformation by microorganisms to methylated species, for example, trimethylantimony (TMSb). In a laboratory incubation experiment, Grob et al. [31] quantified up to 1.35 mg kg^{-1} of TMSb in a shooting range soil with a total antimony concentration of 223.3 mg kg^{-1}. In lesser contaminated soil samples (11.2–34.9 mg Sb kg^{-1}), TMSb concentrations were substantially lower (0.06–0.21 mg kg^{-1}). Under flooded conditions with increased organic matter (2% cow dung amendment), the TMSb ratio increased considerably (making up 7.8% of total antimony in soil solution), illustrating this process dependency on redox conditions [31]. The overall effect of antimony biomethylation on antimony fate in shooting range soil is, however, largely unknown.

8.3.2 Leaching of antimony in shooting range soils

Weathering of bullets under natural climate conditions will lead to antimony leaching to soil porewater and/or surface runoff. In the previously mentioned field study of Barker [24], they found that lead and antimony runoff concentrations depend on soil type. On a controlled trial with a small arms shooting range backstop berm (9.6–281.3 mg kg^{-1}) constructed with different soil types, runoff was sampled for five consecutive summers. The highest antimony runoff concentrations were found in sandy soil (Fig. 8.3), whereas the lowest concentrations were found in silty loam.

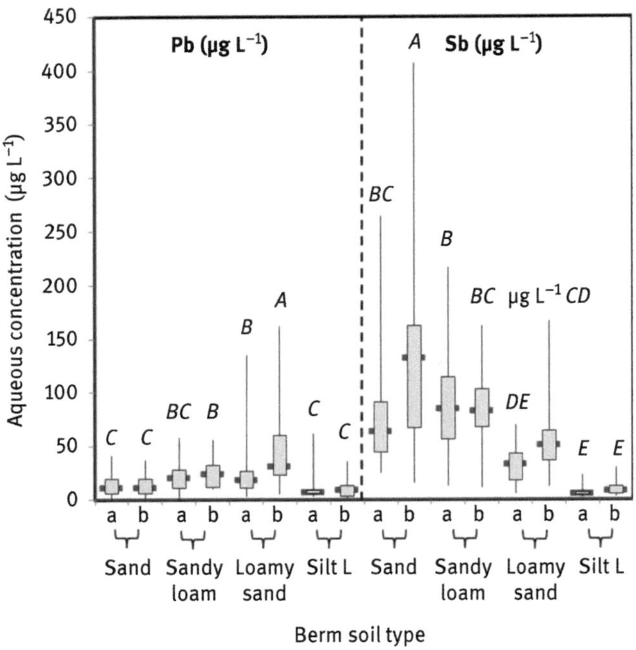

Fig. 8.3: Box plot diagrams of concentrations of lead (left half) and antimony (right half) in runoff from backstop berm as a function of soil type (sandy, sandy loam, loamy sand and silt loam) for the duplicate of contaminated test berms (a and b). Horizontal grey lines indicate median values during four years of samples. The grey boxes depict the lower (25th percentile) and upper (75th percentile) quartiles [24]. Reprinted with permission from Barker et al. (2019).

The sandy soil was assumed to be promoted by favorable conditions for oxidation and a relatively high pH (8.38) yielding the highest concentrations observed. For comparison, the aqueous concentrations of antimony were greater than that of lead in all soil types, indicating potential high antimony mobility, even though the total amount of antimony in bullets is ~2 orders of magnitude lower than that of lead.

Comparable antimony porewater concentrations were found in a four-year field study on leaching from a silty sand backstop berm in Norway [13]. They showed by using suction lysimeters that leaching remained high over four years (19–349 μg L^{-1}) and concluded that this was due to the metallic bullet fragments being a constant source of antimony. Furthermore, they found a seasonal trend in porewater antimony concentrations that were highest in the summer. They attributed this to higher weathering rates of bullets at warmer temperatures along with seasonal differences in soil water content. Hockmann et al. [32] found a similar seasonal trend in antimony leaching under oxic conditions in large field lysimeters filled with alkaline silty loam shooting range soil. Their findings also indicated increased antimony leaching due to increased dissolved organic carbon (DOC) production in the summer competing with antimony for binding sites. Elevated lead and antimony concentrations were also found in in situ porewater from the firing line of six small arms shooting ranges located in a minerotrophic and acidic (pH 3.7–4.9) peatland [16]. Here concentrations of antimony between 3–150 μg L^{-1} were found, showing similar mobility under other conditions than for backstop berms.

As previously described, Sb(V) may reduce to Sb(III) under water logging and reducing conditions affecting antimony leaching behavior. Tandy et al. [33] performed an outdoor lysimeter experiment over several years to compare antimony leaching from a calcareous moderately contaminated shooting range soil (21 mg Sb kg^{-1}; 520 mg Pb kg^{-1}) under drained and prolonged waterlogging (1.5–2.75 y), followed by a 1.5-year period of soil re-oxidation (Fig. 8.4). Short-term waterlogging immobilized antimony, attributed to Sb(V) reduction to Sb(III), which binds more effectively to iron (hydr)oxides at the prevailing pH conditions. However, prolonged waterlogging resulted in a release of antimony into solution as iron (hydr)oxides became reductively dissolved. Furthermore, draining and re-oxidation of the system mobilized antimony, but soil solution and leachate concentrations remained lower than for the control kept under constantly drained conditions (Fig. 8.4). Similar results were found in microcosm experiments on an alkaline shooting range soil [6]. Here, a transition to reducing conditions first led to the immobilization of antimony by sorption to iron (hydr)oxides, as

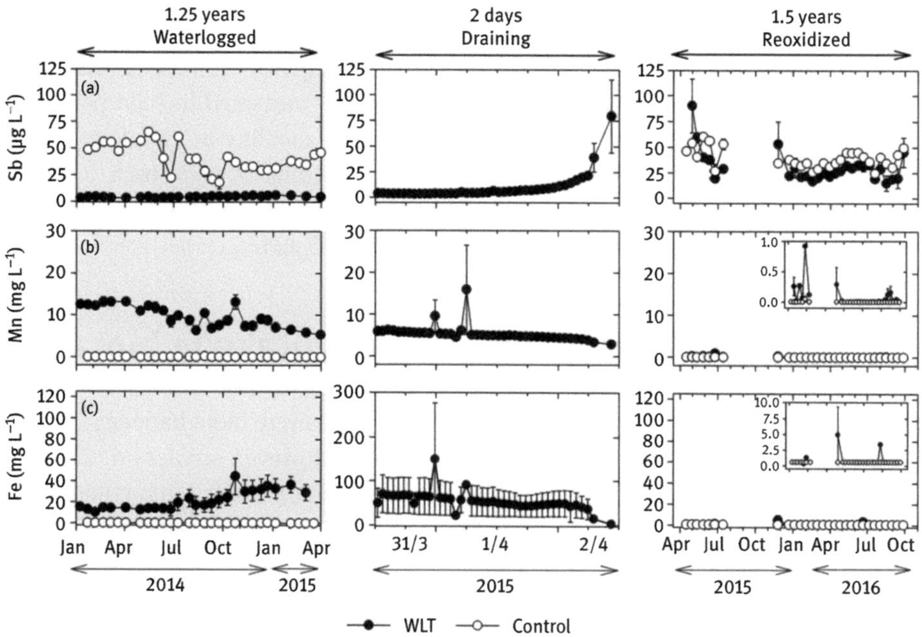

Fig. 8.4: Concentrations of dissolved antimony (a), manganese (b) and iron (c) in leachate samples of the control (open symbols) and water-logged regime, WLT (closed symbols) lysimeter subject to waterlogging and reoxidation. Mean ± standard error, n = 2 [33]. Reprinted with permission from Tandy et al. (2018).

Sb(V) was converted to Sb(III). When reducing conditions continued, the previously sorbed Sb(III) was gradually released into solution due to reductive dissolution of the Fe-oxides. Changes in redox conditions can either reduce antimony porewater concentrations by increased sorption or mobilize antimony by dissolution of Fe-oxides.

Laboratory leaching studies have confirmed high antimony leaching from contaminated shooting range soils. Water extracts (batch tests) on backstop berm soils from various shooting ranges gave substantial antimony eluate concentrations of 450–2,200 µg L^{-1} [10]. Long-term column test carried out on one of the backstop berm samples showed initial leachate concentrations of approximately 700 µg L^{-1}, reducing over time to 300 µg L^{-1}, also reflecting severe antimony pollution [10] on a similar scale as observed in the field.

Metals and antimony may spread from contaminated soils and groundwater to surface water draining the shooting range area. Heier et al. [34] studied the speciation of metals and antimony in a shooting range runoff stream during a period of 21 days using size and charge fractionation systems. While lead concentrations ranged between 15 and 56 µg L^{-1}, mainly present as particulate and colloidal high molecular mass (HMM) species (complexed to dissolved organic matter), antimony was mainly present as low molecular matter (LMM) species in concentrations between <3 and 12 µg L^{-1}. Similar observations were made in a stream draining a shooting range in a peatland with lightly acidic pH (lead and antimony concentrations of 6.9 ± 0.1 µg L^{-1} and 7.4 ± 0.1 µg L^{-1}) [15]. The antimony concentrations measured downstream of the shooting range were high compared with lead, with Pb-Sb ratios of 0.9 (filtered samples) versus a Pb-Sb ratio of 13 measured in solid peat samples in the upper peat layer, indicating a high relative mobility of antimony in this peat system [15]. However, substantially higher concentrations have been found in stream water. In the above-mentioned study of Mariussen et al. [16], they observed lead and antimony concentrations up to 180 µg L^{-1} and 65 µg L^{-1}, respectively.

8.4 Antimony bioavailability and uptake in biota

Exposure of antimony in contaminated shooting range soils to biota happens through uptake of polluted porewater or soil, as well as runoff. However, studies on antimony bioavailability and uptake in biota associated with shooting range contamination are scarce in literature.

8.4.1 Bioavailability

The bioavailable fraction of contaminants can be divided in a direct bioavailable fraction (immediate available) and a bioaccessible (potential available) fraction [35]. Frequently used methods to measure bioavailable metal and metalloid soil fractions are based on both chemical extractions and mechanistic geochemical models, as well as passive samplers such as diffusive gradient in thin films (DGT). In these passive samplers, free and labile ions diffuse from the soil water through a membrane filter and an inert hydrated gel before sorbing to a resin [36]. The hydrated gel acts as a filter by preventing larger molecules and colloids from diffusing through. The sorbent in the DGT, a chelex resin for metal cation binding and an Fe oxide gel for metalloid binding, works as a sink for metals in the porewater and therefore leads to replenishment of labile metals from the solid phase pool. It is generally assumed that the freely dissolved and labile fractions are more bioavailable than particle-bound elements [37].

In a study conducted by Okkenhaug et al. [15] at a small arms shooting range located in a peatland area, the bioavailable fraction of metals and antimony was measured using both chemical extractions and DGTs. The amount of $CaCl_2$-extractable lead (0.1–0.4%) and antimony (0.6–1.5%) was low. Extracting the soil samples at pH 3.75 with NH_4 lactate solution gave also small extractable fractions of antimony (1–5%) relative to total solid concentrations, whereas lead was somewhat higher (6–17%) for comparison. Elements extracted with these methods would include weakly sorbed species, particularly those retained on the soil surface by relatively weak electrostatic interactions and species released by ion exchange processes [38], thus readily mobile. Although chemical extractions showed low antimony bioavailability relative to total solid concentrations, DGT concentrations of antimony in peat soil from the fire line were measured to 11 ± 5 and 23 ± 7 µg L^{-1} in the top layer (0–15 cm). DGT concentrations of lead were significantly lower, 4 ± 2 and 1.5 ± 0.1 µg L^{-1} respectively, emphasizing the differences in availability between these two elements [15].

8.4.2 Uptake of antimony in plants

Small arms shooting ranges are mainly located in remote areas, often used as pasture for sheep and cattle during times of no shooting activity or when the site has been decommissioned [13, 39]. Robinson et al. [40] investigated the accumulation of metal and antimony by vegetation growing on a backstop berm with antimony soil concentrations between 629 and 8,230 mg kg^{-1} (geometric mean). For most of the 10 plant species considered, antimony leaf concentrations were below 2 mg kg^{-1} with a bioaccumulation factor (BAF) < 0.001, indicating low accumulation of antimony from the rhizosphere [40]. Substantial lower soil concentrations of antimony were observed in a field lysimeter experiment carried out on a calcareous and a weakly acidic shooting range soil (approx. 19 mg kg^{-1}); the highest mean shoot concentration was reached by *R. acetosa* (0.3 ± 0.2 mg kg^{-1}) and *P. aviculare* (0.31 ± 0.04 mg kg^{-1}). All plants on the calcareous soil accumulated 1.5–2.6-fold more antimony (0.195–0.275 mg kg^{-1}) than on the acidic soil (0.071–0.128 mg kg^{-1}), most likely due to the higher antimony solubility in soils with high pH value. All the above-ground biomass of various plant species remained below the Swiss threshold value for fodder plants [39].

Hockmann et al. [41] conducted field lysimeter studies with various pasture plants (including *Lolium perenne* and *Trifolium repens, Plantago lanceolate, Rumex obtusifolius*) grown on a calcareous shooting range soil (21 ± 1 mg kg^{-1}) under waterlogged and drained conditions. Antimony uptake showed a high variability among the sampled plant species, and all plant species exhibited increasing trends in concentration in both treatments over the growing period (Fig. 8.5). Shoot antimony concentrations differed by a factor of 60 between plant species but remained in all cases at levels <0.3 mg kg^{-1}. Under waterlogged conditions, a relatively high antimony plant uptake at low total antimony porewater concentrations was observed, indicating a preferential uptake of Sb(III) [41]. This preferential uptake of Sb(III) by the forage grass *Lolium perenne* in a calcareous shooting range soil was also found by Wan et al. [5] in a pot

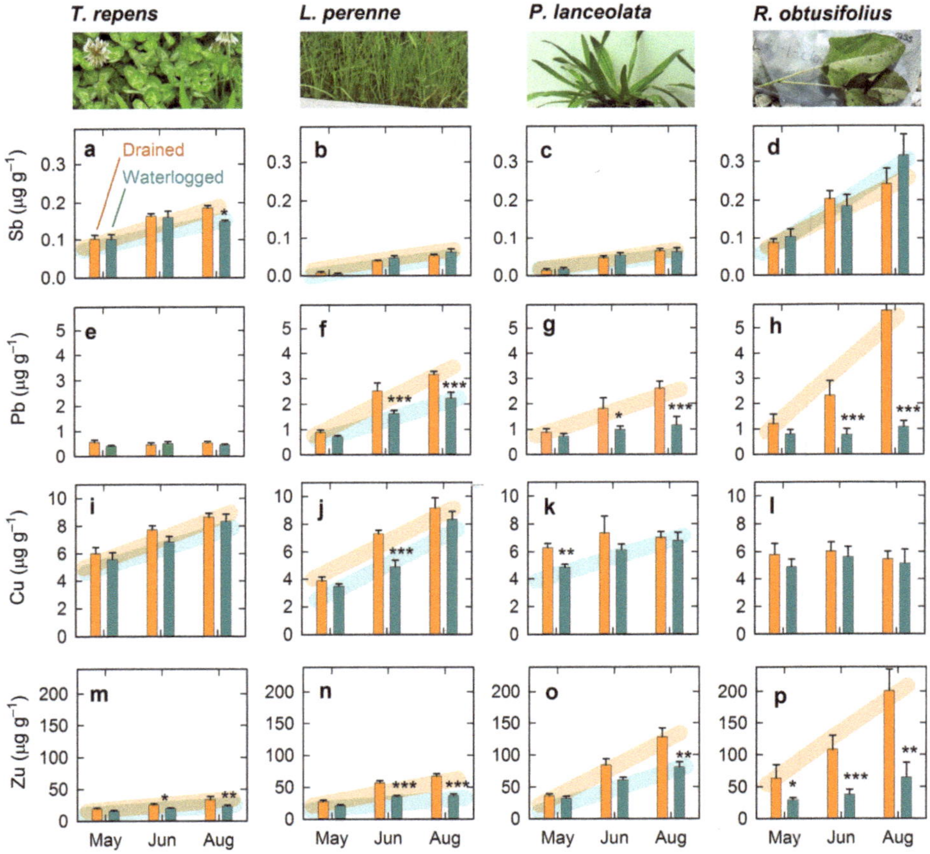

Fig. 8.5: Dry weight concentrations of metals and antimony in shoots sampled from the drained (orange columns) and waterlogged (blue bars) lysimeters in May, June, and August 2013. Error bars represent standard errors (n = 6–8). Linear trends significant to $p < 0.05$ are indicated by orange and blue shading. Asterisks denote significant differences between treatments (* = $p < 0.05$, ** = $p < 0.01$, *** = $p < 0.001$) [41]. Reprinted with permission from Hockmann et al. (2018).

experiment investigating redox speciation. Their study on the uptake of antimony from a shooting range soil by two pasture grass species showed that reducing conditions caused by waterlogging increased antimony shoot concentrations of *L. perenne* by ~tenfold [5]. This was attributed to the preferential uptake of Sb(III) compared to Sb(V) [42], which may have been a major antimony redox species in the solid phase under the waterlogging regime [6]. Based on the study, antimony uptake by plants was considered to be low under both drained and waterlogged conditions, despite the higher bioavailability of Sb(III) to many plant species.

The number of studies addressing the uptake of antimony in plants from shooting range soils is low. However, based on results from the existing studies, the risk of antimony transfer from a shooting range soil into the food chain can be considered as low.

8.5 Remediation options for shooting range soils

Shooting ranges with high environmental load call for remediation measures to prevent spreading and exposure of contaminants. Such remediation measures are often conducted in connection with shooting range closure and land use change. Widespread and heterogeneous contamination, in some cases spread over several square kilometers [15, 43], drives cost and increases complexity. In addition, several shooting ranges are in environments vulnerable to intervention (e.g., peatlands), and it may take several decades up to centuries to restore damaged land to their original state [44].

Management of shooting range soils may include implementation of multiple strategies to reduce contaminant burden and stabilize contaminants. These strategies can broadly be categorized as i) mobilization (e.g., soil washing, enhanced phytoremediation), ii) removal (e.g., separation, excavation), and iii) stabilization (e.g., chemical stabilization) [11]. The mobilization strategy using soil washing is one major physical technique used to reclaim shooting range soil, where liquids (water occasionally with solvents) are used to enhance the removal of contaminants from soils [45]. However, current studies on the effect on removal of antimony from shooting range soils by soil washing are very scarce in literature. Little is also known of the effect on antimony of phytoremediation, which is an established, but poorly documented, method for shooting range soils [46]. Phytoextraction is one phytoremediation in situ technique, where plants that have the capacity to absorb and retain trace elements, hyperaccumulators, are used to extract contaminants from the soil porewater. In a literature review of the antimony soil-plant system, the authors concluded that phytoextraction for antimony contaminated soil is not a realistic option unless high-biomass antimony hyperaccumulator plants are found [47]. Currently, only a limited number of plant species have been tested to salvage shooting range soils, and no studies on the effect of phytoextraction on antimony are found in literature.

The most common remediation of shooting ranges consists of removal by excavation and landfilling the soil as waste. High lead soil concentrations and antimony leaching exceeding the limits of ordinary waste landfills (ANNEX II. Council Directive 1999/31/EC on the landfill of waste) [48] often require disposal at a hazardous waste landfill. To reduce costs, shooting range soil might be treated (e.g., chemical stabilization) prior to landfilling to comply with the leaching criteria for ordinary waste landfills, an alternative that has received increased intention.

Chemical stabilization is by far the remediation strategy that has received most focus in the scientific literature, also including its effects on antimony. Adding phosphate stabilizes contaminated shooting range soils where lead is the main target contaminant by formation of pyromorphite compounds ($Pb_{10}(PO_4)_6X_2$, X = Cl, Br, F, OH) with low solubility [49]. However, such phosphate treatment might mobilize antimony by competitive sorption from phosphate in the soil [50, 51]. Though this is not always the case as shown by Sanderson et al. [52] using rock phosphate to successfully stabilize both lead and antimony in four different shooting range soils.

As an alternative amendment, iron-based sorbents can potentially immobilize both cationic (e.g., lead, copper) and anionic (e.g., antimony) contaminants [13, 43, 53]. In a laboratory batch experiments on slightly acidic and alkaline shooting range soils, Tandy et al. [53] applied two redmud-based amendments consisting mainly of iron and aluminum oxides (2% and 20% addition) alone and in combination with two reducing agents (5% zerovalent iron, Fe(0) and 5% iron sulfate, $FeSO_4 \cdot 7H_2O$). Treatments containing a mixture of iron and aluminum oxides mixed with iron sulfate (VS-IS in Fig. 8.6) showed a substantial decrease in leaching of both lead and antimony (up to ~98%) even at dosages as low as 2%. Iron-based sorbents are thus able to reduce leaching from shooting range soils below the European leaching criteria for landfilling of inert waste landfills.

Effective binding of antimony by iron-based sorbents was also found by Okkenhaug et al. [10] using a commercially available iron oxyhydroxide (CFH-12) and Fe(0) grit to chemically stabilize backstop berm soils in both laboratory batch and saturated column tests. These sorbents mixed with soil (0.5 and 2% w/w) showed sorption of antimony between 46 and 92% based on eluate concentration. To prevent mobilization of lead, copper, and zinc liming was necessary when adding CFH-12, which reacted slightly acidic. Column tests showed significant retention of antimony (89–98%) in soil amended with CFH-12 (2%) mixed with limestone (1%) compared to unamended soil. The sorption capacity of soils amended with Fe(0) (2%) increased steadily up to 72% over the duration period of the column test (64 days), most likely due to the gradual oxidation of Fe(0) to iron oxyhydroxides. To provide a closer simulation of field conditions, Mariussen et al. [43] investigated the effect of the same amendments using unsaturated column test simulation precipitation over a 2.5-year period. CFH-12 (2%) and limestone (1%), in addition to a treatment with preoxidized zerovalent iron grit (2% and 4%), were added to various backstop berm soils. The mixture of CFH-12 and limestone gave, compared to the unamended

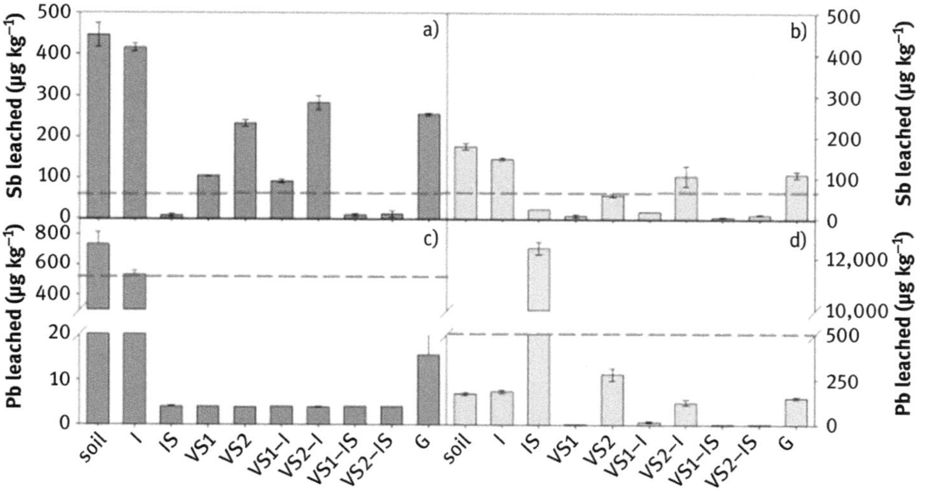

Fig. 8.6: Effect of amendments on leaching of antimony and lead from contaminated shooting range soils after 26 weeks. (a–b) antimony, (c–d) lead. Dashed line represents the regulations for inert wastes (EC, 2002). Treatments, soil = unamended, I = zerovalent iron, IS = $FeSO_4$, VS1 = ViroSoil (iron and aluminum oxides) 1, VS2 = Virosoil 2, VS1-I = VS1 + zerovalent iron, VS1-IS = VS1 + $FeSO_4$, VS2-IS = VS2 + $FeSO_4$, G = goethite. Mean ± SE (n = 3) [53]. Reprinted with permission from Tandy et al. (2017).

soil, a retention up to 89% and 99% for antimony and lead, respectively. Combined soil treatment with phosphate (tricalcium phosphate, CPN, 5%) and iron-oxide (Fe_2O_3) has also shown promising results in laboratory scale, immobilizing both metals and metalloids in a backstop from a military shooting range [54].

Field studies on iron-hydroxide remediated shooting range soil have been carried out by Okkenhaug et al. [13], who investigated the effect of mixing ferric oxyhydroxide powder (CFH-12) (2%) together with limestone (1%) and zerovalent iron (Fe(0); 2%) into contaminated backstop berm soil. This resulted in an average decrease of antimony and lead soil porewater concentrations of 66% and 97%, respectively. The remediation effect was stable over the four-year experimental period, indicating no remobilization. In the field study of Barker et al. [24], $FeCl_3$ was added to various backstop berm soils. Dissolved Fe(III) would precipitate forming iron-(hydr)oxides with active sorption sites. Both lead and antimony concentrations decreased significantly in all studied soil types.

Moderate effect on antimony immobilization in contaminated shooting range soil has been achieved using biochar. Silvani et al. [55] conducted laboratory batch tests with contaminated backstop berm soil with various biochars (BC) based on waste timber and coconut shell, as well as wood shrub and iron-enriched designer biochar. Amending 20% BC led to reductions of 61 and 12% for lead and antimony concentrations. By contrast, amending 20% of the iron-enriched designer biochar, Fe-BC, had a much stronger effect on leaching, with reductions of 98% for lead and 47% for antimony (without a dose effect). In other studies, biochar has also been found to immobilize antimony [50], but in other cases also mobilize antimony [46, 56]. The increase of soil pH and release of phosphorus by biochar amendment can induce antimony mobilization through enhanced electrostatic repulsion from the negatively charged biochar particles [19].

8.6 Lessons and perspectives

Due to widespread use of bullets containing antimony, polluted shooting ranges represent a significant source of antimony in the environment. Weathering of these bullets depends highly on redox conditions, thus affecting their transport and fate. While the deposited bullets contain metallic Sb(0) weathering under oxic conditions yields predominantly more soluble Sb(V), but Sb(III) can be observed under certain reducing conditions. The antimony biogeochemistry cannot be understood without considering the complex interaction with Fe-oxides, important sorption sites for antimony in these soils affecting their mobility. Antimony has been shown to interact with organic matter in soils, serving as both sorption sites and facilitator for oxidation. Shooting range soils, porewater, and surface runoff expose biota to antimony; however, the few available studies indicate low uptake in plants.

Antimony deposited at shooting ranges represents a high environmental load that calls for remediation measures. While the mobilization of antimony by soil washing, for example, has received little attention, the removal of polluted soil has become a common alternative and stabilization by, for example, sorbents is receiving more attention within the scientific community. Phosphates often used to immobilize lead at these sites have been shown to mobilize antimony, at the same time calling for discretion. Other sorbents such as iron-based materials and biochar have shown promising results for both lead and antimony at the same time, although these sorbents' effectiveness might be pH dependent.

Future research should emphasize a better holistic understanding of antimony's role in the biogeochemistry at these polluted sites. At present a knowledge gap still remains on especially Sb bioavailability and the long-term effects of remediation measures.

References

[1] European Commission, Joint Research Centre (JRC). Status of local soil contamination in Europe. JRC Technical Reports. EUR 29124 EN. 2018. https://publications.jrc.ec.europa.eu/repository/bitstream/JRC85913/lbna26376enn.pdf.

[2] European Environment Agency (EEA). Progress in management of contaminated sites. Indicator Assessment. Data and management. 2014.

[3] Randich E, Duerfeldt W, McLendon W, Tobin W. A metallurgical review of the interpretation of bullet lead compositional analysis. Forensic Sci Int 2002, 127, 174–191.
[4] Lewis J, Sjostrom J, Skyllberg U, Hagglund L. Distribution, chemical speciation, and mobility of lead and antimony originating from small arms ammunition in a coarse-grained unsaturated surface sand. J Environ Qual 2010, 39, 863–870.
[5] Wan XM, Tandy S, Hockmann K, Schulin R. Changes in Sb speciation with waterlogging of shooting range soils and impacts on plant uptake. Environ Pollut 2013a, 172, 53–60.
[6] Hockmann K, Tandy S, Lenz M, Nachtegaal M, Janousch M, Schulin R. Release of antimony from contaminated soil induced by redox changes. J Hazard Mater 2014, 275, 215–221.
[7] Ackermann S, Giere R, Newville M, Majzlan J. Antimony sinks in the weathering crust of bullets from Swiss shooting ranges. Sci Total Environ 2009, 407, 1669–1682.
[8] Heier LS, Lien IB, Stromseng AE, Ljones M, Rosseland BO, Tollefsen KE, Salbu B. Speciation of lead, copper, zinc and antimony in water draining a shooting range – Time dependant metal accumulation and biomarker responses in brown trout (Salmo trutta L.). Sci Total Environ 2009, 407, 4047–4055.
[9] Sorvari J, Antikainen R, Pyy O. Environmental contamination at Finnish shooting ranges – the scope of the problem and management options. Sci Total Environ 2006, 366, 21–31.
[10] Okkenhaug G, Amstatter K, Bue HL, Cornelissen G, Breedveld GD, Henriksen T, Mulder J. Antimony (Sb) contaminated shooting range soil: Sb mobility and immobilization by soil amendments. Environ Sci Technol 2013, 47, 6431–6439.
[11] Sanderson P, Fangjie Q, Balaj S, Ayanka W, Ravi N. Contamination, fate and management of metals in shooting range soils – A Review. Curr Pollut Rep 2018, 4, 175–187.
[12] Norwegian Defence Research Establishment, Veileder for undersøkelse, risikovurdering, opprydning og avhending av skytebaner og øvingsfelt (Guideline for mapping, riskassessment, remediation and of shooting ranges). FFI-report 2010/00116, 2010. In Norwegian.
[13] Okkenhaug G, Gebhardt KAG, Amstaetter K, Bue HL, Herzel H, Mariussen E, Almas AR, Cornelissen G, Breedveld GD, Rasmussen G, Mulder J. Antimony (Sb) and lead (Pb) in contaminated shooting range soils: Sb and Pb mobility and immobilization by iron based sorbents, a field study. J Hazard Mater 2016, 307, 336–343.
[14] Sanderson P, Naidu R, Bolan N, Bowman M, McLure S. Effect of soil type on distribution and bioaccessibility of metal contaminants in shooting range soils. Sci Total Environ 2012, 438, 452–462.
[15] Okkenhaug G, Smebye AB, Pabst T, Amundsen CE, Sævarsson H, Breedveld GD. Shooting range contamination: Mobility and transport of lead (Pb), copper (Cu) and antimony (Sb) in contaminated peatland. J Soils Sed 2018, 18, 3310–3323.
[16] Mariussen E, Johnsen IV, Strømseng AE. Distribution and mobility of lead (Pb), copper (Cu), zinc (Zn), and antimony (Sb) from ammunition residues on shooting ranges for small arms located on mires. Environ Sci Pollut Res 2017a, 24, 10182–10196.
[17] Lewinska K, Karczewska A. A release of toxic elements from military shooting range soils as affected by pH and treatment with compost. Geoderma 2019a, 346, 1–10.
[18] Johnson CA, Moench H, Wersin P, Kugler P, Wenger C. Solubility of antimony and other elements in samples taken from shooting ranges. J of Environmental Quality 2005, 34, 248–254.
[19] Rajapaksha AU, Ahmad M, Vithanage M, Kim K-R, Chang JY, Lee SS, Ok YS. The role of biochar, natural iron oxides, and nanomaterials as soil amendments for immobilizing metals in shooting range soil. Environ Geochem Health 2015, 37, 931–942.
[20] Basunia S, Landsberger S. Contents and leachability of heavy metals (Pb, Cu, Sb, Zn, As) in soil at the Pantex firing range, the same range as expected from malachite equilibrium. Amarillo, Texas. J Air Waste Manage 2001, 51, 1428–1435.

[21] Laporte-Saumure M, Martel R, Mercier G. Characterization and metal availability of copper, lead, antimony and zinc contamination at four Canadian small arms firing ranges. Environ Technol 2012, 2011, 32, 767–781.
[22] Clausen JL, Bostick B, Korte N. Migration of lead in surface water, pore water, and groundwater with a focus on firing ranges. Crit Rev Environ Sci Technol 2011, 41, 1397–1448.
[23] Martin WA, Nestler CC, Wynter M, Larson SL. Bullet on bullet fragmentation profile in soils. J Environ Manag 2014, 15, 369–372.
[24] Barker AJ, Douglas TA, Ilgen AG, Trainor TP. Lead and antimony from bullet weathering in newly constructed target berms: Chemical speciation, mobilization, and remediation strategies. Sci Total Environ 2019, 658, 558–569.
[25] Vantelon D, Lanzirotti A, Scheinost AC, Kretzschmar R. Spatial distribution and speciation of lead around corroding bullets in a shooting range soil studied by micro-X-ray fluorescence and absorption spectroscopy. Environ Sci Technol 2005, 39, 4808–4815.
[26] Scheinost AC, Rossberg A, Vantelon D, Xifra I, Kretzschmar R, Leuz AK, Funke H, Johnson CA. Quantitative antimony speciation in shooting-range soils by EXAFS spectroscopy. Geochim Cosmochim Acta 2006, 70, 3299–3312.
[27] Filella M, Williams PA, Belzile N. Antimony in the environment: Knownsand unknowns. Environ Chem 2009, 6, 95–105.
[28] Steely S, Amarasiriwardena D, Xing B. An investigation of inorganic antimony species and antimony associated with soil humic acid molar mass fractions in contaminated soils. Environ Pollut 2007, 148, 590–598.
[29] Ilgen AG, Majs F, Barker AJ, Douglas TA, Trainor TP. Oxidation and mobilization of metallic antimony in aqueous systems with simulated groundwater. Geochim Cosmochim Acta 2014, 132, 16–30.
[30] Lewinska K, Karczewska A, Siepak M, Szopka K, Galka B, Iqbal M. Effects of waterlogging on the solubility of antimony and arsenic in variously treated shooting range soils. Appl Geochem 2019b, 105, 7–16.
[31] Grob M, Wilcke W, Mestrot A. Release and biomethylation of antimony in shooting range soils upon flooding. Soil Systems. 2018, 2, 34.
[32] Hockmann K, Tandy S, Lenz M, Reiser R, Conesa HM, Keller M, Studer B, Schulin R. Antimony retention and release from drained and waterlogged shooting range soil under field conditions. Chemosphere 2015, 134, 536–543.
[33] Tandy S, Hockmann K, Keller M, Studer B, Papritz A, Schulin R. Antimony mobility during prolonged waterlogging and reoxidation of shooting range soil: A field experiment. Sci Tot Environ 2018, 624, 838–844.
[34] Heier LS, Meland S, Ljones M, Salbu B, Stromseng AE. Short-term temporal variations in speciation of Pb, Cu, Zn and Sb in a shooting range runoff stream. Sci Total Environ 2010, 408, 2409–2417.
[35] Reichenberg F, Mayer P. Two complementary sides of bioavailability: Accessibility and chemical activity of organic contaminants in sediments and soils. Environ Toxicol Chem 2006, 25, 1239–1245.
[36] Davison W, Zhang H. In-situ speciation measurements of trace components in natural waters using thin-film gels. Nature 1994, 367, 546–548.
[37] Paquin PR, Gorsuch JW, Apte S, Batley GE, Bowles KC, Campbell PGC, Delos CG, Di Toro DM, Dwyer RL, Galvez F, Gensemer RW, Goss GG, Hostrand C, Janssen CR, McGeer JC, Naddy RB, Playle RC, Santore RC, Schneider U, Stubblefield WA, Wood CM, Wu KB. The biotic ligand model: A historical overview. Comp Biochem Physiol C Toxicol Pharmacol 2002, 133, 3–35.

[38] Gleyzes C, Tellier S, Astruc M. Fractionation studies of trace elements in contaminated soils and sediments: A review of sequential extraction procedures. TrAC-Trends Anal Chem 2002, 21, 451–467.

[39] Evangelou MWH, Hockmann K, Pokharel R, Jakob A, Schulin R. Accumulation of Sb, Pb, Cu, Zn and Cd by various plants species on two different relocated military shooting range soils. J Environ Manage 2012, 108, 102–107.

[40] Robinson BH, Bischofberger S, Stoll A, Schroer D, Furrer G, Roulier S, Gruenwald A, Attinger W, Schulin R. Plant uptake of trace elements on a Swiss military shooting range: Uptake pathways and land management implications. Environ Pollut 2008, 153, 668.

[41] Hockmann K, Tandy S, Studer B, Evangelou MWH, Schulin R. Plant uptake and availability of antimony, lead, copper and zinc in oxic and reduced shooting range soil. Environ Pollut 2018, 238, 255–262.

[42] Wan XM, Tandy S, Hockmann K, Schulin R. Effects of waterlogging on the solubility and redox state of Sb in a shooting range soil and its uptake by grasses: A tank experiment. Plant Soil 2013b, 371, 155–166.

[43] Mariussen E, Johnsen IV, Strømseng AE. Application of sorbents in different soil types from small arms shooting ranges for immobilization of lead (Pb), copper (Cu), zinc (Zn), and antimony (Sb). J Soil Sed 2018, 18, 1558–1568.

[44] Gorham E, Rochefort L. Peatland restoration: A brief assessment with special reference to Sphagnum bogs. Wet Ecol Manag 2003, 11, 109–119.

[45] Khan FI, Husain T, Hejazi R. An overview and analysis of site remediation technologies. J Environ Manage 2004, 71, 95–122.

[46] Bandara T, Vithanage M. Pytoremedation of Shooting Range Soils. In: Ansari AA, et al., eds, Phytoremediation. Management of Environmental Contaminants. Vol. 3, Switzerland, Springer International Publishing, 2016, 469.

[47] Tschan M, Robinson BH, Schulin R. Antimony in the soil–plant system – a review. Environ Chem 2009, 6, 106–115.

[48] Council of the European Union, Council Decision of 19 December 2002 establishing criteria and procedures for the acceptance of waste at landfills pursuant to Article 16 of and Annex II of Directive 1999/31/EC, Official Journal of the European Communities L 11/27 (2003).

[49] Li L, Xing W, Scheckel KG, Xiang G, Ji H, Li H. Lead retention in a calcareous soil influenced by calcium and phosphate amendments. J Hazard Mater 2013, 262, 250–255.

[50] Ahmad M, Lee SS, Lim JE, Lee SE, Cho JS, Moon DH, Hashimoto Y, Ok, YS. Speciation and phytoavailability of lead and antimony in a small arms range soil amended with mussel shell, cow bone and biochar: EXAFS spectroscopy and chemical extractions. Chemosphere 2014, 95, 433–441.

[51] Griggs C S, Martin WA, Larson SL, O'Connnor G, Fabian G, Zynda G, Mackie D. The effect of phosphate application on the mobility of antimony in firing range soils. Sci Total Environ 2011, 409, 2397–2403.

[52] Sanderson P, Naidu R, Bolan N. Effectiveness of chemical amendments for stabilisation of lead and antimony in risk-based land management of soils of shooting ranges. Environ Sci Pollut Res 2015, 22(12), 8942–8956.

[53] Tandy S, Meier N, Schulin R. Use of soil amendments to immobilize antimony and lead in moderately contaminated shooting range soils. J Hazard Mater 2017, 324, 617–625.

[54] Rodriguez-Seijo A, Vega FA, Arenas-Lago D. Assessment of iron-based and calcium-phosphate nanomaterials for immobilisation of potentially toxic elements in soils from a shooting range berm. J Environmental Manag 2020, 267.

[55] Silvani L, Cornelissen G, Smebye AB, Zhang Y, Okkenhaug G, Zimmerman AR, Thune G, Sævarsson H, Hale SE. Can biochar and designer biochar be used to remediate per-and polyfluorinated alkyl substances (PFAS) and lead and antimony contaminated soils?. Sci Total Environ 2019, 694, 133693.
[56] Ahmad M, Lee SS, Lee SE, Al-Wabel MI, Tsang DCW, Ok YS. Biochar-induced changes in soil properties affected immobilization/ mobilization of metals/metalloids in contaminated soils. J Soils Sed 2017, 17, 717–730.

Montserrat Filella and Juan Carlos Rodríguez-Murillo

Chapter 9 From headwaters to oceans: antimony across the aquatic continuum

9.1 Introduction

Interest in the 'environmental life' of antimony started relatively late compared with other chemical elements. The element was overlooked for many years often because the interest in an element is directly related to its perceived 'dangerousness.' Furthermore, in stark contrast with its companion element arsenic, antimony's lower abundance, relative insolubility, and lack of evident (eco)toxic effects did not favor research. Moreover, as for all chemical substances, following antimony's fate in natural systems was dependent on the possibility of measuring it. The first large survey for freshwaters quoted in Filella et al. [1] – which does not preclude the existence of older ones – is the study published by Udodov and Parilov in 1961 [2] where 3,770 measurements of antimony in Siberian rivers were performed by using spectrometry, after preconcentration by sulfide precipitation. It was the emergence of neutron activation analysis (NAA), however, with its multielement analysis capabilities and low detection limits, which fostered measurements in all types of environmental matrices. The first published NAA values, from 1965, concerned freshwaters [3] and oceans [4]. Somewhat later, values obtained by hydride generation techniques followed – coupled with atomic emission spectroscopy [5, 6] and with atomic absorption spectroscopy [7]. Finally, inductively coupled plasma mass spectrometry (ICP-MS) largely took over in the 1990s. Since then, most of the reported values for antimony in the environment have been acquired in the context of multielement studies.

Does 60 years of data collection allow us to understand the behavior of antimony in the environment? This chapter attempts to answer this question in the case of surface waters by following antimony through the aquatic network, understood

Acknowledgments: The authors would like to thank Greg Cutter, Juraj Majzlan, and Roberta Rudnick for enlightening discussions. We are also thankful to Teba Gil-Díaz for making her original data available to us.

Montserrat Filella, Department F.-A. Forel, University of Geneva, Boulevard Carl-Vogt 66, CH-1205 Geneva, Switzerland. montserrat.filella@unige.ch
Juan Carlos Rodríguez-Murillo, Museo Nacional de Ciencias Naturales, CSIC, Serrano 115 dpdo, E-28006, Madrid, Spain. jcmurillo@mncn.csic.es

https://doi.org/10.1515/9783110668711-009

as “a sequence of geological, physical, chemical, and biological systems that transport, transform and deliver elements during their transit from uplands to the ocean coastal zones and possibly into the deep ocean” [8]. Despite the interconnection of all water bodies, most aquatic research is done piece by piece and, often, results are even limited to a single system on a given date. This means that existing data, albeit abundant, are patchy and not necessarily adequate for extracting general conclusions. This chapter is an attempt to integrate existing information in the case of antimony.

Systems with high concentrations of antimony (e.g., supergene environments, acid mine drainage, or geothermal systems) are not included. Such systems are not only characterized by being richer in antimony but also by experiencing different physico-chemical conditions. Thus, although their study is interesting per se and can, no doubt, provide insight into the environmental behavior of the element, observations in such systems do not necessarily apply to ‘non-polluted’ systems (i.e., systems with ‘natural’ concentrations or non-heavily polluted) and, cited out of context, can even be misleading.

The content of this chapter relies strongly on a series of reviews published by one of the authors from 2002 to 2013 [1, 9–16]. Most of the information contained in these articles is still useful and will not be repeated here. These reviews were based on the systematic collection and listing of ‘all that has been published.’ Such an approach was believed necessary at the time of their elaboration in order to set the frame of what was known and unknown about the element. This chapter is also based on the comprehensive collection of published data, though the acritical reproduction of all reported values has been replaced by a thoughtful consideration of their analytical soundness and discussion of the operational character embedded in most values.

Aquatic compartments are open systems and, as such, linked to other environmental compartments, in particular, soils and atmosphere, in which continuous exchanges occur. The complete understanding of the behavior of antimony in aquatic systems, therefore, requires the future integration of this study into the larger frame of the whole biogeochemical cycle of the element. Antimony in soils has been the object of a lot of attention and is discussed in several chapters of this book. No good review devoted to the fate of antimony in the atmosphere exists, and many aspects, especially related to its chemical speciation and processes occurring during antimony atmospheric transport, appear to be little studied. This is particularly worrisome considering that the recent modification of the atmospheric cycle of antimony is the highest ever found for this element over a period of 800,000 years [17].

9.2 Analytical and methodological considerations

From the technical point of view, total 'dissolved' antimony concentrations can now be reliably measured in most aqueous matrices. Too many studies, however, ignore the most elemental analytical chemistry principles, and frequently the reported results have little or no value. Causes are multiple, but, probably, the affordability of quadrupole mass spectrometers and the perception that this type of determination is just 'routine analysis' and that, therefore, it does not demand any specific training can be cited as reasons. The problem is not exclusive for antimony but greatly affects this element. Some of the main issues concerning accuracy, detection limits, and precision are briefly discussed later.

The usual measuring methods require the analyte to be in solution. The total and reliable extraction of antimony from solid matrices such as soils, sediments, and plants is possible but not all digestion methods are equally efficient. For instance, digestion with nitric acid [18] or *aqua regia* [19] is inadequate. Sediments contain a significant amount of silicates that require, for complete dissolution, an acid digestion procedure including HF as a digestion agent. This requirement extends to suspended particulate matter [20]. The use of techniques that do not completely dissolve solid matrices systematically leads to results that are difficult to interpret and impossible to compare.

The measurement of antimony concentrations without using appropriate certified reference materials (CRM) or using CRM with either inadequate concentrations or no certified values for antimony (a problem frequently encountered in multielement studies) is widespread. Blanket statements like "accuracy was checked against standards" with no further details or "values were within X percent" without detailing for which element are common.

At the low concentrations of antimony present in many aquatic systems, avoiding sample contamination is absolutely necessary. As clearly stated by Shotyk et al. [21], "to develop a scientific understanding of the chemical evolution of surface and groundwaters (. . .) requires limits of detection well below the abundance of these elements in natural freshwaters." These limits of detection can only be attained by applying strict protocols during sampling and sample pretreatment; both have a profound effect on detection limits. Filter blanks have been pointed out as sources of uncertainty for many elements when determining their concentrations in suspended particulate material [22]. Digestion procedures are also a critical step. For instance, the microwave acid digestion required in the analysis of antimony in suspended particulate matter by inductively coupled plasma sector field mass spectrometry (ICP-SFMS) implied the shift from an instrumental detection limit of 0.001 ng to a methodological detection limit of 2 ng, 2,000 times higher [20]. Values of procedural blanks need always to be reported but this is rarely the case.

The improvement of detection limits with time is behind reports of decreasing antimony concentrations over the years in different matrices (e.g., seawater [1], biomarkers such as hair [14], urine [15], etc.). This is due to the fact that the manner in which below detection limit (BDL) samples are taken into consideration – or not – when calculating average or median values can lead to the overestimation of reported values. Therefore, improving detection limits nearly automatically leads to reporting lower concentrations. This effect needs to the considered when using or citing old data.

As illustrated in Fig. 9.1(a), measurement precision has improved over the years. This figure shows 20 years of monthly 'dissolved' antimony concentrations in Lake

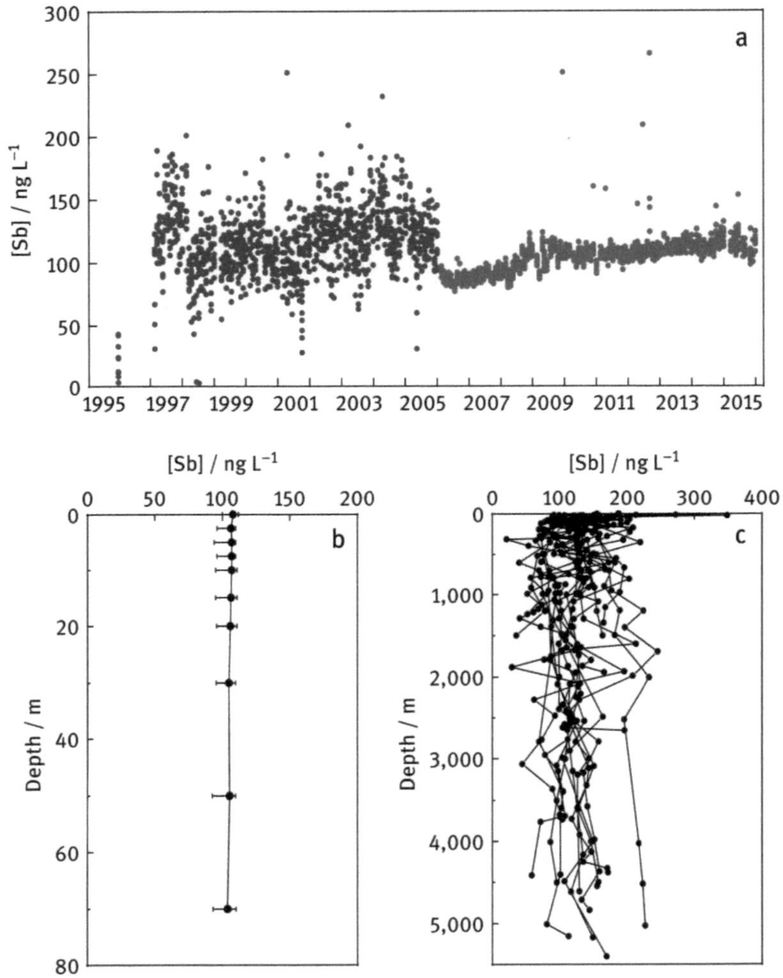

Fig. 9.1: (a) Time series of monthly antimony concentrations for 1995–2015 in Lake Geneva, Switzerland, France (depth at the sampling point, 76 m). Values at all depths for all dates plotted. The ICP-MS instrument was changed in 2006 [23]. (b) Median concentration profiles for antimony in the same lake (2006–2015). Error bars in profile correspond to the interquartile range [23]. (c) Antimony profiles in the Atlantic [24–26] and Pacific Oceans [27, 28].

Geneva (1995–2015) [23]. These concentrations were obtained in the same laboratory by using similar protocols; the change in precision observed in 2006 corresponds to the change of the ICP-MS instrument.

Particular attention needs to be paid to the meaning of data. This aspect is particularly thorny in environmental systems because of the operational character of many of the analytical methods used. In the case of antimony in natural waters, operationality starts with the widespread distinction between 'dissolved' and 'particulate' concentrations based on filtering the samples through 0.45 or 0.22 µm pore size filters. Particle sizes are continuously distributed in natural waters following a power law distribution with a slope often close to −4 [29], thus striding over the 'dissolved' and 'particulate' operational limit. Particles with sizes between 1 nm and 1 µm are usually called colloids because of their behavior in solution [29]; in antimony work, some authors restrict the colloid qualification to particles with sizes between 1 kDa and 0.22 µm (Fig. 9.2) or even below 100 nm [40]. These limits are arbitrary and do not correspond with the behavior of particles as colloids in aqueous media [41]. Not all chemical elements are equally distributed among particle sizes [42]. The assessment of their distribution requires the use of size fractionation techniques, such as cross-flow-filtration (CFF), field-flow-fractionation (FFF), or dialysis, all coupled with a technique that measures the element concentration. Data for antimony in oxic waters are shown in Fig. 9.2 and discussed in the corresponding sections. In general, since the element is mostly present in the 'truly' dissolved fraction (i.e., less than 0.5–1 kDa), it can be safely assumed that the usual filtration procedure really gives the dissolved fraction and that colloids will not play a significant role in its behavior in natural waters.

So-called chemical speciation techniques applied to antimony in natural aquatic samples have mainly focused on three aspects: (i) distinguishing between the two main oxidation states, Sb(III) and Sb(V) [43], and, to a minor extent, measuring alkylantimony species ([44], Chapter 11) (Fig. 9.3); (ii) extracting sediments with reagents having different chemical properties in order to identify the antimony solid binding phases; (iii) using X-ray absorption spectroscopy (XAS) to obtain an insight into the oxidation state and/or about the host phases of antimony. Note the complete lack of CRM with certified values for Sb(III) and Sb(V) which makes it impossible the sound evaluation of published speciation values in all cases. The application and limitations of single and sequential extractions in the case of antimony were extensively discussed in Filella [12]. The reader is advised to consider that earlier publication prior to reading the present chapter and, in particular, Tab. 9.1, which contains data found for non-heavily polluted sediments. XAS techniques are relatively insensitive, and, in practice, only environmental samples containing high antimony concentrations, such as soils close to mines and smelters, shooting ranges or artificial laboratory mixtures, can be studied by these techniques. Thus, no XAS results are included in this chapter.

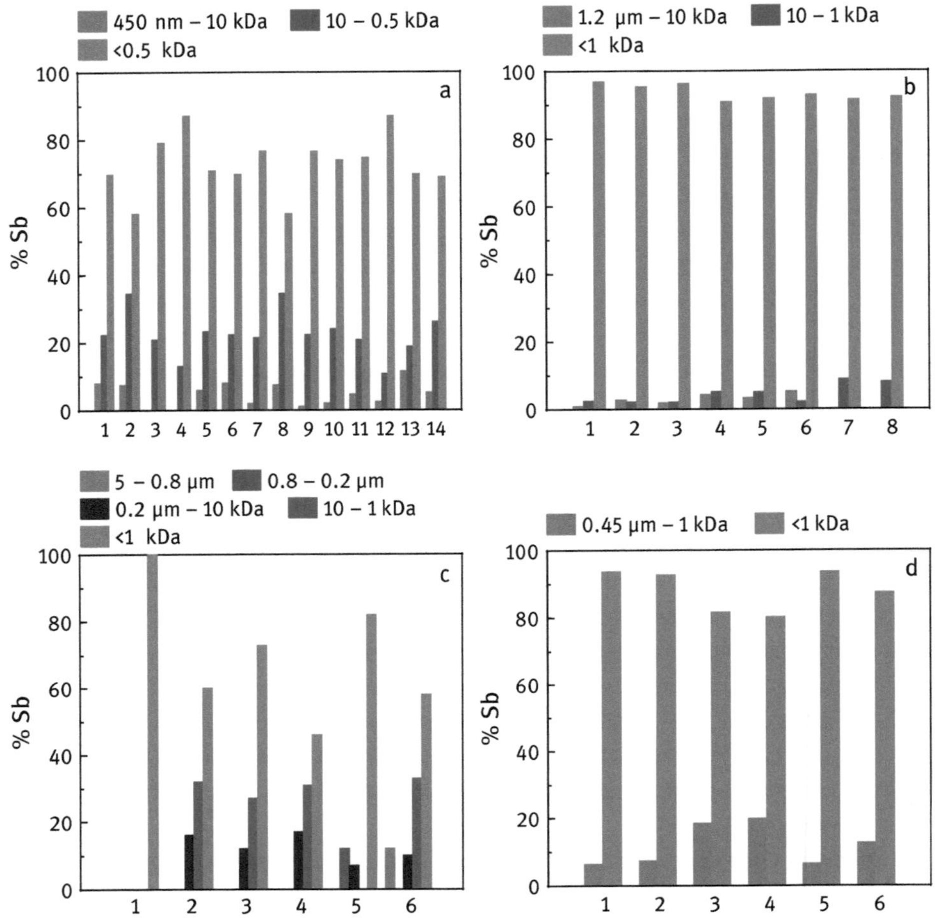

Fig. 9.2: Size distribution of antimony in surface waters. (a) Rivers in Japan [30–32]; (b) 1–3: Po River, Italy, 4–6: Dunajec River basin, Poland [33], 7: Lake Geneva and 8:Versoix River, Switzerland [34]; (c) Pristine Karelian rivers [35]; (d) Arctic and subarctic systems. 1: Lake Vilno [36] (upper limit: 0.22 µm), 2: Lake Svyatoe [37], 3–4: 0–5 ‰ and 5–30 ‰ salinities in the Severnaya Dvina Estuary [38], 5–6: May and July Ob River, Siberia [39]. Data obtained by ultrafiltration in [30–35] and by dialysis in [36–39].

9.3 About time scales

Chemical reactions and transport processes jointly determine the rates of change in total concentration of chemical elements as well as their chemical speciation. In order to clearly understand the fate of antimony in the aquatic continuum, it would be necessary to compare the time scales of transport processes in the different compartments with the characteristic times of chemical reactions. This type of exercise

Tab. 9.1: Antimony in iron and residual fractions in sediment samples obtained by (sequential) extractions. Sediments directly polluted by mining or smelters are not considered[a].

Type of system	System	Experimental details	Sb concentration / $\mu g\ g^{-1}$	% Fe fraction	% residual fraction	Reference
Estuary surface	Puget Sound, USA	NAA CRM n = 23	0.2–1.0	31	not given	[53]
River surface	Garonne River, France	Digestion: tri-acid or alkaline melting attack No CRM Residual by difference n = 4	not given	38	62	[54]
River SPM	Waikato River, New Zealand	Digestion: *aqua regia* CRM Residual?	1.7	63	28	[19]
River surface	Chongming Island, China	Digestion: HF-HNO_3-H_2O_2 No CRM Residual measured n = 54	0.54–4.93	3.1	90.4	[55]

(continued)

Tab. 9.1 (continued)

Type of system	System	Experimental details	Sb concentration / $\mu g\ g^{-1}$	% Fe fraction	% residual fraction	Reference
Lake core	Clearwater (oxic) and McFarlane (anoxic) lakes, Canada	Digestion: mixture of acids CRM? Residual by difference	Clearwater: max 3.75 McFarlane: max 1.44	Clearwater Surface: 50.4 Peak: 46.2 20 cm: 0 McFarlane Surface: 60.5 Peak: 26.5 20 cm: 57.9	Clearwater Surface: 48.8 Peak: 53.8 20 cm: 100 McFarlane Surface: 31.1 Peak: 69.8 20 cm: 32.7	[56]
Anoxic, sulfidic sediments surface	Brown Bay, Antarctica	Digestion: *aqua regia* CRM Residual measured[b] n = 11	Not given	Most 0, when measured (5): 3.2	93.4	[57]
Lake surface	Taihu Lake, China A- 1.8–2.3 m B- < 1 m	Digestion: 65% HNO_3 No CRM Residual measured n = unclear	A: 3.5 B: 3.2	A: 1.2 B: 2.3	A: 96.3 B: 95.4	[58]
Lake core	Nansi Lake, China Mean depth 1.5 m	Digestion unknown No CRM Residual by difference?	~1	5.6	90.4	[59]
Marine surface	Bohai Bay, China	Digestion: HF-HNO_3-$HClO_4$ CRM Residual by difference n = 27	1–3	4.87	78.93	[60]

Estuary surface	Yangtze Estuary, China	Digestion: not given CRM Residual? n = 13	0.36–1.11	~<1 (figure)	96.5	[61]
Estuary surface	Yangtze Estuary, China	Digestion: *aqua regis* CRM Residual by difference n = 28	0.36–1.48	7.22	60.56	[62]
Marine surface	Hangzou Bay, China	Digestion: HF-HNO_3-H_2O_2 CRM Residual by difference n = 20	0.72–1.36	~<1 (figure)	98.7	[63]
Marine surface	Jiaozhou Bay, China	Digestion: HF-HNO_3-$HClO_4$ CRM Residual by difference n = 9	Not given	2.30	85.9	[64]
Marine core	Jiaozhou Bay, China	Digestion: HF-HNO_3-$HClO_4$ CRM Residual measured	<32 µm: 0.16[c] 32–63 µm: 0.47 >63 µm: 0.42	<32 µm: 1.3 32–63 µm: 1.8 >63 µm: 3.5	<32 µm: 96.4 32–63 µm: 95.8 >63 µm: 83.3	[65]

[a] CRM: certified reference material.
[b] Sum of all steps (BCR) agreed with pseudo-total digestion for all elements except Sb, which was considerably lower (30–80%).
[c] Units not given. Probably, $\mu g\ g^{-1}$.

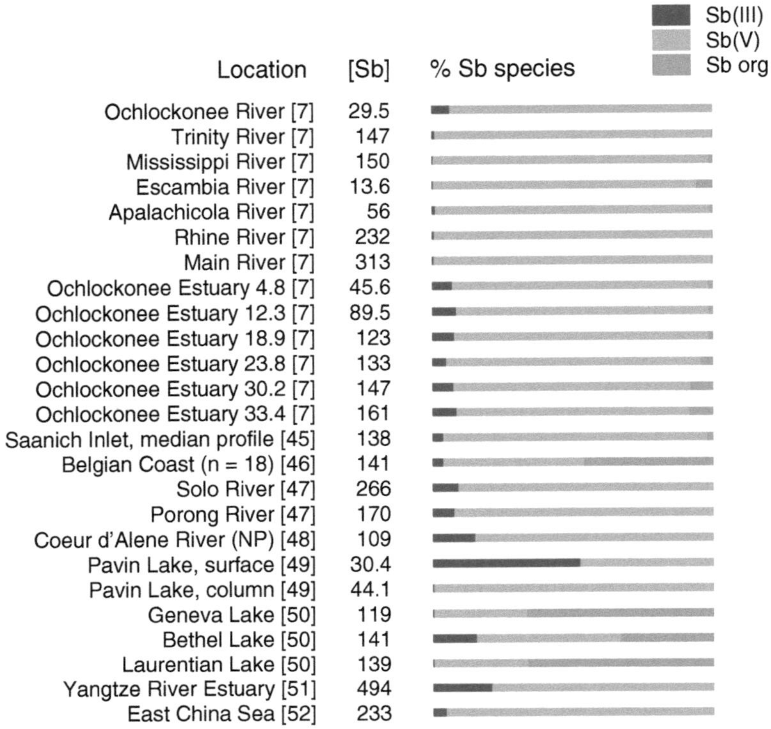

Fig. 9.3: Antimony chemical speciation in oxic surface waters, excluding oceans. Concentrations are 'dissolved' concentrations in ng L^{-1}. The 'Sb org' fraction refers always to alkylated antimony species except in the Belgian coast and Geneva, Bethel and Laurentian lakes where the fraction of antimony is associated to organic matter destroyed by UV irradiation. Values are given in chronological order. Values in Ochlockonee Estuary are salinities in ppt (‰).

has rarely been attempted. In a precursor exercise, Morgan and Stone [66] discussed trace elements in lakes and compared mixing time scales with first- and second-order reaction times of complexation reactions but antimony was not included. The kinetics of ligand exchange reactions for antimony has received scant attention and is a subject entirely absent in common information sources [67, 68], but it has been reported to be slow in a few laboratory studies in biological contexts [69, 70].

The kinetics of antimony redox reactions has received much more attention. Oxidation reactions seem to be much faster than reduction ones but direct estimations in surface waters remain scarce. Cutter and coworkers [71] reported rate constants of 0.008 day^{-1} for Sb(III) oxidation in the upper 100 m in the Black Sea and 1×10^{-6} day^{-1} for Sb(V) reduction in its bottom waters. These extremely low rates of Sb(V) reduction confirmed previous observations by Bertine and Lee [45] who had reported no measurable reduction of Sb(V) to Sb(III) in sulfide-rich waters (Saanich Inlet). Following initial studies by Belzile et al. [72], Sb(III) oxidation by natural and synthetic iron and manganese oxy(hydroxides) has been widely studied in the past few years in the lab, mostly for remediation purposes or oriented toward understanding processes taking place in mine-affected waters. High variation of rates and controlling factors (pH, Fe(III) concentration, light, bacteria, etc.) have been described. Translation of this wealth of information to natural water conditions remains to be rigorously attempted.

Kinetic control of precipitation reactions can also have environmental implications. Majzlan et al. [73] found that tripuhyite – postulated as the ultimate sink of antimony in soils [74] – was grossly supersaturated in all waters for which values could be found in the literature. Even if tripuhyite described in mine wastes is poorly crystalline, extremely fine-grained and with variable Fe/Sb ratio and, thus, expected to be less insoluble than the well-crystallized and stoichiometric form, still no tripuhyite formation was observed in the waters considered. This implies that tripuhyite probably needs decades to nucleate and grow – necessary conditions that might exist in sediments and soils but not in surface waters.

This rapid overview confirms the relevance of considering water residence times in the different chain links that constitute the network of the aquatic continuum. This is not a simple task because, under any general category such as 'rivers' or 'lakes,' a myriad of different systems exist. For this reason, the next sections include short discussions on the diversity and expected range of characteristics in each system category.

9.4 Systems

9.4.1 Rivers

Gaillardet et al. [75], in a much cited work, gave a world average of 70 ng L^{-1} for 'dissolved antimony' (filtered 0.2 µm) in 'natural' rivers discarding data from highly polluted systems. This average was based on values for 14 rivers, 12 of them in America. The average remained unchanged in a second edition of the book, published 11 years later [76]. This world average concentration exactly agrees with the median value of 70 ng L^{-1} (range: < 2–1,210 ng L^{-1}, n = 807) for European rivers obtained in the FOREGS European survey [77]. Extensive surveys over large areas, as for this one, performed with the same analytical approach provide an excellent source of valuable environmental information but no other similar survey is available for river waters. Though not equivalent but nevertheless useful, a group of studies covering rivers discharging in the Arctic provides a comprehensive set of data. They report a similar range of concentrations: a 20–70 ng L^{-1} range in Karelian pristine rivers in a 1,500 km transect N-S from Kola peninsula to Moscow [35], an average value of 73 ng L^{-1} in an extensive survey of the Ob River and flood zone – the Ob is the largest river discharging in the Arctic Ocean [39], values of western Siberian rivers along a permafrost gradient: 30.3 ng L^{-1}

(spring, n = 70), 25.2 ng L^{-1} (summer, n = 67), 49.8 ng L^{-1} (autumn, n = 13), 38.5 ng L^{-1} (winter, n = 55) [78].

Older values, like the ones reported by Byrd [79] for 48 rivers in Asia, USA, and Canada (median: 160 ng L^{-1}), are generally higher. It is impossible to know whether this is due either to contaminated rivers being included, the general decrease in reported values linked to methodological improvements discussed in Section 9.2, or to differences in the lithology of the watersheds. Trace element concentrations in rivers, in the absence of inputs from anthropogenic sources, depend on processes taking place during transport but, mainly, on their abundances in the watershed crust and their mobility during weathering. Linking measured antimony concentrations with the lithology of the watershed was attempted in the FOREGS survey but, given geological internal variability in most watersheds, it is disputable whether a density sampling of one sample per 5,000 km^2 allows for more than establishing patterns in relatively homogenous areas (e.g., observing the presence of lower 'dissolved' values throughout central and northern Fennoscandia on the Precambrian Shield rocks throughout most of Norway and Scotland). In fact, according to some authors [80], the dissolved load is a clear reflection of the weathering of catchment rock types only in few exceptional circumstances and, even in those cases, is only clear for certain elements (e.g., the REE) among which antimony is not included.

Solute concentration-discharge relationships are often used as a tool to understand the behavior of chemical elements because they integrate hydrological and biogeochemical processes. Streamflow may be sustained by water from different sources in catchments. Baseflow represents the discharge of relatively long-lived stores of water, such as groundwater that sustains streamflow between rainfall events. During high rainfall periods, surface runoff may make significant contributions to streamflow. 'Dissolved' antimony has been observed to decrease with discharge in some rivers (Fig. 9.4) following a power law function with slopes −0.566 in the Amazon River [81]; −0.313 and −0.361 in two medium-sized rivers in China, Min River [82], and Beipan River [83]; and −0.043 to −0.287 in several points in the Lot-Garonne river system, France [85]. These observations support a general dilution behavior suggesting that antimony enters the rivers mostly with the baseflow and that flood episodes just have a dilution effect. A more detailed analysis would be needed in each case to evaluate whether soil water, water from the unsaturated zone, or shallow groundwater may also be mobilized as catchments wet up following rainfall. Antimony dilution behavior, however, is not general; the element seems to behave differently depending on the climatic zones or other local or regional factors. For instance, antimony concentrations were found to increase with the discharge in River Ob with maximum concentrations in spring floods [87]; a maximum antimony concentration at the end of the spring flood (52% of the total flux) was also reported in the Severnaya Dvina, the largest European Arctic river [88]. Other observed increases of antimony concentrations with discharge were attributed to strong erosion of surface iron crusts in the Oubangui River [86] or to old mine tailings in a tributary in the Rhône River [84]. Small rivers may also not show a dilution behavior: Matsunaga et al. [89] reported an increase of antimony concentrations with discharge in rainfall events in a rural stream.

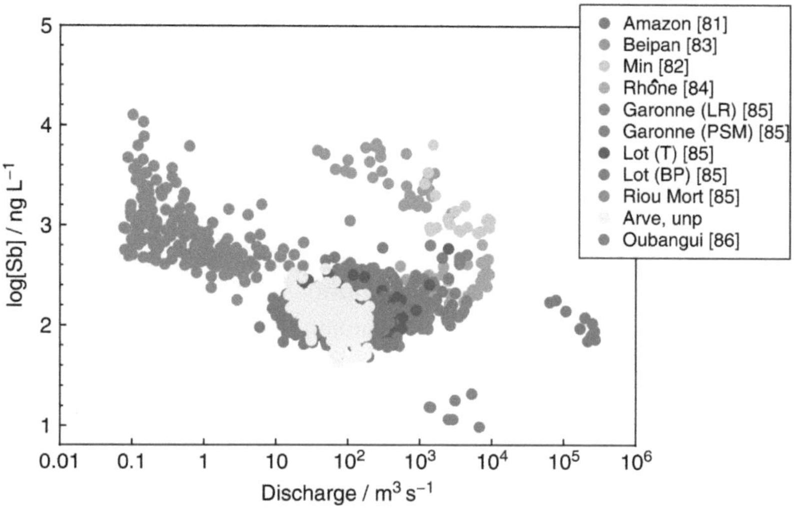

Fig. 9.4: Variation of 'dissolved' antimony concentrations with water discharge in world rivers.

The interpretation of concentration-discharge results might be complicated by the existence of hysteresis. This type of response has been described for antimony in the Lot River ([90] and the Oubangui River [86] but remains clearly understudied for antimony.

Is 'dissolved' antimony in rivers 'truly' dissolved? The scarce published values about the antimony distribution in the colloidal size range in rivers (Fig. 9.2) point to antimony being mostly present in the 'truly' dissolved fraction. Some amount is also found, however, in non-'truly' dissolved fractions depending on systems (and authors?), which might imply a role for colloids, even if minor, in antimony transport in waters. Gil-Diaz et al. [85] reported values of the ratio Sb 0.02 µm to Sb 0.2 µm for the 2003–2007 period for different sites of the Lot-Garonne river system and found it usually to be higher than 0.9, but warned of a high variability of the values measured linked to the contamination of the samples from the filters used. The difficulty of measuring this level of antimony concentrations while avoiding contamination in ultrafiltration systems needs always to be considered.

Concerning antimony redox state, the element is mostly present as Sb(V) in oxic waters, including rivers (Fig. 9.3). Surprisingly, fewer values than expected seem to have been published. Other values have been reported but within the frame of the development of new analytical methods or in the monitoring of highly polluted media.

Viers et al. [91] updated the pioneering work of Martin and Meybeck [92] on the chemical composition of the suspended sediments in the world rivers and reported an average 'particulate' antimony concentration of 2.19 ± 2.41 μg g^{-1} (n = 41). Previous global values, cited by Viers et al. [91], are 2.5 μg g^{-1} [92] and 1.4 μg g^{-1} [93].

The partitioning of an element between the solution and particulate phases in waters, soils, and sediments is often described by using a 'distribution' or 'partition' coefficient (K_d), defined as the quotient between the concentration of the element in the particulate phase (*Cp*), in mass per unit weight of solid phase, and the 'dissolved' concentration (*Cd*) in mass per unit volume of solution. Reported values of K_d (L g^{-1}) for antimony are collected in Tab. 9.2. These K_d values are usually interpreted as antimony having a low affinity for inorganic and organic particles. This might be correct even if it relies on the conceptualization of K_d as being a measure of the sorption on a solid matrix that does not contain the element itself (e.g., the assumption that is based on the existence of an underlying C-type sorption isotherm). However, in the case of antimony, the possibility that *Cp* mostly reflects antimony contents in detrital particles cannot be excluded. Tab. 9.2 details the digestion method used in the determination of the particulate material. This is a key issue because incomplete digestion will give too low K_d values.

In principle, the ratio between the concentration of the compound remaining in the solution and the amount adsorbed on the solid is the same at any concentration, and K_d should not show any dependence on suspended solid particle concentration (SSP) (this is the hypothesis behind the K_d definition). This dependency has been followed in a very few cases for antimony and all possible types of behavior reported with no correlation in two studies on the same river system, two studies with increases and one with a decrease (Tab. 9.2). Decreases have been attributed to the formation of colloids and their subsequent coagulation onto larger filterable particles [102] but also to a change in the particulate origin when high particulate matter concentrations are linked to flood events. This might be the case for antimony given the limited amount found in the colloidal fraction (Fig. 9.2) and the variety of behaviors observed.

Sequential extractions could provide some clue about how antimony is present in the particles. However, almost no values have been reported for suspended sediments (Tab. 9.1). Information for bottom river sediments could also help because resuspension can be very active in rivers with, depending on the stream, continuous exchange of suspended and bottom sediments occurring but, again, there are insufficient data available. No reliable conclusions can be drawn from such limited information.

As shown in Fig. 9.5, the percentage of antimony present in the 'dissolved' fraction depends on K_d values but also on the amount of total particulate matter in the water. This can be extremely variable. Gaillardet et al. [75] estimated the world mean value for big rivers at 350 mg L^{-1}, but many rivers have just a few mg of suspended particles per liter. Given the low K_d values of antimony, antimony will be present

Tab. 9.2: Reported values of partition coefficients, K_d (L g^{-1}), for antimony in surface waters. Values for antimony radioisotopes are not included. Values are in chronological order.

System	K_d / L g^{-1}	Correlation with SPM[a]?	Pore size filtration / μm	'Dissolved' Sb / ng L^{-1}	Digestion particulate matter	Reference
World river average	2.5	–	–	1,000	–	[92]
Thames River, UK	33	–	0.45	270	NAA	[94]
Rhône River, France	5.3	–	0.4	487	Not given	[95]
World river average	31	–	–	70	–	[75, 76]
Waikato River, New Zealand	Summer: 4.5 (median, n = 12) Winter: 15.2 (median, n = 12)	–	0.45	300–900	*Aqua regia*	[19]
Gironde Estuary, France	16.2 (mean, n = 96)	–	0.2	92–292	HF-HCl-HNO_3	[96]
Gironde Estuary, France	12.0 (median, n = 50)	No correlation	0.22	205–245	HF-HCl-HNO_3	[97]
Rhône River, France	6.6 (median, n = 41)	Decreasing values with increasing SPM	0.2	163–644	"Acid digestion"	[84]
Kamo River, Japan	1	–	0.45	160	HF-$HClO_4$-HNO_3	[98]

(continued)

Tab. 9.2 (continued)

System	K_d / L g^{-1}	Correlation with SPM[a]?	Pore size filtration / µm	'Dissolved' Sb / ng L^{-1}	Digestion particulate matter	Reference
Oda River, Japan	1.0 (median, n = 18)	Decreasing values with increasing SPM	0.20	9–39	HF-$HClO_4$-HNO_3	[89]
Gironde Estuary, France	8.0 (median, n = 93)	Increasing values with increasing SPM[b]	0.45	80–400	HF-HNO_3-HNO_3	[99]
Yangtze Estuary, China	0.70	–	0.45	1,119	Not given	[100]
Yangtze Estuary south branch (95% discharge), China	0.43	Increasing values with increasing SPM	0.45	Not given	HF-$HClO_4$-HNO_3	[101]
Lot-Garonne river system, France	15.8 (median, n = 1057)	No correlation	0.45	~200	HF-HNO_3-HNO_3	[85]

[a]SPM: suspended particulate matter.
[b]Not in publication, calculated from values provided by the authors.

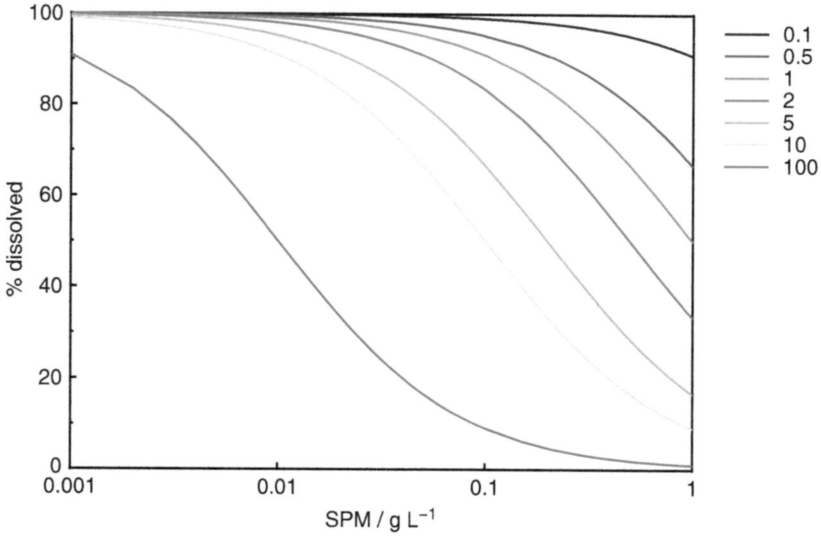

Fig. 9.5: Theoretical percentage of 'dissolved' antimony as a function of the concentration of suspended particulate matter, SPM, for different values of K_d (L g^{-1}).

mostly 'dissolved' in the majority of systems. Globally, however, riverine flux of particulate antimony to the oceans (33 kt y^{-1}, estimated by Viers et al. [91] using a value of 15 Gt y^{-1} for the world suspended flux) exceeds the riverine 'dissolved' antimony flux (2.6 kt y^{-1}; value from Gaillardet et al. [75] considering a total water discharge of 3.74×10^4 km^3 y^{-1} [103]). This is the case for all chemical elements except sodium; antimony is among the elements where the proportion of 'dissolved' is the highest [104]. In practice, up to 90% of suspended particles will settle when river mixes with seawater [105] and, consequently, only a small percentage of the continental material and its associated chemical elements reach the ocean.

Regarding particulate transport, it needs to be mentioned that, although in principle advection is the main transport mechanism in rivers, and sedimentation of particulate matter is not as favored as in lakes or oceans, 63% of the world's long rivers (>1,000 km) are no longer free flowing [106]; they represent 41% of the global river volume with dams being the main contributor. Dams and major upstream hydrological events have shown to affect antimony in estuary sediments [107].

Wastewater treatment plants (WTPs) can significantly contribute to surface water stream flows. An extensive study was performed in Switzerland [108], where effluent and sludge concentrations were measured for 64 WTPs – covering 54% of total ones, estimated a median concentration value of 'dissolved' antimony of 1 µg L^{-1} (min: 0.3, max: 8.1). Other published concentration values for individual WTPs are in the same range [109, 110]. Antimony discharge from WTPs represented 16% of the total 'dissolved' antimony leaving Switzerland in rivers, with water discharge from WTPs being 4,237,000 m^3 day^{-1}, to be compared to 193,622,400 m^3 day for all rivers (all values for 2016). This contribution of WTPs can be considered as representative of an industrialized country with 98% of the population connected to wastewater treatment systems.

9.4.2 Lakes

World lake sizes and volumes follow Pareto's distributions [111]. This means that, in practice, the term 'lake' includes very different water bodies, ranging from few large lakes, such as the Caspian Sea or Lake Baikal, to a large number of small lentic waters. Lakes are generally considered as settling zones in the aquatic continuum but their hydraulic residence times are very variable, ranging from the 400 years of Lake Tanganyika to the median value of 0.1 years of the 20,900 lakes in Brazil. Consequently, their accepted role as settling areas might be not as general as assumed, particularly in the case of nonreactive, essentially dissolved (i.e., conservative) elements. Even if most published concentrations are for 'dissolved' antimony, in the rare cases where 'particulate' antimony has also been reported, the element was found almost exclusively in the so-called dissolved fraction [112]. Size fractionation values for lakes (Fig. 9.3) are similar to rivers and the same considerations apply (Section 9.4.1); that is, antimony is almost exclusively found 'truly' dissolved. A study of four river basins in southern Sweden including 77 inlets and outlets of 36 lakes [113] confirmed the conservative behavior of antimony when the authors found no significant decrease in antimony concentrations when comparing inlets with outlets, and this was not the case for nearly all other elements that gave significant differences, all decreases except calcium and potassium.

Unfortunately, the number of studies of antimony in lakes is much lower than for other compartments such as rivers or estuaries. Reported 'dissolved' antimony concentrations often range between 50 and 200 ng L^{-1} [23, 50, 114–116]. Chemical element concentrations in lakes integrate the signal from the geological characteristics of the watersheds through riverine inputs and direct watershed runoff but, except if small and located in an uniform lithology (e.g., [18, 117],), the possible geological signal is even more blurred than in rivers. It is noteworthy that there is no antimony data for any of the largest lakes in the world except a few values in some US Great Lakes reported in the 1980s [112] and in multielement studies in Lake Baikal [118, 119]. Many concentration values that can be found in the literature are for lakes directly impacted by mining or industrial activities and are point values; they do not contribute to the understanding of antimony behavior.

In non-shallow, temperate lakes, concentration profiles provide key information concerning the behavior of chemical elements, but only very few studies report such profiles for antimony. Vertical – essentially conservative – profiles with a slight depletion in the surface in Lake Pavin (France) [49] and a small surface maximum in Lake Constance (Switzerland, Germany) [115] and Lake Geneva (Switzerland, France) [23] have been reported. Particularly deserving of mention is the robust set of data, with 20 years of monthly 'dissolved' antimony concentration profiles, in Lake Geneva. Profiles were sampled from 1995 to 2016 at the same point (location where the lake is 76 m deep and stratifies in summer) and analyzed by the same laboratory applying good laboratory practice (GLP) protocols (Figs. 9.1(a, b)). This data series shows that 'dissolved' antimony concentrations have been increasing in Lake Geneva at a rate of 1.8 ng L^{-1} per year between 1996 and 2006 and at 3.1 ng L^{-1} per year in the 2006–2015 period. Whether atmospheric deposition, suggested by the slightly higher concentrations measured at the surface, explains that increase is open to question. The impact of atmospheric deposition in 'dissolved' antimony concentrations in lakes has never been studied in detail even if already observed in lakes in Norway many years ago [117].

Lake sediments often record close mining or industrial activities [120, 121], but some antimony concentration peaks in dated core sediments have been attributed to non-local pollution sources with antimony reaching lake bottoms through atmospheric transport and deposition [18, 122–126]. This obviously requires (i) lack of diagenetic processes taking place in the sediments and significantly altering the signal and (ii) atmospheric borne antimony actually reaching the sediments (i.e., no significant dissolution in the lake water column). However, the surface enrichments mentioned earlier, for instance in Lake Geneva, suggest though that some dissolution is taking place. Some solubility measurements exist for seawater [127–130] but not in freshwaters.

Thermal stratification is a central feature of the physical dynamics of temperate lakes. Seasonal thermal stratification is linked to chemical stratification leading to possible anoxic conditions of the hypolimnion. Anoxia is also possible when lakes stratify due to saline conditions or because of high biological productivity, even in shallow ones. In lakes such as Lake Balaton (3.2 m deep) [131] or Lake Taihu, the third largest freshwater lake in China (2 m average depth) [58], the oxic-anoxic interface might just be above the sediment water interface, a situation closer to wetlands.

Antimony is reported to be mostly present as Sb(V) in oxic lake waters (Fig. 9.3) and expected to be as Sb(III) in anoxic conditions. It is well known that oxidation and reduction cycles, in particular associated with iron and manganese cycling, are important components of the biogeochemistry of many lakes and their sediments; therefore, iron oxy(hydroxides) are expected to some extent to control antimony biogeochemical cycle and antimony mobility. This might well be the case for soils but is not as clear for lakes and their sediments. For instance, Palmer et al. [132] found no coupling of antimony with Fe/Mn cycling in small subarctic lakes when studying seasonal variations of arsenic and antimony in surface waters, but other studies have found coupling [56] or mobility linked to the oxidation of Sb(III) to Sb(V) by precipitation of iron oxy(hydroxides) [58]. The vertical conservative profiles mentioned earlier, the low percentage of antimony found bound to iron by sequential extraction (Tab. 9.1) and recent observations in sediment pore waters, mostly obtained by diffusive gradients in thin films (DGT) (Tab. 9.3), suggest decoupling or, at least, a more complex and system-dependent behavior as compared to arsenic.

Tab. 9.3: Antimony in sediment pore waters[a].

	Oxic/anoxic	Technique	Observations	Reference
Baikal Lake	Oxic	Coring and digestion	Layers of Fe/Mn in near surface sediments (mainly of oxy(hydroxides)) Sb shows only slight enrichment (2–5) in the Fe phase, contrasting behavior with As, P, Mo, Cd	[133]
Clearwater Lake, Canada	Oxic, pH 6.6	Peepers, speciation modeling	Sb involved in a dynamic cycle with precipitation-dissolution-diffusion-reprecipitation of Fe and Mn oxy(hydroxides)	[56]
McFarlane Lake, Canada	Seasonally anoxic, pH 7		Retention of S through the formation of iron sulfide and pyrite, which could retain Sb by sorption or favor Sb complexation	
Peruvian near-surface sediments	Within and below oxygen minimum zone	Core squeezing	Sb diffuse into reducing sediments to precipitation depth. "It remains questionable whether Sb is trapped as a sulfide or in association with OM since highest Sb/Al ratios coincide with the TOC maximum."	[134]
Canadian lake	Oxic	Peepers, speciation modeling (WHAM)	Sb supposed present as Sb(V) Sb preferentially associated with authigenic Fe oxy(hydroxides) but in lesser proportions than As and Mo	[135]
Tributary of the Three Gorges Reservoir, China	Not mentioned	DGT (ferrihydrite), not in situ	Overlying water diffuse into the sediments Sb not associated with Fe and Mn; "This might be due to the influence of sulfide, pH and TOC"	[136]
Australian contaminated freshwater wetland sediment	Oxic-anoxic transition	DGT-DET (mercapto-silica, Metsorb-Chelex)	Sb mobilization decoupled from the Fe cycle and, therefore, "more likely linked to sulfur and/or organic carbon"	[137]

Tab. 9.3 (continued)

	Oxic/anoxic	Technique	Observations	Reference
Zrmanja River estuary, Croatia	Transition from oxic to suboxic in the upper few cm	Core centrifugation	Diffusive fluxes (ng cm^{-2} s^{-1}): Velebit channel: 1.12 × 10^{-6}, Karin Sea: −0.09 × 10^{-6}; Novigrad Sea: 0.83 × 10^{-6}	[138]
Taihu Lake, China	Oxic-anoxic depending on month	DGT (ZrO-Chelex) and HR-peeper, not in situ	Higher Sb concentrations of soluble and labile Sb in oxic conditions; Possible oxidation of Sb(III) by plant rhizosphere in July; otherwise, Sb mobility controlled by "Fe and Mn oxides"	[58]
Nansi Lake, China	Anoxic below 4.6 mm	DGT (ZrO-Chelex) and HR-peeper, not in situ	Diffusive flux (µg m^{-2} d^{-1}): −0.56 The release of Sb decoupled from the reductive dissolution of Fe/Mn oxy(hydroxides)	[59]

[a]DET: diffusive equilibration in thin films, DGT: diffusive gradients in thin films, OM: organic matter, TOC: total organic carbon, WHAM: Windermere Humic Aqueous Model.

9.4.3 Transitional systems between land and ocean

Even if river discharge is often understood as occurring in the form of estuarine systems in areas subjected to tidal forcing, the transition zones in the mouths of rivers can take many forms. For instance, rivers can also meet salty water directly, as in the Mediterranean Sea. In all cases, the defining feature of transitional systems between land and ocean is the salinity gradient, a spatial pattern in the mixture of water from two distinctly different sources: seawater and freshwater. When end-member concentrations vary more slowly than the flushing time and there are no intermediate transformations or sources, the spatial structure is determined primarily by mixing of the two end members, and properties vary linearly (conservatively) along the salinity gradient. Nonlinear patterns along the salinity gradient, on the other hand, imply estuarine processes of transformation and non-conservative behavior of the chemical element.

It is generally accepted that "antimony behaviour in estuaries seems to be rather variable depending on the estuary characteristics" [1]. However, this is a blank statement with low information content. With the objective of investigating whether it is possible to extract a general pattern for antimony from published data, all data found for 'dissolved' antimony versus salinity have been plotted in Fig. 9.6. This figure does not include the much cited study from Van der Sloot et al. [142]

because of serious doubts about analytical issues. No East China Sea data [152], where antimony was reported to negatively correlate with salinity (salinity range 23.40 and 34.90, n = 340), are shown because no data were published. Globally, antimony behavior seems to be rather conservative. This would agree with observations in rivers and lakes, but variability in each system is high, most data sets contain a limited number of points, and analytical and temporal reproducibility is never tested. When looking at Fig. 9.6, it is important to remember that deriving any behavior from these representations is based on the assumption that end members do not change over time due to events such as intrusions of upwelled ocean water, changes in river inflow with seasons or annual weather anomalies [143] as well as the fact that there is no source in the estuary. This is the case, for instance, of Tejo Estuary where inputs occur over a broad lateral expanse of the estuary even if they appear on the mixing diagram virtually as point source features [136].

The fate of 'particulate' antimony in estuaries has been much less studied. If conservative behavior of the 'dissolved' fraction is accepted (i.e., exchange with the 'dissolved' phase is excluded), 'particulate' antimony fate is expected to follow the general fate of particulate material, with extensive sedimentation and retention in estuaries and coastal zones. The fate of colloidal antimony was followed in the Severnaya Dvina Estuary and did not change with salinity [38]; thus the authors excluded extensive coagulation happening at the beginning of the mixing zone. Note that this colloidal fraction (<1 kDa) is usually measured as 'dissolved' in most studies.

Atmospheric input rapidly became the predominant input in the Gulf of Lions at the mouth of the Rhône River (Mediterranean Sea) while the riverine influence became negligible in the few tens of kilometers outside the river mouth [95]. More recently, the role of atmospheric input was confirmed in the East China Sea, a marginal sea in western Pacific [52] where box modeling suggested that total antimony input of antimony into the East China Sea is nearly half of the input from the rivers. Even if these calculations were based only on 'dissolved' antimony in the rivers and might also underestimate contributions from rainwater, they clearly confirm the importance of atmospheric contribution to antimony concentrations in coastal areas.

9.4.4 Oceans

The ocean is the last part of the hydrological cycle where chemical elements are removed and inserted into the mantle via subduction. The number of studies of antimony in the oceans is rather limited, and no new values have been measured for nearly 20 years. The first 'dissolved' antimony profiles in oceans were measured in 1969 in the Northeast Pacific Ocean (10–4,000 m) [144] and in the Sargasso Sea (1–5,069 m) [145]. Later, Middleburg and coworkers [146] published nine (unfiltered) profiles in the North Atlantic and, except a profile by Rouxel et al. [147] for the

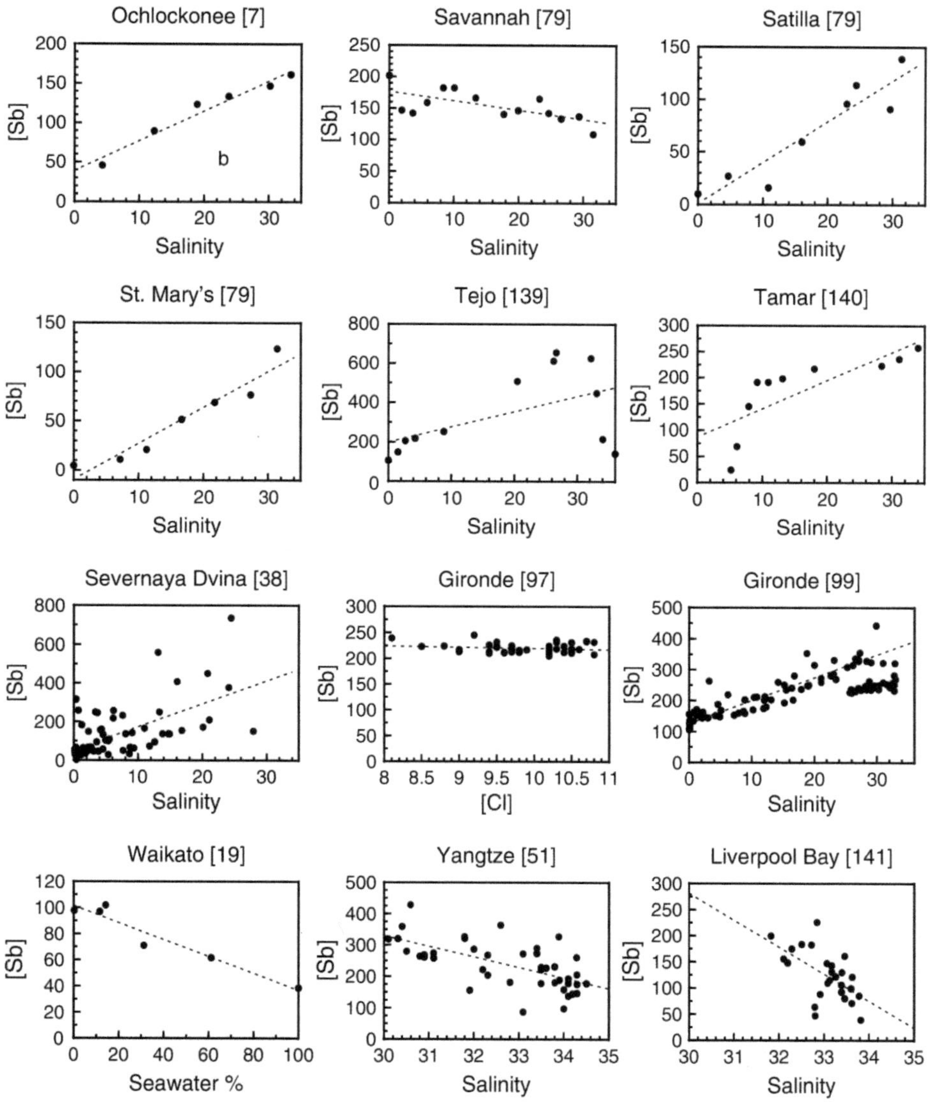

Fig. 9.6: Variation of antimony concentrations with salinity in world estuaries. Concentrations are 'dissolved' concentrations in ng L^{-1}. Salinity in units of parts per thousand (ppt or ‰), except when said otherwise.

Atlantic and a few for the Chatham Rise, New Zealand [28], all other existing profiles have been provided by Cutter and coworkers [24–27] and are associated with those shown in Figs. 9.7(a), (b) for the Pacific and Atlantic Oceans, respectively. The behavior of antimony has been qualified as "conservative" [144, 145], "mildly scavenged element with surface atmospheric. input and strong influences by the mixing of water masses with different Sb" [25, 26], "mildly scavenged" [27]. All profiles from Cutter

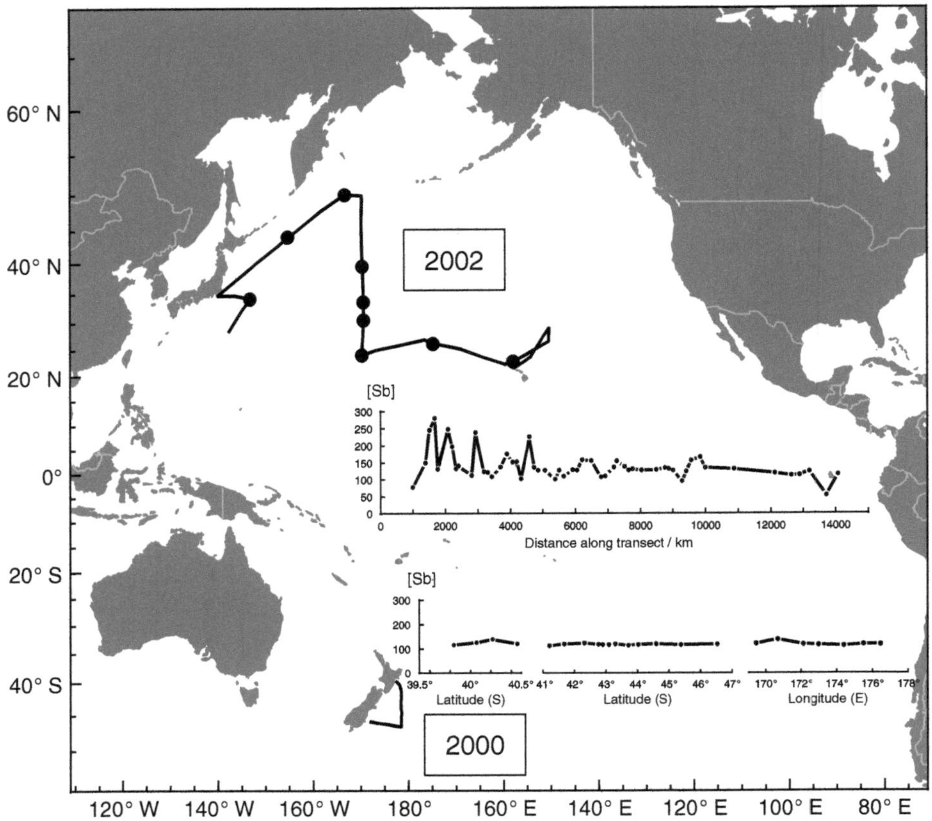

Fig. 9.7: Surface transect values of antimony concentrations in oceans. Concentrations are 'dissolved' concentrations in ng L^{-1}. (above) Pacific Ocean [27, 28]. (next page) Atlantic Ocean [24–26].

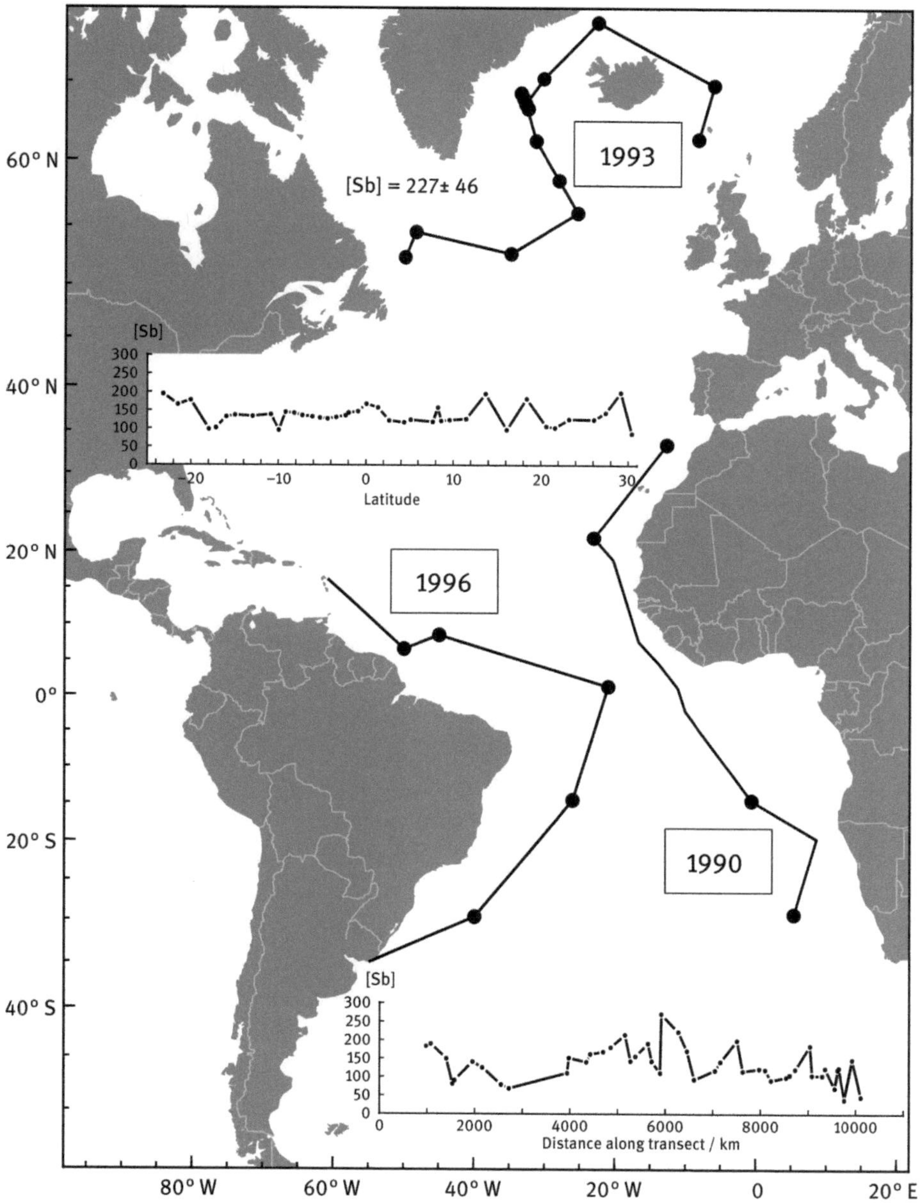

Fig. 9.7 (continued)

and coworkers and Ellwood and Maher campaigns are shown in Fig. 9.3. Profiles in [144, 145] have not been included because we estimate that their concentration values (averages: 210 and 304 ng L^{-1}, respectively) are too high probably due to the analytical methods available at that time (NAA on evaporated seawater). Middleburg et al. [146] data have not been added because samples are unfiltered. The variability of profiles in Fig. 9.3 is high (just compare it with the variability of antimony data in Lake Geneva in Fig. 9.2). This (in and in-between) profile variability has never been considered or discussed. Is this variability natural or is the consequence of poor sampling and measurement precision? Does this variability allow us to state that antimony concentration in Pacific AABW (139 ng L^{-1}) is "much higher" than in South Atlantic AABW (64 ng L^{-1}), confirming the scavenged element character of antimony [27]?

A mean value of 125 ng L^{-1} for 'dissolved' antimony concentrations in oceans can be estimated from profiles in Fig. 9.1(c). This value is lower than the one suggested in Filella et al. [1], 200 ng L^{-1}, and the 195 ng L^{-1} proposed by Bruland and Lohan [148]. Still, antimony concentrations are higher in oceans than in unpolluted river waters. This has also been observed, even at a higher scale, for molybdenum and attributed to the low reactivity of molybdenum's dominant form, an anion [149]. This is also the case for Sb(V) (Section 9.5.1).

A common characteristic in all studies is the mention of the importance of atmospheric deposition as a source of 'dissolved' antimony. This confirms observations in lakes (Section 9.4.2) and coastal seas (Section 9.4.3). Little information is available about the influence of hydrothermal inputs from active ridges on the supply of antimony to the oceans [147]. Metal fluxes in oceans are now being studied by integrating atmospheric deposition, vertical mixing, and upwelling into the mixed layer and into the particle production zone [150, 151]. The behavior of multiple oxidation state elements, including arsenic, has been studied with this approach [152], but antimony is absent from these studies, and our understanding remains based on much older data and simpler methodological approaches.

'Dissolved' antimony is primarily in the pentavalent state in ocean waters. Antimony(III) and methylated antimony, primarily monomethyl antimony (Cutter's studies) but also dimethyl and trimethyl species [28] have been found in surface waters. Conservative behavior of monomethyl antimony in the Pacific Ocean has been reported [27]; this behavior is similar to the one observed for organic species of germanium [153]. More measurements in oceans and similar observations are lacking in freshwaters concerning the presence and behavior of alkylantimony compounds.

9.5 Chemistry at work

The understanding of the environmental behavior of any element requires the integration of knowledge at the macroscopic level – the main subject of this chapter – with the chemical and biological processes at play at the microscopic level. Do observations in Section 9.4 agree with our current comprehension of chemical processes?

9.5.1 Redox and solubility issues

Thermodynamics predicts that (as found in Section 9.4), antimony should be present as Sb(V) in oxic waters with Sb(V) being an anion, octahedral $Sb(OH)_6^-$, in most of the natural pH range [154]. Sb(V) forms $Sb(OH)_5$ (or $HSb(OH)_6$) at acidic pH values (pK 2.8), outside natural pH values. The solubility behavior of Sb(V) was interpreted in the past on the basis of the formation of Sb_2O_5(s), but the formation of this species in environmental conditions has been discredited [155]. Sb(V) solubility is linked to the formation of calcium antimonates (whose dissolution is incongruent and cannot be predicted with a single solubility product) [155, 156] and/or tripuhyite ($FeSbO_4$(s)) [74], but their formation seems unlikely in waters due to either solubility or kinetic considerations (see Section 9.3).

Even if antimony is mostly present as Sb(V) in oxic systems, it is common to find a small percentage of antimony as Sb(III) [1]. The lack of CRM makes it impossible to assess the concentrations reported, but the presence of Sb(III) itself seems indisputable. Possible reasons have been discussed in the past [9], and no conclusive new insights have been provided later. As mentioned in Section 9.3, oxidation and reduction reactions often show slow kinetics, particularly Sb(V) reduction. Note, however, that, irrespective of the causes behind, the concentration levels of Sb(III) reported will not influence the behavior of the element in the aquatic continuum, provided oxic conditions are maintained.

At the adequate redox potential, the uncharged, soluble, tetrahedral $Sb(OH)_3$ species exists over a wide pH range (pK values 1.3 and 11.8). The solubility of Sb_2O_3(s) – with two isomorphic forms: senarmontite, cubic, formed at ambient temperature and valentinite, orthorhombic, formed above 600 °C – is well characterized [154]; Sb(III) concentration in waters does not appear to be limited by Sb_2O_3(s) solubility. Whether this is also the case of atmosphere-borne antimony is an entirely open question.

Note that antimony in hydrothermal systems shows an entirely different behavior with Sb(III) becoming the dominant oxidation state [157, 158].

9.5.2 Influence of natural organic matter

Most existing evidence does not support a significant binding of antimony by natural organic matter (NOM) [9, 12, 13, 88, 159]. However, as shown in Fig. 9.3, a few studies found significant formation of Sb-NOM complexes. UV digestion was applied in these studies but in no other; it is therefore impossible to judge whether this procedure generates an analytical artifact or the claimed complexes exist in natural waters and have been missed in all other studies.

The lack of affinity of Sb(V) and NOM has been explained by both compounds being negatively charged at usual environmental pH values. Laboratory studies of Sb(V) complexation by low molecular mass (l.m.m.) organic ligands [160] only showed complexation at acidic pH values. The possible role of 'anchors' such as Ca and Fe ions has also been evoked to explain some observations [161] but remains a tentative hypothesis.

Observed complexation of Sb(III) by l.m.m. organic ligands in aqueous solution is limited to α-hydroxy-carboxylic acids [162]. Complexation by NOM is stronger for Sb(III) than for Sb(V) and extends over a larger pH range but does not cover all natural waters: complexation studies of Sb(III) by humic-type compounds show an abrupt decrease of binding at pH values above 6 [163]. Since 2012, studies have been published concerning the role of thiol functional groups of solid peat NOM in the sequestration of Sb(III) in anoxic, sulfur-enriched environments [164]. They do not change our current understanding concerning oxic waters but do suggest a possible role of antimony complexation by some types of organic matter in anoxic media like some sediments.

It is necessary to mention that not all NOM in waters are of humic or fulvic types [165]. Binding of antimony by other types of NOM has never been investigated.

9.5.3 Influence of iron oxy(hydroxides)

It is widely written and generally accepted that, under oxic to slightly reducing conditions, the environmental mobility of Sb(V) and Sb(III) is controlled by sorption or co-precipitation reactions with iron oxy(hydroxides). This is supported by many laboratory studies, experimental results based on extractions, and EXAFS observations mostly performed on soils. However, macroscopic observations described in this chapter, pointing to a conservative behavior of antimony in lakes and estuaries and "mildly recycled" in oceans, together with observations in sediments (Tab. 9.1) and generally low K_d values, suggest a more complex picture with clear decoupling of iron and antimony biogeochemical cycles in some systems but not in others. On the basis of existing data, it is impossible to extract any clear conclusion except that the 'generally accepted' role of iron might not be as general as often assumed. Interaction of antimony and iron is not a simple issue; it involves sorption but also redox and precipitation/dissolution reactions, competition with other elements present in larger amounts in waters such as arsenic and phosphorous, simultaneous complexation of antimony and iron by other ligands such as natural organic matter, sulfide, and so on, with all processes subjected to kinetic controls that are often concentration dependent. A systematic evaluation of existing data and their applicability to natural water conditions is pending.

9.5.4 Influence of sulfide

The interaction of antimony with sulfide in sulfide-containing anoxic systems such as sediments and wetlands is not well understood. Sulfide has often been invoked to explain observations in freshwater sediments (Tab. 9.3), but with detailed measurements rarely performed [56]. Contaminated wetlands have received more attention ([166] and references therein). Three, now old, sets of data are systematically cited when discussing the subject: the Baltic Sea ($1 \times 10^{-6} - 5 \times 10^{-5}$ mol kg^{-1} S(-II)) [167], the Black Sea (4×10^{-4} mol kg^{-1} S(-II)) [168], and anoxic brines in the Bannock and Tyro basins in the eastern Mediterranean Sea [169]. Modeling these systems has been attempted using entirely different chemical species datasets [170, 171], never with entirely satisfactory results. Helz et al. [170] modeled Black Sea data considering the existence of $HSb_2S_5^-$, a complex formed by reaction of stibnite with zero-valent sulfur where antimony is simultaneously present as Sb(III) and Sb(V). This chemical species has the advantage of reducing the solubility of stibnite by three orders of magnitude and explaining the unexpected presence of Sb(V) in anoxic waters. Nevertheless, authors failed to explain experimental results entirely and concluded that "we require still better and more complex complete equilibrium information about antimony speciation." Olsen et al. [171] modeled these systems by considering Sb_2S_3(s), $Sb_2S_4^{2-}$, $HSb_2S_4^-$, $H_2Sb_2S_5^{2-}$, and H_3SbS_2O and concluded that their calculations managed to show "the importance of sulfide ligands but Sb(III) interactions with additional ligands are needed." Actually, these conclusions point to the fact that, besides the intrinsic experimental difficulties associated with the study of sulfide-containing natural waters and sediments, the problem of understanding such systems already starts at the level of the characterization of these solution equilibria in the laboratory. Existing constants and the stoichiometry of the corresponding species rely on a very limited number of studies [171–174] and their reinterpretation [175–177], all giving different sets of species and corresponding equilibrium constants. Current incongruities in our current understanding of the solution chemistry of these systems need to be solved at the laboratory level before sound progress will be possible in natural waters.

9.6 Biota at work

The information available on the role of living organisms in antimony environmental fate is very limited. Biological activity has been invoked in the past to explain the presence of thermodynamically unstable Sb(III) in oxygenated waters [10], though as yet the experimental evidence is weak. Very few authors have studied antimony reactions in algae and studies are far from recent [178–181] except for Lehr et al. [182], who showed that a species of eukaryotic, acidothermophilic alga, belonging to the order Cyanidiales, can oxidize Sb(III). Unlike other elements, antimony remained unchanged when studying the diurnal cycle of photosynthesis in small boreal lakes [36, 37]. The contribution of algae to antimony fate in natural waters seems to be, in any case, much more limited than for arsenic [183].

The capacity of *Acidithiobacillus ferroxidans* and *Stibiobacter senarmontii* to oxidize Sb(III) has been known for years (original references in [10]). More recently, a number of studies have been reporting bacteria obtained from highly contaminated soils to be able to oxidize Sb(III) [182, 184–188] and, less often, to reduce Sb(V) [188–190]. The objective of these studies was mostly focused on bioremediation and, in a few cases, they also dealt with systems such as uncontaminated lake sediments [189, 190]. The actual role of bacteria in antimony processes in sediments remains, in practice, unexplored.

9.7 Conclusions

Antimony seems to travel along the aquatic continuum relatively unreactively, provided oxic conditions are maintained. It is somewhat surprising that only limited data are available, which are analytically sound and information-rich in spite of the high number of publications available. The paucity of recent antimony studies in oceans is particularly striking. Many aspects of the environmental chemistry of antimony could benefit from further research, but probably the issues in need of more attention are the understanding of the interplay of antimony and iron (and manganese) processes in aquatic systems and the comprehension of the contribution of air-borne antimony to the cycling of the element in natural waters.

References

[1] Filella M, Belzile N, Chen Y-W. Antimony in the environment: A review focused on natural waters. I. Occurrence. Earth Sci Rev 2002, 57, 125–176.

[2] Udodov PA, Parilov Y. Certain regularities of migration of metals in natural waters. Geochem 1961, 8, 763–771.

[3] Landström O, Wenner CG. Neutron Activation Analysis of Natural Water Applied to Hydrology. Report AE-204. Aktibolaget Atomenergi, Stockholm., 1965 Cited in Onishi H. Antimony. In: Wedepohl KH, ed., Handbook of Geochemistry. Heidelberg, Springer, Vol. 2/4, 1969, 51-B-1–51-O-3.
[4] Schutz DF, Turekian KK. The investigation of the geographical and vertical distribution of several trace elements in sea water using neutron activation analysis. Geochim Cosmochim Acta 1965, 29, 259–313.
[5] Braman RS, Justen LL, Foreback CC. Direct volatilization-spectral emission type detection system for nanogram amounts of arsenic and antimony. Anal Chem 1972, 44, 2195–2199.
[6] Braman RS, Tompkins MA. Atomic emission spectrometric determination of antimony, germanium, and methylgermanium compounds in the environment. Anal Chem 1978, 50, 1088–1093.
[7] Andreae MO, Asmodé J-F, Foster P, Van't Dack L. Determination of antimony(III), antimony(V), and methylantimony species in natural waters by atomic absorption spectrometry with hydride generation. Anal Chem 1981, 53, 1766–1771.
[8] Xenopoulos MA, Downing JA, Kumar MD, Menden-Deuer S, Voss M. Headwaters to oceans: Ecological and biogeochemical contrasts across the aquatic continuum. Limnol Oceanogr 2017, 62, S3–S14.
[9] Filella M, Belzile N, Chen Y-W. Antimony in the environment: A review focused on natural waters. II. Relevant solution chemistry. Earth Sci Rev 2002, 59, 265–285.
[10] Filella M, Belzile N, Lett M-C. Antimony in the environment: A review focused on natural waters. III. Microbiota relevant interactions. Earth Sci Rev 2007, 80, 195–217.
[11] Filella M, Williams PA, Belzile N. Antimony in the environment: Knowns and unknowns. Environ Chem 2009, 6, 95–105.
[12] Filella M. Antimony interactions with heterogeneous complexants in waters, sediments and soils: A review of data obtained in bulk samples. Earth Sci Rev 2011, 107, 325–341.
[13] Filella M, Williams PA. Antimony interactions with heterogeneous complexants in waters, sediments and soils: A review of binding data for homologous compounds. Chem Erde – Geochem 2012, 72(S4), 49–65.
[14] Filella M, Belzile N, Chen Y-W. Human exposure to antimony. II. Contents in some human tissues often used in biomonitoring (hair, nails, teeth). Crit Rev Environ Sci Technol 2012, 42, 1058–1115.
[15] Filella M, Belzile N, Chen Y-W. Human exposure to antimony. III. Contents in some human excreted biofluids (urine, milk, saliva). Crit Rev Environ Sci Technol 2013, 43, 204–254.
[16] Filella M, Belzile N, Chen Y-W. Human exposure to antimony. IV. Contents in human blood. Crit Rev Environ Sci Technol 2013, 43, 2071–2105.
[17] Do Hur S, Soyol-Erdene T-O, Hwang HJ, Han C, Gabrielli P, Barbante C, Boutron CF, Hong S. Climate-related variations in atmospheric Sb and Tl in the EPICA Dome C ice (East Antarctica) during the past 800,000 years. Global Biogeochem Cycles 2013, 27, 930–940.
[18] Grahn E, Karlsson S, Düker A. Sediment reference concentrations of seldom monitored trace elements (Ag, Be, In, Ga, Sb, Tl) in four Swedish boreal lakes – Comparison with commonly monitored elements. Sci Total Environ 2006, 367, 778–790.
[19] Wilson N, Webster-Brown J. The fate of antimony in a major lowland river system, the Waikato River, New Zealand. Appl Geochem 2009, 24, 2283–2292.
[20] Ödman F, Ruth T, Rodushkin I, Pontér C. Validation of a field filtration technique for characterization of suspended particulate matter from freshwater. Part II. Minor, trace and ultra trace elements. Appl Geochem 2006, 21, 2112–2134.
[21] Shotyk W, Bicalho B, Cuss CW, Donner MW, Grant-Weaver I, Haas-Neill S, Javed MB, Krachler M, Noernberg T, Pelletier R, Zaccone C. Trace metals in the dissolved fraction

(< 0.45 μm) of the lower Athabasca River: Analytical challenges and environmental implications. Sci Total Environ 2017, 580, 660–669.

[22] Planquette H, Sherrell RM. Sampling for particulate trace element determination using water sampling bottles: Methodology and comparison to in situ pumps. Limnol Oceanogr: Methods 2012, 10, 367–388.

[23] Rodríguez-Murillo JC, Nirel P, Filella M. Are trace element concentrations increasing in freshwaters? An insight into the last 20 years of Lake Geneva, Switzerland. H2Open J 2018, 1, 87–89.

[24] Cutter GA, Cutter LS. Behaviour of dissolved antimony, arsenic, and selenium in the Atlantic Ocean. Mar Chem 1995, 49, 295–306.

[25] Cutter GA, Cutter LS. Metalloids in the high latitude North Atlantic Ocean: Sources and internal cycling. Mar Chem 1998, 61, 25–36.

[26] Cutter GA, Cutter LS, Featherstone AM, Lohrenz SE. Antimony and arsenic biogeochemistry in the western Atlantic Ocean. Deep-Sea Res II 2001, 48, 2895–2915.

[27] Cutter GA, Cutter LS. Biogeochemistry of arsenic and antimony in the North Pacific Ocean. Geochem Geophys Geosyst 2006, 7, 1–12.

[28] Ellwood MJ, Maher WA. Arsenic and antimony species in surface transects and depth profiles across a frontal zone: The Chatham Rise, New Zealand. Deep-Sea Res I 2002, 49, 1971–1981.

[29] Filella M. Colloidal Properties of submicron particles in natural waters. In: Wilkinson KW, Lead J, eds, Environmental Colloids and Particles: Behaviour, Separation and Characterisation. Chichester, IUPAC-Wiley, 2007, 17–93.

[30] Tanizaki Y, Yamazaki M, Nagatsuka S. Physicochemical speciation of trace elements in river water by means of ultrafiltration. Bull Chem Soc Japan 1985, 58, 2995–3002.

[31] Tanizaki Y, Shimokawa T, Nakamura M. Physicochemical speciation of trace elements in river waters by size fractionation. Environ Sci Technol 1992, 26, 1433–1444.

[32] Tanizaki Y, Shimokawa T, Yamazaki M. Physico-chemical speciation of trace elements in urban streams by size fractionation. Water Res 1992, 26, 55–63.

[33] Vignati D. Trace metal partitioning in freshwater as a function of environmental variables and its implications for metal bioavailability. Terre Environ 2004, 44, 1.

[34] Kottelat R, Vignati DAL, Chanudet V, Dominik J. Comparison of small- and large-scale ultrafiltration systems for organic carbon and metals in freshwater at low concentration factor. Water Air Soil Poll 2007, 187, 343–351.

[35] Pokrovsky OS, Schott J. Iron colloids/organic matter associated transport of major and trace elements in small boreal rivers and their estuaries (NW Russia). Chem Geol 2002, 190, 141–179.

[36] Shirokova LS, Pokrovsky OS, Viers J, Klimov SI, Moreva OY, Zabelina SA, Vorobieva TY, Dupré B. Diurnal variations of trace metals and heterotrophic bacterioplankton concentration in a small boreal lake of the White Sea basin. Ann Limnol – Int J Limnol 2010, 46, 67–75.

[37] Pokrovsky OS, Shirokova LS. Diurnal variations of dissolved and colloidal organic carbon and trace metals in a boreal lake during summer bloom. Water Res 2013, 47, 922–932.

[38] Pokrovsky OS, Shirokova LS, Viers J, Gordeev VV, Shevchenko VP, Chupakov AV, Vorobieva TY, Candaudap F, Causserand C, Lanzanova A, Zouiten C. Fate of colloids during estuarine mixing in the Arctic. Ocean Sci 2014, 10, 107–125.

[39] Vorobyev SN, Pokrovsky OS, Kirpotin SN, Kolesnichenko LG, Shirokova LS, Manasypov RM. Flood zone biogeochemistry of the Ob River middle course. Appl Geochem 2015, 63, 133–145.

[40] Wang Y, Kong L, He M, Ouyang W, Lin C, Liu X. Influences of particles and aquatic colloids on the oxidation of Sb(III) in natural water. ACS Earth Space Chem 2020, 4, 661–671.

[41] Filella M, Buffle J. Factors controlling the stability of submicron colloids in natural waters. Coll Surf 1993, 73, 255–273.

[42] Town RM, Filella M. Size fractionation of trace metal species in freshwaters: Implications for understanding their behaviour and fate. Rev Environ Sci Biotechnol 2002, 1, 277–297.
[43] Daus B, Hansen HR. Analysis of antimony species – lessons learnt from more than two decades of environmental research. Environ Chem 2016, 13, 913–918.
[44] Filella M. Alkyl derivatives of antimony in the environment. In: Sigel A, Sigel H, Sigel RKO, eds., Organometallics Environment and Toxicology. Metal Ions in Life Science. Cambridge, RSC Publishing, Vol. 7, 2010, 267–301.
[45] Bertine KK, Lee DS. Antimony Content and Speciation in the Water Column and Interstitial Waters of Saanich Inlet. In: Wong CS, Boyle E, Bruland KW, Burton JD, Goldberg ED, eds, Trace Metals in Sea Water. New York, NATO Adv. Res. Inst. Plenum Press, 1983, 21–38.
[46] Gillain G, Brihaye C. A routine speciation method for a pollution survey of coastal sea water. Oceanologica Acta 1985, 8, 231–235.
[47] van der Sloot HA, Hoede D, Wijkstra J. Trace oxyanions and their behaviour in the rivers Porong and Solo, the Java Sea and the adjacent Indian Ocean. Neth J Sea Res 1989, 23, 379–386.
[48] Mok W-M, Wai CM. Distribution and mobilization of arsenic and antimony species in the Coeur d'Alene River, Idaho. Environ Sci Technol 1990, 24, 102–108.
[49] Takayanagi K, Cossa D. Vertical distributions of Sb(III) and Sb(V) in Pavin Lake, France. Water Res 1997, 31, 671–674.
[50] Deng T-L, Chen Y-W, Belzile N. Antimony speciation at ultra trace levels using hydride generation atomic fluorescence spectrometry and 8-hydroxyquinoline as an efficient masking agent. Anal Chim Acta 2001, 432, 293–302.
[51] Wu XD, Song JM, Li XG, Yuan HM, Li N. Behaviors of dissolved antimony in the Yangtze River Estuary and its adjacent waters. J Environ Monit 2011, 13, 2292–2303.
[52] Ren J-L, Zhang X-Z, Sun Y-X, Liu S-M, Huang D, Zhang J. Antimony and arsenic biogeochemistry in the East China Sea. Deep Sea Res II 2016, 124, 29–42.
[53] Crecelius EA, Bothner MH, Carpenter R. Geochemistries of arsenic, antimony, mercury, and related elements in sediments of the Puget Sound. Environ Sci Technol 1975, 9, 325–333.
[54] Leleyter L, Probst J-L. A new sequential extraction procedure for the speciation of particulate trace elements in river sediments. Int J Environ Anal Chem 1999, 73, 109–128.
[55] Mao L, Liu L, Yan N, Li F, Tao H, Ye H, Wen H. Factors controlling the accumulation and ecological risk of trace metal(loid)s in river sediments in agricultural field. Chemosphere 2020, 243, 125359.
[56] Chen Y-W, Deng T-L, Filella M, Belzile N. Distribution and early diagenesis of antimony species in sediments and porewaters of freshwater lakes. Environ Sci Technol 2003, 37, 1163–1168.
[57] Larner BL, Palmer AS, Seen AJ, Townsend AT. A comparison of an optimised sequential extraction procedure and dilute acid leaching of elements in anoxic sediments, including the effects of oxidation on sediment metal partitioning. Anal Chim Acta 2008, 608, 147–157.
[58] Ren M, Wang D, Ding S, Yang L, Xu S, Yang C, Wang Y, Zhang C. Seasonal mobility of antimony in sediment-water systems in algae- and macrophyte-dominated zones of Lake Taihu (China). Chemosphere 2019, 223, 108–116.
[59] Ma X, Li C, Yang L, Ding S, Zhang M, Zhang Y, Zhao T. Evaluating the mobility and labile of As and Sb using diffusive gradients in thin-films (DGT) in the sediments of Nansi Lake, China. Sci Total Environ 2020, 713, 136569.
[60] Duan L, Song J, Xu Y, Li X, Zhang Y. The distribution, enrichment and source of potential harmful elements in surface sediments of Bohai Bay, North China. J Hazard Mater 2010, 183, 155–164.

[61] Zhao S, Feng C, Yang Y, Niu J, Shen Z. Risk assessment of sedimentary metals in the Yangtze Estuary: New evidence of the relationships between two typical index methods. J Hazard Mater 2012, 241–242, 164–172.
[62] Duan L-Q, Song J-M, Yuan H-M, Li X-G, Li N, Ma J-K. Distribution, chemical speciation and source of trace elements in surface sediments of the Changjiang Estuary. Environ Earth Sci 2014, 72, 3193–3204.
[63] Mao L, Ye H, Li F, Tao H, Shi L, Chen L, Shi W, Yan G, Chen H. Source-oriented variation in trace metal distribution and fractionation in sediments from developing aquaculture area-A case study in south Hangzhou bay, China. Mar Pollut Bull 2017, 125, 389–398.
[64] Liang X, Song J, Duan L, Yuan H, Li X, Li N, Qu B, Wang Q. Metals in size-fractionated core sediments of Jiaozhou Bay, China: Records of recent anthropogenic activities and risk assessments. Mar Pollut Bull 2018, 127, 198–206.
[65] Liang X, Song J, Duan L, Yuan H, Li X, Li N, Qu B, Wang Q, Xing J. Source identification and risk assessment based on fractionation of heavy metals in surface sediments of Jiaozhou Bay, China. Mar Pollut Bull 2018, 128, 548–556.
[66] Morgan JJ, Stone AT. Kinetics of Chemical Processes of Importance in Lacustrine Environments. In: Stumm W, ed., Chemical Processes in Lakes. New York, Wiley, 1985, 389–426.
[67] Wilkins RG. Kinetics and Mechanisms of Reactions of Transition Metal Complexes, 2nd edition. VCH, Weinheim, Wiley-VCH, 1991.
[68] Richens DT. The Chemistry of Aqua Ions: Synthesis, Structure and Reactivity: A Tour through the Periodic Table of the Elements. Chichester, Wiley, 1997.
[69] Hansen HR, Pergantis SA. Investigating the formation of an Sb(V)–citrate complex by HPLC-ICPMS and HPLC-ES-MS(/MS). J Anal At Spectrom 2006, 21, 1240–1248.
[70] Li Y, Liu J-M, Han F, Jiang Y, Yan X-P. Probing interactions of antimony species with DNA by short column capillary electrophoresis coupled with inductively coupled plasma mass spectrometry. J Anal At Spectrom 2011, 26, 94–99.
[71] Cutter GA. Kinetic controls on metalloid speciation in seawater. Mar Chem 1992, 40, 65–80.
[72] Belzile N, Chen Y-W, Wang Z. Oxidation of antimony (III) by amorphous iron and manganese oxyhydroxides. Chem Geol 2001, 174, 379–387.
[73] Majzlan J, Števko M, Lánczos T. Soluble secondary minerals of antimony in Pezinok and Kremnica (Slovakia) and the question of mobility or immobility of antimony in mine waters. Environ Chem 2016, 13, 927–935.
[74] Leverett P, Reynolds JK, Roper AJ, Williams PA. Tripuhyite and schafarzikite: Two of the ultimate sinks for antimony in the natural environment. Min Mag 2012, 76, 891–902.
[75] Gaillardet J, Viers J, Dupré B. Trace Elements in River Waters. In: Drever JI, Surface and Ground Water, Weathering, and Soils. Treatise on Geochemistry, Vol. 5, 2003, 225–272.
[76] Gaillardet J, Viers J, Dupré B. Trace Elements in River Waters. In: Turekian KK, ed., Treatise on Geochemistry. Oxford, Elsevier, 2nd edn, 2014, 195–123.
[77] Salminen R, Batista MJ, Bidovec M, Demetriades A, De Vivo B, De Vos W, Duris M, Gilucis A, Gregorauskiene V, Halamic J, Heitzmann P, Lima A, Jordan G, Klaver G, Klein P, Lis J, Locutura J, Marsina K, Mazreku A, O'Connor PJ, Olsson SA, Ottesen R-T, Petersell V, Plant JA, Reeder S, Salpeteur I, Sandström H, Siewers U, Steenfelt A, Tarvainen T "Geochemical atlas of Europe. Part 1: Background information, methodology and maps." (Geological Survey of Finland: Espoo) (2005).
[78] Pokrovsky OS, Manasypov RM, Loiko SV, Krickov IA, Kopysov SG, Kolesnichenko LG, Vorobyev SN, Kirpotin SN. Trace element transport in western Siberian rivers across a permafrost gradient. Biogeosciences 2016, 13, 1877–1900.

[79] Byrd JT. Comparative geochemistries of arsenic and antimony in rivers and estuaries. Sci Total Environ 1990, 97(98), 301–314.
[80] Lawrence MG, Greig A, Collerson KD, Kamber BS. Rare earth element and yttrium variability in South East Queensland waterways. Aquat Geochem 2006, 12, 39–72.
[81] Seyler PT, Boaventura GR. Distribution and partition of trace metals in the Amazon basin. Hydrol Process 2003, 17, 1345–1361.
[82] Zhu X, Lang Y-C, Zhong J, Ding H, He H, Yan Z, Li S-L. Variations of trace elements under hydrological conditions in the Min River, Eastern Tibetan Plateau. Acta Geochim 2018, 37, 509–518.
[83] Liu J, Chen B, Xu ZY, Wei Y, Su ZH, Yang R, Ji YX, Wang XD, Zhang LL, An N, Yang F. Tracing solute sources and carbon dynamics under various hydrological conditions in a karst river in southwestern China. Environ Sci Pollut Res 2020, 27, 11375–11386.
[84] Ollivier P, Radakovitch O, Hamelin B. Major and trace element partition and fluxes in the Rhône River. Chem Geol 2011, 285, 15–31.
[85] Gil-Díaz T, Schäfer J, Coynel A, Bossy C, Dutruch L, Blanc G. Antimony in the Lot–Garonne river system: A 14-year record of solid–liquid partitioning and fluxes. Environ Chem 2018, 15, 121–136.
[86] Seyler P, Elbaz-Poulichet F. Biogeochemical control on the temporal variability of trace element concentrations in the Oubangui river (Central African Republic). J Hydrol 1996, 180, 319–332.
[87] Vorobyev SN, Pokrovsky OS, Kolesnichenko LG, Manasypov RM, Shirokova LS, Karlsson J, Kirpotin SN. Biogeochemistry of dissolved carbon, major, and trace elements during spring flood periods on the Ob River. Hydrol Proc 2019, 33, 1579–1594.
[88] Chupakov AV, Pokrovsky OS, Moreva OY, Shirokova LS, Neverova NV, Chupakova AA, Kotova EI, Vorobyeva TY. High resolution multi-annual riverine fluxes of organic carbon, nutrient and trace element from the largest European Arctic river, Severnaya Dvina. Chem Geol 2020, 538, 119491.
[89] Matsunaga T, Tsuduki K, Yanase N, Kritsananuwat R, Ueno T, Hanzawa Y, Naganawa H. Temporal variations in metal enrichment in suspended particulate matter during rainfall events in a rural stream. Limnol 2013, 15, 13–25.
[90] Coynel A, Schäfer J, Dabrin A, Girardot N, Blanc G. Groundwater contributions to metal transport in a small river affected by mining and smelting waste. Water Res 2007, 41, 3420–3428.
[91] Viers J, Dupre B, Gaillardet J. Chemical composition of suspended sediments in World Rivers: New insights from a new database. Sci Total Environ 2009, 407, 853–868.
[92] Martin JM, Meybeck M. Elemental mass balance of material carried by major world rivers. Mar Chem 1979, 7, 173–206.
[93] Savenko VS Chemical composition of World River's suspended matter. GEOS 2006 [175pp. in Russian]. Cited in Viers et al. (2009).
[94] Habib S, Minski MJ. Incidence and variability of some elements in the non tidal region of the River Thames, and River Kennet, U.K. Sci Total Environ 1982, 22, 253–273.
[95] Guieu C, Zhang J, Thomas AJ, Martin JM, Brun-Cottan JC. Significance of atmospheric fallout on the upper layer water chemistry of the North-Western Mediterranean. J Atm Chem 1993, 17, 45–60.
[96] Masson M, Schäfer J, Blanc G, Dabrin A, Castelle S, Lavaux G. Behavior of arsenic and antimony in the surface freshwater reaches of a highly turbid estuary, the Gironde Estuary, France. Appl Geochem 2009, 24, 1747–1756.

[97] Masson M, Lanceleur L, Tercier-Waeber M-L, Schäfer J, Hezard T, Larrose A, Bossy C, Blanc G. Distribution and reactivity of oxyanions (Sb, As, V, Mo) in the surface freshwater reaches of the Gironde Estuary (France). App Geochem 2011, 26, 1222–1230.

[98] Asaoka S, Takahashi Y, Araki Y, Tanimizu M. Comparison of antimony and arsenic behavior in an Ichinokawa River water–sediment system. Chem Geol 2012, 334, 1–8.

[99] Gil-Díaz T, Schäfer J, Pougnet F, Abdou M, Dutruch L, Eyrolle-Boyer F, Coynel A, Blanc G. Distribution and geochemical behaviour of antimony in the Gironde Estuary: A first qualitative approach to regional nuclear accident scenarios. Mar Chem 2016, 185, 65–73.

[100] Yin S, Wu Y, Xu W, Li Y, Shen Z, Feng C. Contribution of the upper river, the estuarine region, and the adjacent sea to the heavy metal pollution in the Yangtze Estuary. Chemosphere 2016, 155, 564–572.

[101] Feng C, Guo X, Yin S, Tian C, Li Y, Shen Z. Heavy metal partitioning of suspended particulate matter-water and sediment-water in the Yangtze Estuary. Chemosphere 2017, 185, 717–725.

[102] Li Y-H, Burkhardt L, Buchholtz M, O'Hara P, Santschi PH. Partition of radiotracers between suspended particles and seawater. Geochim Cosmochim Acta 1984, 48, 2011–2019.

[103] Berner EK, Berner RA. Global Environment: Water, Air, and Geochemical Cycles. Englewood Cliffs, NJ, Prentice-Hall, 1996, 376.

[104] Oelkers EH, Gislason SR, Eiriksdottir ES, Jones M, Pearce CR, Jeandel C. The role of riverine particulate material on the global cycles of the elements. Appl Geochem 2011, 26, S365–S369.

[105] Milliman JD, Farnsworth KL. River Discharge to the Coastal Ocean: A global synthesis. Cambridge, Cambridge University Press, 2013.

[106] Grill G, Lehner B, Thieme M, et al. Mapping the world's free-flowing rivers. Nature 2019, 569, 215–221.

[107] Feng C, Zhao S, Wang D, Niu J, Shen Z. Sedimentary records of metal speciation in the Yangtze Estuary: Role of hydrological events. Chemosphere 2014, 107, 415–422.

[108] Vriens B, Voegelin A, Hug SJ, Kaegi R, Winkel LHE, Buser AM, Berg M. Quantification of element fluxes in wastewaters: A nationwide survey in Switzerland. Environ Sci Technol 2017, 51, 10943–10953.

[109] Yoshida H, Christensen TH, Guildal T, Scheutz C. A comprehensive substance flow analysis of a municipal wastewater and sludge treatment plant. Chemosphere 2015, 138, 874–882.

[110] Hargreaves AJ, Vale P, Whelan J, Constantino C, Dotro G, Cartmell E. Mercury and antimony in wastewater: Fate and treatment. Water Air Soil Pollut 2016, 227, 89.

[111] Messager ML, Lehner B, Grill G, Nedeva I, Schmitt O. Estimating the volume and age of water stored in global lakes using a geo-statistical approach. Nat Commun 2016, 7, 13603.

[112] Rossmann R, Barres J. Trace element concentrations in near-surface waters of the Great Lakes and methods of collection, storage, and analysis. J Great Lakes Res 1988, 14, 188–204.

[113] Temnerud J, Düker A, Karlsson S, Allard B, Bishop K, Fölster J, Köhler S. Spatial patterns of some trace elements in four Swedish stream networks. Biogeosciences 2013, 10, 1407–1423.

[114] Stabel HH, Kleiner J, Merkel P, Sinemus HW. Stoffkreisläufe ausgewählter Spurenelemente im Bodensee. Vom Wasser 1991, 76, 73–91.

[115] Sinemus H-W, Kleiner J, Stabel -H-H, Radziuk B. Combination of flow injection hydride generation and sequestration on a graphite tube for the automated determination of antimony in potable and surface waters. J Anal At Spectrom 1992, 7, 433–437.

[116] Moiseenko TI, Dinu MI, Gashkina NA, Kremleva TA. Aquatic environment and anthropogenic factor effects on distribution of trace elements in surface waters of European Russia and Western Siberia. Environ Res Let 2019, 14. doi: 10.1088/1748-9326/ab17ea

[117] Allen RO, Steinnes E. A contribution to the geochemistry of lakes in Norway. Nor Geol Unders Bull 1987, 409, 35–48.
[118] Suturin AN, Paradina LF, Epov VN, Semenov AR, Lozhkin VI, Petrov LL. Preparation and assessment of a candidate reference sample of Lake Baikal deep water. Spectrochim Acta B 2003, 58, 277–288.
[119] Rahmi D, Zhu Y, Umemura T, Haraguchi H, Itoh A, Chiba K. Determination of 56 elements in Lake Baikal Water by High-Resolution ICP-MS with the aid of a Tandem Preconcentration method. Anal Sci 2008, 24, 1513–1517.
[120] Abbott MB, Wolfe AP. Intensive pre-Incan metallurgy recorded by lake sediments from the Bolivian Andes. Science 2003, 301, 1893–1895.
[121] Andrade CF, Jamieson HE, Kyser TK, Praharaj T, Fortin D. Biogeochemical redox cycling of arsenic in mine-impacted lake sediments and co-existing pore waters near Giant Mine, Yellowknife Bay, Canada. Appl Geochem 2010, 25, 199–211.
[122] Cooke C, Abbott MB, Wolfe AP, Kittleson JL. A millennium of metallurgy recorded by lake sediments from Morococha, Peruvian Andes. Environ Sci Technol 2007, 41, 3469–3474.
[123] Kusunoki K, Sakata M, Tani Y, Seike Y, Ayukawa K. Evaluating the contribution of long-range transport of heavy metals from the Asian continent to their concentrations in sediment cores from Lake Shinji, Western Japan. Water Air Soil Poll 2011, 223, 1151–1160.
[124] Wan D, Song L, Yang J, Jin Z, Zhan C, Mao X, Liu D, Shao Y. Increasing heavy metals in the background atmosphere of central North China since the 1980s: Evidence from a 200-year lake sediment record. Atm Environ 2016, 138, 183–190.
[125] Kusunoki K, Sakata M. Long-range transport of heavy metals from the Asian continent contributes to surface sediments in Lakes Shinji and Nakaumi in western Japan. Limnol 2017, 19, 43–52.
[126] Lin Q, Liu E, Zhang E, Nath B, Shen J, Yuan H, Wang R. Reconstruction of atmospheric trace metals pollution in Southwest China using sediments from a large and deep alpine lake: Historical trends, sources and sediment focusing. Sci Total Environ 2018, 613–614, 331–341.
[127] Crecelius EA. The solubility of coal fly ash and marine aerosols in seawater. Mar Chem 1980, 8, 245–250.
[128] Austin LS, Millward GE. Atmosphere–coastal ocean fluxes of particulate arsenic and antimony. Cont Shelf Res 1986, 6, 459–474.
[129] Kersten M, Kriews M, Förstner U. Partitioning of trace metals released from marine aerosols in coastal seawater. Mar Chem 1991, 36, 165–182.
[130] Hsu SC, Wong GTF, Gong GC, Shiah FK, Huang YT, Kao SJ, Tsai F, Lung SCC, Lin FJ, Lin II, Hung CC, Tseng CM. Sources, solubility and dry deposition of aerosol trace elements over the East China Sea. Mar Chem 2010, 120, 116–127.
[131] Elbaz-Poulichet F, Nagy A, Cserny T. The distribution of redox sensitive elements (U, As, Sb, V and Mo) along a river-wetland-lake system (Balaton Region, Hungary). Aquat Geochem 1997, 3, 267–282.
[132] Palmer MJ, Chetelat J, Richardson M, Jamieson HE, Galloway JM. Seasonal variation of arsenic and antimony in surface waters of small subarctic lakes impacted by legacy mining pollution near Yellowknife, NT, Canada. Sci Total Environ 2019, 684, 326–339.
[133] Müller B, Granina L, Schaller T, Ulrich A, Wehrli BP. As, Sb, Mo, and other elements in sedimentary Fe/Mn layers of Lake Baikal. Environ Sci Technol 2002, 36, 411–420.
[134] Böning P, Brumsack H-J, Böttcher ME, Schnetger B, Kriete C, Kallmeyer J, Borchers SL. Geochemistry of Peruvian near-surface sediments. Geochim Cosmochim Acta 2004, 68, 4429–4451.

[135] Feyte S, Tessier A, Gobeil C, Cossa D. In situ adsorption of mercury, methylmercury and other elements by iron oxyhydroxides and organic matter in lake sediments. Appl Geochem 2010, 25, 984–995.
[136] Gao L, Gao B, Zhou H, Xu D, Wang Q, Yin S. Assessing the remobilization of antimony in sediments by DGT: A case study in a tributary of the Three Gorges Reservoir. Environ Pollut 2016, 214, 600–607.
[137] Arsic M, Teasdale PR, Welsh DT, Johnston SG, Burton ED, Hockmann K, Bennett WW. Diffusive Gradients in Thin Films reveals differences in antimony and arsenic mobility in a contaminated wetland sediment during an oxic-anoxic transition. Environ Sci Technol 2018, 52, 1118–1127.
[138] Fiket Ž, Fiket T, Ivanić M, Mikac N, Kniewald G. Pore water geochemistry and diagenesis of estuary sediments – an example of the Zrmanja River estuary (Adriatic coast, Croatia). J Soils Sediments 2018, 19, 2048–2060.
[139] Andreae MO, Byrd JT, Froelich JPN. Arsenic, antimony, germanium, and tin in the Tejo Estuary. Portugal: Modeling a polluted estuary. Environ Sci Technol 1983, 17, 731–737.
[140] van den Berg CMG. Complex formation and the chemistry of selected trace elements in estuaries. Estuaries 1993, 16, 512–520.
[141] Salaun P, Gibbon-Walsh KB, Alves GM, Soares HM, van den Berg CM. Determination of arsenic and antimony in seawater by voltammetric and chronopotentiometric stripping using a vibrated gold microwire electrode. Anal Chim Acta 2012, 746, 53–62.
[142] van der Sloot HA, Hoede D, Wijkstra J, Duinker JC, Nolting RF. Anionic species of V, As, Se, Mo, Sb, Te and W in the Scheldt and Rhine estuaries and the Southern Bight (North Sea). Estuarine, Coastal Shelf Sci 1985, 21, 633–651.
[143] Cloern JE, Jassby AD, Schraga TS, Nejad E, Martin C. Ecosystem variability along the estuarine salinity gradient: Examples from long-term study of San Francisco Bay. Limnol Oceanog 2017, 62, S272–S291.
[144] Spencer DW, Robertson DE, Turekian KK, Folsom TR. Trace element calibrations and profiles at the Geosecs Test Station in the Northeast Pacific Ocean. J Geophys Res 1970, 75, 7686–7696.
[145] Brewer PG, Spencer DW, Robertson DE. Trace element profiles from the GEOSECS-II test station in the Sargasso Sea. Earth Planet Sci Lett 1972, 16, 111–116.
[146] Middelburg JJ, Hoede D, Van Der Sloot HA, Van Der Weijden CH, Wijkstra J. Arsenic, antimony and vanadium in the North Atlantic Ocean. Geochim Cosmochim Acta 1988, 52, 2871–2878.
[147] Rouxel O, Ludden J, Fouquet Y. Antimony isotope variations in natural systems and implications for their use as geochemical tracers. Chem Geol 2003, 200, 25–40.
[148] Bruland KW, Lohan M. Controls of trace metals in seawater. Treatise Geochem 2003, 6, 23–47.
[149] Smedley PL, Kinniburgh DG. Molybdenum in natural waters: A review of occurrence, distributions and controls. Appl Geochem 2017, 84, 387–432.
[150] Kadko D, Aguilar-Islas A, Buck CS, Fitzsimmons JN, Landing WM, Shiller A, Till CP, Bruland KW, Boyle EA, Anderson RF. Sources, fluxes and residence times of trace elements measured during the U.S. GEOTRACES East Pacific Zonal Transect. Mar Chem 2020, 222, 103781.
[151] Kadko D, Aguilar-Islas A, Bolt C, Buck CS, Fitzsimmons JN, Jensen LT, Landing WM, Marsay CM, Rember R, Shiller AM, Whitmore LM, Anderson RF. The residence times of trace elements determined in the surface Arctic Ocean during the 2015 US Arctic GEOTRACES expedition. Mar Chem 2019, 208, 56–69.
[152] Cutter GA, Moffett JW, Nielsdóttir MC, Sanial V. Multiple oxidation state trace elements in suboxic waters off Peru: In situ redox processes and advective/diffusive horizontal transport. Mar Chem 2018, 201, 77–89.

[153] Lewis BL, Andreae MO, Froelich PN. Sources and sinks of methylgermanium in natural waters. Mar Chem 1989, 27, 179–200.
[154] Filella M, May PM. Computer simulation of the low-molecular-weight inorganic species distribution of antimony(III) and antimony(V) in natural waters. Geochim Cosmochim Acta 2003, 67, 4013–4031.
[155] Diemar GA, Filella M, Leverett P, Williams PA. Dispersion of antimony from oxidizing ore deposits. Pure Appl Chem 2009, 81, 1547–1553.
[156] Cornelis G, Van Gerven T, Snellings R, Verbinnen B, Elsen J, Vandecasteele C. Stability of pyrochlores in alkaline matrices: Solubility of calcium antimonate. Appl Geochem 2011, 26, 809–817.
[157] Oelkers EH, Sherman DM, Ragnarsdottir KV, Collins C. An EXAFS spectroscopic study of aqueous antimony(III)-chloride complexation at temperatures from 25 to 250 °C. Chem Geol 1998, 151(1–4), 21–27.
[158] Zotov AV, Shikina ND, Akinifiyev NN. Thermodynamic properties of the Sb(III) hydroxide complex $Sb(OH)_3(aq)$ at hydrothermal conditions. Geochim Cosmochim Acta 2003, 67, 1821–1836.
[159] Oleinikova OV, Drozdova OY, Lapitskiy SA, Demin VV, Bychkov AY, Pokrovsky OS. Dissolved organic matter degradation by sunlight coagulates organo-mineral colloids and produces low-molecular weight fraction of metals in boreal humic waters. Geochim Cosmochim Acta 2017, 211, 97–114.
[160] Tella M, Pokrovski GS. Stability and structure of pentavalent antimony complexes with aqueous organic ligands. Chem Geol 2012, 292–293, 57–68.
[161] Tighe M, Lockwood P, Wilson S. Adsorption of antimony(V) by floodplain soils, amorphous iron(III) hydroxide and humic acid. J Environ Monitor 2005, 7, 1177–1185.
[162] Tella M, Pokrovski GS. Antimony(III) complexing with O-bearing organic ligands in aqueous solution: An X-ray absorption fine structure spectroscopy and solubility study. Geochim Cosmochim Acta 2009, 73, 268–290.
[163] Buschmann J, Sigg L. Antimony(III) binding to humic substances: Influence of pH and type of humic acid. Environ Sci Technol 2004, 38, 4535–4541.
[164] Besold J, Kumar N, Scheinost AC, Lezama Pacheco J, Fendorf S, Planer-Friedrich B. Antimonite complexation with thiol and carboxyl/phenol groups of peat organic matter. Environ Sci Technol 2019, 53, 5005–5015.
[165] Filella M. Freshwaters: Which NOM matters?. Environ Chem Lett 2009, 7, 21–35.
[166] Hockmann K, Planer-Friedrich B, Johnston SG, Peiffer S, Burton ED. Antimony mobility in sulfidic systems: Coupling with sulfide-induced iron oxide transformations. Geochim Cosmochim Acta 2020, 282, 276–296.
[167] Andreae MO, Froelich PN. Arsenic, antimony, and germanium biogeochemistry in the Baltic Sea. Tellus 1984, 36, 101–117.
[168] Cutter GA. Dissolved arsenic and antimony in the Black Sea. Deep-Sea Res 1991, 38, S825–S843.
[169] Van der Weijden CH, Middelburg JJ, De Lange GJ, van der Sloot HA, Hoede D, Woittiez JRW. Profiles of the redox-sensitive trace elements As, Sb, V, Mo and U in the Tyro and Bannock Basins (eastern Mediterranean). Mar Chem 1990, 31, 171–186.
[170] Helz GR, Valerio MS, Capps NE. Antimony speciation in alkaline sulfide solutions: Role of zerovalent sulfur. Environ Sci Technol 2002, 36, 943–948.
[171] Olsen NJ, Mountain BW, Seward TM. Antimony(III) sulfide complexes in aqueous solutions at 30 °C: A solubility and XAS study. Chem Geol 2018, 476, 233–247.
[172] Akeret R, Über die Löslichkeit von Antimon(3)sulfid. PhD thesis, ETH Zurich (No. 2271) (1953), 77 pp.

[173] Babko AK, Lisetskaya GS. Equilibrium in reactions of formation of thiosalts of tin, antimony, and arsenic in solution. Russian J Inorg Chem 1956, 1, 95–107.

[174] Krupp RE. Solubility of stibnite in hydrogen sulfide solutions, speciation, and equilibrium constants, from 25 to 350 °C. Geochim Cosmochim Acta 1988, 52, 3005–3015.

[175] Kolpakova NN. Laboratory and field studies of ionic equilibria in the Sb_2S_3-H_2O-H_2S system. Geochem Int 1982, 19, 46–54.

[176] Spycher NF, Reed MH. As(III) and Sb(III) sulfide complexes: An evaluation of stoichiometry and stability from existing experimental data. Geochim Cosmochim Acta 1989, 53, 2185–2194.

[177] Akinifiyev NN, Zotov AV, Shikina ND. Experimental studies and self-consistent thermodynamic data in the Sb(III)-S(II)-O-H system. Geochem Int 1994, 31, 27–40.

[178] Kantin R. Chemical speciation of antimony in marine algae. Limnol Oceanogr 1983, 28, 165–168.

[179] Benson AA. Antimony Metabolites in Marine Algae. In: Craig PJ, Glocking F, eds, The Biological Alkylation of Heavy Elements. Spec. Publ. Royal Society of Chemistry, Vol. 66, 1988, 135–137.

[180] Maeda S, Fukuyama H, Yokoyama E, Kuroiwa T, Ohki A, Naka K. Bioaccumulation of antimony by *Chlorella vulgaris* and the association mode of antimony in the cell. Appl Organomet Chem 1997, 11, 393–396.

[181] Maeda S, Ohki A. Bioaccumulation and Biotransformation of Arsenic, Antimony, and Bismuth Compounds by Freshwater Algae. In: Wong Y-S, Tam NFY, eds., Wastewater Treatment with Algae. New York, Springer-Verlag and Landes Bioscience, 1998, 73–92.

[182] Lehr CR, Kashyap DR, McDermott TR. New insights into microbial oxidation of antimony and arsenic. Appl Environ Microbiol 2007, 73, 2386–2389.

[183] Duncan EG, Maher WA, Foster SD. Contribution of arsenic species in unicellular algae to the cycling of arsenic in marine ecosystems. Environ Sci Technol 2015, 49, 33–50.

[184] Li J, Wang Q, Zhang S, Qin D, Wang G. Phylogenetic and genome analyses of antimony-oxidizing bacteria isolated from antimony mined soil. Int Biodeter Biodeg 2013, 76, 76–80.

[185] Hamamura N, Fukushima K, Itai T. Identification of antimony and arsenic-oxidizing bacteria associated with antimony mine tailing. Microbes Environ 2013, 28, 257–263.

[186] Nguyen VK, Lee J-U. Antimony-oxidizing bacteria isolated from antimony-contaminated sediment – A phylogenetic study. Geomicrobiol J 2015, 32, 50–58.

[187] Terry LR, Kulp TR, Wiatrowski H, Miller LG, Oremland RS. Microbiological oxidation of antimony(III) with oxygen or nitrate by bacteria isolated from contaminated mine sediments. Appl Environ Microbiol 2015, 81, 8478–8488.

[188] Nguyen VK, Lee J-U. Isolation and characterization of antimony-reducing bacteria from sediments collected in the vicinity of an antimony factory. Geomicrobiol J 2014, 31, 855–861.

[189] Abin CA, Hollibaugh JT. Dissimilatory antimonate reduction and production of antimony trioxide microcrystals by a novel microorganism. Environ Sci Technol 2014, 48, 681–688.

[190] Kulp TR, Miller LG, Braiotta F, Webb SM, Kocar BD, Blum JS, Oremland RS. Microbiological reduction of Sb(V) in anoxic freshwater sediments. Environ Sci Technol 2013, 48, 218–226.

Teba Gil-Díaz

Chapter 10 Radioactive antimony: known emissions and environmental dispersion

10.1 Introduction

Many radioactive elements are present in the environment both from natural and anthropogenic origins. They are studied worldwide in order to better understand their mobility, bioavailability, and radiotoxicity for organisms and humans. The most followed radionuclides are related to isotopes of iodine and cesium, due to their long half-lives and known biological implications. Nevertheless, radionuclide emissions from nuclear power plant (NPP) accidental events, fuel reprocessing plants, dismantled facilities, or nuclear waste repositories are composed of a cocktail of radioactive elements, presenting all kinds of short-, medium-, and long-term half-lives. For many of these radionuclides, there is still a lack of knowledge regarding their environmental biogeochemical behavior. Such is the case for the radionuclides of antimony, which are followed up by scarce studies and/or included in generalist studies with other multiple radionuclides. Nevertheless, given the current and near future trends of nuclear energy, showing both operating NPPs and dismantling conditions, as well as the international implications of radionuclide dispersion in the environment, nuclear-related environmental risk assessment is a topic of concern that still requires further research. In this chapter, we will see the basic knowledge of radioactive antimony, the emitted activities, the impacted environmental compartments, and the current understanding of its environmental biogeochemical behavior.

10.2 Characteristics of radioactive antimony

All radioactive elements show specific life spans, quantified by the half-life time ($t_{1/2}$), and naturally decay into more stable forms. Sometimes, radionuclides require several decay events constituting characteristic decay chains, each step characterized by its own half-life and decay mode. These changes of element over time together with respective $t_{1/2}$ influence the fate of antimony radionuclides, determining their persistence in a given system.

Teba Gil-Díaz, Friedrich-Schiller-Universität Jena, Institute of Geosciences, Burgweg 11, 07749 Jena, Germany, teba.gil-diaz@uni-jena.de

https://doi.org/10.1515/9783110668711-010

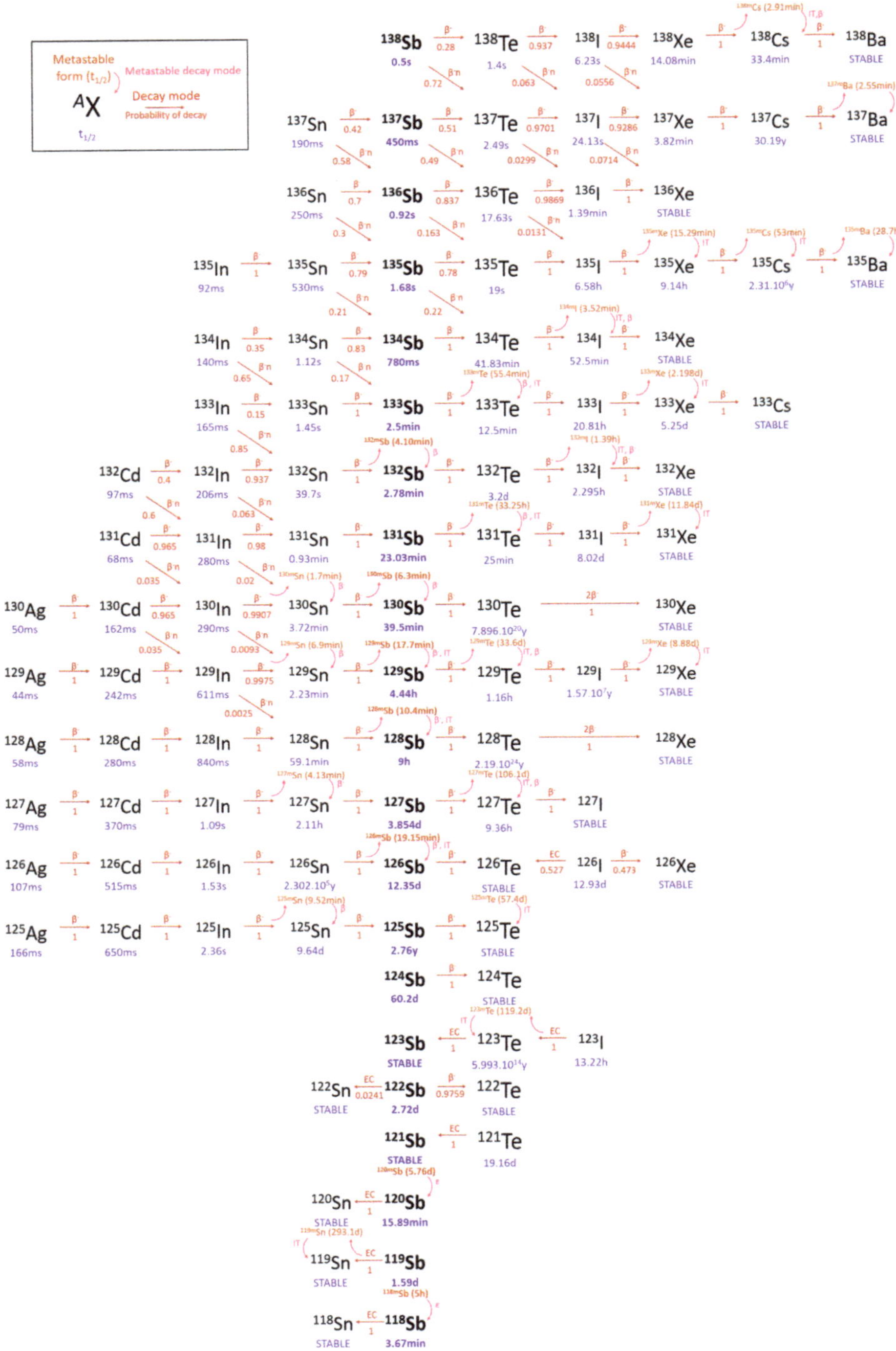

Fig. 10.1: Decay chains of antimony radionuclides. Information on corresponding half-lives (below each isotope), decay modes (EC: electron capture; β⁻: beta decay; IT: isomeric transition; EC: electron

10.2.1 Half-lives and radioactive decay

Antimony presents two stable isotopes (i.e., 57.3% ^{121}Sb and 42.7% ^{123}Sb) and 42 radionuclides. All antimony radionuclides decay by emitting beta (β^-, negatively charged high-energy electrons from the nucleus) and gamma radiation (γ, photon emissions) into several elements: tellurium, the most followed-up radionuclides after accidental events (e.g., ^{131}I, ^{134}Cs, ^{137}Cs), and gaseous xenon, according to the specific radionuclide decay chain (Fig. 10.1). Likewise, other radionuclides of silver, cadmium, indium, and tin can also decay into radioactive antimony, occasionally producing intermediate, longer-lasting, excited states (i.e., metastable forms) before decaying into the daughter antimony radionuclides. However, metastable forms are generally not a relevant feature for antimony given their short $t_{1/2}$ (Fig. 10.1). The contribution of the parent radionuclides to the budget of antimony may be non-negligible for specific isotopes given the fact that some, especially those of tin, may show higher fission yields than the radionuclides of antimony themselves (data not shown). Out of all antimony radionuclides, the most probable outcomes from nuclear fission reactions based on known fission yields are ^{131}Sb, ^{132}Sb, and ^{133}Sb. However, there are only three relevant half-lives rendering antimony radionuclides persistent in the environment: ^{126}Sb ($t_{1/2}$ = 12.4 d; β^-decay energy of 3,670 keV), ^{124}Sb ($t_{1/2}$ = 60 d; β^-decay energy of 2,904 keV), and ^{125}Sb ($t_{1/2}$ = 2.76 y; β^-decay energy of 767 keV). From these, only ^{125}Sb is listed in the International Commission on Radiological Protection (ICRP) as a radionuclide of potential importance in occupational exposure [1]. More detailed information about the general decay radiation, emission energies, and gamma lines of antimony radionuclides can be found in ICRP [2].

10.2.2 Unique anthropogenic origin

All antimony radionuclides have an anthropogenic origin, mainly as primary fission products of nuclear reactions, implying the disintegration of fissionable, heavy atoms such as those of ^{235}U or ^{239}Pu. Its building up within the nuclear fuel and reactor core in NPPs depends on the fissionable fuel content and on the fuel burnup, also known as fuel utilization or the fraction of fuel atoms spent during fission. Nevertheless, already low-fuel burnups of 1% produce enough fission products to detect antimony radionuclides [5]. Other origins of antimony radionuclides include activated tin in the zirconium alloy within the reactor core structures, becoming radioactive ^{125}Sb [6], as

---

Fig. 10.1 (continued)
capture), decay products (red arrows), and decay probability (from 0 to 1) are given. Metastable nuclides (in orange) and respective half-lives are also shown where appropriate. The above decay chains have been assembled using available data [2, 3], modified from [4].

well as Sb/Be clusters or stops and bearings of certain pumps containing Sb, producing $^{122}Sb/^{124}Sb$ by activation of stable $^{121}Sb/^{123}Sb$ [7].

10.3 Sources and emitted activities to the environment

The major sources of antimony radionuclides to the environment are related to past nuclear bomb tests, current nuclear fuel reprocessing plants, and NPP activities, including sporadic NPP accidental events [8]. Other minor sources are related to the fields of analytical chemistry (e.g., calibrations of high-resolution detectors [9]) or nuclear medical theragnostics [10].

The chemical forms in which antimony radionuclides are released into the environment are versatile, according to the source and the specific conditions during the discharge. For this reason, radionuclides of antimony have been classified as (i) mainly *dissolved* within the liquid effluents of nuclear installations [11], (ii) *refractory*, that is, expected to be released to the environment following the fate of the fuel or reactor core structures after NPP accidental releases, and (iii) *volatile* (like cesium), based on both NPP accidental releases and experimental data representing various realistic fuel compositions in pressurized water reactors (i.e., VERCORS program [12]).

10.3.1 Historical releases from punctual nuclear bomb tests

Between 1945 and 1980, more than 540 atmospheric nuclear tests were carried out mostly in the northern hemisphere by several countries (EEUU, Russia, UK, France, and China), and since 1961, these have been gradually replaced by underground tests. These atmospheric tests released radioactive particles into the troposphere and/or the stratosphere according to the loaded mass of the bomb, showing variable dispersion from near range (<1,000 km) to worldwide homogenization, depositing by gravity and/or wet deposition as debris of nuclear fission, spent fuel, or construction materials. Atmospheric deposition occurs within 30 days for tropospheric transport (i.e., relevant for short- to medium-lived radionuclides) and between 2 and 12 months in polar regions or 8 and 24 months in equatorial regions (i.e., containing longer-lived radionuclides) for stratospheric dispersion [13]. The total activities released for ^{125}Sb, excluding local deposition, were estimated to be 0.52 HBq (Hexa-Becquerel, HBq = 1×10^{18} Bq), similar to those of ^{137}Cs ($t_{1/2}$ of 30.14 y, 0.91 HBq) and lower than other short-/medium-lived radionuclides like tritium (^{3}H, $t_{1/2}$ of 12.32 y, 240 HBq) and ^{131}I ($t_{1/2}$ of 8.02 d, 651 HBq [14]). This radioactivity was deposited in both terrestrial and aquatic systems, though nowadays it is probably no longer present in the environment for the case of ^{125}Sb ($t_{1/2}$ of 2.76 y). Nevertheless, past in situ measurements of biological accumulation from exposure to ^{125}Sb from nuclear tests exist for the Enewetak and Bikini Atolls, evidencing high accumulations in plankton (8.2 pCi g^{-1} dry weight, dw, equivalent to ~303 Bq kg^{-1} dw), algae (0.34–0.91 pCi g^{-1} dw, equivalent to ~13–34 Bq kg^{-1} dw), and invertebrates (0.013–0.060 pCi g^{-1} dw, equivalent to ~0.5–2 Bq kg^{-1} dw [15]), with no evidences in higher levels of the trophic chain.

10.3.2 Continuous releases from nuclear installations: NPPs and fuel reprocessing plants

Nuclear installations are generally linked to aquatic systems and are allowed to release in a controlled manner certain amounts of radionuclides in their liquid discharges, within the regulatory safety limits regarding radioactivity and dose (<1 mSv y^{-1}). Alongside with other radionuclides, radioactive levels of both ^{124}Sb and ^{125}Sb are generally monitored in these effluents, though ^{122}Sb and ^{127}Sb were also requested to enter the monitoring lists due to their potential environmental impacts [16]. Liquid effluents from the most known nuclear fuel reprocessing plants (La Hague in France and Sellafield in UK) used to release 1×10^{-3} TBq y^{-1} of ^{124}Sb and 1–20 TBq y^{-1} of ^{125}Sb (Tera-Becquerel, TBq = 1×10^{12} Bq) and decreased to 0.50 TBq y^{-1} in the mid-1990s (i.e., La Hague [17]), representing respective activities of <100 mBq L^{-1} and ~2 Bq kg^{-1} [18]. Long-term monitoring in fluvial systems hosting a high density of both NPP and fuel reprocessing plants (e.g., four NPPs and one reprocessing plant in the Rhône River in France) has registered total activities of 30 ± 10 mBq L^{-1} in filtered water and 14 ± 7 Bq kg^{-1} in dried sediments between 1980 and 1986 for ^{125}Sb [19], decreasing to ~0.1 mBq L^{-1} for both ^{124}Sb and ^{125}Sb between 2002 and 2012 (equivalent to 5–10 GBq in 2008 [20]), with no detectable levels in the suspended particle matter [21]. Monitoring from NPP liquid effluents in Korea between 2011 and 2015 have registered releases of ^{122}Sb, ^{124}Sb, and ^{125}Sb in particulate forms (not detected in dissolved or gaseous forms), reporting total annual radiation doses to the public living nearby ranging between 3.20×10^{-5} and 2.06×10^{-3} mSv in particulate form (e.g., including radionuclides of Cs, Co, Sr, Zr, and others [22]). Other aquatic systems hosting a lower density of nuclear installations (e.g., two NPPs in the Gironde Estuary in France) recently showed emissions of ^{124}Sb and ^{125}Sb below "detection limits" in all sampled matrices (i.e., water samples <0.05 Bq L^{-1}, soils and sediments <2 Bq kg^{-1} dw, aquatic and terrestrial biological samples <11 Bq kg^{-1} dw [23]).

Aquatic organisms also accumulate ^{125}Sb in the coastal areas exposed to the effluents of La Hague. Some examples include macroalgae (i.e., varying concentrations between 1 and 60 Bq kg^{-1} dw between 1980 and 1990 [24, 25]) and bivalves/gastropods (i.e., 0.20–2.50 Bq kg^{-1} fresh weight, fw, and up to 5 Bq kg^{-1} in mussels [17, 24]), showing concentrations below detection limits since 2000.

These observations suggest that continuous environmental releases of antimony radionuclides are in both dissolved and particle phases, though the dissolved phase seems to be dominant. However, it is not clear if this implies a transport within the colloidal or truly dissolved phase. In any case, ^{125}Sb shows detectable biotransfer and bioaccumulation in aquatic organisms, which are of the same order of magnitude than those quantified after punctual releases from atmospheric bomb tests (Section 10.3.1). The reported activities in liquid (between 0.1 and 100 mBq L^{-1}, equivalent to concentrations <3 fg L^{-1}) and particle forms (~2 Bq kg^{-1}, equivalent to ~0.05 fg kg^{-1}) are several orders of magnitude below environmental levels compared to stable antimony (e.g., average ~200 ng L^{-1} in seawater, and crustal concentrations 0.2×10^6–2×10^6 ng kg^{-1} [26]).

10.3.3 Registered accidental releases from NPPs: cases of Chernobyl and Fukushima

There have been two major nuclear power plant accidental events in human history that, according to the International Nuclear and Radiological Event Scale (INES), have rated the highest in severity and radiological exposure to living organisms: the NPP accidental releases of Chernobyl (CNPP, Ukraine, April 26, 1986, INES 7) and Fukushima Daiichi (FDNPP, Japan, March 11, 2011, INES 7). These accidents released non-negligible amounts of several radionuclides involving important radiation doses and widespread environmental dispersion [27], monitored in greater (e.g., ^{137}Cs) or lesser extent (e.g., ^{125}Sb) worldwide in all environmental compartments. Especially after the FDNPP, attention has been drawn to radionuclide transport in aquatic systems given the direct, unexpected liquid discharges to coastal waters, though still widely understudied nowadays [28].

Out of all registered atmospheric releases, the total output of ^{125}Sb was only computed from the CNPP accident, accounting for 0.23 PBq (Peta-Becquerel, PBq = 1×10^{15} Bq) compared to emitted 1,200–1,700 PBq of ^{131}I and 74–85 PBq of ^{137}Cs [27]. Lower releases from the FDNPP are assumed given the reported evidences in atmospheric and soil samples [29, 30]. The ^{125}Sb from CNPP was widely dispersed worldwide, with comparable deposition rates to those of ^{137}Cs [31]. Measured activities in continental surface samples all over Europe include examples in Greek grass and soils (1×10^{-5} Bq kg^{-1} and 4.4 Bq kg^{-1}, respectively [32]), as well as Finnish lichen (~2,000 Bq m^{-2} in 1987, 24% still present in 1990 [31]). Over time, activities were also detected in the first 4 cm of Swedish soils 6.5 years after the CNPP accident (max. 23.2 ± 3.1 kBq m^{-2}, for an estimated original ground deposition of 60 kBq ^{125}Sb m^{-2} compared to 900 kBq ^{134}Cs m^{-2} and 1,600 kBq ^{137}Cs m^{-2} [33]), and even low levels were detected 15 years later in the first 3 cm of Finnish forest soils (1.4–14 Bq kg^{-1} 1999–2000 [34]).

Despite the direct discharges of radionuclides to coastal waters after the FDNPP accident, very few studies reported activities of ^{125}Sb, only focusing on marine bottom sediments along the coast of Japan with activities ranging between 1 and 10 Bq kg^{-1} from May 2011 to February 2012 [35]. In fact, there are more evidences of atmosphere-water interactions after the CNPP accident, reporting the presence of dissolved ^{125}Sb despite the lack of direct discharges. A transfer from radioactive debris emitted by CNPP to reactive natural particles in Lake Constance in 1986 showed 300 Bq m^{-2} of ^{125}Sb sediment loading [36]. Contrastingly, deposition of ^{125}Sb into the Scheldt Estuary (the Netherlands) did not impact the relatively high background levels as the marine input entering the estuary already contained ^{125}Sb in the particle load due to the influence of discharges from La Hague and Sellafield [37]. Soft tissues of aquatic organisms such as marine bivalves and gastropods in the coast of Monaco also registered ^{125}Sb at least one month after CNPP accident, presenting activities ranging between 0.48 and 4.2 Bq kg^{-1} fw in *M. galloprovincialis* and up to 22 Bq kg^{-1} fw in *P. lusitanica* [38].

These accidental, punctual events (acute exposure) suggest a widespread reach of ^{125}Sb, potentially soluble and transferable between environmental compartments. These releases are bioavailable to aquatic organisms to a similar extent than those from nuclear bomb tests (Section 10.3.1) and continuous releases (Section 10.3.2), despite the variability in the total emitted activities (e.g., punctual 0.52 HBq vs. 0.23 PBq vs. continuous 0.50 TBq y^{-1}).

10.4 Environmental reactivity and biogeochemical behavior

In general, environmental biogeochemical processes do not distinguish between stable and radioactive isotopes. For this reason, it is generally assumed that the biogeochemical behavior of radioactive elements will follow that of their analogue stable isotopes. This hypothesis is especially applied for radionuclides with short to medium half-lives, where relevant environmental persistence allowing studies on the elemental biogeochemical behavior and toxicity is only possible when the sources release high amounts of radionuclides or in a continuous manner. Radiolabeled studies could potentially present more reliable results compared to environmental and laboratory spikes of classical stable isotopes as they circumvent the analytical challenges of environmentally relevant trace to ultra-trace levels in complex matrices like seawater. However, the number of studies using radiolabeled elements (especially laboratory approaches using adequate spikes) is not abundant due to the radiological implications requiring, for example, adequate laboratory facilities, appropriate waste management, and radioprotection measures for the scientific staff according to the spiking activities. Nevertheless, the examples below will focus mostly on studies of ^{125}Sb behavior, compared to studies on stable antimony when applicable.

10.4.1 Solid fractionation and solid/liquid partitioning (K_d)

Several studies suggest that geologically derived (inherited) antimony is mostly present in the refractory ("residual") fraction of sediments/soils, and the iron oxide fraction represents the next relevant carrier phase [39]. Few studies have gone beyond reporting radioactive levels released from nuclear installations (from accidental or continuous discharges), with the aim of actually understanding the solid fractionation and mobility of radionuclides, particularly for ^{125}Sb. Among the rare cases, Carbol et al. [40] performed sequential selective extractions for ^{125}Sb and ^{137}Cs on Swedish soils after the CNPP accidental event. Despite potential side effects of applied reagents during selective extraction schemes concerning Sb solid fractionation [41] and the low radioactive levels in the study, ~50% of ^{125}Sb was associated to the ion-exchangeable fraction and ~25% to the organic matter of the O_h-horizon, concluding that ^{125}Sb was loosely bound compared to ^{137}Cs (<30% labile [40]). Higher mobility of added stable antimony has also been observed for several solid fractions of coastal sediments compared to the inherited antimony fraction [41, 42]. These observations may point out a higher mobility of the added/anthropogenic antimony (stable or radioactive) compared to the inherited forms.

In fact, adsorption experiments suggest a slow but overall high solubility of added ^{125}Sb. For instance, ~83% of ^{125}Sb remained in the <0.45 µm fraction after a 5 days exposure to coastal solid and water matrices [43]. Likewise, >85% of ^{125}Sb was found in solution after 50 h of exposure to limestone, dolomite, quartz, and marine sediments with solid/liquid ratios ranging between 1 and 100 g L^{-1} [44], and >80% was also in solution for spikes of 6.1 µg L^{-1} of stable Sb(V) in contact with 10 g L^{-1} of gibbsite ($Al(OH)_3$) at pH 8 [45]. It has also been observed that 45%–85% of ^{125}Sb remains in the dissolved phase after contact with synthetic mixtures of chemisorbents containing MnO_2, iron, and thiol functional groups [46].

Reported solid/liquid partitioning coefficients (i.e., logK_d values, used for transport models in nuclear risk assessment), also reflect a general preference of added Sb in the dissolved phase. Results range between 0.3 and 2.0 L kg^{-1} for marine sediments of the North Adriatic [44], between 1.4 and 3.1 L kg^{-1} for sediments near La Hague between 1982 and 1992 [17], and between 1.1 and 2.9 L kg^{-1} for stable Sb(V) spikes versus single minerals [45]. These values are relatively lower than the recommended guidelines reported by IAEA [47] based on environmental stable antimony (i.e., logK_ds of 3.3–3.6 kg L^{-1} for continental margins and deep-sea sediments), lower than values from highly antimony enriched sites (e.g., abandoned mines, logK_d ~3.0–5.0 L kg^{-1} [48, 49]), and within the lower range of estimated logK_d values for radioisotopes in freshwater, based on overall published data (i.e., 2.7–4.7 L kg^{-1}, 5th–95th percentiles) [50]. Nevertheless, K_d values are site specific and an inter-comparison between punctual measurements in different areas/conditions and a general application worldwide should be taken with care. For example, logK_d values may range from 1.2 to 2.2 L kg^{-1} according to the nature of the soil (e.g., from sandy to clay or organic rich soils [51]). In addition, many studies on K_d values lack certain details, nuancing the understanding of antimony transport, such as (i) unspecified spike concentrations in experimental conditions, (ii) the applied definition of "dissolved antimony" (e.g., <0.45 µm, occasionally <0.20 µm, vs. "truly dissolved" phases <0.02 µm), (iii) the characteristic sorption kinetics of different solids in specific matrices (e.g., desorption rates in seawater may vary between 27% in 2 h for urban particles and 83% in 100 h polluted coastal aerosols [52, 53]) influencing the strict definition and potential misuse of K_d in environmental samples (adsorption/desorption equilibrium) versus distribution ratio (R_d) (i.e., any solid/liquid interaction before reaching equilibrium [54]), and (iv) the unknown similarities between field- and laboratory-based K_d values [55], where field-based K_d values integrate inherited (refractory) trace elements and laboratory-based K_d values address solid/liquid partitioning of spiked elements (better simulating anthropogenic discharges).

The role of the colloidal phase in the transport of added antimony is also not conclusive and/or site dependent. The few studies verifying the colloidal forms of ^{125}Sb from both experimental spikes (filtering at 0.45 µm and 0.025 µm [19]) and environmental ^{125}Sb releases (i.e., coastal water from Sellafield subjected to ultrafiltration, <10 kDa [56]) suggest no role of the colloidal fraction. Nevertheless, studies on groundwater from underground nuclear tests showed $\log K_d$ of 2.9–3.3 L kg^{-1} within the 0.05–0.003 µm colloidal phase [57], in line with evidences of Sb transport associated to river colloids [58]. Furthermore, the importance of the particle phase in ^{125}Sb transport may not be discarded as ^{125}Sb is present in marine sediments [59] and can be fixed to the particle phase after discharges from La Hague and Sellafield [37] or be transferred from radioactive debris to reactive natural particles [36].

10.4.2 Transfer rates, bioaccumulation factors, and radiotoxicity

Radionuclides can be toxic due to both chemical properties and radioactive dose. Ecotoxicological studies with stable antimony are described in Chapter 14. Surprisingly, some radiotoxic experiments have actually been performed at very high concentrations (e.g., 1–8 µCi ^{125}Sb L^{-1}, equivalent to 37–296 kBq L^{-1} and 1–10 ng L^{-1} [60]) compared to emitted environmental activities, and relatively lower than average concentrations of stable antimony (Sections 10.3.1–10.3.3). One would assume that such initial levels of exposure would certainly influence the obtained results on biological transfer rates, accumulation, and radiotoxicity. However, the accumulations registered in the different tested organisms (algae, crustaceans, molluscs, and teleosts) were lower than those found in coastal environments for ^{125}Sb releases from nuclear installations and for stable antimony [60]. These seemingly mismatching/divergent results highlight several points applicable in ecological studies: (i) the importance of speciation, thus, the chemical forms in which the elements are released into the environment versus those present during experimental conditions, (ii) the synergic/antagonist effect of accumulation mechanisms when the organisms are exposed to a cocktail of radionuclides (i.e., similar and/or higher spikes of ^{110m}Ag and ^{60}Co, biologically more relevant than antimony [60]), (iii) the influence of other physico-chemical conditions in elemental transformations and also in organism welfare, and, most importantly, (iv) the exposure pathway, both direct (dissolved) and indirect (via the trophic chain) for organisms in the environment, potentially rendering higher accumulations than in experimental setups.

The lack of in situ (environmental) and ex situ (laboratory) studies, together with the need for completing/increasing the representativeness of databases for modeling/predictive purposes in nuclear risk assessments (e.g., ERICA/FREDERICA Tool [61, 62]) may be the reason why lists of accumulation factors report indistinctly values from laboratory experiments and environmental studies based on both stable and ^{125}Sb exposures (Tab. 10.1). Concentration factors vary among trophic levels and exposure conditions between <0.17 and average 1,400 L kg^{-1}, meaning that antimony is somehow accumulated within the organisms. There is a general trend towards lower accumulations for radioactive antimony, though this may reflect the absolute, lower concentrations of ^{125}Sb compared to stable antimony. Nevertheless, these databases contain high uncertainties (see standard deviations in Tab. 10.1 when specified by the reference study), potentially related to the interplay of (i) low sampled population (when specified), (ii) the intrinsic bias due to generalizing concentration factors from different exposure conditions (equilibrium vs. non-equilibrium conditions, element behavior in environment vs. laboratory, acute vs. chronic exposure, concentration levels, etc.), and/or (iii) an innate high variability in Sb accumulation within the same trophic level. In any case, reported ^{125}Sb bioaccumulation factors are actually really low compared to other contaminants of concern (e.g., ~10^6 for stable cadmium and silver [63]) and other radionuclides (e.g., up to 10^4 for ^{110m}Ag [60]). Furthermore, there seems to be no biomagnification of ^{125}Sb along the food chain within an artificial biocoenosis [58], nor from environmental observations (e.g., Sections 10.3.1–10.3.3 and Strohal et al. [44]). More detailed information on bioaccumulation factors of stable antimony in aquatic and terrestrial organisms can be found in Chapter 14.

Exposure routes of concern for antimony to humans are mainly related to ingestion and inhalation pathways [66]. Concerning the ingestion pathway, ^{125}Sb seems to accumulate within the edible parts of molluscs (body) and fish (muscles, accounting <10% of the total fixed quantity [19]). Elimination periods from the body of aquatic organisms range between 10 days for prawns (*Crangon crangon*), 21–58 days for mussels and clams (*Mytilus edulis, Tapes japónica*), 43 days for fish contaminated by trophic pathways, 27–70 days by direct pathways, and 72 days for crabs (*Carcinus maenas* [8]). Despite these relatively short biological retention times, marine organisms still register ^{125}Sb released from past radioactive waste deep-sea dumping sites, several years later (i.e., continuous source [67]). Accidental inhalation of ^{125}Sb oxide aerosols by humans has shown long in-body (lung) retention times of 850 ± 250 days in non-smoker males and 2,700 ± 1,000 days for smokers [68]. Nevertheless, the radiotoxic

Tab. 10.1: Concentration factors of ^{125}Sb in several organisms based on experimental setups (ex situ) and/or after exposure to environmental releases (in situ). Concentration factors account for the relative radioactivity in the organism compared to that in the water (direct) and/or food (trophic pathway). The standard deviation (SD), number of samples (n), and fresh weight (fw) versus dry weight (dw) are noted when specified in the original reference.

Organisms	Conditions	Concentration factor	Reference
Concentration factors in freshwater systems (L kg^{-1})			
Phytoplankton	In situ	300 (fw) (n = 1)	[61][a]
Zooplankton	In situ	74 (fw) (n = 1)	[61][a]
Fish (general)	In situ/Ex situ	290 ± 540* (fw) (n = 7)	[61][a]
	In situ	33 ± 97* (n = 141)	[64]
Benthic feeding	In situ	20 ± 21 (n = 22)	[64]
Piscivorous	In situ	36 ± 110* (n = 113)	[64]
Aquatic moss: *Platyhypnidium sp.*	In situ	30 (fw)	[65]
Vascular plants	In situ	1,500** (fw) (n = 1)	[61][a]
	In situ	25 ± 18** (fw) (n = 6)	[64]
Molluscs: gastropods	In situ	49** (fw) (n = 1)	[64]
Insect larvae	In situ	82 ± 69** (fw) (n = 14)	[64]
Reptiles	In situ	2,300** (fw) (n = 2)	[64]
Concentration factors in brackish systems (L kg^{-1})			
Molluscs	In situ	160 (fw) (n = 1)	[64]
Concentration factors in marine systems (L kg^{-1})			
Phytoplankton	In situ	1,000	[47][a]
Zooplankton	In situ	1,300 ± 2,500** (fw) (n = 13)	[61][a]
	In situ	80** (dw)	[47][a]
Macroalgae (general)	In situ	20	[47][a]
	In situ	220 ± 490* (fw) (n = 44)	[64]
Red algae: *Corallina officinalis*	Ex situ	~17 (fw) direct exposure	[60][b]
Chondrus crispus	In situ	23 ± 9 (fw)	[17][c]

Tab. 10.1 (continued)

Organisms	Conditions	Concentration factor	Reference
Concentration factors in marine systems (L kg^{-1})			
Brown algae:	In situ	13 ± 5	[17][c]
Fucus serratus	In situ	9 ± 14	[17][c]
Laminaria digitata *Pelvetia canaliculata*	In situ	19 ± 20 (fw)	[17][c]
Cnidaria (anemones)	Ex situ	1.20 (fw)	[60][b]
	In situ	90 (fw) (n = 1)	[61][a]
Molluscs (general)	In situ	20	[18]
	In situ	300* (20** fw for cephalopods)	[47][a]
	In situ	470 ± 860* (fw) (n = 7)	[64]
	Ex situ	1 – 1.3 (fw) direct exposure	[60][b]
	Ex situ	~3.5 (fw) trophic pathway	[60][b]
Bivalves (general)	In situ/Ex situ	530 ± 920* (fw) (n = 6)	[61][a]
Ostrea edulis	In situ	21 ± 21 Patella	[17][c]
Mytilus edulis	In situ	17 ± 6 scallops	[17][c]
Patella vulgata	In situ	<10 (fw) mussels	[17][c]
Crustaceans (general)	In situ	10	[18]
	In situ	300* (dw)	[47][a]
	In situ	1,400 ± 2,600** (fw) (n = 4)	[61][a]
	Ex situ	1 – 1.3 (fw) direct exposure	[60][b]
	Ex situ	~3.5 (fw) trophic pathway	[60][b]
	In situ	25 ± 7 crabs	[17][c]
Fish (general)	In situ	20	[18]
	In situ	600**	[47][a]
	In situ/Ex situ	220 ± 760* (fw) (n = 31)	[61][a]
Acanthogobius flavimanus	Ex situ	< 1 (fw) direct exposure	[60][b]
Blennius pholis	Ex situ	0.3 – 1.2 (fw) trophic pathway	[60][b]
Cyprinus carpio	Ex situ	< 0.17 direct exposure	[19][d]

Tab. 10.1 (continued)

Organisms	Conditions	Concentration factor	Reference
Concentration factors in terrestrial systems (kg dry soil per kg of fresh weight organism)			
Annelids	Ex situ	0.006** (n = 1)	[64]
Lichens and bryophytes	In situ/Ex situ	0.39 ± 0.24* (n = 4)	[64]
Molluscs: gastropods	In situ	0.25 ± 0.24* (n = 7)	[64]
Soil invertebrates	Ex situ	0.00593**	[62]
Grasses and herbs	In situ	0.025**	[62]
Shrubs	In situ/Ex situ In situ	0.092 ± 0.055* (n = 3) 0.00239**	[64] [62]
Mammals eating shrubs	In situ	0.00000215**	[62]

*Not clearly stated if measured with stable or ^{125}Sb and/or both indistinctly; **Based on stable Sb; [a]Database of both stable and radioactive exposures in environmental and laboratory conditions; [b]After 20–40 days of exposure to 37–296 kBq L^{-1} of ^{125}Sb (experimental); [c]Environmental dispersion of ^{125}Sb, 5 km away from La Hague; [d]After 20 days of exposure to ~1,087 kBq L^{-1} of ^{125}Sb (experimental).

consequences of an internal exposure and retention of ^{125}Sb are yet not clear. Accumulation of ^{125}Sb within the cytosol of animal or plant cells [69] and/or in specific storage/detoxification organs could expose the organ to intense irradiation from low energy beta radiation (e.g., ^{125}Sb) compared to other radionuclides with higher energies (i.e., ~2,000 keV, like ^{110m}Ag and ^{60}Co), where both beta and gamma radiation eventually leave both organ and organism. Nevertheless, comparative studies suggest that ^{125}Sb is potentially less harmful than other radionuclides presenting higher biological roles (e.g., ^{110m}Ag for aquatic organisms [60]) and higher activities (e.g., ^{60}Co, ^{137}Cs, ^{226}Ra, ^{239}Pu, and ^{241}Am, based on numerical models calculating the relative doses for a human population exposed on a daily basis to contaminated sediments through all potential pathways [11]).

10.4.3 Environmental dispersion and fate

Overall, the previous sections suggest that the environmental releases of anthropogenic antimony may affect and transfer between all environmental compartments. Sporadic studies on atmospheric depositions from accidental releases of "hot" particles and condensed matter on terrestrial systems or aquatic environments suggest releases of solvable ^{125}Sb. The dissolution rates and/or retention times in the solid fraction may be variable according to the specific characteristics of the released material and the physico-chemical properties of the receiving medium. For instance, the vertical dissolution of ^{125}Sb (0.15–0.23 cm y^{-1}) and its diffusion speed (0.04–0.13 cm^2 y^{-1}) may be faster than ^{137}Cs [70] or show longer retention times in different soil horizons of forest ecosystems [11]. Long-term environmental monitoring of direct liquid discharges of ^{125}Sb from La Hague into the English Channel and the North Sea between 1983 and 1994 suggests a non-negligible dispersion of dissolved ^{125}Sb along the northern coasts of western and central Europe. These studies were originally intended to evaluate the environmental impact from La Hague but turned out to use ^{125}Sb as a conservative radiotracer for validating medium- to long-term hydrodynamic models. These models forecast controlled or accidental releases of soluble substances for simulations of several months (^{125}Sb) to several years (^{137}Cs) for support in decision-making processes based on K_d values [71–73]. This dispersion of ^{125}Sb is in accordance with the few studies on stable antimony in coastal waters and oceans, highlighting its conservative to mildly scavenged behavior, with non-negligible atmospheric inputs [74]. Despite this known behavior of ^{125}Sb, it is surprising that no similar studies followed after the direct liquid discharges from FDNPP into the coastal waters, showing higher influences from atmospheric depositions. This means that there are no standard/unique forms for environmental releases of Sb radionuclides from nuclear installations.

Few comprehensive studies aiming at understanding the biogeochemical cycle for both added (radioactive or stable) and inherited antimony exist, difficulting further conclusive remarks. Dispersion scenarios aiming at evaluating the risk of radionuclide releases into the environment for currently operating NPPs, future potential accidental releases, and dismantling strategies should account for the seemingly site-specific, soluble character of added ^{125}Sb into the environment. This is of particular interest for aquatic systems (rivers, estuaries, and coastal areas), where radionuclides of antimony (mainly ^{124}Sb, ^{125}Sb, and ^{126}Sb) may be both bioavailable and highly dispersive, subjected to local and coastal hydrodynamics [75], eventually decaying into stable isotopes of tellurium (Fig. 10.1). The environmental persistence of these antimony radionuclides may range from days to years according to the specific source (i.e., forms of the involved isotopes and specific half-lives), the released activities and the duration of the discharge (punctual vs. continuous). In order to understand and accurately predict the environmental transport, distribution, and fate of antimony radionuclides, on-site K_d values should be determined from experimental results based on environmental matrices, and the aforementioned temporal scales should be coupled to antimony sorption kinetics (both R_d and K_d) as well as to on-site hydrological timescales such as water/sediment residence times, particularly in highly dynamic systems presenting variable particle loads such as estuaries [75]. It seems likely that the emitted activities will be potentially more relevant from the radiotoxic point of view (exposure to γ and β^- radiation) than from elemental toxicity (e.g., examples on antimony toxicity in aquatic and terrestrial ecosystems can be found in Chapters 7 and 14), but further studies should clarify the radiotoxic relevance of antimony, including data on effective doses. All this information, complemented by a better understanding of the environmental biogeochemistry of stable antimony (still lacking), is essential for the development of post-accidental scenarios of the fate and impact of radioactive Sb, fundamental for medium- to long-term nuclear risk assessment and necessary for upcoming, future management strategies such as dismantling nuclear facilities.

References

[1] Sonzogni AA. Chart of Nuclides NuDat 2.6 – National Nuclear Data Center. Brookhaven National Laboratory, 2013. Available in http://www.nndc.bnl.gov/nudat2/reCenter.jsp?z=51&n=68.

[2] International Commission on Radiological Protection. Individual monitoring for intakes of radionuclides by workers: Design and interpretation (Ann ICRP 19 (1/3)). ICRP Publication Oxford (1988), 54 pp.

[3] Gray T, Mann N, Whitby M. Periodic Table of Isotopes, 2017. Available in http://periodictable.com/Isotopes/051.123/index.html.

[4] Gil-Díaz T. Tellurium radionuclides produced by major accidental events in nuclear power plants. Environ Chem 2019, 16, 296–302.

[5] Kleykamp H. The chemical state of the fission products in oxide fuels. J Nuc Mat 1985, 131, 221–246.

[6] Iguchi Y, Baba T, Kawakami H. Study for nuclide transfer ratio of particles generated by thermal cutting. In: 14th International Conference on Nuclear Engineering. American Society of Mechanical Engineers, Vol. 42460, 2006, 443–449.

[7] Garcier Y, Guers R, Bidard F, Colson P. Memento de la radioprotection en exploitation. EDF Report, Paris (2014), 363 pp.

[8] Beaugelin-Seiller K, Garnier-Laplace J. Fiche Radionucléide, Antimoine 125 et environnement. IRSN report, Paris (2002), 25 pp.

[9] Rajput MU, Ali N, Hussain S, Mujahid SA, MacMahon D. Beta decay of the fission product ^{125}Sb and a new complete evaluation of absolute gamma ray transition intensities. Radiat Phys Chem 2012, 81, 370–378.

[10] Uddin MS, Hermanne A, Sudár S, Aslam MN, Scholten B, Coenen HH, Qaim SM. Excitation functions of α-particle induced reactions on enriched ^{123}Sb and ^{nat}Sb for production of ^{124}I. Appl Radiat Isotopes 2011, 69, 699–704.

[11] International Atomic Energy Agency. Environmental impact of radioactive releases. In: International Symposium on Environmental Impact of Radioactive Releases. IAEA, Vienna, 1995.

[12] Pontillon Y, Ducros G, Malgouyres PP. Behavior of fission products under severe PWR accident conditions VERCORS experimental programme – Part 1: General description of the programme. Nucl Eng Des 2010, 240, 1843–1852.

[13] Renaud P. Les essais atmosphériques d'armes nucléaires: des retombées radioactives à l'échelle planétaire. Fiche information, 2008. Available in https://www.irsn.fr/FR/connaissances/Environnement/surveillance-environnement/resultats/retombees-tirs-armes-nucleaires/Pages/2-essais-nucleaires-atmospheriques.aspx#.X0zpg4vtZaQ.

[14] González AJ. Résidus radioactifs de la guerre froide: l'héritage radiologique. IAEA Bulletin 1998, 40, 1–10.

[15] Welander AD. Distribution of radionuclides in the environment of Eniwetok and Bikini Atolls, August 1964 (No. CONF-670503). In: Nelson DJ, Gunad FC, eds., Symposium on Radioecology. Washington, 1969, 346–354.
[16] Booth RS. Compendium of radionuclides found in liquid effluents of nuclear power stations (No. ORNL-TM–3801). U.S. Atomic Energy Commission, Oak Ridge National Laboratory (1975), 17 pp.
[17] Gandon R, Du Bois PB, Baron Y. Caractère conservatif de l'antimoine 125 dans les eaux marines soumises à l'influence des rejets de l'usine de retraitement des combustibles irradiés de La Hague. Radioprotection 1998, 33, 457–482.
[18] GRNC Groupe Radioécologie Nord Cotentin. Revue critique des mesures dans l'environnement – mesures GT2. Rapport final, volume 2. IRSN, Fontenay aux Roses (1999).
[19] Baudin JP, Garnier J, Lambrechts A, Foulquier L. Experimental Data related to physico-chemical forms of antimony 125 and silver 110m and to their transfers to a freshwater fish, the carp (No. CEA-R–5476). CEA Centre d'Etudes Nucléaires de Cadarache (1989), 77 pp.
[20] Antonelli C, Boullier V, Eyrolle F, Gurriaran R, Cagnat X, Mekhlouche D, Raimbault P, Fornier M. Flux de radioactivité exportés par le Rhône en Méditerranée en 2008. (Report n°2010–04). IRSN Cadarache (2010), 41 pp.
[21] Eyrolle-Boyer F, Antonelli C, Renaud P, Tournieux D. Origins and trend of radionuclides within the lower Rhône River over the last decades. Radioprotection 2015, 50, 27–34.
[22] Kong TY, Kim S, Lee Y, Son JK, Maeng SJ. Radioactive effluents released from Korean nuclear power plants and the resulting radiation doses to members of the public. Nucl Eng Technol 2017, 49, 1772–1777.
[23] Barbey P, Boilley D, Josset M, Migeon A, Bernollin A, Dunand E. Etude de la radioactivite dans l'environnement autour du CNPE du Blayais. ACRO (2012), 98 pp.
[24] Germain P, Masson M, Baron Y. Contribution aux recherches sur les mouvements des éléments à l'état de traces dans les eaux côtières de la Manche par l'étude de la distribution spatiale de traceurs radioactifs industriels dans les moules et les fucus (CEA-R-5534). CEN Saclay (1990).
[25] Masson M, Van Weers AW, Groothuis REJ, Dahlgaard H, Ibbett RD, Leonard KS. Time series for seawater and seaweed of ^{99}Tc and ^{125}Sb originating from releases of La Hague. J Marine Syst 1995, 6, 397–413.
[26] Filella M, Belzile N, Chen YW. Antimony in the environment: A review focused on natural waters: I. Occurrence. Earth-Sci Rev 2002, 57, 125–176.
[27] Steinhauser G, Brandl A, Johnson TE. Comparison of the Chernobyl and Fukushima nuclear accidents: A review of the environmental impacts. Sci Total Environ 2014, 470, 800–817.
[28] Buesseler K, Dai M, Aoyama M, Benitez-Nelson C, Charmasson S, Highley K, Maderich V, Masqué P, Morris PJ, Oughton D, Smith JN. Fukushima Daiichi–derived radionuclides in the ocean: Transport, fate, and impacts. Annu Rev Mar Sci 2017, 9, 173–203.
[29] Kojima S, Arinobu T, Kosuda C, Kato M, Furukawa M. Radiochemical analysis of fission and neutron activation products released from the Fukushima Daiichi Nuclear Power Plant accident (2). In: Proceeding of annual meeting 2012 and the 56th symposium on radiochemistry of the Japan Society of Nuclear and Radiochemical Sciences. Japan, 2012, 172 pp.
[30] Thakur P, Ballard S, Nelson R. An overview of Fukushima radionuclides measured in the northern hemisphere. Sci Total Environ 2013, 458, 577–613.
[31] Lehto J, Paatero J, Pehrman R, Kulmala S, Suksi J, Koivula T, Jaakkola T. Deposition of gamma emitters from Chernobyl accident and their transfer in lichen–soil columns. J Environ Radioactiv 2008, 99, 1656–1664.

[32] Papastefanou C, Manolopoulou M, Charalambous S. Radiation measurements and radioecological aspects of fallout from the cherbonyl reactor accident. J Environ Radioactiv 1988, 7, 49–64.
[33] Carbol P, Solatie D, Erdmann N, Nylén T, Betti M. Deposition and distribution of Chernobyl fallout fission products and actinides in a Russian soil profile. J Environ Radioactiv 2003, 68, 27–46.
[34] Ylipieti J, Rissanen K, Kostiainen E, Salminen R, Tomilina O, Täht K, Gilucis A, Gregorauskiene V. Chernobyl fallout in the uppermost (0–3 cm) humus layer of forest soil in Finland, North East Russia and the Baltic countries in 2000–2003. Sci Total Environ 2008, 407, 315–323.
[35] Kusakabe M, Oikawa S, Takata H, Misonoo J. Spatiotemporal distributions of Fukushima-derived radionuclides in nearby marine surface sediments. Biogeosciences 2013, 10, 5019–5030.
[36] Robbins JA, Lindner G, Pfeiffer W, Kleiner J, Stabel HH, Frenzel P. Epilimnetic scavenging of Chernobyl radionuclides in Lake Constance. Geochim Cosmochim Acta 1992, 56, 2339–2361.
[37] Martin JM, Wollast R, Loijens M, Thomas A, Mouchel JM, Nieuwenhuize J. Origin and fate of artificial radionuclides in the Scheldt estuary. Mar Chem 1994, 46, 189–202.
[38] Whitehead NE, Ballestra S, Holm E, Huynh-Ngoc L. Chernobyl radionuclides in shellfish. J Environ Radioactiv 1988, 7, 107–121.
[39] Filella M. Antimony interactions with heterogeneous complexants in waters, sediments and soils: A review of data obtained in bulk samples. Earth-Sci Rev 2011, 107, 325–341.
[40] Carbol P, Skålberg M, Skarnemark G. Cesium and antimony behaviour in water and forest soil after the Chernobyl accident. Radiochim Acta 1995, 69, 259–270.
[41] Gil-Díaz T, Schäfer J, Filella M, Dutruch L, Bossy C. Fractionation of inherited and spiked antimony (Sb) in fluvial/estuarine bulk sediments: Unexpected anomalies in parallel selective extraction protocols. Appl Geochem 2019, 108, 104386.
[42] Brannon JM, Patrick JWH. Fixation and mobilization of antimony in sediments. Environ Pollut B 1985, 107–126.
[43] McCauley JE. Ecological studies of radioactivity in the Columbia River estuary and adjacent Pacific Ocean (No. RLO-1750–54). U.S. Atomic Energy Commission, United States (1970), 199 pp.
[44] Strohal P, Huljev D, Lulić S, Picer M. Antimony in the coastal marine environment, North Adriatic. Estuar Coast Mar Sci 1975, 3, 119–123.
[45] Essington ME. Antimony (V) adsorption by variable-charge minerals. Tennessee Univ Knoxville (2013), 170 pp.
[46] Johnson BE, Santschi PH, Addleman RS, Douglas M, Davidson J, Fryxell GE, Schwantes JM. Optimization and evaluation of mixed-bed chemisorbents for extracting fission and activation products from marine and fresh waters. Anal Chim Acta 2011, 708, 52–60.
[47] International Atomic Energy Agency. Sediment distribution coefficients and concentration factors for biota in the marine environment. Technical Reports Series No. 422, Vienna (2004), 95 pp.
[48] Craw D, Wilson N, Ashley PM. Geochemical controls on the environmental mobility of Sb and As at mesothermal antimony and gold deposits. Trans Inst Min Metall B 2004, 113, B3–B10.
[49] Manaka M, Yanase N, Sato T, Fukushi K. Natural attenuation of antimony in mine drainage water. Geochem J 2007, 41, 17–27.
[50] Ciffroy P, Durrieu G, Garnier JM. Probabilistic distribution coefficients (K_ds) in freshwater for radioisotopes of Ag, Am, Ba, Be, Ce, Co, Cs, I, Mn, Pu, Ra, Ru, Sb, Sr and Th—implications for uncertainty analysis of models simulating the transport of radionuclides in rivers. J Environ Radioactiv 2009, 100, 785–794.

[51] Gil-García C, Tagami K, Uchida S, Rigol A, Vidal M. New best estimates for radionuclide solid–liquid distribution coefficients in soils. Part 3: Miscellany of radionuclides (Cd, Co, Ni, Zn, I, Se, Sb, Pu, Am, and others). J Environ Radioactiv 2009, 100, 704–715.

[52] Austin LS, Millward GE. Atmosphere–coastal ocean fluxes of particulate arsenic and antimony. Cont Shelf Res 1986, 6, 459–474.

[53] Kersten M, Kriews M, Forstner U. Partitioning of trace metals released from marine aerosols in coastal seawater. Mar Chem 1991, 36, 165–182.

[54] McKinley IG, Alexander WR. Constraints on the applicability of 'in-situ distribution coefficient' values. J Environ Radioactiv 1992, 15, 19–34.

[55] Ciceri G, Traversi AL, Martinotti W, Queirazza G. Radionuclide partitioning between water and suspended matter: Comparison of different methodologies. In: Studies in Environmental Science. Elsevier, 1988, Vol. 34, 353–375.

[56] Salbu B, Bjørnstad HE, Svaren I, Prosser SL, Bulman RA, Harvey BR, Lovett MB. Size distribution of radionuclides in nuclear fuel reprocessing liquids after mixing with seawater. Sci Total Environ 1993, 130, 51–63.

[57] Buddemeier RW, Hunt JR. Transport of colloidal contaminants in groundwater: Radionuclide migration at the Nevada Test Site. App Geochem 1988, 3, 535–548.

[58] Filella M, Belzile N, Chen YW. Antimony in the environment: A review focused on natural waters II. Relevant solution chemistry. Earth-Sci Rev 2002, 59, 265–285.

[59] Tassi Pelati L, Triulzi C. Radioactivity contents in some plankton and sea water samples collected during the period between 1960–68 and their correlations (No. CISE-E–712). Energ. Nucl. (Milan) 1969, 16, 17–26.

[60] Amiard JC. A study of the uptake and toxicity of some stable and radioactive pollutants in marine organisms: Antimony, silver, cobalt and strontium in mollusks, crustaceans and teleosts. PhD thesis, Paris (1978), 139 pp.

[61] Hosseini A, Thørring H, Brown JE, Saxén R, Ilus E. Transfer of radionuclides in aquatic ecosystems–default concentration ratios for aquatic biota in the Erica Tool. J Environ Radioactiv 2008, 99, 1408–1429.

[62] Beresford NA, Barnett CL, Howard BJ, Scott WA, Brown JE, Copplestone D. Derivation of transfer parameters for use within the ERICA Tool and the default concentration ratios for terrestrial biota. J Environ Radioactiv 2008, 99, 1393–1407.

[63] Lanceleur L, Schäfer J, Chiffoleau JF, Blanc G, Auger D, Renault S, Baudrimont M, Audry S. Long-term records of cadmium and silver contamination in sediments and oysters from the Gironde fluvial–estuarine continuum–Evidence of changing silver sources. Chemosphere 2011, 85, 1299–1305.

[64] International Atomic Energy Agency. Handbook of Parameter Values for the Prediction of Radionuclide Transfer to Wildlife. Technical Reports Series No. 479. Vienna (2014), 211 pp.

[65] Baudin JP, Lambrechts A, Pally M. Utilisation des mousses aquatiques comme indicateurs de contamination radioactive. Hydroécol Appl 1991, 3, 209–240.

[66] Dieter HH, Fawell JK, Ohanian E., Giddings M, Toft P, Magara Y, Jackson P, Bartram J, Callan P, Hashizume H, Vickers C, Sheffer M. Antimony in Drinking-water, Background document for development of WHO Guidelines for Drinking-water Quality (WHO/SDE/WSH/03.04/74). World Health Organization, 2003. Available in http://www.who.int/water_sanitation_health/dwq/chemicals/antimony.pdf.

[67] Feldt W, Kanish G, Lauer R. Radioactive contamination of the NEA dumping sites. In: Impacts of radionuclide releases into the marine environment. IAEA, 1981.

[68] Garg SP, Singh IS, Sharma RC. Long term lung retention studies of ^{125}Sb aerosols in humans. Health Phys 2003, 84, 457–468.

[69] Bonotto S, Bossus A, Nuyts G, et al. Experimental uptake study of ^{60}Co, ^{137}Cs, ^{125}Sb, and ^{65}Zn in four marine algae. In: Park PK, eds., Wastes in the ocean. Wiley, New York, Vol. 3, 1983, 287–300.
[70] Bossew P, Gastberger M, Gohla H, Hofer P, Hubmer A. Vertical distribution of radionuclides in soil of a grassland site in Chernobyl exclusion zone. J Environ Radioactiv 2004, 73, 87–99.
[71] Du Bois PB, Guéguéniat P. Quantitative assessment of dissolved radiotracers in the English Channel: Sources, average impact of la Hague reprocessing plant and conservative behaviour (1983, 1986, 1988, 1994). Cont Shelf Res 1999, 19, 1977–2002.
[72] Du Bois PB, Dumas F. Fast hydrodynamic model for medium-and long-term dispersion in seawater in the English Channel and southern North Sea, qualitative and quantitative validation by radionuclide tracers. Ocean Model 2005, 9, 169–210.
[73] Periáñez R, Miró C. Characteristic times in the English Channel from numerical modelling: Supporting decision-making. J Radiol Prot 2009, 29, 219–237.
[74] Cutter GA, Cutter LS, Featherstone AM, Lohrenz SE. Antimony and arsenic biogeochemistry in the western Atlantic Ocean. Deep Sea Res II 2001, 48, 2895–2915.
[75] Gil-Díaz T, Schäfer J, Pougnet F, Abdou M, Dutruch L, Eyrolle-Boyer F, Coynel A, Blanc G. Distribution and geochemical behaviour of antimony in the Gironde Estuary: A first qualitative approach to regional nuclear accident scenarios. Mar Chem 2016, 185, 65–73.

Jaime N. Caplette and Adrien Mestrot

Chapter 11 Biomethylation and biovolatilization of antimony

11.1 Introduction

Geogenic or anthropogenic antimony can be released into the environment and subjected to microbial-facilitated biomethylation and biovolatilization. The initial pioneering into the field of antimony biomethylation and biovolatilization was due to a proposed linkage of sudden infant death syndrome (SIDS) and toxic gases produced as by-products from microbial biovolatilization (e.g., *Scopularipsus brevicaulis*) in infant mattress coverings, where antimony is added as a flame retardant [1]. This hypothesis was later refuted to be the cause of SIDS as appreciable concentrations of stibines, arsines, and/or phosphines were not found, and there was no direct causal relationship between urine and blood antimony concentrations and infant deaths [2–6].

A majority of the work on antimony biomethylation and biovolatilization has focused on pure microbial cultures with some minor emphasis on the aquatic environment, soils and sediments, biota, sewage sludge fermenters, landfill sites, and geothermal hot spring environments [7–29]. Since the most recent extensive reviews on the topic in 2011 very little work has been done [30–32]. This chapter focuses on a basic introduction to the chemistry of environmentally relevant methyl-Sb species, the methylation mechanism, advantages and limitations of measurement techniques, occurrence and behavior in environmental compartments, and toxicology. When possible, this chapter focuses on discussing the recent developments in the past decade and current limitations in the field.

11.1.1 Terminology

The nomenclature listed in Tab. 11.1 will be used throughout the chapter. If the oxidation state of the methylated antimony species is unknown or not mentioned in the cited literature, they are referred to herein as monomethylantimony (MMSb), dimethylantimony (DMSb), and trimethylantimony (TMSb) with a descriptive modifier (e.g., non-volatile).

Jaime N. Caplette, Adrien Mestrot, Institute of Geography, University of Bern, Hallerstrasse 12, CH-3012 Bern, Switzerland. adrien.mestrot@giub.unibe.ch

https://doi.org/10.1515/9783110668711-011

Tab. 11.1: Describes the antimony species, chemical formula, abbreviations, and corresponding oxidation state (when known). Due to the lack of standard nomenclature and for the sake of clarity, we chose to use the full formulas as abbreviations for the volatile antimony species.

Name	Chemical formula	Abbreviation	Oxidation
Antimonite	$Sb(OH)_3$	Sb(III)	III
Antimonate	$Sb(OH)_6^-$	Sb(V)	V
Inorganic antimony		Sb_{IO}	III/V
Volatile			
Stibine	SbH_3	SbH_3	III
Monomethylstibine	$(CH_3)SbH_2$	$(CH_3)SbH_2$	III
Dimethylstibine	$(CH_3)_2SbH$	$(CH_3)_2SbH$	III
Trimethylstibine	$(CH_3)_3Sb$	$(CH_3)_3Sb$	III
Non-volatile			
Monomethylstibonic acid	$CH_3SbO(OH)_2$	MMSb(V)	V
Dimethylstibinic acid	$(CH_3)_2SbO(OH)$	DMSb(V)	V
Trimethylantimony oxide	$(CH_3)_3SbO$	TMSb(V)	V
Unspecified			
Monomethylantimony		MMSb	III/V
Dimethylantimony		DMSb	III/V
Trimethylantimony		TMSb	III/V

11.1.2 Basic chemical properties

The most common oxidation states for antimony in the environment are the trivalent (III) and pentavalent (V) forms. Sb(III) is likely present under naturally occurring conditions as $Sb(OH)_3$, and Sb(V) is likely as $Sb(OH)_6^-$.

The trivalent methylated species of antimony are more volatile than the pentavalent species. The known volatile species are the hydrides stibine (SbH_3), a toxic gas, and species containing 1–3 methyl groups: monomethylstibine ($(CH_3)SbH_2$), dimethylstibine ($(CH_3)_2SbH$), and trimethylstibine ($(CH_3)_3Sb$) (Tab. 11.1). Methylated stibines behave as weak Lewis bases, and $(CH_3)_3Sb$ has low solubility in water [36, 37].

The most volatile species is SbH_3 (Tab. 11.2). A higher volatility, relative to $(CH_3)_3Sb$, would be expected for $(CH_3)SbH_2$ and $(CH_3)_2SbH$ due to lower boiling points (Tab. 11.2) and higher stability in synthetic air [38]. $(CH_3)_3Sb$ should be the least volatile and most unstable [14, 38, 39]. Oxidation of $(CH_3)_3Sb$ produces oxidation products of $(CH_3)_3SbO$ as well as cyclic and linear oligomers [40]. Parris and Brinckman calculated the rates of oxidation for $(CH_3)_3Sb$ as $2 \times 10^3\ M^{-1}\ s^{-1}$, yielding a half-life of 50 s [39]. Haas and Feldmann showed that $(CH_3)_3Sb$ is more stable than that Parris and Brinckman have previously reported [38, 39]. The authors found that

Tab. 11.2: Basic properties of the volatile stibines.

Species	Boiling point / °C	Vapor pressure (25 °C) / kPa	Half-life in dark Synthetic air (20 °C) / h	Half-life in UV light Synthetic air (30 °C) / h
SbH_3	−17[a]	4.2 x 10^{59} [b]	1250[a]	1[a]
$(CH_3)SbH_2$	41[c]	*NA*	1100[a]	0.15[a]
$(CH_3)_2SbH$	60.7[c]	*NA*	110[a]	0.1[a]
$(CH_3)_3Sb$	81[a]	13.8[d]	75[a]	0.1[a]

[a]Source: [14].
[b]Calculated from [33].
[c]Source: [34].
[d]Calculated from [35].

$(CH_3)_3Sb$ was the least stable species in the vapor phase and a loss of 24.6% of the species was observed within 24 hours [38]. Also, the presence of $(CH_3)_3Sb$ in landfill sites, measured hours after emission, indicates that $(CH_3)_3Sb$ may be more stable in the atmosphere than previously believed (Tab. 11.2) [14].

Monomethylstibonic acid (MMSb(V), $(CH_3)SbO(OH)_2$), and dimethylstibinic acid (DMSb(V), $(CH_3)_2SbO(OH)$), behave as weak acids, whereas trimethylantimony oxide (TMSb(V), $(CH_3)_3SbO$)) behaves as a weak base in solution [31, 41]. DMSb(V) has low solubility in water and common organic solvents [31, 42]. Studies using ESI-MS showed $TMSbCl_2$ and $TMSb(OH)_2$, at neutral pH, hydrolyze easily and likely occur as a cation, $(CH_3)_3Sb(OH)^+$ [41, 43].

11.2 Biomethylation pathway

Antimony biomethylation may follow the enzymatically controlled pathway proposed by Challenger and coworkers (i.e., the Challenger Pathway). This pathway was originally proposed for arsenic but now is largely disproven [44–50]. It consists of an initial reduction of Sb(V) to Sb(III); Sb(III) then undergoes an oxidative methylation step from the carbocation (CH_3^+) to the nucleophilic Sb [24, 26, 28, 45]. This is followed by a reductive two-electron transfer step facilitated by the thiol groups on glutathione (reducing methylated Sb(V) species to methylated Sb(III) species) [24, 26, 28, 45]. This mechanism continues in a stepwise fashion until the final product, $(CH_3)_3Sb$, is produced [5, 24, 26, 28, 45]. The methyl donor is most likely *S*-adenosylmethionine (SAM), a common methyl donor in biological systems [24, 26, 28, 45].

This work was supported by the identification of non-volatile intermediate products, such as MMSb and DMSb, and their volatile counterparts in various incubations of aerobic and anaerobic microorganisms and the incorporation of $^{13}CD_3$ into methyl-

Sb species (non-volatile DMSb and TMSb) from $^{13}CD_3$-labelled L-methionine (a precursor to SAM) [7]. This suggests that SAM is the methyl donor for antimony methylation for methylated species formed by *S. brevicaulis* from a potassium antimony(III) tartrate substrate (PAT) [7].

Isotopically enriched $^{123}Sb(V)$ using methylcobalamin (CH_3-B_{12}), a common methyl donor for Hg, as the methyl donor and glutathione as the reductant showed that CH_3-B_{12} is unlikely the sole methyl-donor in the inoculum of methanogenic archaea derived from sewage sludge fermenters [27, 28, 51]. Methylated antimony species (MMSb and DMSb) were produced by the methanogens, but the sole species from sewage sludge fermenters (non-volatile TMSb and $(CH_3)_3Sb$) was not produced [28]. An Sb- and methyl-Sb-glutathione complex was identified indicating the role of glutathione with antimony reduction during the methylation process [28]. Glutathione complexes may be stabilizing methylated Sb(III) species, being intermediates during the methylation process [28].

11.2.1 Factors affecting the biomethylation of antimony

Biomethylation of antimony appears to be a fortuitous process due to the low yields of methylated (0.8%) and volatilized (0.001–0.03%) antimony rather than a detoxification mechanism [5, 11, 26, 28, 52, 53]. The substrate speciation has an important role in antimony biomethylation. It has been reported that 5X less methylated species are formed when an Sb(V) substrate ($KSb(OH)_6$) is used compared to PAT for an aerobic microbe (*Cryptococcus humicolous*) or no production of methylated species in *S. brevicaulis* inoculum [52, 54]. *C. humicolous* produced the most methylated Sb species when using a PAT substrate [52]. Other studies have shown that Sb(V) can be biomethylated, but substrates with Sb(V) did not produce MMSb and produced more non-volatile TMSb (10.69 $\mu g\ L^{-1}$) [52].

The relationship between substrate Sb(III) concentration and biomethylation has been reported to be positively correlated in aerobic microbes [5, 11, 52]. In anaerobic microbes, no apparent relationship between substrate concentration and biomethylation or biovolatilization was observed [26, 55]. *Flavobacterium* sp. also showed a dependence on the concentration of antimony in the substrate and the methylation potential, for example, 10–20 mg L^{-1} PAT produced up to 4 ± 0.5 $\mu g\ L^{-1}$ non-volatile methylated species but at PAT concentrations of 30 mg L^{-1} the non-volatile methylated species production decreased (0.9 ± 0.1 $\mu g\ L^{-1}$) [11]. At low PAT substrate concentrations (e.g., 10 or 20 mg L^{-1}), non-volatile MMSb and DMSb were observed in the supernatant of *Flavobacterium* sp. but with a higher concentration (e.g., 30 mg L^{-1}) non-volatile TMSb was the dominant species (81%) [11]. By using a PAT substrate, the highest production of methylated antimony species was formed by *C. humicolous* but when the concentration of PAT exceeded 500 mg L^{-1} the production of methylated species was reduced [52]. In headspace gases of *M. formicicium* low substrate concentrations (0.005 nM Sb) produced volatile SbH_3 (17.3 ng), $(CH_3)SbH_2$ (1.7 ng), $(CH_3)_2SbH$ (1.9 ng), and $(CH_3)_3Sb$ (16.1 ng), whereas at concentrations of 0.05 nM Sb or higher only $(CH_3)_3Sb$ was produced [55].

The introduction of inorganic arsenic showed an enhancing effect on the production of methylated antimony species, and arsenic methylation was more efficient than antimony [11, 52]. Comparisons between arsenic treated and untreated Sb(III) substrates also showed an influence on antimony speciation in *Flavobacterium* sp. [11]. In substrates without arsenic, non-volatile MMSb (42%) and DMSb (58%) were detected, whereas with arsenic treatment antimony was formed earlier and MMSb (14%), DMSb (61%), and TMSb (25%) were observed in the culture supernatants [11]. Hartmann and coworkers also observed that the presence of trivalent arsenic influenced the speciation of antimony; in these treatments, only non-volatile TMSb was detected [52]. The addition of antimony to the media did not enhance or change antimony speciation in *C. humicolous* [52]. Selenium may also enhance antimony methylation [28].

11.3 Analytical techniques

A representative overview showing the types of sampling, method detection limits (LOD) (when possible), separation systems, and coupling typically used for organoantimony speciation is in Tab. 11.3. For more detailed information, refer to the chapters and reviews by Filella, Andrewes, and Miravet [30, 31, 56].

11.3.1 Standards and certified reference materials

Due to a lack of commercially available standards, quantification of SbH_3, $(CH_3)SbH_2$, $(CH_3)_2SbH$, and $(CH_3)_3Sb$ requires derivatization from the non-volatile trimethylantimony standards. This technique may suffer from molecular rearrangements (see Section 11.3.3.1 Hydride Generation below). MMSb(V) and DMSb(V) have been synthesized in the lab but this procedure is complicated and subject to skepticism due to a lack of purity of the material and the synthesis has not been reproduced [42, 57, 58]. $TMSb(V)Br_2$ is the only standard commercially available and is used to produce methylated hydrides for gas chromatography (GC) speciation analysis and is used as a standard for TMSb(V) in liquid chromatography [14, 20, 26, 54, 59–62].

There are a lack of certified reference materials (CRM) available for inorganic antimony (Sb_{IO}) species, let alone methyl-Sb species. Most studies focus on sample spiking as a method validation step, although important, there is an expected difference in behavior in spiked samples and natural samples. Species-specific

Tab. 11.3: Examples of the sampling types, separation techniques, and detection system coupling possibilities for methyl-Sb speciation.

Sampling / Preservation	Separation method	Detection system	Species identified[a]	Quantitative	LOD (methyl-Sb)	References
	(HG)-GC	MS	All NV	Y	15 ng	[21]
	(HG)-GC	AAS	All NV All V	Y	20 ng dm^{-3} 1 ng	[8,24,52,53,67]
SPME	GC	MS	All V	X	0.4 ng g^{-1}	[53]
SPME	(HG)-GC	AAS	Sb_{IO}, DMSb(V), TMSb(V)	X	0.1 µg L^{-1}	[7,24,54]
CT-(CF)	(HG)-GC	AAS	$(CH_3)_3Sb$	Y	5 ng	[72]
CT-(CF)	(HG)-GC	ICP-MS	All V	Y	0.15 pg–5 ng	[13,14,19,26,28, 54,67,73,74]
	HPLC	HG-(UV)-AFS	Sb(III), Sb(V), TMSb(V)	Y	0.13 µg L^{-1}	[64,75]
	HPLC	ICP-MS	Sb(III), Sb(V), TMSb(V)	Y	0.05–0.2 µg L^{-1}	[20,43,62,76]

[a]V indicates all volatile species measured and identified (SbH_3, $(CH_3)SbH_2$, $(CH_3)_2SbH$, and $(CH_3)_3Sb$). NV indicates all non-volatile species measured from matrix (Sb_{IO}, MMSb, DMSb, and TMSb). HG is used to measure non-volatile species.

CRMs exist for methyl-Hg, organo-Sn species, inorganic, and methylated arsenic species in biological, plant, and soil matrices yet there is a lack of CRMs for antimony speciation in any matrix.

11.3.2 Sample preservation

The stability of methylated volatile stibines is summarized in Section 11.1.2, "Basic chemical properties," and Tab. 11.2 [14, 38, 63]. Much less work has been done to date on the stability of methylated species in biota, soils, sediments, and in solution. For example, there is a lack of information on the stability of TMSb(V) in solution and proper storage practices. Filtered seawater samples showed TMSb(V) was stable at 4 °C with and without the addition of EDTA after 7 days of sampling [64]. It is generally accepted that freeze-drying is the best preservation technique for speciation analysis [65].

11.3.3 Sampling and preservation of volatile and non-volatile species

Several techniques are used to sample and preserve the integrity of volatile methylated antimony species. The direct analysis of volatile stibines (or those produced from dissolved species though hydride generation (HG)) is possible when the concentration of methylated antimony species is high enough. This technique may preserve the integrity of the methyl-Sb species and is quantitative but lacks adaptability for the field. Polymer bags are an efficient sampling technique as they do not obstruct the speciation of the compounds and the gases can be directly analyzed or cryogenically trapped (CT) and analyzed [38, 66]. It is quantitative and qualitative, yet volume limitations restrict this technique to high-emission point sources, and samples should be measured within 24 hours [13, 39, 57, 58].

CT is a non-destructive technique that kinetically stabilizes SbH_3, $(CH_3)SbH_2$, $(CH_3)_2SbH$, and $(CH_3)_3Sb$ [13, 38]. Typically, the volatile antimony species (or derivatized non-volatile species) are trapped on a stationary phase packing [8, 14, 19, 26–28, 56, 67]. This technique may be impractical in remote field areas due to limitations on the mobility of the method.

Solid phase microextraction (SPME) is typically coupled with GC and is commonly used to separate the methylated species from Sb_{IO} [53, 68]. It has the advantage that it is a cost-effective, fast, and simple technique that combines sampling, pre-concentration, and separation into one-step. SPME can be used in both lab and field-based studies. One of the major disadvantages of this technique is the competition between analytes on the solid phase causing a misrepresentation of the analyte of interest on the SPME fiber [47].

The headspace gases of *S. brevicaulis* were collected using nitric acid traps for total volatile antimony (which was identified with MS as $(CH_3)_3Sb$) by Jenkins and coworkers [69]). This method was never validated, but the authors reported that only 51% of the antimony was recovered from the traps [69]. Recently, a method developed originally for volatile arsines was used to measure volatile stibines emanating from rice paddies sampled from the Xikuangshan antimony mining area in Lengshuijiang, China [70, 71]. This method has not been validated for antimony (only for arsenic) and may represent an underestimation of volatile antimony [70, 71].

11.3.4 Solid sample extraction and stability of extracts

Ultra-sonication, heat, time, and extracts are used for assistance with speciation analysis in soils, biota, and plants. Typically, extracts with chelating properties such as citric acid and EDTA have the highest recoveries of TMSb in soils and biota. MeOH:water, acetic acid, and water extractions have had poor recoveries of TMSb in soils and biota (<1%), but Mueller and coworkers recovered 5–20% TMSb(V) from

Pteris vittate using water extraction [54, 76–79]. An oxalic–ascorbic acid extraction worked well for TMSb(V) extraction in plants (98.8 ± 2.5%) and shooting range soils (78.9 ± 11%), whereas a two-step extraction from mining-impacted soils recovered 100% of the spiked TMSb(V) standard [20, 62, 76]. Extraction of TMSb(V) from spiked algae samples with EDTA was suggested as it was the most efficient extract for algae and mollusks [75]. Extraction studies on chitin using acetic acid and MeOH:water preserved the speciation of TMSb(V) [80].

11.3.5 Speciation analysis

11.3.5.1 Hydride generation

HG is commonly used for speciation analysis of metal(loid)s that can produce volatile hydrides. The original oxidation state of the species cannot be determined using HG. HG has been subject of debate for antimony speciation, especially for the speciation of non-volatile MMSb, DMSb, and TMSb. It has been shown that significant molecular rearrangements can occur during HG giving rise to artifacts during analysis [67, 81]. During the derivatization of $(CH_3)_3SbCl_2$, $(CH_3)_3Sb(OH)_2$, and synthesized MMSb, DMSb, and TMSb standards using HG-GC-AAS and HG-GC-MS, four peaks corresponding to SbH_3, $(CH_3)SbH_2$, $(CH_3)_2SbH$, and $(CH_3)_3Sb$ were identified [21, 58]. These effects are enhanced at lower pH equating up to 10% in natural water samples [67]. By using acetic acid, preconditioning the reactor vessel, and having a semi-continuous flow with rapid O_2 exclusion during the HG process, molecular rearrangements were reduced and only one peak was detected from the $(CH_3)_3SbCl_2$ standard corresponding to $(CH_3)_3{}^{121/123}Sb$ [21, 67, 80]. Although these arrangements have not been observed for arsenic speciation, they may contribute to misinterpretation of antimony speciation in the environment.

11.3.5.2 Gas chromatography

GC coupled with various detectors is a highly useful technique for determining molecular information, identification, and quantification of volatile stibines (Tab. 11.3). One major disadvantage of GC is its lack of available antimony standards and that it relies solely on the production of volatile stibines from the HG technique (dissolved methylated species). SbH_3, $(CH_3)SbH_2$, $(CH_3)_2SbH$, and $(CH_3)_3Sb$ are produced as artifacts from a $TMSbBr_2$ solution standard and used for species identification during measurements. GC has been used to identify, for example, volatile and non-volatile methylated antimony species from headspaces gases, sewage sludge fermenters, and incubated alluvial soil headspaces [26, 28, 53, 82].

11.3.5.3 High-performance liquid chromatography

One of the advantages of high-performance liquid chromatography (HPLC) is that it does not suffer from molecular rearrangement issues inherent to HG, and depending on the type of matrix, no prior preparation steps may be required before injection. One of the major disadvantages is the lack of commercially available standards. Separation of Sb species has been successful mostly using the anion exchange (ANEX) column [77]. When a cation exchange column was used, TMSb(V) was not detected, and with reverse-phase chromatography, TMSb(V) was only eluted with the solvent front in alkaline mobile phases [77]. The separation of TMSb(V) was shown to be complicated using an isocratic elution program as Sb(III) and TMSb(V) were co-eluting but the addition of an organic modifier reduced this effect [20, 62, 76, 83]. Sb(III), Sb(V), and TMSb(V) were successfully separated in seawaters, algae, mollusks, marine sediments, and airborne particulate matter using isocratic and gradient elution programs and ANEX [59, 64, 75, 84, 85].

11.3.6 Detector systems

11.3.6.1 Atomic absorption spectrometer

Atomic absorption spectrometry (AAS) has the advantage of being a user-friendly and cheap technique but suffers from higher detection limits (μg L^{-1} range) [31]. AAS has been used for the detection of non-volatile DMSb and TMSb in pure culture inocula [53, 54].

11.3.6.2 Atomic fluorescence spectrometer

Atomic fluorescence spectrometry (AFS) is a technique that offers a high level of sensitivity, in the ng L^{-1} range (higher than AAS) and cheaper to operate than ICP-MS. Sb_{IO} and TMSb(V) were separated using HPLC and detected in waters and contaminated mining-impacted soils, tailings, and sediments [85, 86].

11.3.6.3 Inductively coupled plasma mass spectrometer

Inductively coupled plasma mass spectrometry (ICP-MS) has the added benefit of low detection limits typically in the ng L^{-1} range but the range of pg L^{-1} using GC-ICP-MS for volatile antimony has been accomplished [87, 88]. Sb_{IO} and TMSb(V) have been successfully separated and detected using HPLC-ICP-MS in soils, waters, and biota [62, 78, 79, 89].

11.3.6.4 Mass spectrometer

Mass spectrometry (MS) has been used to identify molecular information for volatile stibines produced from the HG process and to identify the speciation of the $TMSbCl_2$ and/or $TMSb(OH)_2$ standards in solution. Antimony has two stable isotopes: 121 (57.2%) and 123 (42.8%) [90]. These masses are used to identify the fragmentation of antimony compounds and subsequently identify species. For example, antimony artifacts during the HG process were confirmed using HG-GC-MS and their corresponding masses [21]. TMSb solution standards were identified with electrospray time of flight mass spectrometer and electrospray ionization mass spectrometer [41, 43].

11.4 Occurrence and behavior

11.4.1 Aerobic microorganisms

A majority of the research for aerobic microbial antimony biomethylation has focused on the filamentous fungus, *S. brevicaulis*. The most common detected volatile methylated species in *S. brevicaulis* headspace gases and bioreactors is $(CH_3)_3Sb$ [5, 7, 19, 54, 69]. Trace amounts of SbH_3, $(CH_3)SbH_2$, and $(CH_3)_2SbH$ were also detected in the headspace gases of aerobic bioreactors [54]. Numerous reports of methylated antimony species formed by *S. brevicaulis* indicate relatively high fluxes of methylated antimony species. For example, Craig and coworkers reported 25–50 ng of $(CH_3)_3Sb$ produced from PAT substrate (60 mg dm^{-3}), and Jenkins and colleagues reported up to 4,107 ± 981 ng of $(CH_3)_3Sb$ produced from a PAT substrate (400 mg) [5, 72]. In the study by Andrewes and colleagues, only 5 of the 63 inoculums produced volatile antimony which was in the pg range, but in the large bioreactor experiment, the authors observed up to 800 pg of $(CH_3)_3Sb$ after 8 days and all methylated species were detected [54]. For non-volatile methylated species, methyl-Sb in the cell extracts in *S. brevicaulis* cultures was not detected, but the saturated Sb_2O_3 substrate produced up to 2 μg L^{-1} of non-volatile DMSb and 0.8 μg L^{-1} of non-volatile TMSb in the media after 1 month [54]. After 2 months with 1,000 mg L^{-1} PAT substrate, non-volatile DMSb (7.1 μg L^{-1}) and TMSb (3.5 μg L^{-1}) were produced in the media [54]. In a 1,000 μg L^{-1} substrate after 12 hours, 60–538 pg of $(CH_3)_3Sb$ was detected in the headspace gases [19]. After 18 days of this experiment 10 ng $(CH_3)_3Sb$ was observed with the most produced during the beginning of the experiment [19].

Flavobacterium sp. (aerobic chemotroph), *Cryptococcus humicolous* (isolated from raw sewage), and *Phaeolus schweinitzii* can biomethylate antimony from Sb_{IO} substrates [11, 23, 52]. Non-volatile MMSb, DMSb, and TMSb have been detected in the inoculum supernatants [11, 23, 52]. In *Flavobacterium* sp. cultures a maximum of 4 ± 0.5 μg L^{-1} non-volatile methylated species was formed in substrates of 20 μg L^{-1} PAT where MMSb and DMSb were equal in proportion [11]. A maximum of 183.6 μg L^{-1} non-volatile methylated Sb species was detected in the *C. humicolous* inoculum (100 mg L^{-1} Sb) and non-volatile DMSb (49.3%) was the major species, but MMSb (14%) and TMSb (36.7%) were also detected in the treatments of *C. humicolous* [52]. Non-volatile TMSb was the dominant species produced (up to 520 μg L^{-1} in 40 days), but DMSb was also detected, and up to 6% of the substrate was methylated by the wood-rotting fungus, *P. schweinitzii* [23].

11.4.2 Anaerobic microorganisms

Anaerobic microorganisms are efficient at biomethylating antimony. There are numerous reports of methylation of antimony substrates by sewage sludge fermenters, anaerobic soil enrichment cultures, and from landfill sites [12, 14, 26–28, 55, 72, 82]. $(CH_3)_3Sb$ was consistently the most abundant volatile species detected in the headspace gases of anaerobic soil enrichment cultures, *Clostridium* spp., *P. fluorescens*, peptolytic bacteria, methanogenic archaea, and sewage sludge enrichment cultures [6, 12, 55, 72, 91]. In the headspace gases of *M. formicicium* SbH_3 (17.3 ng), $(CH_3)SbH_2$ (1.7 ng), and $(CH_3)_2SbH$ (1.9 ng) were also detected. The SRB, *D. gigas* was not capable of producing volatile Sb [55]. *Clostridium* spp. (*C. acetobutylicum, C. butryricum,* and *C. cochlearium*), also produced non-volatile MMSb, DMSb, and TMSb (total species up to 21.3 μg dm^{-3}) where MMSb was the main species (76 ± 27.9%) [12]. The maximum yield of non-volatile methylated species was 0.043% of the PAT substrate by *C. acetobutylicum* after 28 days [12].

In sewage sludge fermenters $(CH_3)_3Sb$ was the sole volatile product measured (77.5 ng L^{-1}) after 1 week of incubation even though the substrate had a low antimony concentration (1.7 mg kg^{-1}) [55]. The methanogenic archaea, *M. formicicium*, was the most efficient at volatilizing antimony, forming up to three orders of magnitude more $(CH_3)_3Sb$ than other cultures [55]. In garden and garage soil enrichment cultures, stimulated for nitrate reducers, 40–3,450 ng of $(CH_3)_3Sb$ (2.3% methylation rate) was produced, and $(CH_3)_3Sb$ was the only volatile species detected in 12 of 104 cultures and suggests that nitrate reducing conditions may be favorable for $(CH_3)_3Sb$ formation [72]. The microbe *P. fluorescens* showed an indication of volatile $(CH_3)_3Sb$ (566–4,897 μg L^{-1}) in the headspace gases of 24 of 48 tested incubations [91].

11.4.3 Aquatic environments

Methylated antimony species may represent around 10% of the total antimony and even up to 21% in surface waters from the Atlantic Ocean has been reported [15, 92–96]. Andreae and coworkers were the first to report methylated antimony (MMSb(V) and DMSb(V)) species in waters from the Gulf of Mexico, polluted rivers from Germany, and estuary waters from Florida (Ochlockonee River) [92]. The speciation of methylated antimony in these earlier studies is likely inaccurate due to issues with the HG technique [92]. Nevertheless, methylated antimony species are present in the waters but only the degree of methylation cannot be distinguished [92]. DMSb and TMSb have also been detected in the estuary and polluted river waters, the surface waters of Chatham Rise, and the Baltic Sea but are typically in lower abundance than MMSb [8, 53, 64, 92, 93, 96]. Methyl-Sb showed conservative behavior in vertical profiles of the Pacific Ocean and long residence time in surficial waters of Atlantic Ocean studies – a different behavior than observed for methyl-As species, which tend to have a strong depth dependence [15, 95].

In most studies, methylated antimony species are restricted to the upper sections of the vertical water profiles [15, 93, 95]. In the Baltic Sea MMSb (maximum 0.13 nM) showed enrichment in the photic and anoxic zones in one transect, whereas in other transects, MMSb tends to have high proportions in the surface waters than a consistent distribution with depth [96].

Andreae and coauthors reported methyl-Sb species in the Baltic Sea to be formed by bacteria because there were no methyl-Sb species in the marine plankton or algae [96]. Other workers have also suggested the formation of these species by bacteria [15, 16, 93, 95]. The lack of methylated species at depth, in Chatham Rise and the North Pacific Ocean, could indicate the importance of microbiota for methylation of antimony (e.g., phytoplankton) and that they are not particularly stable and are recycled with upwelling and downwelling at these zones [93, 95].

11.4.4 Soils and sediments

Antimony biomethylation in the soil and sediment environment focuses on highly contaminated sites and reducing environments. The main methylated species detected in sediment pore waters from the Ruhr River and a redox-controlled contaminated floodplain soil was MMSb (max 0.61 $\mu g\ L^{-1}$ MMSb), TMSb(V) in the pore waters of shooting range soils (3.5 $\mu g\ L^{-1}$ TMSb(V) or 7.8% of the total antimony), and DMSb in the pore waters of industrially contaminated soil ($0.424 \pm 0.018\ \mu g\ L^{-1}$) [10, 62, 83, 86, 97–99]. MMSb ($56.03 \pm 11.77\ \mu g\ kg^{-1}$) was the major methylated species in soil extracts from an arable soil [9]. TMSb(V) was the only methylated species detected in extracts of mining-impacted soils (0.006–2.3 $mg\ kg^{-1}$), industrially polluted soils ($4.3 \pm 1.1\ \mu g\ L^{-1}$), and after flooding in shooting range soils (1.35 $mg\ kg^{-1}$, 0.6% of total antimony). TMSb (max 0.6 $\mu g\ L^{-1}$) was the major methylated species observed in the surface waters of flooded industrially contaminated sediment but MMSb and DMSb were also detected [10, 62, 78, 83, 100]. Acidified, eutrophic sediments spiked with arsenic and antimony produced up to 2.6 $\mu g\ L^{-1}$ DMSb in the surface water column after 36 days [10]. The most abundant non-volatile methylated species detected in an arable soil (20–40 cm depth) was MMSb, but DMSb and TMSb were also present in the soil [9].

Enhancing reducing conditions by flooding microcosms has been shown to increase the methylation of antimony [10, 62, 94, 98]. The production of TMSb(V) in soil pore waters and the production of volatile antimony (up to 20%) were enhanced when adding a nutrient source, such as the amendment of manure or rice straw [62, 71]. This enhancement indicates that biomethylation and biovolatilization may be a microbial-facilitated process in soils and sediments [9, 10, 62, 71]. Other workers have reported that the methylation of antimony did not appear to follow productivity cycles in river sediments and was controlled more by suspended sediments during heavy rain and winter flooding events [97].

Adsorption studies on soils surrounding the Xikuangshan mine indicate that TMSb(V) may be less mobile in soils with Fe- and Al-rich phases but more mobile than organo-As species [86]. For example, adsorption of TMSb(V) in the Ferrasols was lower than observed for monomethylarsenic (97%) and dimethylarsenic (87%) species, indicating a higher mobilization potential of TMSb(V) [86]. In the pore waters of a floodplain soil, a higher proportion of methylated antimony species than methylated arsenic species was observed [98]. This suggests antimony may be more susceptible to methylation and more mobile than arsenic in these sediments [98].

Other workers have also identified $(CH_3)_3Sb$ in the headspace gases of an anaerobically incubated alluvial soil with a low antimony concentration (1.3 mg kg^{-1}) [82]. Over 3 months, approximately 1 nmol of $(CH_3)_3Sb$ was produced as the only detected volatile species [82]. $(CH_3)_3Sb$ was also detected in emissions from alpine wetland bogs [66].

11.4.5 Biota

MMSb, DMSb, and TMSb have been detected in mosses, liverwort, pondweed, ferns, and other plants surrounding mining-impacted areas, and the uptake of TMSb(V) has been evaluated in hydroponic solutions in ryegrass, rye, fescue, and wheat [20, 21, 80, 89]. DMSb and trace MMSb were detected in the extracts of liverwort (181–186 ng g^{-1} DMSb) and mosses (maximum 0.17 ± 0.01 mg kg^{-1} DMSb in *Drepanocladus* sp.) impacted from historical and modern mining activities [8, 80]. Pondweed extracts, influenced by mining effluent in the Northwest Territories, Canada, had four peaks corresponding to SbH_3, MMSb, DMSb, and TMSb [21]. TMSb was detected as the dominant methylated species in extracts of fern (*Pteris vittata*) leaves (0.6 ± 0.01 µg kg^{-1}) and roots (1 ± 0.6 µg kg^{-1}) surrounding contaminated sites, in leaf extracts of perennial herbs and ferns surrounding the Xikuangshan mine, China (10–99% of the speciation), and in *Hydnum cupressiforme Hedw.* (2.87 ± 0.32 µg g^{-1}) from the Eastern Pyrenees, Spain [61, 79, 101].

In TMSb(V) spiked hydroponic solutions, the highest amount of antimony uptake was in the shoots of fescue plants and the least in plant roots [20, 89]. The authors observed 150–240 μg kg^{-1} TMSb(V) in all roots of Sb(V) treatments as well as a range of 50–150 μg kg^{-1} TMSb(V) in plant shoots of all the Sb_{IO} treatments [89]. In TMSb(V) hydroponic solutions, TMSb(V) was the dominant species in the roots and shoots [89]. TMSb(V) was taken up the least in plant roots in all treatments [89]. Hydroponic experiments in ryegrass, rye, fescue, and wheat plants in hydroponic solutions showed TMSb(V) nutrient spiked solutions having the highest mobility and easy uptake in plant shoots [89]. The results of TMSb(V) uptake in the plants were counter-intuitive as TMSb(V) should be present either as a neutral or positively charged ion in solution at pH 6, whereas Sb(V) should be negatively charged [89]. This indicates different uptake mechanisms in the plant for Sb(V) and TMSb(V) in plant roots potentially using glycerol transports by aquaporins [89].

Sb_{IO} may be methylated in plant shoots as TMSb has been detected (up to 6%) in the shoots of ryegrass, rye, wheat, and fescue and plants surrounding the Xikuangshan Mine [20, 61, 89]. Demethylation may also be occurring in the shoots of ryegrass, rye, wheat, and fescue as Sb_{IO} was detected in the shoots of the plants in TMSb(V) treated hydroponic solutions [20, 89].

Methyl-Sb species in organisms remain ambiguous. There has only been one report of methyl-Sb species in animals to the author's knowledge. De Gregori and coworkers did not detect TMSb in mollusks or algae off the Chilean coast. DMSb and TMSb were detected by Koch and coworkers in snails nearby a Au mining-impacted region in the Northwest Territories, Canada [8].

TMSb has also been detected in human spot urine samples from individuals working in the battery industry and people that consumed Sb-containing fish [102, 103]. In addition to TMSb, the people who consumed Sb-containing fish also had MMSb and DMSb in their urine [102]. In the fish ingestion study, after 24 h, MMSb (3 ng L^{-1}), DMSb (1 ng L^{-1}), and TMSb (2 ng L^{-1}) were detected [102]. The highest observed methylated species was TMSb after 14 hours of ingestion [102]. In the urine, Sb_{IO} was dominant and the methylated antimony species are lower than observed for arsenic analogs [102].

11.4.6 Sewage sludge fermenters

$(CH_3)_3Sb$ is the only detected volatile antimony species in the headspace gases of sewage sludge fermenters, making up to 0.1% of the total antimony species [13, 26, 28, 38, 55, 73]. Up to 14.72 μg m^{-3} Sb has been detected in sewage gases [13]. In wastewater incubations with low substrate antimony, 77.5 ± 4.6 ng L^{-1} $(CH_3)_3Sb$ was produced in the headspace gases [55]. The high production of $(CH_3)_3Sb$ in these experiments is suggested by the authors due to 1) antimony being either more easily volatilized or 2) taken up by microbial cells easier than other metal(loid)s [55].

All non-volatile species, and an ethyl-Sb species in one study, have been detected in the filtrates of sewage sludge and wastewaters [26, 28, 73]. MMSb (2.2 ± 0.3 ng g^{-1}), DMSb (0.14 ± 0.02 ng g^{-1}), and TMSb (0.96 ± 0.4 ng g^{-1}) were detected in the filtrates of sewage sludge and can equate up to 0.6% of the total antimony [26]. Typically, MMSb was the dominant methylated species in the non-volatile sewage sludge extracts and filtrates [26, 28]. In abiotic treatments of sewage sludge, Sb_{IO} was produced as well as minor DMSb (0.2–0.43 ng) and MMSb (2.9–3.8 ng) [28]. A majority of the non-volatile methylated antimony species were observed within the cell, suggesting that methylation may be occurring in the cell, and only 1/10th was in the sludge filtrate with a faster exclusion of TMSb than MMSb from the cell [26].

In sewage sludge samples, methanogenic archaea may be the major methylators [26]. After 7 days, archaea produced more $(CH_3)_3Sb$ than SRB treatments [26]. The methanogenic archaea were up to 6X more efficient at Sb-methylation [26]. This indicates that SRB may not be the major methylators in sewage sludge [26].

11.4.7 Landfills

Methylated and volatile antimony have been detected in the standing waters that landfill gases percolate through and in the landfill gases [13, 73, 74, 104]. $(CH_3)_3Sb$ was the dominant volatile species (up to 71.6 µg m^{-3}) observed in all landfill gases from Canada and Germany, and can make up to 98% of the speciation; SbH_3 was also detected [13, 73, 74, 104]. Two waste gases from Germany had 40 to 2,400 ng m^{-3} Sb [74].

In the condensed water of the waste disposal in Hessen, Germany, and wetland waters that the gases percolate through Sb_{IO}, MMSb(V), DMSb(V), and TMSb(V) were detected [14, 73]. When stibines percolate through wetland waters, $(CH_3)_3Sb$ is oxidized to the pentavalent TMSb(V) in solutions and more easily demethylated than the arsenic counterparts [14]. For example, higher proportions of MMSb(V) and DMSb(V) were observed than their As analogs, suggesting that $(CH_3)SbH_2$ and $(CH_3)_2SbH$ may be relatively more unstable in water [14]. Inlet and outlet gases (which percolate through the water phase) were compared to test how efficient the immobilization of these gases is by the water and showed no significant difference in composition [14]. This suggests that in wetland landfill sites the contact time between solution and gas was not sufficient for immobilization and the subsequent gases are released into the atmosphere [14].

11.4.8 Geothermal hot springs

Geothermal hot springs are capable of dissolving higher amounts of metals due to their extreme conditions such as high temperature and alkaline or acidic conditions and have specialized microbial communities [22]. $(CH_3)_3Sb$ has been the sole volatile Sb species detected in these environments and is likely a product of microbial activity from algal mats [8, 22]. Non-volatile MMSb, DMSb, and TMSb were identified in the hot spring waters in Meager Creek, Canada, and Tokkanu, New Zealand

and can account up to 0.12% of the total Sb [8, 22]. Volatile antimony was detected in Yellowstone National Park where an acidic, high temperature (270–300 °C) hot spring produced the most volatile Sb (0.1–29 mg m^{-3}) and $(CH_3)_3Sb$ was detected above an algal mat in Meager Creek [8, 22, 105].

11.5 Toxicity

There have been a limited number of studies evaluating the toxicity of methylated antimony species. $(CH_3)_3Sb$ and SbH_3 have been identified as genotoxins, although the estimated amount of $(CH_3)_3Sb$ to cause damage is high, 10,000 mg m^{-3} $(CH_3)_3Sb$, and unrealistic in the environment [36]. $(CH_3)_3SbCl_2$ caused genotoxic effects in hamster ovary cells but it was not easily membrane-permeable and did not bind to the cell membrane [106]. $(CH_3)_3SbCl_2$ only caused DNA damage when forcibly taken up [106]. It suggests that resistance to $(CH_3)_3SbCl_2$ is implemented at the uptake channels and not intracellularly. There may be an active efflux mechanism or a metabolite for further cellular resistance [106]. There is a huge gap in knowledge for antimony toxicity, let alone methyl-Sb toxicity.

11.6 Future outlooks

Currently, it is unclear if the biomethylation and biovolatilization of antimony is a mechanism of detoxification or a fortuitous process, and the exact mechanism of methylation may still need to be further elucidated (e.g., Challenger mechanism is unlikely true for arsenic methylation). It is still unclear how stable volatile methylated antimony species are in the atmosphere; the attempted studies were done in synthetic air experiments and are contradictory, casting doubt on their environmental mobility, which could be underestimated. A major concern with the volatilization of metals entails how stable they are in the atmosphere. As TMSb(V) has been detected bound with airborne particulate matter, methylated species may be mobilized further than their original point source. Possibly, $(CH_3)_3Sb$, once emitted to the atmosphere, could be oxidized by UV or radicals to TMSb(V) and/or bind to particulate matter. Uptake mechanisms and conversion of antimony and methyl-Sb species in biota are currently unknown. There is a lack of information relating to the influence of these conversions on the mobility and toxicity of antimony in the environment. As previous studies have indicated that a wide array of microorganisms are capable of biomethylating and biovolatilizing antimony, it may be an important section of the biogeochemical cycle of antimony but it currently remains ambiguous. This lack of knowledge could be attributed to inadequate methods of measurements, which either require a high level of sophistication (e.g., CT-GC-ICP-MS) or suffer from molecular rearrangements (e.g., HG-based techniques) giving rise to artifacts during analysis. With the advancement of speciation methods, these unknowns could be addressed.

Abbreviations

CH_3-B_{12}	Methyl-cobalamin
$(CH_3)SbH_2$	Monomethylstibine
$(CH_3)_2SbH$	Dimethylstibine
$(CH_3)_3Sb$	Trimethylstibine
ANEX	Anion exchange
CT	Cryogenic trapping
DMSb	Dimethylantimony species
DMSb(V)	Dimethylstibinic acid
ESI	Electrospray ionization
GC	Gas chromatography
HG	Hydride generation
HPLC	High-performance liquid chromatography
ICP	Inductively coupled plasma
MMSb	Monomethylantimony species
MMSb(V)	Methylstibonic acid
MS	Mass spectrometry
PAT	Potassium antimony(III) tartrate $KSbOC_4H_4O_6 \cdot ½H_2O$
SAM	S-Adenosylmethionine
SbH_3	Stibine
Sb(III)	Antimonite $Sb(OH)_3$
Sb_{IO}	Inorganic Sb species
Sb(V)	Antimonate $Sb(OH)_6^-$
SPME	Solid phase microextraction
SRB	Sulfate reducing bacteria
TMSb	Trimethylantimony species
TMSb(V)	Trimethylantimony oxide

References

[1] Richardson BA. Sudden Infant Death Syndrome: A possible primary cause. J Forensic Sci Soc 1994, 34(3), 199–204.

[2] Pearce RB, Callow ME, Macaskie LE. Fungal volatilization of arsenic and antimony and the sudden infant death syndrome. FEMS Microbiol Lett 1998, 158(2), 261–265.

[3] Jenkins RO, Morris TA, Craig PJ, Goessler W, Ostah N, Wills KM. Evaluation of cot mattress inner foam as a potential site for microbial generation of toxic gases. Hum Exp Toxicol 2000, 19, 693–702.

[4] Cullen A, Shannon W, Thornton L, et al. Concentrations of antimony in infants dying from SIDS and infants dying from other causes. Arch Dis Child 2000, 82(3), 244–247.

[5] Craig PJ, Jenkins RO, Dewick R, Miller DP. Trimethylantimony generation by Scopulariopsis brevicaulis during aerobic growth. Sci Total Environ 1999, 229(1–2), 83–88.

[6] Gates P, Harrop H, Pridham J, Smethurst B. Can microorganisms convert antimony trioxide or potassium antimony tartrate to methylated stibines?. Sci Total Environ 1997, 205(2–3), 215–221.
[7] Andrewes P, Cullen WR, Feldmann J, Koch I, Polishchuk E. Methylantimony compound formation in the medium of Scopulariopsis brevicaulis cultures: 13CD3-L-methionine as a source of the methyl group. Appl Organomet Chem 1999, 13(10), 681–687.
[8] Koch I, Wang L, Feldmann J, Andrewes P, Reimer KJ, Cullen WR. Antimony species in Environmental Samples. Int J Environ Anal Chem 2000, 77(2), 111–131.
[9] Duester L, Diaz-Bone RA, Kosters J, Hirner AV. Methylated arsenic, antimony and tin species in soils. J Environ Monit 2005, 7(12), 1186–1193.
[10] Duester L, Vink JP, Hirner AV. Methylantimony and -arsenic species in sediment pore water tested with the sediment or fauna incubation experiment. Environ Sci Technol 2008, 42(16), 5866–5871.
[11] Jenkins RO, Forster SN, Craig PJ. Formation of methylantimony species by an aerobic prokaryote: Flavobacterium sp. Arch Microbiol 2002, 178(4), 274–278.
[12] Smith LM, Craig PJ, Jenkins RO. Formation of involatile methylantimony species by Clostridium spp. Chemosphere 2002, 47(4), 401–407.
[13] Feldmann J, Hirner A. Occurrence of volatile metal and metalloid species in landfill and sewage gases. Int J Environ Anal Chem 1995, 60(2), 339–359.
[14] Feldmann J. Volatilization of metals from a landfill site: Generation and immobilization of volatile species of tin, antimony, bismuth, mercury, arsenic, and tellurium on a municipal waste deposit in Delta, British Columbia. In: Biogeochemistry of Environmentally Important Trace Elements. Washington, DC, 2002, 128–140.
[15] Cutter GA, Cutter LS. Biogeochemistry of arsenic and antimony in the North Pacific Ocean. Geochemistry, Geophys Geosystems 2006, 7(5), 1–12.
[16] Cutter GA. Dissolved arsenic and antimony in the Black Sea. Deep Res Part I Oceanogr Res Pap 1991, 38(Supplement 2), S825–43.
[17] Cutter GA, Cutter LS. Behavior of dissolved antimony, arsenic, and selenium in the Atlantic Ocean. Mar Chem 1995, 49(4), 295–306.
[18] Ji Y, Sarret G, Schulin R, Tandy S. Fate and chemical speciation of antimony (Sb) during uptake, translocation and storage by rye grass using XANES spectroscopy. Environ Pollut 2017, 231(2), 1322–1329.
[19] Andrewes P, Cullen W, Polishchuk E. Confirmation of the aerobic production of trimethylstibine by Scopulariopsis brevicaulis. Appl Organomet Chem 1999, 13(9), 659–664.
[20] Mestrot A, Ji Y, Tandy S, Wilcke W. A novel method to determine trimethylantimony concentrations in plant tissue. Environ Chem 2016, 13(6), 919–926.
[21] Dodd M, Pergantis SA, Cullen WR, Li H, Eigendorf GK, Reimer KJ. Antimony speciation in freshwater plant extracts by using hydride generation-gas chromatography-mass spectrometry. Analyst 1996, 121(2), 223–228.
[22] Hirner AV, Feldmann J, Krupp E, Grümping R, Goguel R, Cullen WR. Metal(loid)organic compounds in geothermal gases and waters. Org Geochem 1998, 29(5–7), 1765–1778.
[23] Andrewes P, Cullen WR, Polishchuk E, Reimer KJ. Antimony biomethylation by the wood rotting fungus Phaeolus schweinitzii. Appl Organomet Chem 2001, 15(6), 473–480.
[24] Andrewes P, Cullen WR, Polishchuk E. Arsenic and antimony biomethylation by Scopulariopsis brevicaulis: Interaction of arsenic and antimony compounds. Environ Sci Technol 2000, 34(11), 2249–2253.
[25] Andrewes P, Cullen WR, Polishchuk E. Antimony biomethylation by scopulariopsis brevicaulis: Characterization of intermediates and the methyl donor. Chemosphere 2000, 41(11), 1717–1725.

[26] Wehmeier S, Feldmann J. Investigation into antimony mobility in sewage sludge fermentation. J Environ Monit 2005, 7, 1194–1199.
[27] Wehmeier S, Ellam R, Feldmann J. Isotope ratio determination of antimony from the transient signal of trimethylstibine by GC-MC-ICP-MS and GC-ICP-TOF-MS. J Anal At Spectrom 2003, 18(9), 1001–1007.
[28] Wehmeier S, Raab A, Feldmann J. Investigations into the role of methylcobalamin and glutathione for the methylation of antimony using isotopically enriched antimony(V). Appl Organomet Chem 2004, 18(12), 631–639.
[29] Meyer J, Michalke K, Kouril T, Hensel R. Volatilisation of metals and metalloids: An inherent feature of methanoarchaea?. Syst Appl Microbiol 2008, 31(2), 81–87.
[30] Filella M. Alkyl Derivatives of Antimony in the Environment. In: Sigel A, Sigel H, Sigel RK, editors, Organometallics in Environment and Toxicology. 2010, 261–301.
[31] Andrewes P, Cullen WR. Organoantimony compounds in the Environment. In: Craig PJ, editor, Organometallic compounds in the Environment. John Wiley & Sons, 2003, 277–351.
[32] Jenkins RO. Biomethylation of Arsenic, Antimony and Bismuth. In: Sun H, editor, Biological Chemistry of Arsenic, Antimony and Bismuth. 2011, 145–180.
[33] Yaws C. Inorganic Compounds and Elements, 4th Edition. Handbook of Vapor Pressure, 1995.
[34] Burg AB, Grant L. The methylstibines and the monomer dimethylstibinoborine. J Am Chem Soc 1959, 81(1), 1–5.
[35] Rosenbaum EJ, Sandberg R. Vapor pressures of trimethylphosphine trimethylarsine and trimethylstibine. J Am Chem Soc 1940, 62(6), 1622–1623.
[36] Andrewes P, Kitchin KT, Wallace K. Plasmid DNA damage caused by stibine and trimethylstibine. Toxicol Appl Pharmacol 2004, 194(1), 41–48.
[37] Atkins P, Overton T, J.r W, Armstrong M. Inorganic Chemistry, 5th Edition. Oxford, W.H. Freeman and Company, 2010.
[38] Haas K, Feldmann J. Sampling of trace volatile metal(loid) compounds in ambient air using polymer bags: A convenient method. Anal Chem 2000, 72(17), 4205–4211.
[39] Parris GE, Brinckman FE. Reactions Which Relate to Environmental Mobility of Arsenic and Antimony. II. Oxidation of Trimethylarsine and Trimethylstibine. Environ Sci Technol 1976, 10(12), 1128–1134.
[40] Craig PJ, Forster SA, Jenkins RO, Lawson G, Miller D, Ostah N. Use of mass spectroscopic techniques to elucidate the nature of the products of the oxidation of trimethylstibine in air. Appl Organomet Chem 2001, 15(6), 527–532.
[41] Zheng J, Takeda A, Furuta N. Investigating the electrospray mass spectra of inorganic and organic antimony compounds. J Anal At Spectrom 2001, 16(1), 62–67.
[42] Meinema HA, Noltes JG. Investigations on organoantimony compounds. VI. Preparation and properties of thermally stable dialkylantimony(V) compounds of the types R2Sb(OR′)3, R2Sb (OAc)3 and R2Sb(O)OH. J Organomet Chem 1972, 36(2), 313–322.
[43] Lintschinger J, Schramel O, Kettrup A. The analysis of antimony species by using ESI-MS and HPLC-ICP-MS. Fresenius J Anal Chem 1998, 361, 96–102.
[44] Challenger F. Biological methylation. Chem Rev 1945, 36(3), 315–361.
[45] Bentley R, Chasteen TG. Microbial Methylation of Metalloids: Arsenic, Antimony, and Bismuth. Microbiol Mol Biol Rev 2002, 66(2), 250–271.
[46] Hayakawa T, Kobayashi Y, Cui X, Hirano S. A new metabolic pathway of arsenite: Arsenic-glutathione complexes are substrates for human arsenic methyltransferase Cyt19. Arch Toxicol 2005, 79(4), 183–191.
[47] Mestrot A, Planer-Friedrich B, Feldmann J. Biovolatilisation: A poorly studied pathway of the arsenic biogeochemical cycle. Environ Sci Process Impacts 2013, 15(9), 1639–1651.

[48] Cullen WR. Chemical mechanism of arsenic biomethylation. Chem Res Toxicol 2014, 27(4), 457–461.
[49] Marapakala K, Qin J, Rosen BP. Identification of catalytic residues in the As(III) S-adenosylmethionine methyltransferase. Biochemistry 2012, 51(5), 944–951.
[50] Marapakala K, Packianathan C, Ajees AA, et al. A disulfide-bond cascade mechanism for arsenic(III) S-adenosylmethionine methyltransferase. Acta Crystallogr Sect D Biol Crystallogr 2015, 71(3), 505–515.
[51] Wood JM, Kennedy FS, Rosen CG. Synthesis of methylmercury compounds by extracts of a mehanorganic bacterium. Nature 1968, 220, 173–174.
[52] Hartmann LM, Craig PJ, Jenkins RO. Influence of arsenic on antimony methylation by the aerobic yeast Cryptococcus humicolus. Arch Microbiol 2003, 180(5), 347–352.
[53] Smith LM, Maher WA, Craig PJ, Jenkins RO. Speciation of volatile antimony compounds in culture headspace gases of Cryptococcus humicolus using solid phase microextraction and gas chromatography-mass spectrometry. Appl Organomet Chem 2002, 16, 287–293.
[54] Andrewes P, Cullen WR, Feldmann J, Koch I, Polishchuk E, Reimer E. The production of methylated organoantimony compounds by Scopulariopsis brevicaulis. Appl Organomet Chem 1998, 12, 827–842.
[55] Michalke K, Wickenheiser EB, Mehring M, Hirner AV, Hensel R. Production of Volatile Derivatives of Metal(loid)s by Microflora Involved in Anaerobic Digestion of Sewage Sludge. Appl Environ Microbiol 2000, 66(7), 2791–2796.
[56] Miravet R, Hernández-Nataren E, Sahuquillo A, Rubio R, López-Sánchez JF. Speciation of antimony in environmental matrices by coupled techniques. Trends Anal Chem 2010, 29(1), 28–39.
[57] Wieber M, Waly J. Methanestibonic Acid. Zeitschrift für Naturforsch B 1990, 45(12), 1615–1617.
[58] Dodd M, Grundy SL, Reimer KJ, Cullen WR, Methylated antimony(V) compounds: Synthesis, hydride generation properties and implications for aquatic speciation. Appl Organomet Chem 1992, 6, 207–211.
[59] Zheng J, Ohata M, Furuta N. Studies on the speciation of inorganic and organic antimony compounds in airborne particulate matter by HPLC-ICP-MS. Analyst 2000, 125(6), 1025–1028.
[60] Quiroz W, Arias H, Bravo M, Pinto M, Lobos MG, Cortés M. Development of analytical method for determination of Sb(V), Sb(III) and TMSb(V) in occupationally exposed human urine samples by HPLC-HG-AFS. Microchem J 2011, 97(1), 78–84.
[61] Wei C, Ge Z, Chu W, Feng R. Speciation of antimony and arsenic in the soils and plants in an old antimony mine. Environ Exp Bot 2015, 109, 31–39.
[62] Grob M, Mestrot A, Wilcke W. Release and Biomethylation of Antimony in Shooting Range Soils upon Flooding. Soil Syst 2018, 2(34), 1–34.
[63] Parris GE, Brinckman FE. Reactions which relate to the environmental mobility of arsenic and antimony. I. Quaternization of trimethylarsine and trimethylstibine. J Org Chem 1975, 40(25), 3801–3803.
[64] De Gregori I, Quiroz W, Pinochet H, Pannier F, Potin-Gautier M. Simultaneous speciation analysis of Sb(III), Sb(V) and (CH 3)3SbCl2 by high performance liquid chromatography-hydride generation-atomic fluorescence spectrometry detection (HPLC-HG-AFS): Application to antimony speciation in sea water. J Chromatogr A 2005, 1091(1–2), 94–101.
[65] Zhang S, Wang S, Shan X. Effect of sample pretreatment upon the metal speciation in sediments by a sequential extraction procedure. Chem Speciat Bioavailab 2001, 13(3), 69–74.
[66] Wickenheiser EB, Michalke K, Hensel R, et al. Volatile Compounds in Gases Emitted from Wetland Bogs Near Lake Cadagno. 1998.

[67] Koch I, Feldmann J, Lintschinger J, Serves SV, Cullen WR, Reimer KJ. Demethylation of Trimethylantimony Species in Aqueous Solution during Analysis by Hydride Generation/Gas Chromatography with AAS and ICP MS Detection. Appl Organometal Chem 1998, 12, 129–136.
[68] Díez S, Bayona JM. Trace element determination by combining solid-phase microextraction hyphenated to elemental and molecular detection techniques. J Chromatogr Sci 2006, 44(7), 458–471.
[69] Jenkins R, Irgolic K. Biomethylation of inorganic antimony compounds by an aerobic fungus: Scopulariopsis brevicaulis. Environ Sci Technol 1998, 32(7), 882–885.
[70] Mestrot A, Uroic MK. Quantitative and qualitative trapping of arsines deployed to assess loss of volatile arsenic from paddy soil. Environ Sci Technol 2009, 43(21), 8270–8275.
[71] Yan H, Wang X, Yang Y, Duan G, Zhang H, Cheng W. The effect of straw-returning on antimony and arsenic volatilization from paddy soil and accumulation in rice grains. Environ Pollut 2020, 263(PartA), 1–8.
[72] Jenkins R, Craig P, Miller D, Stoop LCAM, Ostah N, Morris T. Antimony biomethylation by mixed cultures of micro-organisms under anaerobic conditions. Appl Organomet Chem 1998, 12, 449–455.
[73] Feldmann J, Grümping R, Hirner AV. Determination of volatile metal and metalloid compounds in gases from domestic waste deposits with GC/ICP-MS. Fresenius J Anal Chem 1994, 350, 228–234.
[74] Hirner AV, Feldmann J, Goguel R, Rapsomanikis S, Fischer R, Andreae MO. Volatile metal and metalloid species in gases from municipal waste deposits. Appl Organomet Chem 1994, 8(1), 65–69.
[75] De Gregori I, Quiroz W, Pinochet H, Pannier F, Potin-Gautier M. Speciation analysis of antimony in marine biota by HPLC-(UV)-HG-AFS: Extraction procedures and stability of antimony species. Talanta 2007, 30(3), 458–465.
[76] Ge Z, Wei C. Simultaneous analysis of SbIII, SbV and TMSb by high performance liquid chromatography-inductively coupled plasma-mass spectrometry detection: Application to antimony speciation in soil samples. J Chromatogr Sci 2013, 51(5), 391–399.
[77] Lintschinger J, Koch I, Serves S, Feldmann J, Cullen WR. Determination of antimony species with high-performance liquid chromatography using element specific detection. Fresenius J Anal Chem 1997, 359(6), 484–491.
[78] Ulrich N. Speciation of antimony(III), antimony(V) and trimethylstiboxide by ion chromatography with inductively coupled plasma atomic emission spectrometric and mass spectrometric detection. Anal Chim Acta 1998, 359, 245–253.
[79] Müller K, Daus B, Mattusch J, Stärk H, Wennrich R. Simultaneous determination of inorganic and organic antimony species by using anion exchange phases for HPLC-ICP-MS and their application to plant extracts of Pteris vittata. Talanta 2009, 78(3), 820–826.
[80] Craig PJ, Forster SN, Miller D, Jenkins RO. An analytical method for the detection of methylantimony species in environmental matrices: Methylantimony levels in some UK plant material. Analyst 1999, 124(8), 1243–1248.
[81] Dodd M, Grundy SL, Reimer KJ, Cullen WR. Methylated antimony(V) compounds: Synthesis, hydride generation properties and implications for aquatic speciation. Appl Organomet Chem 1992, 6(2), 207–211.
[82] Meyer J, Schmidt A, Michalke K, Hensel R. Volatilisation of metals and metalloids by the microbial population of an alluvial soil. Syst Appl Microbiol 2007, 30(3), 229–238.
[83] Yang H, He M. Speciation of Antimony in Soils and Sediments by Liquid Chromatography–Hydride Generation–Atomic Fluorescence Spectrometry. Anal Lett 2015, 48(12), 1941–1953.

[84] Bellido-Martín A, Gómez-Ariza JL, Smichowsky P, Sánchez-Rodas D. Speciation of antimony in airborne particulate matter using ultrasound probe fast extraction and analysis by HPLC-HG-AFS. Anal Chim Acta [Internet] 2009, 649, 191–195.
[85] Sayago A, Beltrán R, Recamales MAF, Gómez-Ariza JL. Optimization of an HPLC-HG-AFS method for screening Sb(V), Sb(III), and Me3SbBr2 in water samples. J Anal At Spectrom 2002, 17(10), 1400–1404.
[86] Yang H, He M. Adsorption of methylantimony and methylarsenic on soils, sediments, and mine tailings from antimony mine area. Microchem J 2015, 123, 158–163.
[87] Feldmann J. Summary of a calibration method for the determination of volatile metal(loid) compounds in environmental gas samples by using gas chromatography-inductively coupled plasma mass spectrometry. J Anal At Spectrom 1997, 12(9), 1069–1076.
[88] Raja PMV, Barron AR Physical Methods in Chemistry and Nano Science. LibreTexts; 2020.
[89] Ji Y, Mestrot A, Schulin R, Tandy S. Uptake and Transformation of Methylated and Inorganic Antimony in Plants. Front Plant Sci 2018, 9, 1–10.
[90] Chang T, Qian Q, Zhao M, Wang J. The isotopic abundance of antimony. Int J Mass Spectrom Ion Process 1993, 123(1), 77–82.
[91] Gürleyük H, Van Fleet-Stalder V, Chasteen TG. Confirmation of the Biomethylation of Antimony Compounds. Appl Organomet Chem 1997, 11(6), 471–483.
[92] Andreae MO, Asmodé J, Foster P, Van't Deck L. Determination of antimony(III), antimony(V), and methylantimony species in natural waters by atomic absorption spectrometry with hydride generation. Anal Chem 1981, 53(12), 1766–1771.
[93] Ellwood M, Maher W. Arsenic and antimony species in surface transects and depth profiles across a frontal zone: The Chatham Rise, New Zealand. Deep Res Part I Oceanogr Res Pap 2002, 49(11), 1971–1981.
[94] Bertine K, Lee D. Antimony Content and Speciation in the Water Column and Interstitial Waters of Saanich Inlet. NATO Conf Ser 4 Mar Sci 1983, 9, 21–38.
[95] Cutter G, Cutter L, Featherstone A, Lohrenz SE. Antimony and arsenic biogeochemistry in the western Atlantic Ocean. Deep Res Part II Top Stud Oceanogr 2001, 48(13), 2895–2915.
[96] Andreae MO, Froelich PN. Arsenic, antimony, and germanium biogeochemistry in the Baltic Sea. Chem Phys Meteorol 1984, 36(2), 101–117.
[97] Duester L, Hartmann L, Hirner A. Methylated arsenic and antimony species in suspended matter of the river Ruhr, Germany. Biogeosciences Discuss 2008, 5(2), 1361–1378.
[98] Frohne T, Rinklebe J, Diaz-Bone RA, Du Laing G. Controlled variation of redox conditions in a floodplain soil: Impact on metal mobilization and biomethylation of arsenic and antimony. Geoderma 2011, 160(3–4), 414–424.
[99] Yang H, He M. Distribution and Speciation of Selenium, Antimony, and Arsenic in Soils and Sediments Around the Area of Xikuangshan (China). Clean – Soil, Air, Water 2016, 44(11), 1429–1598.
[100] Ulrich N. Determination of antimony species with fluoride as modifier and flow injection hydride generation inductively-coupled plasma emission spectrometry. Anal Chim Acta 2000, 417(2), 201–209.
[101] Miravet R, Bonilla E, López-Sánchez J, Rubio R. Antimony speciation in terrestrial plants. Comparative studies on extraction methods. J Environ Monit 2005, 7(12), 1207–1213.
[102] Kresimon J, Grüter UM, Hirner AV. HG/LT-GC/ICP-MS coupling for identification of metal(loid) species in human urine after fish consumption. Fresenius J Anal Chem 2001, 371(5), 586–590.
[103] Krachler M, Emons H. Urinary antimony speciation by HPLC-ICP-MS. J Anal At Spectrom 2001, 16(1), 20–25.

[104] Feldmann J, Koch I, Cullen WR. Complementary use of capillary gas chromatography-mass spectrometry (ion trap) and gas chromatography-inductively coupled plasma mass spectrometry for the speciation of volatile antimony, tin and bismuth compounds in landfill and fermentation gases. Analyst 1998, 123(5), 815–820.
[105] Planer-friedrich B, Merkel B. Volatile Metals and Metalloids in Hydrothermal Gases. Environ Sci Technol 2006, 40(10), 3181–3187.
[106] Dopp E, Hartmann LM, Florea AM, et al. Trimethylantimony dichloride causes genotoxic effects in Chinese hamster ovary cells after forced uptake. Toxicol Vitr 2006, 20(6), 1060–1065.

Robert Wysocki

Chapter 12 Molecular mechanisms of antimony transport and detoxification

12.1 Introduction

Antimony is a toxic metalloid, and nearly all organisms possess various mechanisms for its detoxification. Biologically relevant forms of antimony, antimonite [Sb(III)] and antimonate [Sb(V)], are widespread in nature, but usually in very low concentrations [1, 2]. Thus, organisms are not likely to have developed specific resistance mechanisms of antimony. However, antimony often occurs together with arsenic in the earth's crust [3, 4] and has several chemical and biological properties similar to arsenic [3, 5, 6]. Under reducing conditions and at biologically relevant pH values, both metalloids exist in the form of hydroxide molecules, $Sb(OH)_3$ and $As(OH)_3$, and show high affinity for the thiol groups of peptides and proteins [3, 5–7]. Hence, Sb(III) resistance is conferred by proteins that have evolved to allow the growth of early organisms in the presence of arsenite [As(III)], which has always been present in much higher concentrations in the environment compared to Sb(III) and was a dominant form in primordial oxygen-free oceans [8–10]. Consequently, detoxification transporters specific for both As(III) and Sb(III) or only for As(III) have been identified to date [11–13]. As(III) and Sb(III) also share common entry pathways into cells via the channel proteins called aquaglyceroporins [11–14]. Importantly, downregulation or inactivation of As(III)/Sb(III) uptake proteins is often observed in organisms with acquired metalloid resistance [15–18]. Arsenate [As(V)] structurally mimics phosphate and uses the plasma membrane transporters involved in phosphate uptake to enter the cells [11]. In contrast, Sb(V) exists as a monovalent anion $Sb(OH)_6^-$ in solutions, and it remains to be established how pentavalent antimony is taken up by the cells [4, 5]. Once in the cytoplasm, Sb(V) has to be reduced to the more toxic trivalent state to allow its export out of the cell via the plasma membrane transporters specific for As(III)/Sb(III) or the aquaglyceroporins [11–14, 17]. Alternatively, Sb(III) can be complexed with thiol-containing peptides, such as glutathione (GSH), trypanothione ($T(SH)_2$), and phytochelatins (PCs), followed by transport of resulting Sb(III)-thiol

Acknowledgments: The author's work on mechanisms of antimony toxicity and transport is supported by grants from the National Science Centre, Poland (2015/19/B/NZ1/00327; 2016/21/B/NZ2/01751; 2019/35/B/NZ3/00379).

Robert Wysocki, Institute of Experimental Biology, University of Wroclaw, Kanonia 6/8, 50-328, Wroclaw, Poland. robert.wysocki@uwr.edu.pl

https://doi.org/10.1515/9783110668711-012

conjugates into the intracellular compartments, such as the vacuole, or out of the cell [11–13, 17–19]. Oxidation of Sb(III) to less toxic Sb(V) is another example of antimony detoxification utilized by bacteria isolated from antimony-contaminated sites [20]. The presence of methylated antimony species in environmental samples has also been reported [20–22]. However, the significance of antimony biomethylation for detoxification processes in bacteria, archaea, or fungi has been questioned due to the low rate of antimony biomethylation by unspecified enzymes, trace quantities of methylated antimony compounds generated by organisms, and possible inaccurate measurements in published studies [22, 23].

12.2 Uptake and release of antimony via the aquaglyceroporins

Up to the mid-1990s, the knowledge of how antimony is accumulated by organisms was scarce. No transporters responsible for antimony uptake were known, and at that time it was suggested that antimony enters the cells by diffusion across the plasma membrane. In the breakthrough work from Barry P. Rosen's group, a classical genetic approach using transposon-mediated random mutagenesis made it possible to determine that inactivation of the *Escherichia coli glpF* gene encoding the glycerol channel results in increased resistance to Sb(III) [24]. Solely based on this phenotype, it was concluded that GlpF mediates uptake of Sb(III) in the form of $Sb(OH)_3$ due to molecular resemblance to glycerol (Fig. 12.1). Surprisingly, the *glpF* mutant was not more resistant to As(III) and an alternative route for As(III) accumulation in bacteria was suggested [24]. Inspired by this work, Wysocki et al. [25] showed for the first time in eukaryotes that Fps1, the closest orthologue of GlpF in budding yeast, is involved in accumulation of both As(III) and Sb(III) (Fig. 12.2). Then, it was confirmed by direct transport measurements in *E. coli* that GlpF is permeable to both As(III) and Sb(III) [26]. GlpF and Fps1 belong to the aquaglyceroporin subfamily of the major intrinsic protein (MIP) family of channel proteins that are ubiquitously present in all kingdoms of life [27]. Aquaglyceroporins mediate transport of small uncharged molecules such as glycerol or urea in an energy-independent manner [28]. Discovery of Fps1 as the uptake route for metalloids in eukaryotes and the usage of *fps1* deletion mutant in heterologous complementation assays facilitated identification of accumulation pathways for As(III) and Sb(III) in other model organisms [29–32]. It was first reported for AqpS from the legume symbiont bacterium *Sinorhizobium meliloti* that aquaglyceroporins may also play a physiological function in metalloid efflux and resistance [33]. Functional characterization of plant aquaglyceroporins [30] and Fps1 [34] revealed that these channels indeed contribute to metalloid tolerance by transport of As(III) and Sb(III) out of the cell down the concentration gradient.

The MIPs are characterized by the common topology of two duplicated segments, each consisting of three transmembrane α-helical regions and one half-helix containing

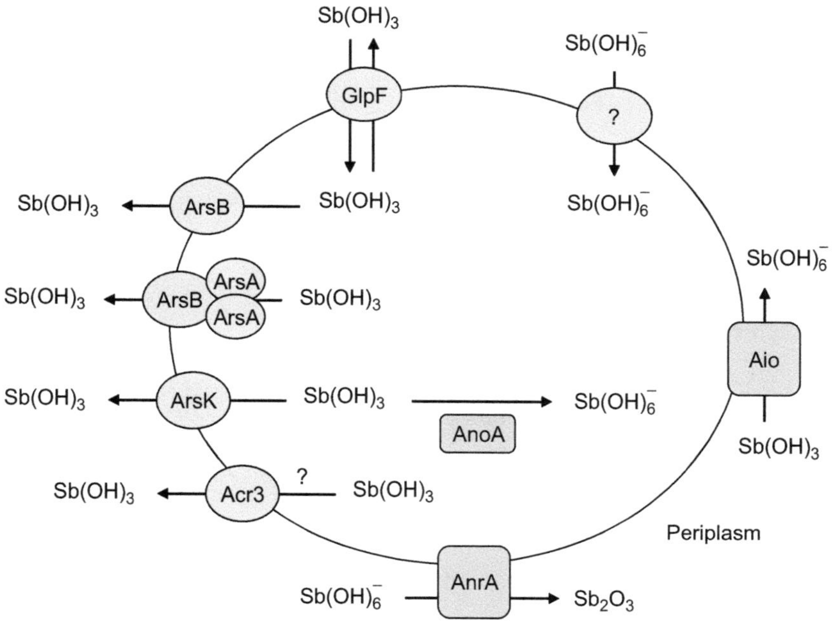

Fig. 12.1: Antimony tolerance pathways in bacteria.
Sb(III) ($Sb(OH)_3$) is accumulated by the bidirectional aquaglyceroporin GlpF. Sb(V) ($Sb(OH)_6^-$) uptake pathway is not known. Sb(III) can be detoxified by three types of As(III)/Sb(III)-specific transporters, the H^+/metalloid antiporter ArsB, the ATP-dependent ArsA/ArsB pump, or the ArsK permease. In some bacteria, such as the cyanobacterium *Synechocystis* sp. and the soil bacterium *Agrobacterium tumefaciens*, Sb(III) might also be extruded by the H^+/metalloid antiporter Acr3 but this requires direct transport evidence. Alternatively, Sb(III) is oxidized to less-toxic Sb(V) by antimonite oxidase AnoA or the respiratory antimonite oxidase complex Aio (AioA/AioB) in *A. tumefaciens*. In the periplasm of *Desulfuribacillus stibiiarsenatis*, the respiratory antimonate reductase AnrA converts Sb(V) to Sb_2O_3 followed by its precipitation.

the conserved Asn-Pro-Ala (NPA) motif [28]. Structural studies of the animal aquaporins [35–37] and *E. coli* GlpF [38] revealed that they exist as homotetramers in the plasma membrane and six transmembrane α-helical regions of each monomer fold into the hourglass structure. In the center of each monomer, two-half helices align by interacting with each other via the NPA motifs to form a selective pore, which is ~1 Å wider in GlpF than in the water-permeable aquaporins. This allows glycerol and other larger molecules to pass through the GlpF channel, while the passage of protons and other charged molecules is still restricted due to the presence of NPA motifs and the aromatic/arginine constriction region at the outer site of the membrane, respectively. $As(OH)_3$ and $Sb(OH)_3$ readily pass through aquaglyceroporins as they exhibit a pyramidal structure with charge distribution very similar to glycerol conformation [39]. Interestingly, aquaglyceroporins are also permeable to other metalloids, such as boron and silicon [40]. However, since borate [$B(OH)_3$] and silicate [$Si(OH)_4$] have distinct structures and more uniform charge distribution, it was suggested that the volume of

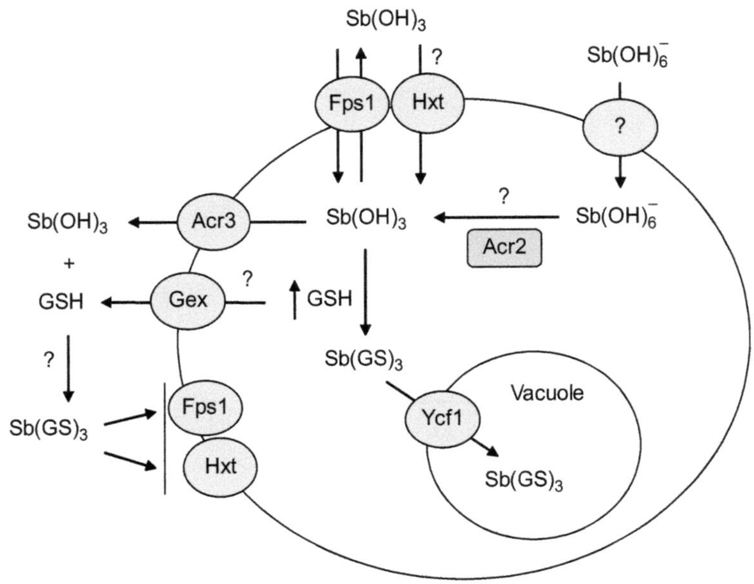

Fig. 12.2: Routes of antimony uptake and detoxification in the yeast *Saccharomyces cerevisiae*. Sb(III) ($Sb(OH)_3$) uptake is mediated by the aquaglyceroporin Fps1. Hexose transporters (Hxts) have been shown to contribute to As(III) accumulation but their role in Sb(III) uptake has not been tested yet. Sb(III) is exported by the H^+/metalloid antiporter Acr3 or diffuses via the aquaglyceroporin Fps1 down the concentration gradient. In response to Sb(III) stress, yeast cells upregulate synthesis of glutathione (GSH), which is used for Sb(III) complexation in the cytosol. The resulting Sb(III)-glutathione complexes ($Sb(GS)_3$) are then sequestrated in the vacuole by the ATP-binding cassette (ABC) transporter Ycf1. Similarly to As(III) stress, free glutathione might be extruded by Gex1 and Gex2 transporters to form Sb(III)-glutathione complexes in the external medium and thereby limit Sb(III) uptake through Fps1 and hexose transporters. Sb(V) ($Sb(OH)_6^-$) uptake pathway is not known. The hypothetical role of arsenate reductase Acr2 in Sb(V) reduction has not been established.

metalloid compounds might play a major role in their ability to enter cells via aquaglyceroporins [41].

12.2.1 The yeast aquaglyceroporin Fps1

The budding yeast *Saccharomyces cerevisiae* Fps1 is an unusual member of the MIP family. Fps1 plays a physiological role in osmoregulation by regulating glycerol efflux rather than uptake [42, 43]. In response to hyperosmotic stress to compensate water loss, yeast cells increase the intracellular concentration of glycerol by upregulating its synthesis and closing the Fps1 channel to prevent glycerol efflux [43, 44]. In contrast, when yeast cells are moved from high to low osmolarity conditions, Fps1 is rapidly opened to release glycerol [43]. Such tight regulation of Fps1 is possible due to the presence of unusual long extensions of N- and C-terminal cytoplasmic regions that are subjected to post-translational modifications by mitogen-activated protein kinases (MAPKs) [43–47]. Phosphorylation of the N-terminus at Thr231 by MAPK Hog1 and possibly other MAPKs triggers closure of the Fps1 channel under hyperosmotic [43, 48] and metalloid stress [45] or its removal from the plasma membrane in response to acetic acid stress [46]. In addition, Fps1 activity is positively regulated by redundantly acting cytosolic paralog proteins, Rgc1 and Rgc2/Ask10, which belong to a family of pleckstrin homology domain proteins [49, 50]. It was shown that Rgc2 keeps Fps1 in the opened state by interacting with the C-terminus of Fps1, while the stress-activated Hog1 phosphorylates Rgc2 at multiple sites, which triggers eviction of Rgc1/2 from Fps1 and contributes to closing of the channel [50]. Finally, it was demonstrated that phosphorylation of Ser537 in the C-terminus of Fps1 by MAPK Slt2 is also required for Fps1 opening [47]. However, it is not known how mechanistically such post-translational modifications or interactions with Rgc1/2 contribute to the functional state of Fps1.

Such unique properties of Fps1 facilitated not only identification of Fps1 as a As(III)/Sb(III) uptake route but also understanding of its regulation under metalloid stress (Fig. 12.2). Cells lacking the *FPS1* gene or a pair of *RGC1* and *RGC2* genes become highly resistant to As(III) and Sb(III) [25, 49]. Similarly, exposing cells to a high concentration of NaCl, which causes closure of Fps1, renders cells more resistant to both metalloids [25]. On the other hand, expression of the constitutively open Fps1 variants lacking either the N-terminal extension (*fps1-Δ1*) or the MAPK phosphorylation site (*fps1-T231A*) greatly increases sensitivity of yeast cells to As(III) and Sb(III) [25]. Moreover, cells lacking Hog1 are highly sensitive to both metalloids and accumulate more As(III) compared to wild-type cells [45]. Fps1 is also regulated at the expression level, as we observed a decrease of Fps1 mRNA levels upon addition of As(III) [25].

Interestingly, the *fps1Δ* mutant is sensitive to As(V) [25], suggesting that the physiological role of Fps1 may be the efflux of As(III) formed by reduction of As(V) in the cytoplasm (Fig. 12.2). To our surprise, we found that overexpression of *FPS1* in *hog1Δ* or wild-type cells confers increased resistance to As(III) and Sb(III) instead of elevated sensitivity as one might have expected [34]. This suggested that Fps1 contributes to both As(V) and As(III)/Sb(III) resistance. Consequently, after the initial drop of *FPS1* mRNA synthesis, transcription of the *FPS1* gene was increased during long-term exposure to As(III), while overexpression of *FPS1* reduced As(III) accumulation in yeast cells [34]. Moreover, overexpression of the *fps1-S537A* variant lacking the Slt2-dependent phosphorylation site or overexpression of wild-type *FPS1* in cells lacking Slt2 did not result in increased resistance to As(III) or elevated As(III) efflux [47]. Thus, the role of Fps1 in As(III), and most likely Sb(III) export, requires Slt2-dependent opening of the channel. The question remains how Fps1 as a bidirectional channel is able to mediate As(III)/Sb(III) export against the concentration gradient of metalloids (Fig. 12.2). We postulated that after initial closing of Fps1 to prevent As(III)Sb(III) uptake, Fps1 is reopened to allow As(III)/Sb(III) export after reversing the concentration gradient due to complexation of metalloids by GSH outside the cells (see 12.3.2.1) [47, 51].

12.2.2 Plant aquaglyceroporins

Plant genomes encode a high number of MIPs (e.g., 35 in the model plant *Arabidopsis thaliana*), which are localized not only in the plasma membrane but also in intracellular membranes [52, 53]. Based on the phylogenetic analysis, plant members of the MIP family are currently classified into five subfamilies: the plasma membrane intrinsic proteins (PIPs), tonoplast intrinsic proteins (TIPs), Nodulin 26-like intrinsic proteins (NIPs), small and basic intrinsic proteins (SIPs), and X intrinsic proteins (XIPs) [52–54]. As NIPs are believed to be the functional equivalents of aquaglyceroporins, they were chosen by Bienert et al. [30] for heterologous expression in yeast cells to identify plant proteins capable of As(III)/Sb(III) transport. Expression of six NIPs, AtNIP5;1 and AtNIP6;1 from *A. thaliana*; OsNIP2;1 (Lsi1) and OsNIP3;2 from the rice *Oryza sativa*; LjNIP5;1 and LjNIP6;1 from *Lotus japonicus*, clearly sensitized the *fps1*Δ mutant to As(III) and increased both tolerance to As(V) and As(III) uptake in yeast, indicating that they are involved in bidirectional transport of As(III) (Fig. 12.3) [30]. Complementation tests of Sb(III) tolerance suggested

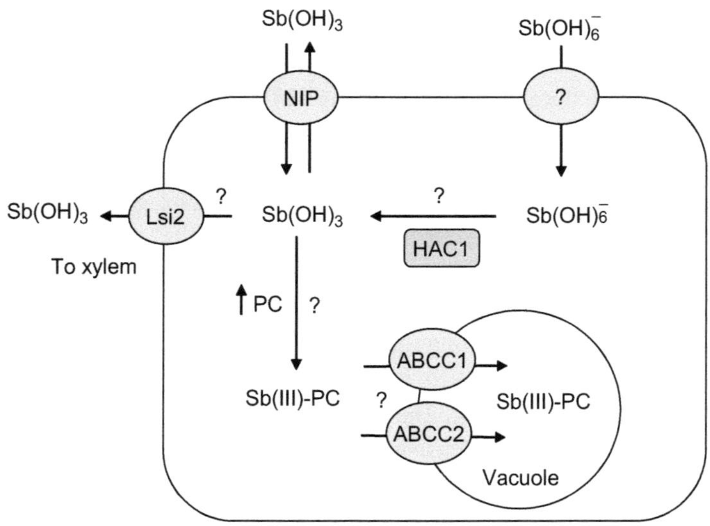

Fig. 12.3: Pathways of antimony uptake and detoxification in the flowering plant *Arabidopsis thaliana*. In roots, Sb(III) ($Sb(OH)_3$) is accumulated by aquaglyceroporins of the Nodulin 26-like intrinsic protein (NIP) subfamily, such as AtNIP1;1 and AtNIP7;1. Sb(V) ($Sb(OH)_6^-$) uptake pathway is not known. It has not been tested whether arsenate reductase HAC1 also recognizes Sb(V) as a substrate. As NIP proteins act as a bidirectional channel, they can also contribute to Sb(III) translocation from roots to shoots or Sb(III) export to the environment down the concentration gradient. Upregulation of phytochelatin (PC) synthesis by Sb(III) and transport of Sb(III)-PC complexes by ATP-binding cassette (ABC) transporters of the ABCC subfamily (ABCC1 and ABCC2) from the cytosol to the vacuole require experimental evidence. Similarly, it remains to be established whether the silicon and arsenite transporter Lsi2 contributes to Sb(III) translocation from root cells to xylem.

that AtNIP7;1, OsNIP2;1, and LjNIP6;1 are efficient transporters of Sb(III), while AtNIP5;1, AtNIP6;1, OsNIP3;2, and LjNIP5;1 are moderately permeable to Sb(III) [30].

In separate studies, involvement in As(III) transport was also shown for OsNIP1;1, OsNIP2;2 (Lsi6), OsNIP3;1 [32], OsNIP3;3 [55], AtNIP1;1, AtNIP1;2, AtNIP5;1 [56], AtNIP7;1 [57], VvNIP1;1, and VvNIP6;1 from grapevine (*Vitis vinifera*) [58], and HvNIP1;2 from barley (*Hordeum vulgare*) [55]. Functional analysis of NIPs in plants revealed that only OsNIP2;1 (Lsi1) [32] and AtNIP1;1 [56] exhibit significant physiological importance for As(III) bidirectional transport and tolerance in whole plants [56]. Both OsNIP2;1 (Lsi1) and AtNIP1;1 are highly expressed and localized in the plasma membrane of root cells, contributing to As(III) uptake from the environment [32, 56]. However, other NIP proteins may contribute to As(III) translocation from roots to shoots, as it was demonstrated for AtNIP3;1 [59] and AtNIP7;1 [60] in *A. thaliana*. A few recent reports implicated that members of other subfamilies of plant MIP proteins are also permeable to As(III), including PIP proteins from the rice *O. sativa* (OsPIP2;4, OsPIP2;6, and OsPIP2;7) [61], and one member of the TIP subfamily (PvTIP4;1) in the fern *Pteris vittata* [62]. Unfortunately, in none of the studies mentioned earlier was the role of plant MIPs in Sb(III) transport analyzed. However, Kamiya and Fuiiwara [63] clearly showed that AtNIP1;1 is a major uptake route of Sb(III). Inactivation of AtNIP1;1 resulted in 50% reduction of Sb(III) uptake and increased Sb(III) tolerance in *A. thaliana*. Importantly, disruption mutants of AtNIP1;2 and AtNIP5;1, which were shown to accumulate As(III), showed no effect on Sb(III) accumulation and tolerance [63]. Thus, further functional characterization of various plant MIPs is needed to establish their role in Sb(III) transport.

12.2.3 Protozoan aquaglyceroporins

Pathways of antimony transport are intensively studied in protozoan *Leishmania* species due to the usage of Sb(V)-based drugs in the treatment of human diseases caused by these intracellular parasites and emerging Sb(V) resistance in field isolates of *Leishmania* (see also Chapter 13) [17]. Although Sb(V) uptake transporters are not known, it is generally accepted that Sb(III) generated by biological reduction of Sb(V) is the active form of the drug. It was first demonstrated by Gourbal et al. [64] that the aquaglyceroporin AQP1 constitutes a major As(III) and Sb(III) uptake pathway route in *Leishmania* (Fig. 12.4). AQP1 is also permeable to water, glycerol, and several other compounds and is localized to the flagellum of *Leishmania* promastigotes (an extracellular form of parasite) and in the flagellar pocket membrane and contractile vacuole of amastigotes (an intracellular form of parasite) [65]. Overexpression of *Leishmania major* AQP1 (LmAQP1) highly sensitized wild-type and Sb(V)-resistant *Leishmania* strains to As(III) and Sb(III) [65], while disruption of the *LmAQP1* gene conferred increased As(III) and Sb(III) resistance [66]. Other studies revealed that

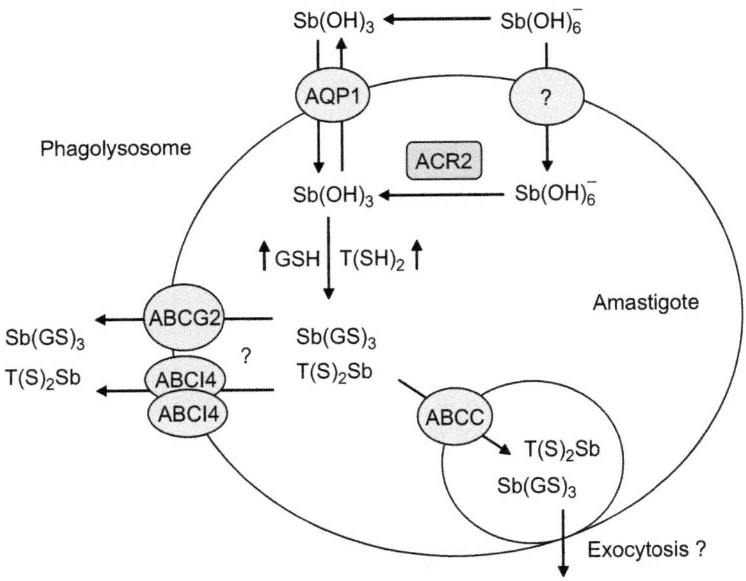

Fig. 12.4: Antimony resistance pathways in the *Leishmania* amastigote.
Sb(III) ($Sb(OH)_3$) uptake is mediated by the aquaglyceroporin AQP1. Sb(V) uptake pathway is not known. Sb(V) is enzymatically or non-enzymatically reduced to Sb(III) in host cells or in the cytosol of *Leishmania* amastigotes by antimonate reductase ACR2. Sb(III) formed by Sb(V) reduction can be released by AQP1 down the concentration gradient. In response to antimony treatment, *Leishmania* cells increase synthesis of thiol-rich peptides, such as glutathione (GSH) and trypanothione ($T(SH)_2$), in order to chelate Sb(III) in the cytosol. The resulting Sb(III)-thiol complexes ($Sb(GS)_3$ and $T(S)_2Sb$) are sequestrated in the intracellular vesicles by ATP-binding cassette (ABC) transporters of the ABCC subfamily, such as MRPA (ABCC3), or probably exported by ABC transporters, ABCG2 and ABCI4, into the phagolysosome lumen of host cells. It has also been suggested that Sb(III)-thiol complexes accumulated in the intracellular vesicles might be released into the phagolysosome by means of exocytosis.

downregulation of LmAQP1 at the mRNA level and point mutations in the *AQP1* genes are associated with acquired resistance to antimonials [67–70].

Remarkably, similar to the yeast Fps1, LmAQP1 is post-translationally regulated by the Hog1 functional orthologue LmMPK2 [71]. However, in contrast to the negative regulation of Fps1 by Hog1 in yeast, LmMPK2 activity does not alter permeability of the LmAQP1 channel but instead increases its stability and triggers uniform localization of AQP1 in the plasma membrane, as it was shown in promastigotes of *L. donovani*. Moreover, mutation of the MPK2-dependent phosphorylation site in LmAQP1 or inactivation of LmMPK2 kinase activity resulted in decreased Sb(III) uptake in *Leishmania* cells. Consequently, deletion of *MPK2* in *L. mexicana* conferred Sb(III) resistance. Recently, it has been shown that AQP1 is subjected to proteasomal degradation, probably depending on ubiquitination of the N-terminal Lys12 residue [72]. Interestingly, expression of the kinase-dead MPK2 variant destabilized the wild-type AQP1 protein but not AQP1-K12A or AQP1-K12R variants. Mutation of Lys12 not only improved stability of the AQP1 protein but also restricted its localization to the flagellum. It has been suggested that Mpk2-dependent phosphorylation of AQP1 at Thr197 in the cytoplasmic loop might interfere with ubiquitination of Lys12 in the N-terminus and subsequent protein degradation [72].

12.2.4 Vertebrate aquaglyceroporins

The human genome encodes 11 members of the MIP family [28]. The involvement of human AQP7 and AQP9 aquaglyceroporins in As(III) and Sb(III) uptake was demonstrated by heterologous expression in yeast cells and *X. laevis* oocytes (Fig. 12.5) [29]. AQP3 and AQP10 showed little or no permeability to As(III) when expressed in oocytes [73]. Among mammalian aquaglyceroporins, AQP9 proved to be the most efficient transporter of As(III) [73]. Although human AQP3 showed little permeability to As(III) in a heterologous system, overexpression of AQP3 in human cell lines led to an increase in As(III) uptake and sensitivity, whereas decreased expression of AQP3 correlated with As(III) resistance [14]. Thus, in addition to AQP7 and AQP9, AQP3 may contribute to As(III) accumulation in certain cell types. However, it is not known whether AQP3 has any role in Sb(III) uptake. The role of five zebrafish orthologues of

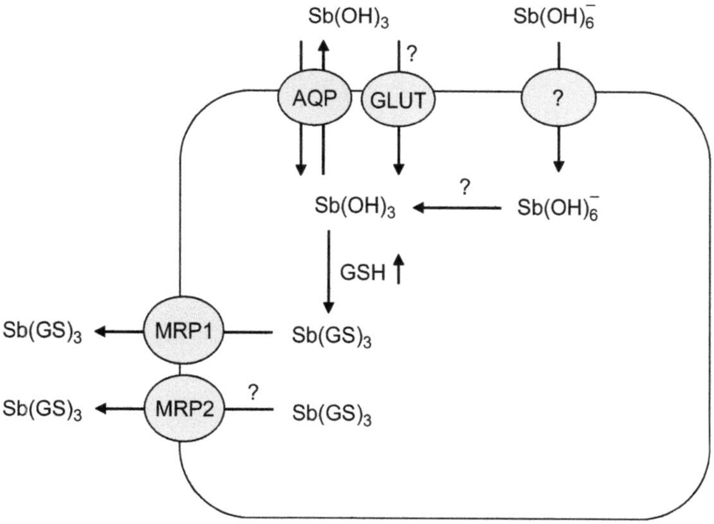

Fig. 12.5: Antimony uptake and efflux pathways in mammalian cells.
Sb(III) ($Sb(OH)_3$) is accumulated by aquaglyceroporins, such as AQP7 and AQP9. Sb(III) uptake via hexose transporters (GLUTs) has not been analyzed yet. Sb(V) ($Sb(OH)_6^-$) uptake and reduction pathways are not known. In the cytosol, Sb(III) is complexed by glutathione (GSH) followed by extrusion of Sb(III)-glutathione conjugates ($Sb(GS)_3$) by the ATP-binding cassette (ABC) transporter MRP1. The role of MRP2 in $Sb(GS)_3$ efflux requires direct transport evidence.

human aquaglyceroporins (Aqp3, Aqp3l, Aqp9a, Aqp9b, and Aqp10) in metalloid transport was also investigated [74]. All five were equally permeable to As(III) when expressed in *X. laevis* oocytes. Aqp9b and Aqp10 showed a high rate of Sb(III) transport, Aqp3l and Aqp9a showed some permeability to Sb(III), while Aqp3 did not transport this metalloid.

Inorganic arsenicals are subjected to methylation in the liver [75, 76]. AQP9 is highly expressed in hepatocytes, and it was found that AQP9 also facilitates translocation of methylated forms of As(III) and As(V) across the membranes [77, 78]. This led to the suggestion that AQP9 plays an important role in detoxification of arsenic in hepatocytes by extruding methylated forms of arsenic to the blood stream down a concentration gradient. The analysis of AQP9-null mice confirmed this notion as lack of AQP9 expression led to accumulation of arsenic in the liver and other organs, 90% reduction of arsenic present in urine and feces, and increased toxicity in mice [79]. In contrast, enzymatic methylation of antimony has not yet been demonstrated in human cells.

In yeast, plant, and human cells, As(III) is accumulated not only by aquaglyceroporins but also by hexose transporters [80–82]. Importantly, yeast cells respond to As (III) stress by triggering ubiquitin-dependent endocytosis and degradation of hexose transporters to reduce As(III) uptake [83]. It is tempting to speculate that Sb(III) is also recognized by hexose transporters, but it has not been demonstrated yet (Fig. 12.2).

12.3 Pathways of antimony detoxification

12.3.1 Sb(III)/As(III)-specific transporters

A high level of Sb(III) resistance is conferred by As(III)/Sb(III)-specific exporters, which are exclusively expressed in the presence of their substrates. Up to now, three families of As(III)/Sb(III) transporters have been identified: ArsB, Acr3, and ArsK. Members of ArsB and ArsK families are widespread in prokaryotes, whereas Acr3-type transporters are found in both prokaryotes and eukaryotes [11–13]. Interestingly, protozoans, animals, and higher plants lack As(III)/Sb(III)-specific transporters.

12.3.1.1 ArsB

The *E. coli* ArsB is the best characterized member of the ArsB family, which belongs to the ion transporter superfamily [84]. ArsB exhibits a unique dual mode of metalloid transport depending on the presence of its cytosolic subunit ArsA [85]. The integral inner membrane protein ArsB mediates an $H^+/As(OH)_3$ or $H^+/Sb(OH)_3$ antiport coupled to the electrochemical proton gradient (Fig. 12.1) [26, 86]. ArsB contains 12 membrane-spanning α-helices [87] and a single cysteine residue (Cys369) located in the 11th transmembrane region (TM11) [88]. However, mutation of Cys369 had no effect on the transport activity of ArsB, suggesting that thiol chemistry is not involved in metalloid/proton exchange across the membrane [88]. However, when ArsB is associated with the ArsA homodimer exhibiting As(III)/Sb(III)-dependent ATPase activity [89, 90], the ArsA/ArsB complex acts as an obligatory ATP-powered pump that transports metalloid with much higher efficiency compared to ArsB (Fig. 12.1) [91]. Three cysteine residues (Cys113, Cys172, and Cys422) were shown to be essential for ArsA activity [92]. The crystal structure of ArsA revealed that these cysteine residues are involved in Sb(III) binding that results in allosteric modulation of ATPase activity [92, 93]. The third protein, ArsD, acts as a metallochaperone, which physically interacts with ArsA and directly transfers metalloids to this subunit, thereby stimulating its ATPase activity and the rate of transport at a lower concentration of As(III)/Sb(III) [94]. However, in the absence of crystal structure of ArsB, the exact mechanism of As(III) and Sb(III) translocation via ArsA/ArsB or ArsB remains unknown.

Genes encoding ArsA, ArsB, and ArsD are localized to *ars* operons inducible by As(III) and Sb(III). The model *ars* operon of *E. coli* plasmid R773 consists of five genes: *arsR*, *arsD*, *arsA*, *arsB*, and *arsC* [95, 96]. In contrast, the chromosomal copy of the *ars* operon in *E. coli* contains only three genes: *arsR*, *arsB*, and *arsC* [97]. ArsC is an arsenate reductase that converts As(V) to As(III) to allow arsenic extrusion by ArsA/ArsB or ArsB [98, 99]. ArsR is a negative regulator of the *ars* operon expression [100, 101]. In the absence of metalloids, the ArsR homodimer is bound to the promoter of the *ars* operon to inhibit its transcription as a single unit [101, 102]. Binding of As(III) to cysteine residues in ArsR triggers a conformation change in the repressor complex and its release from the promoter [103, 104]. Thus, ArsR plays a key role in sensing the presence of toxic metalloids in bacteria.

Members of the ArsB family were thought to be restricted to prokaryotes. However, it was demonstrated that a distant orthologue of ArsB called Lsi2 exists in rice and mediates transport of silicon as well as As(III) driven by the proton gradient [32]. Due to its localization at the proximal side of both exodermis and endodermis cells of roots, Lsi2 is responsible for the movement of As(III) from roots to shoots. Thus, Lsi2 contributes to As(III) accumulation instead of detoxification. It is not known whether Lsi2 displays activity toward Sb(III) (Fig. 12.3).

12.3.1.2 Acr3

The first *ACR3* gene encoding a novel type of As(III) transporter was cloned and characterized in *S. cerevisiae* [105, 106]. The yeast Acr3 became a founder member of the Acr3 family, which belongs to the bile/arsenite/riboflavin transporter (BART) superfamily [107]. Acr3 proteins are ubiquitously present in bacteria, archaea, fungi, and lower plants [108–110]. Acr3 proteins studied so far confer high resistance to

As(III) [105, 106, 109–112], share a topology of 10 transmembrane spans [109, 113, 114], and catalyze the $H^+/As(OH)_3$ antiport [115, 116]. The common feature of Acr3 proteins is the presence of a cysteine residue in the middle of TM4 [109, 117]. Mutation of this residue in Acr3 from the bacteria *Alkaliphilus metalliredigens* (Cys138) and *Corynebacterium glutamicum* (Cys129) [109] or *S. cerevisiae* (Cys151) [117] resulted in loss of function, strongly suggesting that thiol chemistry is involved in As(III) translocation across the membrane. Mutagenetic analysis of conserved residues located in transmembrane regions of Acr3 proteins led to identification of a glutamic acid residue in TM9 (Glu353 in the *S. cerevisiae* Acr3 and Glu305 in the *C. glutamicum* Acr3) as essential for transport activity [116, 118].

We generated a hypothetical three-dimensional model of the yeast Acr3 protein using the crystal structure of the sodium bile acid symporter from *Yersinia frederiksenii* as a template [114]. The model predicts that four transmembrane helices (TM1, TM2, TM6, and TM7) form a still panel domain, while six transmembrane helices (TM3-TM5, TM8-10) form a motile core domain [114]. TM4 and TM9 are probably discontinuous helices that run in opposite directions and cross each other within unwound regions in the center of the Acr3 structure. These regions are highly conservative in Acr3 proteins and contain Cys151 and Glu353, which were suggested to serve as a binding site for As(III) and H^+, respectively [109, 116–118]. According to the alternating access mechanism proposed for secondary active transporters [119], we suggested that when As(III) binds to the TM4-TM9 crossover on the cytoplasmic side of the membrane, it triggers rotation of the core domain, allowing release of As(III) outside the cell and simultaneously opening of the H^+ binding site to the exterior [114]. Upon H^+ binding, the core domain of Acr3 rotates again to release H^+ in the cytoplasm and make the As(III) binding site accessible again in the interior. However, the actual structure of Acr3 is needed to test this hypothesis.

Although Acr3 transporters are regarded as As(III) specific [105, 106, 109–112, 116], we found that the *S. cerevisiae* Acr3 is also able to transport Sb(III) out of the cell [120] but with three times lower efficiency compared to As(III) (Fig. 12.2) [117]. In support of this notion, the yeast *acr3* mutants exhibited moderate sensitivity to Sb(III), whereas overexpression of the *ACR3* gene fully complemented Sb(III) sensitivity of the *ycf1Δ* mutant lacking a major Sb(III) detoxification pathway in budding yeast [117, 120]. Moreover, transcription of the *ACR3* gene is strongly induced not only by As(III) and As(V) but also Sb(III) [121, 122]. Interestingly, disruption of *ACR3* genes in the cyanobacterium *Synechocystis* sp. [123] and in the soil bacterium *Agrobacterium tumefaciens* [124] conferred increased sensitivity to both As(III) and Sb(III). This suggests that some bacterial Acr3 proteins can mediate translocation of Sb(III), but this requires confirmation by direct in vivo and in vitro transport studies (Fig. 12.1).

Similar to *arsB*, bacterial *ACR3* genes are located in *ars* operons under the control of ArsR-type repressors [108, 125]. Based on the coexistence of *ACR3* and *arsA* genes in many *ars* operons and lack of *arsB*, it was suggested that the ArsA ATPase may also bind to Acr3 to form an ATP-driven pump, analogous to the ArsA/ArsB complex [126, 127]. However, in eukaryotes orthologues of the ArsA ATPase, Get3 in budding yeast and ASNA-1 in mammalian cells, play an unrelated function in post-transcriptional targeting proteins with a single transmembrane region to the endoplasmic reticulum [128, 129].

In *S. cerevisiae* the *ACR3* gene is under the control of the transcriptional activator Yap8, which is constantly bound to the *ACR3* promoter but in an inactive form in the absence of an As(III)/Sb(III) inducer [122]. Kumar et al. [130] showed that Yap8 acts as a direct sensor of metalloids that undergoes conformation changes and activation upon As(III) or Sb(III) binding to three cysteine residues of Yap8. In addition, As(III) treatment leads to stabilization of the Yap8 protein, which is expressed at very low levels [131]. In contrast, the group of Claudina Rodrigues-Pousada postulated that Yap8 is regulated at the level of subcellular localization and accumulates in the nucleus during As(III) stress only [132].

Many transporters localized to the plasma membrane are regulated post-transcriptionally at the level of endocytosis and subsequent vacuolar/lysosomal degradation [133], including hexose transporters during As(III) stress [83]. Although the yeast Acr3 is subjected to ubiquitin-dependent internalization followed by degradation in the vacuole, this transporter is a relatively long-lived plasma membrane protein both in the presence and absence of its substrates [134]. It is important to mention that Acr3 proteins are not always localized to the plasma membrane as the fern Acr3 is targeted to the vacuolar membrane [110].

12.3.1.3 ArsK

The *Agrobacterium tumefaciens* ArsK has been recently identified as a novel type of As(III)/Sb(III) transporter belonging to the major facilitator superfamily (MFS) (Fig. 12.1) [135]. Orthologues of ArsK are widespread in bacterial species resistant to As(III) but absent in archaea and eukaryotes. ArsK possesses a topology of 12 transmembrane regions and is permeable to a wide range of substrates, including inorganic As(III) and Sb(III) as well as organoarsenic compounds, such as methylarsenite and trivalent roxarsone. Expression of the *arsK* gene is under the control of the ArsR2 repressor and is induced by all substrates transported by ArsK [135].

12.3.2 Sb(III) complexation and sequestration via the ABC transporters

The toxicity of As(III) and Sb(III) is primarily due to their high affinity to thiol groups of cysteine residues resulting in inactivation of proteins and/or protein misfolding and aggregation [7, 136, 137]. However, cells took advantage of this feature of trivalent metalloids to sense their presence and activate thiol-based detoxification mechanisms. As described earlier, As(III)/Sb(III) binding to proteins via cysteine residues regulates activity of transcription factors (e.g., ArsR and Yap8) to induce expression of genes involved in metalloid detoxification (e.g., *ars* operon and *ACR3*) [104, 130], stimulates the ATPase activity of the ArsA/ArsB pump in bacteria [93], and may be the basis for As(III)/Sb(III) translocation via the Acr3 transporter [109, 117]. Another defense mechanism used by most eukaryotes involves upregulation of thiol-containing ligand synthesis, such as the tripeptide glutathione (γ-Glu-Cys-Gly, GSH), a conjugate of spermidine and GSH (trypanothione, $T(SH)_2$), and oligomers of GSH called phytochelatins (PCs), to chelate free molecules of As(III)/Sb(III) in the cytoplasm and prevent their binding to proteins [11, 12, 17–19]. The resulting As(III)/Sb (III)-thiol conjugates are not stable at neutral pH [138, 139] and thus are sequestrated into the intracellular compartments of lower pH, such as the vacuole, or exported out of the cell [11–13, 18, 19]. The sequestration of As(III)/Sb(III)-thiol conjugates is mediated by members of the ABCC subfamily of the ATP-binding cassette (ABC) transporter superfamily, including Ycf1 in *S. cerevisiae*, MRP1 in mammals, MRPA in *Leishmania*, and ABCC1 and ABCC2 in *A. thaliana* [11–13, 18, 19].

ABC transporters are ubiquitously present in all domains of life and use ATP hydrolysis to mediate unidirectional transport of a broad range of structurally diverse substrates [140]. The canonical ABC transporters are composed of four subunits, two nucleotide-binding domains (NBD) and two transmembrane domains (TMDs), each consisting of 6 to 10 α-helical regions. Structural data of exporter-type ABC transporters suggest that the ATP-driven dimerization of NBDs results in switching of the substrate-bound TMDs from the inward-facing to the outward-facing conformation followed by ATP hydrolysis to release the substrate and reset the transporter conformation to the inward position [140].

12.3.2.1 Ycf1 and glutathione in *S. cerevisiae*

In budding yeast, the ABC transporter Ycf1 confers resistance to several heavy metals and metalloids by catalyzing their transport as glutathione *S*-conjugates from the cytosol to the vacuole [111, 141–144]. Interestingly, the *ycf1Δ* mutant exhibits much higher sensitivity to Sb(III) compared to the *acr3Δ* mutant, while the double *acr3Δ ycf1Δ* mutant is slightly more sensitive than the single *ycf1Δ* mutant [111, 122]. This strongly suggests that Ycf1 is the main route of Sb(III) detoxification in budding yeast (Fig. 12.2). Ycf1 is constitutively expressed and the activity of *YCF1* promoter is not increased by addition of its substrates [122]. In contrast, expression of enzymes involved in sulfur assimilation and GSH biosynthesis are considerably elevated by cadmium and As(III) [145], suggesting that GSH synthesis is a rate-limiting factor for Ycf1-mediated accumulation of glutathione *S*-conjugates in the vacuole (Fig. 12.2). During Sb(III) treatment, we observed a two-fold increase of promoter activity of the *GSH1* gene encoding γ-glutamylcysteine synthetase, which catalyzes the first step of GSH synthesis [122]. Moreover, deletion of *YAP1* encoding a transcriptional activator of many genes involved in the stress response, including *GSH1*, conferred Sb(III) sensitivity, while accumulation of Yap1 in the nucleus was induced by Sb(III) [122].

Interestingly, Thorsen et al. [51] demonstrated that in response to As(III), yeast cells export GSH via the MFS transporters Gex1 and Gex2 as well as ABC transporters to chelate As(III) in the external medium (Fig. 12.2). The double *gex1Δ gex2Δ* mutant was more sensitive to As(III) compared to the wild type. Moreover, transport studies revealed that yeast cells accumulated little $As(GS)_3$, leading to the conclusion that GSH efflux and formation of As(III)-glutathione conjugates outside the cell decrease As(III) uptake and contribute to resistance. However, no such response was observed in the presence of cadmium, and the role of GSH export in Sb(III) tolerance has not been investigated (Fig. 12.2).

12.3.2.2 MRP1 and glutathione in mammals

Formation of As(III)-glutathione conjugates followed by ATP-dependent transport out of the cell constitutes a major As(III) detoxification pathway in mammalian cells [11–13, 146]. Two members of the C subfamily of ABC transporters, MRP1 (ABCC1) and MRP2 (ABCC2), are the best characterized efflux pathways of inorganic and organic As(III)-glutathione conjugates that clearly contribute to As(III) resistance and clearance from the body [146]. MRP1 is ubiquitously expressed in the human body with the exception of the liver. In contrast, MRP2 is abundant in the liver and plays a key role in secretion of As(III) into the bile. Several lines of evidence suggest that MRP1 is also involved in Sb(III) resistance, possibly by GSH-dependent efflux of Sb(III) (Fig. 12.5). In vitro studies showed that overexpression of MRP1 confers increased resistance to As(III) and Sb(III) in human cell lines [147, 148], while MRP1-deficient murine cells are hypersensitive to As(III) and Sb(III) [149]. Importantly, the human lung cancer GLC4/Sb30 cell line selected for Sb(III) resistance exhibited elevated Sb(III) efflux due to overexpression of MRP1 [150]. GLC4/Sb30 cells are cross-resistant to As(III), and it was shown that chemical depletion of GSH resulted in reduced As(III) efflux [151]. It was later confirmed that MRP1 mediates transport of $As(GS)_3$ [152]. Collectively, these results suggest that MRP1 also transports Sb(III) in a GSH-conjugated form, but this assumption requires direct evidence. In contrast to MRP1, the role of MRP2 in Sb(III) detoxification has not been investigated in detail. An early study reported that Sb(III) decreased GSH level by 30% in hepatocytes and provided evidence that the biliary excretion of Sb(III) is GSH-dependent [153]. Moreover, expression of MRP2 in rat and human hepatocytes is increased in response to both As(III) and Sb(III) [154]. Thus, it is tempting to speculate that MRP2 also mediates $Sb(GS)_3$ efflux in hepatocytes (Fig. 12.5).

12.3.2.3 ABC transporters and trypanothione in *Leishmania*

Early studies on genetic determinants of acquired cross-resistance to As(III) and antimonials in *Leishmania* revealed that amplification of three genes, *GSH1*, *ODC*, and *MRPA/ABCC3* (formerly known as *PGPA*), is often associated with this phenomenon [15, 17, 18, 155–157]. As mentioned earlier, *GSH1* encodes for γ-glutamylcysteine synthetase, catalyzing a rate-limiting step in glutathione biosynthesis. *ODC* encodes for ornithine decarboxylase, which is required for spermidine biosynthesis. Both molecules are used to form a bis-glutathione-spermidine conjugate called trypanothione, which is a major source of reduced thiols in *Leishmania* and *Trypanosoma* species [158]. *MRPA* encodes an ABC transporter, an orthologue of Ycf1 and Mrp1 transporters, and its mutation was reported to reverse As(III) and Sb(III) resistance [159, 160]. In sum, these observations led to a model in which complexation of As(III)/Sb(III) by trypanothione and MRPA-dependent transport of resulting conjugates from the cytosol into the intracellular compartment are responsible for increased resistance to arsenicals and antimonials in *Leishmania* (Fig. 12.4) [157, 159]. Later on, direct transport studies using membrane vesicles isolated from *Leishmania* confirmed that MRPA mediates ATP-dependent transport of As(III)-glutathione conjugates [160]. Interestingly, MRPA was shown to be localized to the membrane of intracellular vesicles situated near the flagellar pocket, which is the site of endocytosis and exocytosis in *Leishmania* [161]. This suggested that after sequestration into a lumen of the intracellular vesicles Sb(III)-trypanothione conjugates might be further secreted out of the cell by exocytosis (Fig. 12.4). However, overexpression of MRPA did not decrease As(III) accumulation in *Leishmania* cells, leading to the hypothesis that another type of transporter localized to the plasma membrane is responsible for extrusion of metalloid-thiol conjugates [159, 162]. In search of this hypothetical transporter, Leprohon et al. [163] functionally analyzed the complete set of *Leishmania* ABC transporters from the C subfamily. It was found that all ABCC transporters are localized to the intracellular membrane compartments and three of them, ABCC4, ABCC5, and ABCC7, contribute to Sb(III) resistance [163].

Recently, two additional ABC transporters, ABCG2 and ABCI4, were shown to confer Sb(III) tolerance in *Leishmania* species (Fig. 12.4) [164, 165]. Importantly, ABCG2 is localized not only in the intracellular vesicles but also partially in the plasma membrane, and its overexpression decreases Sb(III) accumulation in the cells. As ABCG2 mediates efflux of glutathione and trypanothione, it was proposed that ABCG2 may export Sb(III)-thiol conjugates either by exocytosis or directly from the cytosol to the external medium (Fig. 12.4) [164, 166]. ABCI4 is a half-type ABC transporter that forms homodimers in the membrane and exhibits dual localization in mitochondria and the plasma membrane [165]. Overexpression of ABCI4 confers increased resistance to Sb(III), As(III), heavy metals, and several other toxic compounds by mediating their efflux. Similarly to ABCG2, ABCI4 extrudes thiols, suggesting that it may act as an exporter of Sb(III)-thiol conjugates (Fig. 12.4) [165].

12.3.2.4 ABC transporters and phytochelatins in plants and fungi

Phytochelatins (PCs) are thiol-rich oligopeptides, $(\gamma\text{-Glu-Cys})_n$-Gly (n = 2–5), which are enzymatically synthesized from GSH by phytochelatin synthase (PCS) [167]. PCSs belong to the papain-like cysteine protease superfamily and are ubiquitously present in land plants; however, *PCS* genes were also identified in other eukaryotic kingdoms, including fungi and animals, and a few cyanobacteria [167, 168]. PCSs are constitutively expressed, but the activity of phytochelatin synthase is induced by the presence of heavy metals and As(III) [168]. The involvement of PCs in arsenic tolerance has been demonstrated in several plant species [169–172], the fission yeast *Schizosaccharomyces pombe* [169], and the nematode *Caenorhabditis elegans* [173]. In *A. thaliana* and *S. pombe*, PC-metal(loid) complexes are sequestrated into the vacuole by C-type ABC transporters: AtABCC1, AtABCC2, and SpABCC2 (Fig. 12.3) [172, 174]. Little is known about the role of PCs in Sb(III) tolerance. Our group demonstrated that heterologous expression of *PCS* genes from *S. pombe* and *Triticum aestivum* (wheat) in *S. cerevisiae* fully complements the Sb(III) sensitivity of the *ycf1Δ* mutant [175]. Moreover, Kamiya and Fujiwara [176] reported that mutation of the *A. thaliana PCS1* gene results in high sensitivity to both As(III) and Sb(III), indicating the role of PCs in Sb(III) tolerance (Fig. 12.3).

12.3.3 Antimony biotransformation as a means of detoxification

In most organisms tolerance to As(V) requires its reduction to a trivalent form, which is then removed from the cytoplasm by various membrane transporters. Reduction of As(V) for detoxification purposes is catalyzed by several classes of arsenate reductases, such as ArsC in prokaryotes, and Acr2 and HAC1 in eukaryotes [177, 178]. Sb(V) reduction is also an initial step in antimony biotransformation. So far, the enzymatic reduction of Sb(V) has been demonstrated for Acr2 from *L. major* (Fig. 12.4) [179]. Acr2 proteins belong to the rhodanese/Cdc25 phosphatase superfamily [180], and LmACR2 possesses dual activity of tyrosine phosphatase and antimonate reductase [181]. Interestingly, expression of *LmACR2* complemented As(V) sensitivity of the *E. coli* mutant lacking the ArsC arsenate reductase. Moreover, the purified LmACR2 was shown to catalyze reduction of both As(V) and Sb(V), suggesting that other members of the Acr2 family may also exhibit antimonate reductase activity [179].

It was recently discovered that the anaerobic bacterium *Desulfuribacillus stibiiarsenatis* is capable of using Sb(V) as a terminal electron acceptor for anaerobic respiration followed by precipitation of resulting Sb(III) in the form of antimony trioxide microcrystals (Fig. 12.1) [182–184]. Enzymes of the dimethyl sulfoxide reductase (DMSOR) family of complex iron–sulfur molybdoenzymes (CISMs) are involved in respiratory reduction and oxidation of several substrates, including As(III) and As(V) [185, 186]. Based on strong transcriptional activation in the presence of Sb(V), the member of the DMSOR family named AnrA was proposed to function as a catalytic subunit of antimonate reductase, likely localized in the periplasm (Fig. 12.1) [184].

Oxidation of As(III) to less-toxic As(V) is commonly observed in prokaryotes and provides not only a detoxification mechanism but also energy for growth [187]. Several Sb(III)-oxidizing bacterial species have been recently isolated from antimony-contaminated sites [188, 189]. Wang et al. [190] reported that the *A. tumefaciens* arsenite oxidase Aio consisting of two subunits, the iron-sulfur and molybdenum-containing subunit AioA and the small Rieske subunit AioB, can also convert Sb(III) to Sb(V) (Fig. 12.1). However, inactivation of the *aioA* gene slightly increased Sb(III) sensitivity and reduced the Sb(III) oxidation rate by only 25%, suggesting the existence of additional Sb(III) oxidase [190]. Indeed, a novel type of antimonite oxidase AnoA was shown to contribute to Sb(III) oxidation in *A. tumefaciens* (Fig. 12.1) [191]. Inactivation of the *anoA* gene resulted in a decrease of both Sb(III) resistance and Sb(III) oxidation capacity. The expression of *anoA* was specifically induced by Sb(III), and overexpression increased the rate of Sb(III) oxidation. AnoA belongs to the short-chain dehydrogenase/reductase (SDR) superfamily of ubiquitously present proteins, which catalase $NAD(P)H/NAD(P)^+$-dependent reduction/oxidation reactions of a wide range of substrates, including polyols [192, 193]. In vitro studies confirmed that AnoA oxidizes Sb(III) to Sb(V) in an NADP+-dependent manner [191]. Thus, AnoA represents a novel type of antimonite oxidase in bacteria.

12.4 Challenges and perspectives

Antimony is a neglected metalloid compared to extensively studied arsenic in biological systems. With the exception of *Leishmania* research, mechanisms of antimony transport and resistance are investigated within studies primarily focused on arsenic, but this is not always the case. For example, transport specificity of plant aquaglyceroporins toward Sb(III) and the role of phytochelatins in Sb(III) tolerance in plants have barely been investigated. Similarly, it has not yet been demonstrated whether hexose transporters facilitate uptake of Sb(III). Other major challenges in antimony biological research include identification of Sb(V) uptake routes, determination of which members of the Acr3 family and the MRP family are able to transport Sb(III), and revealing the mechanism of ArsB- and Acr3-mediated metalloid transport across the membranes at the structural level. More studies are also required to understand transcriptional and post-transcriptional regulation mechanisms of Sb(III) transporters in order to design strategies for modulation of Sb(III) accumulation in organisms. This could help to overcome antimony resistance in *Leishmania*, reduce accumulation of antimony in crops cultivated in contaminated areas, or improve phytoremediation processes.

References

[1] Filella M, Belzile N, Chen YW. Antimony in the environment: A review focused on natural waters: I. Occurrence. Earth Sci Rev 2002, 57, 125–176.
[2] Filella M, Williams PA, Belzile N. Antimony in the environment: Knowns and unknowns. Environ Chem 2009, 6, 95–105.
[3] Wilson SC, Lockwood PV, Ashley PM, Tighe M. The chemistry and behaviour of antimony in the soil environment with comparisons to arsenic: A critical review. Environ Pollut 2010, 158, 1169–1181.
[4] Fawcett SE, Jamieson HE, Nordstrom DK, McCleskey RB. Arsenic and antimony geochemistry of mine wastes, associated waters and sediments at the Giant mine, yellow knife, northwest territories, Canada. Appl Geochem 2015, 62, 3–17.
[5] Filella M, Belzile N, Chen YW. Antimony in the environment: A review focused on natural waters: II. Relevant solution chemistry. Earth Sci Rev 2002, 59, 265–285.
[6] Filella M, Belzile N, Lett MC. Antimony in the environment: A review focused on natural waters. III. Microbiota relevant interactions. Earth Sci Rev 2007, 80, 195–217.
[7] Shen S, Li XF, Cullen WR, Weinfeld M, Le XC. Arsenic binding to proteins. Chem Rev 2013, 113, 7769–7792.
[8] Sforna M, Philippot P, Somogyi A, van Zuilen MA, Medjoubi K, Schoepp-Cothenet B, Nitschke W, Visscher PT. Evidence for arsenic metabolism and cycling by microorganisms 2.7 billion years ago. Nature Geosci 2014, 7, 811–815.
[9] Mukhopadhyay R, Rosen BP, Pung LT, Silver S. Microbial arsenic: From geocycles to genes and enzymes. FEMS Microbiol Rev 2002, 26, 311–325.
[10] Chen SC, Sun GX, Rosen BP, Zhang SY, Deng Y, Zhu BK, Rensing C, Zhu YG. Recurrent horizontal transfer of arsenite methyltransferase genes facilitated adaptation of life to arsenic. Sci Rep 2017, 7, 7741.
[11] Maciaszczyk-Dziubinska E, Wawrzycka D, Wysocki R. Arsenic and antimony transporters in eukaryotes. Int J Mol Sci 2012, 13, 3527–3548.
[12] Tamás MJ. Cellular and molecular mechanisms of antimony transport, toxicity and resistance. Environ Chem 2016, 13, 955–962.
[13] Garbinski LD, Rosen BP, Chen J. Pathways of arsenic uptake and efflux. Environ Int 2019, 126, 585–597.
[14] Mukhopadhyay R, Bhattacharjee H, Rosen BP. Aquaglyceroporins: Generalized metalloid channels. Biochim Biophys Acta 2014, 1840, 1583–1591.
[15] Croft SL, Sundar S, Fairlamb AH. Drug resistance in leishmaniasis. Clin Microbiol Rev 2006, 19, 111–126.
[16] Gjuvsland AB, Zörgö E, Samy JK, Stenberg S, Demirsoy IH, Roque F, Maciaszczyk-Dziubinska E, Migocka M, Alonso-Perez E, Zackrisson M, Wysocki R, Tamás MJ, Jonassen I, Omholt SW, Warringer J. Disentangling genetic and epigenetic determinants of ultrafast adaptation. Mol Syst Biol 2016, 12, 892.
[17] Ponte-Sucre A, Gamarro F, Dujardin JC, Barrett MP, López-Vélez R, García-Hernández R, Pountain AW, Mwenechanya R, Papadopoulou B. Drug resistance and treatment failure in leishmaniasis: A twenty-first century challenge. PLoS Negl Trop Dis 2017, 11, e0006052.
[18] Capela R, Moreira R, Lopes F. An overview of drug resistance in protozoal diseases. Int J Mol Sci 2019, 20, 5748.
[19] Feng R, Wei C, Tu S, Ding Y, Wang R, Guo J. The uptake and detoxification of antimony by plants: A review. Environ Exp Bot 2013, 96, 28–34.

[20] Li J, Wang Q, Oremland RS, Kulp TR, Rensing C, Wang G. Microbial antimony biogeochemistry: Enzymes, regulation, and related metabolic pathways. Appl Environ Microbiol 2016, 82, 5482–5495.
[21] Bentley R, Chasteen TG. Microbial methylation of metalloids: Arsenic, antimony, and bismuth. Microbiol Mol Biol Rev 2002, 66, 250–271.
[22] Filella M. Alkyl derivatives of antimony in the environment. Met Ions Life Sci 2010, 7, 267–301.
[23] Filella M, Williams PA. Antimony biomethylation in culture media revisited in the light of solubility and chemical speciation considerations. Environ Toxicol 2010, 25, 431–439.
[24] Sanders OI, Rensing C, Kuroda M, Mitra B, Rosen BP. Antimonite is accumulated by the glycerol facilitator GlpF in *Escherichia coli*. J Bacteriol 1997, 179, 3365–3367.
[25] Wysocki R, Chéry CC, Wawrzycka D, Van Hulle M, Cornelis R, Thevelein JM, Tamás MJ. The glycerol channel Fps1p mediates the uptake of arsenite and antimonite in *Saccharomyces cerevisiae*. Mol Microbiol 2001, 40, 1391–1401.
[26] Meng YL, Liu Z, Rosen BP. As(III) and Sb(III) uptake by GlpF and efflux by ArsB in *Escherichia coli*. J Biol Chem 2004, 279, 18334–18341.
[27] Park JH, Saier MH Jr. Phylogenetic characterization of the MIP family of transmembrane channel proteins. J Membr Biol 1996, 153, 171–180.
[28] King LS, Kozono D, Agre P. From structure to disease: The evolving tale of aquaporin biology. Nat Rev Mol Cell Biol 2004, 5, 687–698.
[29] Liu Z, Shen J, Carbrey JM, Mukhopadhyay R, Agre P, Rosen BP. Arsenite transport by mammalian aquaglyceroporins AQP7 and AQP9. Proc Natl Acad Sci USA 2002, 99, 6053–6058.
[30] Bienert GP, Thorsen M, Schüssler MD, Nilsson HR, Wagner A, Tamás MJ, Jahn TP. A subgroup of plant aquaporins facilitate the bi-directional diffusion of $As(OH)_3$ and $Sb(OH)_3$ across membranes. BMC Biol 2008, 6, 26.
[31] Isayenkov S, Maathuis FJM. The *Arabidopsis thaliana* aquaglyceroporin AtNIP7;1 is a pathway for arsenite uptake. FEBS Lett 2008, 582, 1625–1628.
[32] Ma JF, Yamaji N, Mitani N, Xu XY, Su YH, McGrath SP, Zhao FJ. Transporters of arsenite in rice and their role in arsenic accumulation in rice grain. Proc Natl Acad Sci USA 2008, 105, 9931–9935.
[33] Yang HC, Cheng J, Finan TM, Rosen BP, Bhattacharjee H. Novel pathway for arsenic detoxification in the legume symbiont *Sinorhizobium meliloti*. J Bacteriol 2005, 187, 6991–6997.
[34] Maciaszczyk-Dziubinska E, Migdal I, Migocka M, Bocer T, Wysocki R. The yeast aquaglyceroporin Fps1p is a bidirectional arsenite channel. FEBS Lett 2010, 584, 726–732.
[35] Murata K, Mitsuoka K, Hirai T, Walz T, Agre P, Heymann JB, Engel A, Fujiyoshi Y. Structural determinants of water permeation through aquaporin-1. Nature 2000, 407, 599–605.
[36] Ren G, Reddy VS, Cheng A, Melnyk P, Mitra AK. Visualization of a water-selective pore by electron crystallography in vitreous ice. Proc Natl Acad Sci USA 2001, 98, 1398–1403.
[37] Sui H, Han BG, Lee JK, Walian P, Jap BK. Structural basis of water-specific transport through the AQP1 water channel. Nature 2001, 414, 872–878.
[38] Fu D, Libson A, Miercke LJ, Weitzman C, Nollert P, Krucinski J, Stroud RM. Structure of a glycerol-conducting channel and the basis for its selectivity. Science 2000, 290, 481–486.
[39] Porquet A, Filella M. Structural evidence of the similarity of $Sb(OH)_3$ and $As(OH)_3$ with glycerol: Implications for their uptake. Chem Res Toxicol 2007, 20, 1269–1276.
[40] Bienert GP, Schüssler MD, Jahn TP. Metalloids: Essential, beneficial or toxic? Major intrinsic proteins sort it out. Trends Biochem Sci 2008, 33, 20–26.
[41] Bhattacharjee H, Mukhopadhyay R, Thiyagarajan S, Rosen BP. Aquaglyceroporins: Ancient channels for metalloids. J Biol 2008, 7, 33.
[42] Luyten K, Albertyn J, Skibbe WF, Prior BA, Ramos J, Thevelein JM, Hohmann S. Fps1, a yeast member of the MIP family of channel proteins, is a facilitator for glycerol uptake and efflux and is inactive under osmotic stress. EMBO J 1995, 14, 1360–1371.

[43] Tamás MJ, Luyten K, Sutherland FC, Hernandez A, Albertyn J, Valadi H, Li H, Prior BA, Kilian SG, Ramos J, Gustafsson L, Thevelein JM, Hohmann S. Fps1p controls the accumulation and release of the compatible solute glycerol in yeast osmoregulation. Mol Microbiol 1999, 31, 1087–1104.
[44] Ahmadpour D, Geijer C, Tamás MJ, Lindkvist-Petersson K, Hohmann S. Yeast reveals unexpected roles and regulatory features of aquaporins and aquaglyceroporins. Biochim Biophys Acta 2014, 1840, 1482–1491.
[45] Thorsen M, Di Y, Tängemo C, Morillas M, Ahmadpour D, Van der Does C, Wagner A, Johansson E, Boman J, Posas F, Wysocki R, Tamás MJ. The MAPK Hog1p modulates Fps1p-dependent arsenite uptake and tolerance in yeast. Mol Biol Cell 2006, 17, 4400–4410.
[46] Mollapour M, Piper PW. Hog1 mitogen-activated protein kinase phosphorylation targets the yeast Fps1 aquaglyceroporin for endocytosis, thereby rendering cells resistant to acetic acid. Mol Cell Biol 2007, 27, 6446–6456.
[47] Ahmadpour D, Maciaszczyk-Dziubinska E, Babazadeh R, Dahal S, Migocka M, Andersson M, Wysocki R, Tamás MJ, Hohmann S. The mitogen-activated protein kinase Slt2 modulates arsenite transport through the aquaglyceroporin Fps1. FEBS Lett 2016, 590, 3649–3659.
[48] Hedfalk K, Bill RM, Mullins JG, Karlgren S, Filipsson C, Bergstrom J, Tamás MJ, Rydström J, Hohmann S. A regulatory domain in the C-terminal extension of the yeast glycerol channel Fps1p. J Biol Chem 2004, 279, 14954–14960.
[49] Beese SE, Negishi T, Levin DE. Identification of positive regulators of the yeast Fps1 glycerol channel. PLoS Genet 2009, 5, e1000738.
[50] Lee J, Reiter W, Dohnal I, Gregori C, Beese-Sims S, Kuchler K, Ammerer G, Levin DE. MAPK Hog1 closes the *S. cerevisiae* glycerol channel Fps1 by phosphorylating and displacing its positive regulators. Genes Dev 2013, 27, 2590–2601.
[51] Thorsen M, Jacobson T, Vooijs R, Navarrete C, Bliek T, Schat H, Tamás MJ. Glutathione serves an extracellular defence function to decrease arsenite accumulation and toxicity in yeast. Mol Microbiol 2012, 84, 1177–1188.
[52] Chaumont F, Barrieu F, Wojcik E, Chrispeels MJ, Jung R. Aquaporins constitute a large and highly divergent protein family in maize. Plant Physiol 2001, 125, 1206–1215.
[53] Johanson U, Karlsson M, Johansson I, Gustavsson S, Sjövall S, Fraysse L, Weig AR, Kjellbom P. The complete set of genes encoding major intrinsic proteins in Arabidopsis provides a framework for a new nomenclature for major intrinsic proteins in plants. Plant Physiol 2001, 126, 1358–1369.
[54] Danielson JA, Johanson U. Unexpected complexity of the aquaporin gene family in the moss *Physcomitrella patens*. BMC Plant Biol 2008, 8, 45.
[55] Katsuhara M, Sasano S, Horie T, Matsumoto T, Rhee J, Shibasaka M. Functional and molecular characteristics of rice and barley NIP aquaporins transporting water, hydrogen peroxide and arsenite. Plant Biotechnol 2014, 31, 213–219.
[56] Kamiya T, Tanaka M, Mitani N, Ma JF, Maeshima M, Fujiwara T. NIP1;1, an aquaporin homolog, determines the arsenite sensitivity of *Arabidopsis thaliana*. J Biol Chem 2009, 284, 2114–2120.
[57] Isayenkov SV, Maathuis FJ. The *Arabidopsis thaliana* aquaglyceroporin AtNIP7;1 is a pathway for arsenite uptake. FEBS Lett 2008, 582, 1625–1628.
[58] Sabir F, Gomes S, Loureiro-Dias MC, Soveral G, Prista C. Molecular and functional characterization of grapevine NIPs through heterologous expression in aqy-null *Saccharomyces cerevisiae*. Int J Mol Sci 2020, 21, 663.
[59] Xu W, Dai W, Yan H, Li S, Shen H, Chen Y, Xu H, Sun Y, He Z, Ma M. Arabidopsis NIP3;1 plays an important role in arsenic uptake and root-to-shoot translocation under arsenite stress conditions. Mol Plant 2015, 8, 722–733.

[60] Lindsay ER, Maathuis FJ. *Arabidopsis thaliana* NIP7;1 is involved in tissue arsenic distribution and tolerance in response to arsenate. FEBS Lett 2016, 590, 779–786.
[61] Mosa KA, Kumar K, Chhikara S, Mcdermott J, Liu Z, Musante C, White JC, Dhankher OP. Members of rice plasma membrane intrinsic proteins subfamily are involved in arsenite permeability and tolerance in plants. Transgenic Res 2012, 21, 1265–1277.
[62] He Z, Yan H, Chen Y, Shen H, Xu W, Zhang H, Shi L, Zhu YG, Ma M. An aquaporin PvTIP4;1 from *Pteris vittata* may mediate arsenite uptake. New Phytol 2016, 209, 746–761.
[63] Kamiya T, Fujiwara T. Arabidopsis NIP1;1 transports antimonite and determines antimonite sensitivity. Plant Cell Physiol 2009, 50, 1977–1981.
[64] Gourbal B, Sonuc N, Bhattacharjee H, Legare D, Sundar S, Ouellette M, Rosen BP, Mukhopadhyay R. Drug uptake and modulation of drug resistance in *Leishmania* by an aquaglyceroporin. J Biol Chem 2004, 279, 31010–31017.
[65] Figarella K, Uzcategui NL, Zhou Y, LeFurgey A, Ouellette M, Bhattacharjee H, Mukhopadhyay R. Biochemical characterization of *Leishmania major* aquaglyceroporin LmAQP1: Possible role in volume regulation and osmotaxis. Mol Microbiol 2007, 65, 1006–1017.
[66] Plourde M, Ubeda JM, Mandal G, Monte-Neto RL, Mukhopadhyay R, Ouellette M. Generation of an aquaglyceroporin AQP1 null mutant in *Leishmania major*. Mol Biochem Parasitol 2015, 201, 108–111.
[67] Marquis N, Gourbal B, Rosen BP, Mukhopadhyay R, Ouellette M. Modulation in aquaglyceroporin *AQP1* gene transcript levels in drug-resistant *Leishmania*. Mol Microbiol 2005, 57, 1690–1699.
[68] Mandal G, Mandal S, Sharma M, Charret KS, Papadopoulou B, Bhattacharjee H, Mukhopadhyay R. Species-specific antimonial sensitivity in *Leishmania* is driven by post-transcriptional regulation of AQP1. PLoS Negl Trop Dis 2015, 9, e0003500.
[69] Mandal S, Maharjan M, Singh S, Chatterjee M, Madhubala R. Assessing aquaglyceroporin gene status and expression profile in antimony-susceptible and -resistant clinical isolates of *Leishmania donovani* from India. J Antimicrob Chemother 2010, 65, 496–507.
[70] Potvin JE, Leprohon P, Queffeulou M, Sundar S, Ouellette M. Mutations in an aquaglyceroporin as a proven marker of antimony clinical resistance in the parasite *Leishmania donovani*. Clin Infect Dis 2020, ciaa1236. doi: 10.1093/cid/ciaa1236
[71] Mandal G, Sharma M, Kruse M, Sander-Juelch C, Munro LA, Wang Y, Vilg JV, Tamás MJ, Bhattacharjee H, Wiese M, Mukhopadhyay R. Modulation of *Leishmania major* aquaglyceroporin activity by a mitogen-activated protein kinase. Mol Microbiol 2012, 85, 1204–1218.
[72] Sharma M, Mandal G, Mandal S, Bhattacharjee H, Mukhopadhyay R. Functional role of lysine 12 in *Leishmania major* AQP1. Mol Biochem Parasitol 2015, 201, 139–145.
[73] Liu Z, Carbrey JM, Agre P, Rosen BP. Arsenic trioxide uptake by human and rat aquaglyceroporins. Biochem Biophys Res Commun 2004, 316, 1178–1185.
[74] Hamdi M, Sanchez MA, Beene LC, Liu Q, Landfear SM, Rosen BP, Liu Z. Arsenic transport by zebrafish aquaglyceroporins. BMC Mol Biol 2009, 10, 104.
[75] Drobna Z, Styblo M, Thomas DJ. An overview of arsenic metabolism and toxicity. Curr Protoc Toxicol 2009, 42, 4.31.1–4.31.6.
[76] Khairul I, Wang QQ, Jiang YH, Wang C, Naranmandura H. Metabolism, toxicity and anticancer activities of arsenic compounds. Oncotarget 2017, 8, 23905–23926.
[77] Liu Z, Styblo M, Rosen BP. Methylarsonous acid transport by aquaglyceroporins. Environ Health Perspect 2006, 114, 527–531.
[78] McDermott JR, Jiang X, Beene LC, Rosen BP, Liu Z. Pentavalent methylated arsenicals are substrates of human AQP9. Biometals 2010, 23, 119–127.

[79] Carbrey JM, Song L, Zhou Y, Yoshinaga M, Rojek A, Wang Y, Liu Y, Lujan HL, DiCarlo SE, Nielsen S, Rosen BP, Agre P, Mukhopadhyay R. Reduced arsenic clearance and increased toxicity in aquaglyceroporin-9-null mice. Proc Natl Acad Sci USA 2009, 106, 15956–15960.

[80] Liu Z, Boles E, Rosen BP. Arsenic trioxide uptake by hexose permeases in *Saccharomyces cerevisiae*. J Biol Chem 2004, 279, 17312–17318.

[81] Liu Z, Sanchez MA, Jiang X, Boles E, Landfear SM, Rosen BP. Mammalian glucose permease GLUT1 facilitates transport of arsenic trioxide and methylarsonous acid. Biochem Biophys Res Commun 2006, 351, 424–430.

[82] Duan GL, Hu Y, Schneider S, McDermott J, Chen J, Sauer N, Rosen BP, Daus B, Liu Z, Zhu YG. Inositol transporters AtINT2 and AtINT4 regulate arsenic accumulation in Arabidopsis seeds. Nat Plants 2015, 2, 15202.

[83] Jochem M, Ende L, Isasa M, Ang J, Schnell H, Guerra-Moreno A, Micoogullari Y, Bhanu M, Gygi SP, Hanna J. Targeted degradation of glucose transporters protects against arsenic toxicity. Mol Cell Biol 2019, 39, e00559–18.

[84] Prakash S, Cooper G, Singhi S, Saier MH Jr. The ion transporter superfamily. Biochim Biophys Acta 2003, 1618, 79–92.

[85] Dey S, Rosen BP. Dual mode of energy coupling by the oxyanion-translocating ArsB protein. J Bacteriol 1995, 177, 385–389.

[86] Kuroda M, Dey S, Sanders OI, Rosen BP. Alternate energy coupling of ArsB, the membrane subunit of the Ars anion-translocating ATPase. J Biol Chem 1997, 272, 326–331.

[87] Wu J, Tisa LS, Rosen BP. Membrane topology of the ArsB protein, the membrane subunit of an anion-translocating ATPase. J Biol Chem 1992, 267, 12570–12576.

[88] Chen Y, Dey S, Rosen BP. Soft metal thiol chemistry is not involved in the transport of arsenite by the Ars pump. J Bacteriol 1996, 178, 911–913.

[89] Tisa LS, Rosen BP. Molecular characterization of an anion pump. The ArsB protein is the membrane anchor for the ArsA protein. J Biol Chem 1990, 265, 190–194.

[90] Rosen BP, Weigel U, Karkaria C, Gangola P. Molecular characterization of an anion pump. The *arsA* gene product is an arsenite(antimonate)-stimulated ATPase. J Biol Chem 1988, 263, 3067–3070.

[91] Dey S, Dou D, Rosen BP. ATP-dependent arsenite transport in everted membrane vesicles of *Escherichia coli*. J Biol Chem 1994, 269, 25442–25446.

[92] Bhattacharjee H, Li J, Ksenzenko MY, Rosen BP. Role of cysteinyl residues in metalloactivation of the oxyanion-translocating ArsA ATPase. J Biol Chem 1995, 270, 11245–11250.

[93] Zhou T, Radaev S, Rosen BP, Gatti DL. Structure of the ArsA ATPase: The catalytic subunit of a heavy metal resistance pump. EMBO J 2000, 19, 4838–4845.

[94] Lin YF, Walmsley AR, Rosen BP. An arsenic metallochaperone for an arsenic detoxification pump. Proc Natl Acad Sci USA 2006, 103, 15617–15622.

[95] Chen CM, Misra TK, Silver S, Rosen BP. Nucleotide sequence of the structural genes for an anion pump. The plasmid-encoded arsenical resistance operon. J Biol Chem 1986, 261, 15030–15038.

[96] Wu J, Rosen BP. The *arsD* gene encodes a second trans-acting regulatory protein of the plasmid-encoded arsenical resistance operon. Mol Microbiol 1993, 8, 615–623.

[97] Carlin A, Shi W, Dey S, Rosen BP. The *ars* operon of *Escherichia coli* confers arsenical and antimonial resistance. J Bacteriol 1995, 177, 981–986.

[98] Ji G, Silver S. Reduction of arsenate to arsenite by the ArsC protein of the arsenic resistance operon of *Staphylococcus aureus* plasmid pI258. Proc Natl Acad Sci USA 1992, 89, 9474–9478.

[99] Oden KL, Gladysheva TB, Rosen BP. Arsenate reduction mediated by the plasmid-encoded ArsC protein is coupled to glutathione. Mol Microbiol 1994, 12, 301–306.

[100] Wu J, Rosen BP. Metalloregulated expression of the *ars* operon. J Biol Chem 1993, 268, 52–58.
[101] Xu C, Shi W, Rosen BP. The chromosomal *arsR* gene of *Escherichia coli* encodes a trans-acting metalloregulatory protein. J Biol Chem 1996, 271, 2427–2432.
[102] Xu C, Rosen BP. Dimerization is essential for DNA binding and repression by the ArsR metalloregulatory protein of *Escherichia coli*. J Biol Chem 1997, 272, 15734–15738.
[103] Shi W, Wu J, Rosen BP. Identification of a putative metal binding site in a new family of metalloregulatory proteins. J Biol Chem 1994, 269, 19826–19829.
[104] Shi W, Dong J, Scott RA, Ksenzenko MY, Rosen BP. The role of arsenic-thiol interactions in metalloregulation of the *ars* operon. J Biol Chem 1996, 271, 9291–9297.
[105] Bobrowicz P, Wysocki R, Owsianik G, Goffeau A, Ułaszewski S. Isolation of three contiguous genes, *ACR1, ACR2* and *ACR3*, involved in resistance to arsenic compounds in the yeast *Saccharomyces cerevisiae*. Yeast 1997, 13, 819–828.
[106] Wysocki R, Bobrowicz P, Ułaszewski S. The *Saccharomyces cerevisiae ACR3* gene encodes a putative membrane protein involved in arsenite transport. J Biol Chem 1997, 272, 30061–30066.
[107] Mansour NM, Sawhney M, Tamang DG, Vogl C, Saier MH Jr. The bile/arsenite/riboflavin transporter (BART) superfamily. FEBS J 2007, 274, 612–629.
[108] Yang Y, Wu S, Lilley RM, Zhang R. The diversity of membrane transporters encoded in bacterial arsenic-resistance operons. PeerJ 2015, 3, e943.
[109] Fu HL, Meng Y, Ordóñez E, Villadangos AF, Bhattacharjee H, Gil JA, Mateos LM, Rosen BP. Properties of arsenite efflux permeases (Acr3) from *Alkaliphilus metalliredigens* and *Corynebacterium glutamicum*. J Biol Chem 2009, 284, 19887–19895.
[110] Indriolo E, Na G, Ellis D, Salt DE, Banks JA. A vacuolar arsenite transporter necessary for arsenic tolerance in the arsenic hyperaccumulating fern *Pteris vittata* is missing in flowering plants. Plant Cell 2010, 22, 2045–2057.
[111] Ghosh M, Shen J, Rosen BP. Pathways of As(III) detoxification in *Saccharomyces cerevisiae*. Proc Natl Acad Sci USA 1999, 96, 5001–5006.
[112] Sato T, Kobayashi Y. The *ars* operon in the *skin* element of *Bacillus subtilis* confers resistance to arsenate and arsenite. J Bacteriol 1998, 180, 1655–1661.
[113] Aaltonen EK, Silow M. Transmembrane topology of the Acr3 family arsenite transporter from *Bacillus subtilis*. Biochim Biophys Acta 2008, 1778, 963–973.
[114] Wawrzycka D, Markowska K, Maciaszczyk-Dziubinska E, Migocka M, Wysocki R. Transmembrane topology of the arsenite permease Acr3 from *Saccharomyces cerevisiae*. Biochim Biophys Acta Biomembr 2017, 1859, 117–125.
[115] Maciaszczyk-Dziubinska E, Migocka M, Wysocki R. Acr3p is a plasma membrane antiporter that catalyzes As(III)/H^+ and Sb(III)/H^+ exchange in *Saccharomyces cerevisiae*. Biochim Biophys Acta 2011, 1808, 1855–1859.
[116] Villadangos AF, Fu HL, Gil JA, Messens J, Rosen BP, Mateos LM. Efflux permease CgAcr3-1 of *Corynebacterium glutamicum* is an arsenite-specific antiporter. J Biol Chem 2012, 287, 723–735.
[117] Maciaszczyk-Dziubinska E, Migocka M, Wawrzycka D, Markowska K, Wysocki R. Multiple cysteine residues are necessary for sorting and transport activity of the arsenite permease Acr3p from *Saccharomyces cerevisiae*. Biochim Biophys Acta 2014, 1838, 747–755.
[118] Markowska K, Maciaszczyk-Dziubinska E, Migocka M, Wawrzycka D, Wysocki R. Identification of critical residues for transport activity of Acr3p, the *Saccharomyces cerevisiae* As(III)/H^+ antiporter. Mol Microbiol 2015, 98, 162–174.
[119] Shi Y. Common folds and transport mechanisms of secondary active transporters. Annu Rev Biophys 2013, 42, 51–72.
[120] Maciaszczyk-Dziubinska E, Wawrzycka D, Sloma E, Migocka M, Wysocki R. The yeast permease Acr3p is a dual arsenite and antimonite plasma membrane transporter. Biochim Biophys Acta 2010, 1798, 2170–2175.

[121] Bobrowicz P, Ulaszewski S. Arsenical – induced transcriptional activation of the yeast *Saccharomyces cerevisiae ACR2* and *ACR3* genes requires the presence of the *ACR1* gene product. Cell Mol Biol Lett 1998, 3, 13–20.
[122] Wysocki R, Fortier PK, Maciaszczyk E, Thorsen M, Leduc A, Odhagen A, Owsianik G, Ulaszewski S, Ramotar D, Tamás MJ. Transcriptional activation of metalloid tolerance genes in *Saccharomyces cerevisiae* requires the AP-1-like proteins Yap1p and Yap8p. Mol Biol Cell 2004, 15, 2049–2060.
[123] López-Maury L, Florencio FJ, Reyes JC. Arsenic sensing and resistance system in the cyanobacterium *Synechocystis* sp. strain PCC 6803. J Bacteriol 2003, 185, 5363–5371.
[124] Kang YS, Shi Z, Bothner B, Wang G, McDermott TR. Involvement of the Acr3 and DctA anti-porters in arsenite oxidation in *Agrobacterium tumefaciens* 5A. Environ Microbiol 2015, 17, 1950–1962.
[125] Ben Fekih I, Zhang C, Li YP, Zhao Y, Alwathnani HA, Saquib Q, Rensing C, Cervantes C. Distribution of arsenic resistance genes in prokaryotes. Front Microbiol 2018, 9, 2473.
[126] Castillo R, Saier MH. Functional promiscuity of homologues of the bacterial ArsA ATPases. Int J Microbiol 2010, 2010, 187373.
[127] Fu HL, Rosen BP, Bhattacharjee H. Biochemical characterization of a novel ArsA ATPase complex from *Alkaliphilus metalliredigens* QYMF. FEBS Lett 2010, 584, 3089–3094.
[128] Stefanovic S, Hegde RS. Identification of a targeting factor for posttranslational membrane protein insertion into the ER. Cell 2007, 128, 1147–1159.
[129] Schuldiner M, Metz J, Schmid V, Denic V, Rakwalska M, Schmitt HD, Schwappach B, Weissman JS. The GET complex mediates insertion of tail-anchored proteins into the ER membrane. Cell 2008, 134, 634–645.
[130] Kumar NV, Yang J, Pillai JK, Rawat S, Solano C, Kumar A, Grøtli M, Stemmler TL, Rosen BP, Tamás MJ. Arsenic directly binds to and activates the yeast AP-1-Like transcription factor Yap8. Mol Cell Biol 2015, 36, 913–922.
[131] Di Y, Tamás MJ. Regulation of the arsenic-responsive transcription factor Yap8p involves the ubiquitin-proteasome pathway. J Cell Sci 2007, 120, 256–264.
[132] Menezes RA, Amaral C, Delaunay A, Toledano M, Rodrigues-Pousada C. Yap8p activation in *Saccharomyces cerevisiae* under arsenic conditions. FEBS Lett 2004, 566, 141–146.
[133] Babst M. Regulation of nutrient transporters by metabolic and environmental stresses. Curr Opin Cell Biol 2020, 65, 35–41.
[134] Wawrzycka D, Sadlak J, Maciaszczyk-Dziubinska E, Wysocki R. Rsp5-dependent endocytosis and degradation of the arsenite transporter Acr3 requires its N-terminal acidic tail as an endocytic sorting signal and arrestin-related ubiquitin-ligase adaptors. Biochim Biophys Acta Biomembr 2019, 1861, 916–925.
[135] Shi K, Li C, Rensing C, Dai X, Fan X, Wang G. Efflux transporter ArsK is responsible for bacterial resistance to arsenite, antimonite, trivalent roxarsone, and methylarsenite. Appl Environ Microbiol 2018, 84, e01842–18.
[136] Ge R, Sun H. Bioinorganic chemistry of bismuth and antimony: Target sites of metallodrugs. Acc Chem Res 2007, 40, 267–274.
[137] Jacobson T, Navarrete C, Sharma SK, Sideri TC, Ibstedt S, Priya S, Grant CM, Christen P, Goloubinoff P, Tamás MJ. Arsenite interferes with protein folding and triggers formation of protein aggregates in yeast. J Cell Sci 2012, 125, 5073–5083.
[138] Sun H, Yan SC, Cheng WS. Interaction of antimony tartrate with the tripeptide glutathione. Implication for its mode of action. Eur J Biochem 2000, 267, 5450–5457.
[139] Yan S, Li F, Ding K, Sun H. Reduction of pentavalent antimony by trypanothione and formation of a binary and ternary complex of antimony(III) and trypanothione. J Biol Inorg Chem 2003, 8, 689–697.

[140] Wilkens S. Structure and mechanism of ABC transporters. F1000Prime Rep 2015, 7, 14.
[141] Szczypka MS, Wemmie JA, Moye-Rowley WS, Thiele DJ. A yeast metal resistance protein similar to human cystic fibrosis transmembrane conductance regulator (CFTR) and multidrug resistance-associated protein. J Biol Chem 1994, 269, 22853–22857.
[142] Li ZS, Szczypka M, Lu YP, Thiele DJ, Rea PA. The yeast cadmium factor protein (YCF1) is a vacuolar glutathione S-conjugate pump. J Biol Chem 1996, 271, 6509–6517.
[143] Gueldry O, Lazard M, Delort F, Dauplais M, Grigoras I, Blanquet S, Plateau P. Ycf1p-dependent Hg(II) detoxification in *Saccharomyces cerevisiae*. Eur J Biochem 2003, 270, 2486–2496.
[144] Préveral S, Ansoborlo E, Mari S, Vavasseur A. Forestier C. Metal(loid)s and radionuclides cytotoxicity in *Saccharomyces cerevisiae*. Role of YCF1, glutathione and effect of buthionine sulfoximine. Biochimie 2006, 88, 1651–1663.
[145] Thorsen M, Perrone GG, Kristiansson E, Traini M, Ye T, Dawes IW, Nerman O, Tamás MJ. Genetic basis of arsenite and cadmium tolerance in *Saccharomyces cerevisiae*. BMC Genomics 2009, 10, 105.
[146] Leslie EM. Arsenic-glutathione conjugate transport by the human multidrug resistance proteins (MRPs/ABCCs). J Inorg Biochem 2012, 108, 141–149.
[147] Cole SP, Sparks KE, Fraser K, Loe DW, Grant CE, Wilson GM, Deeley RG. Pharmacological characterization of multidrug resistant MRP-transfected human tumor cells. Cancer Res 1994, 54, 5902–5910.
[148] Lorico A, Bertola A, Baum C, Fodstad O, Rappa G. Role of the Multidrug Resistance Protein 1 in protection from heavy metal oxyanions: Investigations *in vitro* and in MRP1-deficient mice. Biochem Biophys Res Commun 2002, 291, 617–622.
[149] Rappa G, Lorico A, Flavell RA, Sartorelli AC. Evidence that the multidrug resistance protein (MRP) functions as a co-transporter of glutathione and natural product toxins. Cancer Res 1997, 57, 5232–5237.
[150] Vernhet L, Courtois A, Allain N, Payen L, Anger JP, Guillouzo A, Fardel O. Overexpression of the multidrug resistance-associated protein (MRP1) in human heavy metal-selected tumor cells. FEBS Lett 1999, 443, 321–325.
[151] Vernhet L, Allain N, Payen L, Anger JP, Guillouzo A, Fardel O. Resistance of human multidrug resistance-associated protein 1-overexpressing lung tumor cells to the anticancer drug arsenic trioxide. Biochem Pharmacol 2001, 61, 1387–1391.
[152] Leslie EM, Haimeur A, Waalkes MP. Arsenic transport by the human multidrug resistance protein 1 (MRP1/ABCC1). Evidence that a tri-glutathione conjugate is required. J Biol Chem 2004, 279, 32700–32708.
[153] Gyurasics A, Koszorús L, Varga F, Gregus Z. Increased biliary excretion of glutathione is generated by the glutathione-dependent hepatobiliary transport of antimony and bismuth. Biochem Pharmacol 1992, 44, 1275–1281.
[154] Vernhet L, Séité MP, Allain N, Guillouzo A, Fardel O. Arsenic induces expression of the multidrug resistance-associated protein 2 (MRP2) gene in primary rat and human hepatocytes. J Pharmacol Exp Ther 2001, 298, 234–239.
[155] Grondin K, Haimeur A, Mukhopadhyay R, Rosen BP, Ouellette M. Co-amplification of the gamma-glutamylcysteine synthetase gene *gsh1* and of the ABC transporter gene *pgpA* in arsenite-resistant *Leishmania tarentolae*. EMBO J 1997, 6, 3057–3065.
[156] Haimeur A, Guimond C, Pilote S, Mukhopadhyay R, Rosen BP, Poulin R, Ouellette M. Elevated levels of polyamines and trypanothione resulting from overexpression of the ornithine decarboxylase gene in arsenite-resistant *Leishmania*. Mol Microbiol 1999, 34, 726–735.
[157] Légaré D, Papadopoulou B, Roy G, Mukhopadhyay R, Haimeur A, Dey S, Grondin K, Brochu C, Rosen BP. Efflux systems and increased trypanothione levels in arsenite-resistant *Leishmania*. Exp Parasitol 1997, 87, 275–282.

[158] Fairlamb AH, Cerami A. Metabolism and functions of trypanothione in the Kinetoplastida. Annu Rev Microbiol 1992, 46, 695–729.

[159] Papadopoulou B, Roy G, Dey S, Rosen BP, Ouellette M. Contribution of the *Leishmania* P-glycoprotein-related gene *ltpgpA* to oxyanion resistance. J Biol Chem 1994, 269, 11980–11986.

[160] Papadopoulou B, Roy G, Dey S, Rosen BP, Olivier M, Ouellette M. Gene disruption of the P-glycoprotein related gene *pgpa* of *Leishmania tarentolae*. Biochem Biophys Res Commun 1996, 224, 772–778.

[161] Légaré D, Richard D, Mukhopadhyay R, Stierhof YD, Rosen BP, Haimeur A, Papadopoulou B, Ouellette M. The *Leishmania* ATP-binding cassette protein PGPA is an intracellular metal-thiol transporter ATPase. J Biol Chem 2001, 276, 26301–26307.

[162] Dey S, Papadopoulou B, Haimeur A, Roy G, Grondin K, Dou D, Rosen BP, Ouellette M. High level arsenite resistance in *Leishmania tarentolae* is mediated by an active extrusion system. Mol Biochem Parasitol 1994, 67, 49–57.

[163] Leprohon P, Légaré D, Ouellette M. Intracellular localization of the ABCC proteins of *Leishmania* and their role in resistance to antimonials. Antimicrob Agents Chemother 2009, 53, 2646–2649.

[164] Perea A, Manzano JI, Castanys S, Gamarro F. The LABCG2 transporter from the protozoan parasite *Leishmania* is involved in antimony resistance. Antimicrob Agents Chemother 2016, 60, 3489–3496.

[165] Manzano JI, García-Hernández R, Castanys S, Gamarro F. A new ABC half-transporter in *Leishmania major* is involved in resistance to antimony. Antimicrob Agents Chemother 2013, 57, 3719–3730.

[166] Perea A, Manzano JI, Kimura Y, Ueda K, Castanys S, Gamarro F. Leishmania LABCG2 transporter is involved in ATP-dependent transport of thiols. Biochem J 2018, 475, 87–97.

[167] Clemens S, Peršoh D. Multi-tasking phytochelatin synthases. Plant Sci 2009, 177, 266–271.

[168] Rea PA. Phytochelatin synthase: Of a protease a peptide polymerase made. Physiol Plant 2012, 145, 154–164.

[169] Ha SB, Smith AP, Howden R, Dietrich WM, Bugg S, O'Connell MJ, Goldsbrough PB, Cobbett CS. Phytochelatin synthase genes from Arabidopsis and the yeast *Schizosaccharomyces pombe*. Plant Cell 1999, 11, 1153–1164.

[170] Schmöger ME, Oven M, Grill E. Detoxification of arsenic by phytochelatins in plants. Plant Physiol 2000, 122, 793–801.

[171] Hartley-Whitaker J, Ainsworth G, Vooijs R, Ten Bookum W, Schat H, Meharg AA. Phytochelatins are involved in differential arsenate tolerance in *Holcus lanatus*. Plant Physiol 2001, 126, 299–306.

[172] Song WY, Park J, Mendoza-Cózatl DG, Suter-Grotemeyer M, Shim D, Hörtensteiner S, Geisler M, Weder B, Rea PA, Rentsch D, Schroeder JI, Lee Y, Martinoia E. Arsenic tolerance in Arabidopsis is mediated by two ABCC-type phytochelatin transporters. Proc Natl Acad Sci USA 2010, 107, 21187–21192.

[173] Vatamaniuk OK, Bucher EA, Ward JT, Rea PA. A new pathway for heavy metal detoxification in animals. Phytochelatin synthase is required for cadmium tolerance in *Caenorhabditis elegans*. J Biol Chem 2001, 276, 20817–20820.

[174] Mendoza-Cózatl DG, Zhai Z, Jobe TO, Akmakjian GZ, Song WY, Limbo O, Russell MR, Kozlovskyy VI, Martinoia E, Vatamaniuk OK, Russell P, Schroeder JI. Tonoplast-localized Abc2 transporter mediates phytochelatin accumulation in vacuoles and confers cadmium tolerance. J Biol Chem 2010, 285, 40416–40426.

[175] Wysocki R, Clemens S, Augustyniak D, Golik P, Maciaszczyk E, Tamás MJ, Dziadkowiec D. Metalloid tolerance based on phytochelatins is not functionally equivalent to the arsenite transporter Acr3p. Biochem Biophys Res Commun 2003, 304, 293–300.

[176] Kamiya T, Fujiwara T. A novel allele of the Arabidopsis phytochelatin synthase 1 gene conferring high sensitivity to arsenic and antimony. Soil Sci Plant Nutr 2011, 57, 272–278.
[177] Messens J, Silver S. Arsenate reduction: Thiol cascade chemistry with convergent evolution. J Mol Biol 2006, 362, 1–17.
[178] Chao DY, Chen Y, Chen J, Shi S, Chen Z, Wang C, Danku JM, Zhao FJ, Salt DE. Genome-wide association mapping identifies a new arsenate reductase enzyme critical for limiting arsenic accumulation in plants. PLoS Biol 2014, 12, e1002009.
[179] Zhou Y, Messier N, Ouellette M, Rosen BP, Mukhopadhyay R. *Leishmania major* LmACR2 is a pentavalent antimony reductase that confers sensitivity to the drug pentostam. J Biol Chem 2004, 279, 37445–37451.
[180] Bordo D, Bork P. The rhodanese/Cdc25 phosphatase superfamily. Sequence-structure-function relations. EMBO Rep 2002, 3, 741–746.
[181] Zhou Y, Bhattacharjee H, Mukhopadhyay R. Bifunctional role of the leishmanial antimonate reductase LmACR2 as a protein tyrosine phosphatase. Mol Biochem Parasitol 2006, 148, 161–168.
[182] Abin CA, Hollibaugh JT. Dissimilatory antimonate reduction and production of antimony trioxide microcrystals by a novel microorganism. Environ Sci Technol 2014, 48, 681–688.
[183] Abin CA, Hollibaugh JT. *Desulfuribacillus stibiiarsenatis sp. nov.*, an obligately anaerobic, dissimilatory antimonate- and arsenate-reducing bacterium isolated from anoxic sediments, and emended description of the genus *Desulfuribacillus*. Int J Syst Evol Microbiol 2017, 67, 1011–1017.
[184] Abin CA, Hollibaugh JT. Transcriptional response of the obligate anaerobe *Desulfuribacillus stibiiarsenatis* MLFW-2^T to growth on antimonate and other terminal electron acceptors. Environ Microbiol 2019, 21, 618–630.
[185] McEwan AG, Ridge JP, McDevitt CA, Hugenholtz P. The DMSO reductase family of microbial molybdenum enzymes; molecular properties and role in the dissimilatory reduction of toxic elements. Geomicrobiol J 2002, 19, 3–21.
[186] Grimaldi S, Schoepp-Cothenet B, Ceccaldi P, Guigliarelli B, Magalon A. The prokaryotic Mo/W-bisPGD enzymes family: A catalytic workhorse in bioenergetic. Biochim Biophys Acta 2013, 1827, 1048–1085.
[187] Silver S, Phung LT. Genes and enzymes involved in bacterial oxidation and reduction of inorganic arsenic. Appl Environ Microbiol 2005, 71, 599–608.
[188] Lehr CR, Kashyap DR, McDermott TR. New insights into microbial oxidation of antimony and arsenic. Appl Environ Microbiol 2007, 73, 2386–2389.
[189] Li J, Yu H, Wu X, Shen L, Liu Y, Qiu G, Zeng W, Yu R. Novel hyper antimony-oxidizing bacteria isolated from contaminated mine soils in China. Geomicrobiol J 2018, 35, 713–720.
[190] Wang Q, Warelow TP, Kang Y-S, Romano C, Osborne TH, Lehr CR, Bothner B, McDermott TR, Santini JM, Wang G. Arsenite oxidase also functions as an antimonite oxidase. Appl Environ Microbiol 2015, 81, 1959–1965.
[191] Li J, Wang Q, Li M, Yang B, Shi M, Guo W, McDermott TR, Rensing C, Wang G. Proteomics and genetics for identification of a bacterial antimonite oxidase in *Agrobacterium tumefaciens*. Environ Sci Technol 2015, 49, 5980–5989.
[192] Kallberg Y, Oppermann U, Jörnvall H, Persson B. Short-chain dehydrogenase/reductase (SDR) relationships: A large family with eight clusters common to human, animal, and plant genomes. Protein Sci 2002, 11, 636–641.
[193] Kavanagh KL, Jörnvall H, Persson B, Oppermann U. Medium- and short-chain dehydrogenase/reductase gene and protein families: The SDR superfamily: Functional and structural diversity within a family of metabolic and regulatory enzymes. Cell Mol Life Sci 2008, 65, 3895–3896.

Jose Antonio Ruiz-Postigo and Saurabh Jain

Chapter 13
Pentavalent antimonials in the treatment of human leishmaniasis

13.1 Introduction

The different forms of leishmaniasis in humans are caused by a protozoan parasite from over 20 *Leishmania* species. Parasites are transmitted from reservoir hosts, domestic or sylvatic mammals to humans (zoonotic leishmaniasis) or from human to human (anthroponotic leishmaniasis), or by the bite of infected female phlebotomine sand flies (vector-borne transmission).

Leishmaniasis can be transmitted by shared syringes, by blood transfusion, and congenitally from mother to infant, but these modes of transmission are rarer than the vector-borne transmission.

Among tropical diseases, leishmaniasis ranks seventh in terms of disability adjusted life years (DALYs), a measure of disease burden, and second in terms of mortality [1, 2].

Leishmaniasis is considered endemic in 97 countries and territories worldwide causing an estimated 700,000 to one million new cases annually.

Pentavalent antimonials are the main treatment for cutaneous leishmaniasis in Africa, the Americas, and Asia, and for visceral leishmaniasis in Africa (in combination with other drugs in East Africa) and the Americas.

13.2 Leishmaniasis in humans

There are three main forms of leishmaniasis in humans: cutaneous, the most frequent; mucocutaneous, which can be mutilating and fatal; and visceral, which is almost always fatal without treatment.

Two other forms of leishmaniasis require special attention: post-kala-azar dermal leishmaniasis and *Leishmania*–HIV coinfection.

Jose Antonio Ruiz-Postigo, Saurabh Jain, Department of Control of Neglected Tropical Diseases, WHO, Geneva, Switzerland. postigoj@who.int

https://doi.org/10.1515/9783110668711-013

13.2.1 Cutaneous leishmaniasis (CL)

The clinical features of cutaneous leishmaniasis vary between and within regions, reflecting different species of parasite, immunological status, and the response of the patient.

A "classical" lesion starts as a papule or nodule at the site of the inoculation; it grows slowly, taking at least one week to reach its final size. A crust develops centrally and if it falls away there will be a painless ulcer with raised edge and variable surrounding induration, which heals gradually over months or years, leaving a depressed scar with altered pigmentation. There is, however, a wide variety of clinical presentations such as single nodules with little inflammation, multiple ulcers, lesions severely inflamed, or concomitant lymphadenopathies.

13.2.2 Mucocutaneous leishmaniasis (MCL)

The salient feature of *Leishmania* species that cause mucocutaneous leishmaniasis is that they cause metastasis to the mucosal tissues of the mouth and upper respiratory tract. It can present from several months to 20 or more years after a cutaneous lesion. Nasal lesions are always leading to obstruction of the nostril and, later, perforation of the septum with collapse and broadening of the nose. The pharynx, palate, larynx, trachea, and upper lip may also be affected. In the final stage there is severe mutilation and it almost never heals spontaneously. Intercurrent pneumonia is the commonest cause of death.

13.2.3 Visceral leishmaniasis (VL)

In endemic areas, the disease tends to be relatively chronic. The incubation period ranges from 10 days to over 1 year, and the onset of the disease is usually gradual. The common symptoms are fever, malaise, chills, weight loss, anorexia, and discomfort in the left hypochondrium. Intercurrent infections are common. Darkening of the skin is typically found in India (the Hindi name *kala-azar* means "black fever").

Sporadic visceral leishmaniasis in nonindigenous people who enter an endemic area tends to be acute, with abrupt onset of fever beginning three weeks to two years after exposure. The disease may progress rapidly.

Malnutrition and immunosuppression, notably HIV infection, predispose to clinical disease in people infected by *L. donovani–L. infantum* complex. The disease is almost always fatal without treatment.

13.2.4 Post-kala-azar dermal leishmaniasis (PKDL)

It is commonest in East Africa and on the Indian subcontinent. PKDL usually appears six months to one or more years after apparent cure of visceral leishmaniasis. Hypopigmented or erythematous macules on any part of the body may later become papular or nodular and infiltrative. Lesions are often confused with vitiligo or leprosy. PKDL may heal spontaneously in Africa but rarely, if ever, in India. The diagnosis is usually clinical by the triad of typical rash, its distribution, and the previous history of VL.

13.2.5 *Leishmania* and HIV coinfection

HIV and *Leishmania* reinforce each other in a detrimental manner. Visceral leishmaniasis is more likely to develop in HIV-infected patients and impairs their response to antiretroviral treatment. In profoundly immunosuppressed patients, atypical sites such as the gastrointestinal tract, lungs, and skin may be infected.

Cutaneous leishmaniasis in AIDS patients shows multiple, polymorphic, and relapsing lesions.

13.3 Epidemiological aspects

13.3.1 Burden of leishmaniasis

Leishmaniasis is considered endemic in 97 countries and territories worldwide, causing an estimated 700,000 to one million new cases annually.

In 2018, more than 95% of new visceral leishmaniasis cases reported to WHO occurred in 10 countries: Brazil, China, Ethiopia, India, Iraq, Kenya, Nepal, Somalia, South Sudan, and Sudan (Fig. 13.1). Over 85% of new cutaneous leishmaniasis cases occurred in 10 countries: Afghanistan, Algeria, Bolivia, Brazil, Colombia, Iran (Islamic Republic of), Iraq, Pakistan, the Syrian Arab Republic, and Tunisia (Fig. 13.2) [3].

13.3.2 Major foci and human behavior

13.3.2.1 Mediterranean Basin

In the Mediterranean Basin, visceral leishmaniasis is a zoonosis caused by *L. infantum*. Houses and their surroundings form the microfoci where human–fly contact occurs and where transmission takes place. Dogs are the proven reservoir of *L. infantum*, but other animals attract sand flies into the houses as well.

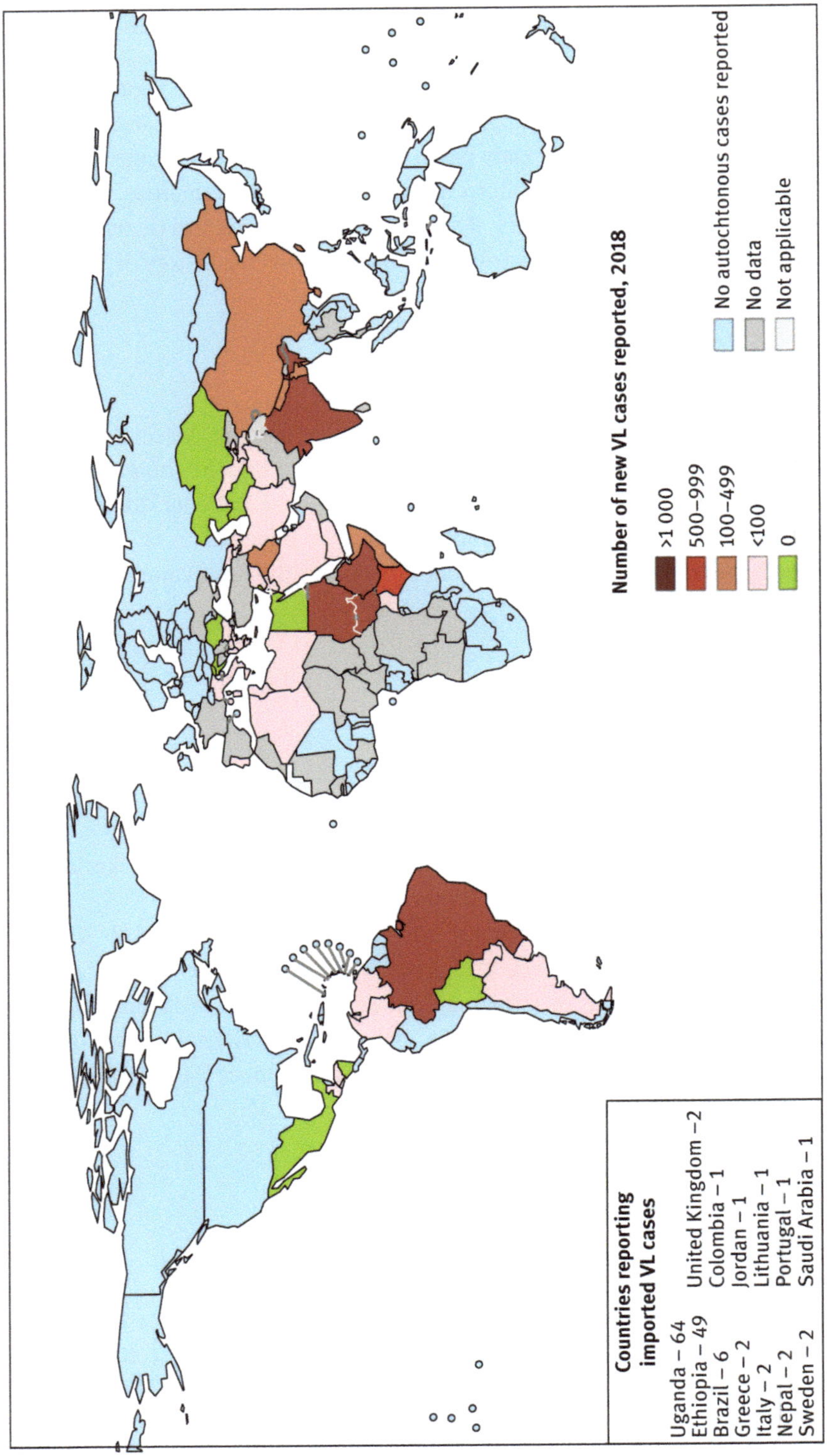

Fig. 13.1: Status of endemicity of visceral leishmaniasis worldwide, 2018.

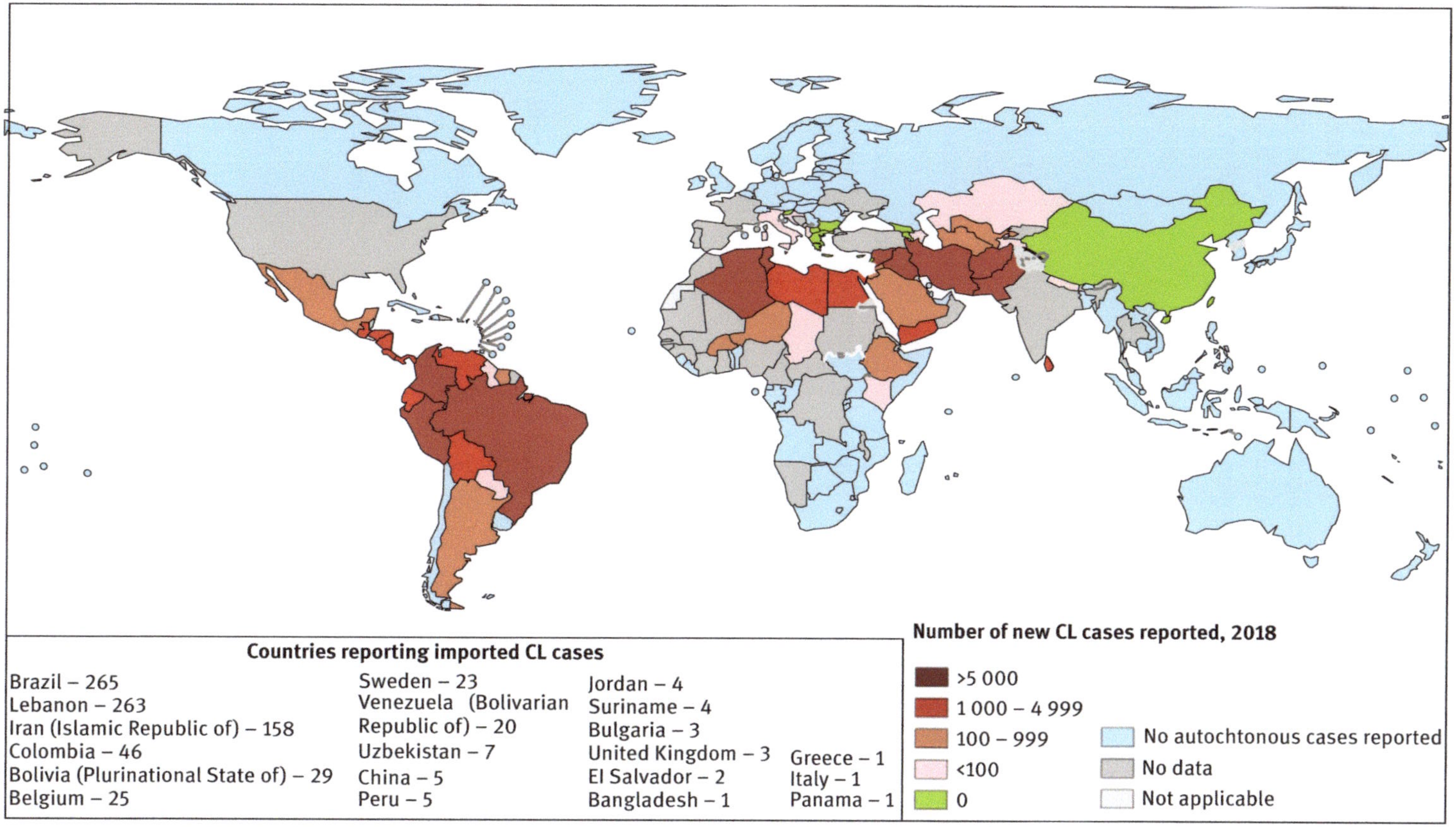

Fig. 13.2: Status of endemicity of cutaneous leishmaniasis worldwide, 2018.

L. infantum transmission foci are also found in Afghanistan, the Islamic Republic of Iran, Pakistan, and Central Asia.

L. major and *L. tropica* are the main causative agents of zoonotic and anthroponotic cutaneous leishmaniasis, respectively. The risk may be increased due to man-made ecological changes (e.g., agricultural projects or irrigation systems extended). Transmission to humans is favored by sleeping outside without a bednet. *L. major* is usually found in rural areas, while *L. tropica* is found in urban and periurban areas.

13.3.2.2 Southeast Asia

In the Indian subcontinent, there are anthroponotic foci of visceral leishmaniasis due to *L. donovani* that occurs in Bangladesh, India, and Nepal. The transmission takes place in agricultural villages where houses are frequently constructed with mud walls and earthen floors, and cattle and other livestock are kept close to human dwellings. Sleeping outside or on the ground may increase the risk of infection. Sporadic cases of local transmission have been reported from Bhutan and Thailand.

In Sri Lanka, where cutaneous leishmaniasis is the most prevalent form of the disease, cattle breeding and living near paddy fields were associated with increased transmission of *L. donovani* [4].

13.3.2.3 East Africa

L. donovani is the species causing visceral leishmaniasis. While sporadic sylvatic zoonotic visceral leishmaniasis transmission is well recognized, sustained peridomestic and domestic cycles in villages have also been demonstrated. Large outbreaks are thought to be due to anthroponotic transmission [5].

L. aethiopica causes zoonotic cutaneous leishmaniasis mainly in areas where there is increased human–fly contact in the proximity to the hyraxes [6].

13.3.2.4 The Americas

Visceral leishmaniasis in the Americas results from zoonotic transmission of *L. infantum*. The habit of keeping dogs and other domestic animals inside the house is thought to promote human infection.

The epidemiology of cutaneous leishmaniasis in the Americas is complex. The transmission cycles are predominantly zoonotic, but the reservoir hosts vary by species and location and in many cases are not fully known. Occupational exposure, such as rubber tapping, military operations, or road constructions, remains important to acquire the infection. More recently widespread deforestation has led to a rapid increase in cases and periurban and even urban transmission.

13.3.3 Socioeconomic factors

Poverty increases the risk for leishmaniasis. Poor housing, lack of waste management, or open sewerage may increase sand fly breeding and resting sites, as well as their access to humans [7].

Economically or social security–driven migration may result in nonimmune individuals entering areas with transmission [8].

13.3.4 Malnutrition

Poor protein-energy, iron, vitamin A, and zinc nutritional status increases the risk that an infection will progress to clinically manifest visceral leishmaniasis or mucocutaneous leishmaniasis.

13.3.5 Population movements

Epidemics of both visceral and cutaneous leishmaniasis are often associated with migration and the introduction of nonimmune people into areas with existing transmission cycles.

13.3.6 Environmental changes

In most endemic regions, leishmaniasis is characterized by a patchy distribution with discrete transmission foci. Natural changes or man-made modifications can result in either an increase or a decrease in the incidence of the disease.

Urbanization, incursion of agricultural farms, and settlements into forested areas or suburbs extending to areas with a high density of rodent reservoir hosts take people close to the transmission zones where increased incidences of cutaneous leishmaniasis have been reported [9, 10].

13.3.7 Climate change

Changes in temperature, rainfall, and humidity can have strong effects on the ecology of *Leishmania* vectors and reservoir hosts by altering their distribution and influencing their survival and population sizes.

Drought, famine, and floods could lead to massive displacement and migration of people to areas with transmission of leishmaniasis.

13.3.8 Periodic fluctuations in incidence of disease

There are marked seasonal and interannual fluctuations in incidence of leishmaniasis in many settings. The predominant source of fluctuation in transmission is the annual cycle of sand flies. The peak transmission rate is the product of vector abundance and infection rate.

Because the incubation period varies widely, from less than one month to more than two years, human incidence may not be clearly linked to the seasonal transmission rate.

The WHO Global Health Observatory [11] and country profiles [12] on leishmaniasis provide examples of interannual and seasonal fluctuations from highly endemic countries.

13.4 Pentavalent antimonials and human leishmaniasis

13.4.1 Overview

Pentavalent antimonials have been extensively used for the treatment of both cutaneous and visceral leishmaniasis for over seven decades. The two pentavalent antimonials currently in use in humans are meglumine antimoniate and sodium stibogluconate. They are chemically similar, and their toxicity and efficacy are related to their antimonial content. Meglumine antimoniate (N-methyl glucamine antimoniate) solution contains 8.1% Sb(V) (81 mg mL^{-1}), whereas sodium stibogluconate contains 10% Sb(V) (100 mg mL^{-1}). The efficacy of these compounds varies depending on the endemic region, the clinical form, and the *Leishmania* species. Due to high-level toxicity and other side effects, quality assurance of antimonials is a critical concern as substandard medicines can cause severe toxicity and death.

Common side effects are anorexia, vomiting, nausea, abdominal pain, malaise, myalgia, arthralgia, headache, metallic taste, and lethargy [13]. Electrocardiographic changes, which are generally reversible, depend on the dose and duration of treatment, the commonest being T-wave inversion, a prolonged Q–T interval, and arrhythmia. Cardiotoxicity and sudden death are serious, but uncommon, side effects. Prolongation of a corrected Q–T interval (>0.5 s) signals the likely onset of serious and fatal cardiac arrhythmia. Pancreatic enzyme concentrations are commonly raised, but clinical pancreatitis is uncommon. Elevated liver enzyme concentrations, leukopenia, anemia, and thrombopenia are not uncommon. Patients should be monitored by serum chemistry, complete blood counts, and electrocardiography. If serious side effects arise (in most cases hepato- or cardiotoxicity), the drug should be changed [14].

Pentavalent antimonials are not recommended in pregnant women [15], HIV-*Leishmania* coinfected patients [16], and patients above 50 years old. According to some manufacturers, they should also be avoided in patients with significant renal, cardiac, and hepatic impairments and in patients who have experienced a serious adverse reaction to a previous dose [17].

Health personnel intending to use pentavalent antimonials should follow manufacturer's instructions before administering the product.

13.4.2 Overdose

According to manufacturer product information, excessive overdose may lead to liver damage (serious jaundice), kidney damage (acute renal insufficiency), heart disorders (bradycardia, Q–T interval prolongation, T-wave inversion or flattening), hematopoietic disorders (anemia, agranulocytosis), and neurological disorders (polyneuritis) [18].

13.4.3 Interactions

No interactions have been reported. However, there have been some reports of cardiac arrhythmias and sudden deaths when amphotericin B deoxycholate was administered soon after sodium stibogluconate for retreatment of visceral leishmaniasis.

13.4.4 Mechanism of action and resistance

Pentavalent antimonials require a competent immune status of leishmaniasis patients for drug efficacy. Immunosuppressed patients due to HIV [19], transplantation, and rheumatology-, hematology-, and oncology-related conditions [20, 21] present higher relapse rates.

It is now generally accepted that all pentavalent antimonials are prodrugs that require biological reduction to the trivalent form [Sb(III)] for antileishmanial activity [22, 23].

Although resistance has been reported from Africa [24, 25] and Brazil [26], the most relevant context where resistance to pentavalent antimonials made them to be abandoned was the development of parasite resistance to these compounds in the Indian subcontinent for the treatment of visceral leishmaniasis. Starting in the 1990s, up to a 65% relapse rate led to a change in the treatment policy to other antileishmanial medicines in the Southeast Asia region [27].

Leishmania resistance to pentavalent antimonials has been attributed to several mechanisms such as a single amino acid substitution (H451Y) in *Leishmania* calcium-dependent kinase SCAMK [28], aquaglyceroporin 1-encoding gene (AQP1) related to the antimony resistance marker (ARM) 56/ARM58 upregulation [29], and increased thiol levels [30].

The presence of resistance has led to the exploration of combination therapies and alternative medicines for the treatment of leishmaniasis.

13.4.5 Pharmacokinetics

Following intramuscular or intravenous administration, pentavalent antimonials are excreted rapidly through kidneys, bringing the blood antimony levels to less than 10% of the peak levels within the first 12 hours.

13.4.6 Combination therapy

Pentavalent antimonials have been used in combination with other medicines (e.g., pentoxifylline, allopurinol, paromomycin, immunomodulators) and physical methods (e.g., cryotherapy) for the treatment of cutaneous leishmaniasis. There is still, however, a need for more robust evidence as to which combination to strongly recommend depending on the *Leishmania* species [31, 32].

For visceral leishmaniasis, the combination of pentavalent antimonials with paromomycin has been tested for over 30 years in East Africa [33, 34] and it was recommended as the first-line treatment in that region at the last WHO Expert Committee in 2010 [35].

13.4.7 Use of pentavalent antimonials by leishmaniasis type

13.4.7.1 Visceral leishmaniasis caused by *L. donovani* in East Africa and Yemen

The first-line treatment is the combination of pentavalent antimonials (20 mg Sb(V) per kg body weight per day intramuscularly or intravenously) plus paromomycin (15 mg [11 mg base] per kg body weight per day intramuscularly) for 17 days [36].

This combination is considered effective and safe, with an initial cure rate in primary VL patients of 95%, 2% with serious adverse event, and a mortality of 1% [37]. The relapse rate in treated patients with visceral leishmaniasis is 10% [38].

When paromomycin is not available pentavalent antimonials can be used as monotherapy at the same dose but for 30 days instead of 17 days [39].

In East Africa, pentavalent antimonials are contraindicated in relapse cases, severely ill patients, pregnant women, HIV-*Leishmania* coinfected patients, or when toxicity is not tolerated by the patient. In these situations, the second-line treatment should be used.

13.4.7.2 Post-kala-azar dermal leishmaniasis in East Africa

The WHO Expert Committee in 2010 recommended pentavalent antimonials for 30–60 days [40]. However, a more recent retrospective study has shown more favorable outcomes when pentavalent antimonials were given in combination with paromomycin for 17 days [41].

13.4.7.3 Visceral leishmaniasis caused by *L. infantum* in the Mediterranean Basin, Middle East, Central Asia, and South America

Although the first-line treatment recommended by WHO is the liposomal amphotericin B [35], the difficulties for many countries in Africa, Asia, and South America in obtaining that medicine make pentavalent antimonials the actual treatment in that situation [42, 43].

13.4.7.4 Cutaneous leishmaniasis caused by *L. major*, *L. tropica*, *L. aethiopia*, and *L. infantum*

Countries where CL due to those species is highly endemic, that is, in Africa and Asia, are mainly using pentavalent antimonials as the first-line treatment [44].

In general terms, for mild infections the medicine is used as intralesional injections, 1–5 mL per session every 3–7 days, 1–5 sessions. For moderate or severe infections, intramuscularly or intravenously, 20 mg Sb(V) per kg body weight per day for 10–20 days.

13.4.7.5 Cutaneous leishmaniasis in the Americas

Most *Leishmania* infections are treated with pentavalent antimonials, 10–20 mg Sb(V) per kg body weight per day for 20 days, intramuscularly or intravenously [45]. Recently, the intralesional route has been explored and it seems to be cost-effective [46].

13.4.8 Access to quality-assured pentavalent antimonials

Access to quality-assured pentavalent antimonials is sometimes difficult due to challenges in the manufacturing processes, the supply chain including forecast, the availability of funds for procurement, or the importation regulations and administrative steps required in the countries. There is only one manufacturer for each of the two compounds currently used, and the demand is not steady, which poses difficulties when there are sudden increases in the needs due to outbreaks or when a batch must be recalled. Meglumine antimoniate (Glucantime) is produced by Sanofi and sodium stibogluconate is a generic produced by Albert David.

Although meglumine antimoniate and sodium stibogluconate are supposedly interchangeable, countries tend to have only one of them in their essential medicines list [43, 47] and cannot easily import the other one in case the medicine they usually import is not available. Additionally, some countries depend on external donors in order to access pentavalent antimonials, which may cause fluctuations to place orders and therefore challenges in the manufacturing process.

13.4.9 Way forward

Since they were first used, pentavalent antimonials have saved over four million lives in the treatment of visceral leishmaniasis. Despite their toxicity and the challenges in ensuring access to them, they will still be used for several years because it does not seem that there will be a real alternative to replace them in the short term.

For cutaneous leishmaniasis, most countries cannot implement alternative treatments such as thermotherapy or cryotherapy. Therefore, pentavalent antimonials will also continue to be a pillar to treat tens of thousands of cases every year.

Researchers are investing efforts to find oral treatments and new ways to use pentavalent antimonials to provide better therapies for patients [48–51].

References

[1] GBD. 2015 DALYs and HALE Collaborators. Global, regional, and national disability-adjusted life-years (DALYs) for 315 diseases and injuries and healthy life expectancy (HALE), 1990–2015: A systematic analysis for the Global Burden of Disease Study 2015. Lancet 2016, 388(10053), 1603–1658.

[2] GBD. 2015 Mortality and Causes of Death Collaborators. Global, regional, and national life expectancy, all-cause mortality, and cause-specific mortality for 249 causes of death, 1980–2015: A systematic analysis for the Global Burden of Disease Study 2015. Lancet 2016, 388(10053), 1459–1544.

[3] Ruiz-Postigo JA, Grout L, Jain S. Global leishmaniasis surveillance, 2017–2018, and first report on 5 additional indicators. Wkly Epidemiol Rec 2020, 95, 265–280.

[4] Amarasinghe A, Wickramasinghe SA. Comprehensive review of cutaneous Leishmaniasis in Sri Lanka and identification of existing knowledge gaps. Acta Parasit 2020, 65, 300–309.

[5] Abubakar A, Ruiz-Postigo JA, Pita J, et al Visceral leishmaniasis outbreak in South Sudan 2009–2012: Epidemiological assessment and impact of a multisectoral response. PLoS Negl Trop Dis 2014, 8, e2720.

[6] Henten SV, Adriaensen W, Fikre H, et al. Cutaneous leishmaniasis due to *Leishmania aethiopica*. EClinical Medicine 2019, 6, 69–81.

[7] Okwor I, Uzonna J. Social and economic burden of human leishmaniasis. Am J Trop Med Hyg 2016, 94, 489–493.

[8] Coulborn RM, Gebrehiwot TG, Schneider M, et al. Barriers to access to visceral leishmaniasis diagnosis and care among seasonal mobile workers in Western Tigray, Northern Ethiopia: A qualitative study. PLoS Negl Trop Dis 2018, 12, e0006778.

[9] Gebremichael D. Zoonotic impact and epidemiological changes of leishmaniasis in Ethiopia. Open Vet J 2018, 8, 432–440.

[10] Arce A, Estirado A, Ordobas M, et al. Re-emergence of leishmaniasis in Spain: Community outbreak in Madrid, Spain, 2009 to 2012. Euro Surveill 2013, 18, 20546.

[11] Global Health Observatory (GHO) data on Leishmaniasis (Accessed October 13, 2020, at https://www.who.int/gho/neglected_diseases/leishmaniasis/en/)

[12] Leishmaniasis country profile-Priority countries (Accessed October 13, 2020, at https://www.who.int/leishmaniasis/burden/endemic-priority-alphabetical/en/)

[13] Oliveira LF, Schubach AO, Martins MM, et al. Systematic review of the adverse effects of cutaneous leishmaniasis treatment in the New World. Acta Trop 2011, 118, 87–96.

[14] Marques SA, Merlotto MR, Ramos PM, Marques MEA. American tegumentary leishmaniasis: Severe side effects of pentavalent antimonial in a patient with chronic renal failure. An Bras Dermatol 2019, 94, 355–357.

[15] Mueller M, Balasegaram M, Koummuki Y, Ritmeijer K, Santana MR, Davidson R. A comparison of liposomal amphotericin B with sodium stibogluconate for the treatment of visceral leishmaniasis in pregnancy in Sudan. J Antimicrob Chemother 2006, 58, 811–815.

[16] Diro E, Lynen L, Mohammed R, Boelaert M, Hailu A, van Griensven J. High parasitological failure rate of visceral leishmaniasis to sodium stibogluconate among HIV co-infected adults in Ethiopia. PLoS Negl Trop Dis 2014, 8, e2875.

[17] Pentostam™ (Accessed October 15, 2020, at http://www.pharmaline.co.il/images/newsletter registration/gsk/2013/25.07.13/pentostam_dr.pdf)

[18] Glucantime® summary of product characteristics (Accessed October 15, 2020, at https://www.wipo.int/export/sites/www/research/en/data/sanofi/marketed_products/Glucantime.pdf)

[19] Silva de Lima UR, Vanolli L, Moraes EC, Ithamar JS, Azevedo CMPES. Visceral leishmaniasis in Northeast Brazil: What is the impact of HIV on this protozoan infection? PLoS One 2019, 14, e0225875.
[20] van Griensven J, Carrillo E, López-Vélez R, Lynen L, Moreno J. Leishmaniasis in immunosuppressed individuals. Clin Microbiol Infect 2014, 20, 286–299.
[21] Schwartz T, Jensenius M, Blomberg B, Fladeby C, Mæland A, Pettersen FO. Imported visceral leishmaniasis and immunosuppression in seven Norwegian patients. Trop Dis Travel Med Vaccines 2019, 5, 16.
[22] Simon L, Croft SL, Sundar S, Fairlamb AH. Drug Resistance in Leishmaniasis. Clin Microbiol Rev 2006, 19, 111–126.
[23] Frézard F, Demicheli C, Ribeiro RRR. Pentavalent antimonials: New perspectives for old drugs. Molecules 2009, 14, 2317–2336.
[24] Ritmeijer K, Davidson RN. Royal Society of Tropical Medicine and Hygiene joint meeting with Médecins Sans Frontières at Manson House, London, 20 March 2003: Field research in humanitarian medical programmes. Médecins Sans Frontières interventions against kala-azar in the Sudan, 1989–2003. Trans R Soc Trop Med Hyg 2003, 97, 609–613.
[25] Ben Salah AB, Ben Ismail R, Amri F, et al. Investigation of the spread of human visceral leishmaniasis in central Tunisia. Trans R Soc Trop Med Hyg 2000, 94, 382–386.
[26] Bryceson AD, Chulay JD, Ho M, et al. Visceral leishmaniasis unresponsive to antimonial drugs: I, clinical and immunological studies. Trans R Soc Trop Med Hyg 1985, 79, 700–704.
[27] Sundar S, More DK, Singh MK, et al. Failure of Pentavalent Antimony in Visceral Leishmaniasis in India: Report from the Center of the Indian Epidemic. Clin Infect Dis 2000, 31, 1104–1107.
[28] Vergnes B, Gazanion E, Mariac C, et al. A single amino acid substitution (H451Y) in Leishmania calcium-dependent kinase SCAMK confers high tolerance and resistance to antimony. J Antimicrob Chemother 2019, 74, 3231–3239.
[29] Rugani JN, Gontijo CMF, Frézard F, Soares RP, Monte-Neto RLD. Antimony resistance in Leishmania (Viannia) braziliensis clinical isolates from atypical lesions associates with increased ARM56/ARM58 transcripts and reduced drug uptake. Mem Inst Oswaldo Cruz 2019, 114, e190111.
[30] Magalhães LS, Bomfim LG, Mota SG, et al. Increased thiol levels in antimony-resistant Leishmania infantum isolated from treatment-refractory visceral leishmaniasis in Brazil. Mem Inst Oswaldo Cruz 2018, 113, 119–125.
[31] Berbert TRN, de Mello TFP, Wolf Nassif P, et al. Pentavalent antimonials combined with other therapeutic alternatives for the treatment of cutaneous and mucocutaneous Leishmaniasis: A systematic review. Dermatol Res Pract 2018, 9014726.
[32] Husein-ElAhmed H, Gieler U, Steinhoff M. Evidence supporting the enhanced efficacy of pentavalent antimonials with adjuvant therapy for cutaneous leishmaniasis: A systematic review and meta-analysis. J Eur Acad Dermatol Venereol 2020, 34, 2216–2228.
[33] Melaku Y, Collin SM, Keus K, Gatluak F, Ritmeijer K, Davidson RN. Treatment of kala-azar in southern Sudan using a 17-day regimen of sodium stibogluconate combined with paromomycin: A retrospective comparison with 30-day sodium stibogluconate monotherapy. Am J Trop Med Hyg 2007, 77, 89–94.
[34] Chunge CN, Owate J, Pamba HO, Donno L. Treatment of visceral leishmaniasis in Kenya by aminosidine alone or combined with sodium stibogluconate. Trans R Soc Trop Med Hyg 1990, 84, 221–225.
[35] WHO/Expert Committee on the Control of Leishmaniases. Control of the Leishmaniases: Report of a Meeting of the WHO Expert Committee on the Control of Leishmaniases. Geneva, Switzerland, WHO, 2010.

[36] Atia AM, Mumina A, Tayler-Smith K, et al. Sodium stibogluconate and paromomycin for treating visceral leishmaniasis under routine conditions in eastern Sudan. Trop Med Int Health 2015, 20, 1674–1684.
[37] Kimutai R, Musa AM, Njoroge S, et al. Safety and effectiveness of sodium stibogluconate and paromomycin combination for the treatment of visceral Leishmaniasis in Eastern Africa: Results from a pharmacovigilance programme. Clin Drug Investig 2017, 37, 259–272.
[38] Musa A, Khalil E, Hailu A, et al. Sodium stibogluconate (SSG) & paromomycin combination compared to SSG for visceral leishmaniasis in East Africa: A randomised controlled trial. PLoS Negl Trop Dis 2012, 6, e1674.
[39] Gebreyohannes EA, Bhagvathula AS, Abegaz TM, Seid MA. Treatment outcomes of visceral leishmaniasis in Ethiopia from 2001 to 2017: A systematic review and meta-analysis. Infect Dis Poverty 2018, 7, 108.
[40] Zijlstra EE, Musa AM, Khalil EA, el-Hassan IM, el-Hassan AM. Post-kala-azar dermal leishmaniasis. Lancet Infect Dis 2003, 3, 87–98.
[41] Abongomera C, Gatluak F, Buyze J, Ritmeijer KA. Comparison of the effectiveness of sodium stibogluconate monotherapy to sodium stibogluconate and paromomycin combination for the treatment of severe Post Kala Azar dermal Leishmaniasis in South Sudan – A retrospective cohort study. PLoS One 2016, 11, e0163047.
[42] Romero GAS, Costa DL, Costa CHN, et al. Efficacy and safety of available treatments for visceral leishmaniasis in Brazil: A multicenter, randomized, open label trial. PLoS Negl Trop Dis 2017, 11, e0005706.
[43] Adel A, Boughoufalah A, Saegerman C, et al. Epidemiology of visceral leishmaniasis in Algeria: An update. PLoS One 2014, 9, e99207.
[44] WHO/Regional Office for the Eastern Mediterranean. Manual for Case Management of Cutaneous Leishmaniasis in the WHO Eastern Mediterranean Region. Cairo, Egypt, WHO, 2014.
[45] Pan American Health Organization. Leishmaniasis in the Americas: Treatment Recommendations. Washington, D.C., PAHO, 2018.
[46] Brito NC, Machado de Assis TS, Rabello A, Cota G. Intralesional infiltration versus parenteral use of meglumine antimoniate for treatment of cutaneous leishmaniasis: A cost-effectiveness analysis. PLoS Negl Trop Dis 2019, 13, e0007856.
[47] Essential Medicines selection: National Medicines List/Formulary/Standard Treatment Guidelines (Accessed October 14, 2020, at https://www.who.int/selection_medicines/country_lists/en/)
[48] Carvalho SH, Frézard F, Pereira NP, et al. American tegumentary leishmaniasis in Brazil: A critical review of the current therapeutic approach with systemic meglumine antimoniate and short-term possibilities for an alternative treatment. Trop Med Int Health 2019, 24, 380–391.
[49] Alves F, Bilbe G, Blesson S, et al. Recent development of visceral Leishmaniasis treatments: Successes, pitfalls, and perspectives. Clin Microbiol Rev 2018, 31(4), e00048–18.
[50] Serban G. Future Prospects in the treatment of parasitic diseases: 2-Amino-1,3,4-thiadiazoles in Leishmaniasis. Molecules 2019, 24, 1557.
[51] No JH. Visceral leishmaniasis: Revisiting current treatments and approaches for future discoveries. Acta Trop 2016, 155, 113–123.

Koen Oorts

Chapter 14
Ecotoxicity of antimony

14.1 Introduction

This chapter deals with the adverse effects of antimony on biota and ecosystems, with a focus on the aquatic, sediment, and terrestrial environmental compartments. Antimony is a naturally occurring trace element with no known biological function. The ecotoxicity of antimony is less studied compared to other metal(loid)s such as arsenic, cadmium, copper, niquel, lead, selenium, or zinc and only a few relevant and reliable ecotoxicity data are published. Apart from some data published in research journals, most of the ecotoxicity data were generated in the framework of the European chemical regulations (Regulation (EC) No. 1907/2006 of the European Parliament and of the Council on the Registration, Evaluation, Authorisation and Restriction of Chemicals (REACH) and its precursor the European Existing Substances Regulation (793/93/EEC)). Under these regulations, a risk assessment is required to ensure absence of adverse effects on humans and the environment during production and use of chemicals, and a minimum dataset is required to determine the hazard potential of the substances in the environment. For more information on these risk assessments, see chapter 16 on legislating antimony and its compounds.

For metals and metalloids, it is generally accepted that the dissolved metal(loid) ion is the driver for the observed effects [1]. The ecotoxicity assessment of antimony is therefore based on all available data where observed effects can be related to the common dissolved inorganic antimony ion. This approach can be considered valid when (i) differences in solubility among antimony compounds are taken into account, (ii) (eco)toxicity is only affected by the antimony species and not by the counter ions, and (iii) the biological and physico-chemical properties of the receiving environmental or physiological media determine the speciation of the antimony moiety, regardless of the speciation of the compounds released, or there are no major differences in toxicity of different species. A good understanding of the environmental chemistry of antimony in the exposure media is therefore critical for a proper evaluation of the ecotoxicity data and assigning results to the common antimony moiety. In natural systems, antimony is mainly present as Sb(V) in oxic environments and as Sb(III) in anoxic systems [2]. Within the environmental relevant pH range (pH 3 to 8), both Sb(III) and Sb(V) are present as their hydrolyzed species $Sb(OH)_3$ and $Sb(OH)_6^-$, respectively [2]. Oxidation of Sb(III) to Sb(V) is expected to be fast in both natural water and soil in the presence of amorphous iron and manganese oxyhydroxides and

Koen Oorts, ARCHE Consulting, Belgium. koen.oorts@arche-consulting.be

https://doi.org/10.1515/9783110668711-014

is usually completed within 1 week [3, 4]. A summary of the ecotoxicity data identified for various inorganic antimony substances is presented together with a discussion on the relevance of these data for the toxicity of the common antimony moiety.

14.2 Quality criteria for ecotoxicological data

A first step in the ecotoxicological assessment of substances is a thorough quality screening of the data available. This screening covers both the relevance of the data for toxicity of the antimony ion and the quality and reliability of the test methods and results. A detailed description of the test methodology, measurements, and data analysis performed is therefore required before data can be considered for determining the ecotoxicological hazard profile of a substance. This step is especially important for metal(loid)s because the test conditions and physicochemical properties of the test media strongly affect the bioavailability and hence the ecotoxicity results [1]. In summary, the following criteria were considered for assessing the relevance and reliability of test results for ecotoxicity of antimony:

- The test substance must be reported and must be a high-purity substance without presence of impurities affecting the effects. There should be no indications for combined toxicity from other substances present in the test medium. The test substance must be well mixed with the test medium, with sufficient equilibration time in sediment and soil to ensure proper partitioning between the solid and liquid phases before the start of the bioassays. Tests for metal(loid)s should generally be conducted with high-purity soluble metal salts. If tests are being conducted with sparingly soluble compounds (e.g., Sb_2O_3)[1], preparation of exposure solutions should incorporate a proper equilibration time after spiking. Toxicity studies from various inorganic antimony substances are accepted provided that there is evidence that there are no confounding factors (change in pH, ionic strength, toxicity of other constituents) and that all effects could be attributed to antimony alone and there is no concern on differences in bioavailability and ecotoxicity of the antimony ion. For this reason, data generated with the soluble organic antimony salts (e.g., antimony potassium tartrate) are not retained for assessment of the hazard properties of the antimony moiety because complexing of antimony by the organic moiety is likely to alter its environmental fate, bioavailability, and toxicity [5].
- Test must be performed in natural or standard artificial (e.g., OECD) waters, sediments, and soils. The physico-chemical properties (e.g., pH, hardness, dissolved organic carbon content for water, pH, clay content and organic matter content for sediment and soil) of the test media can have a significant impact on

1 In this chapter, Sb_2O_3, $SbCl_3$ and $SbCl_5$ refer to solid substances.

the speciation of antimony, and therefore, they must be properly described and be within the natural boundaries of the environment under study.
- As mentioned earlier, the availability of measured dissolved antimony concentrations is critical in the data selection for toxicity to aquatic organisms in order to correct for potential differences in solubility of the test substances (e.g., $SbCl_3$ versus Sb_2O_3) and to account for potential precipitation reactions making Sb less available for aquatic organisms. Measured dissolved antimony concentrations will avoid overestimation of the dissolved antimony concentration (= bioavailable fraction) and hence underestimation of the antimony toxicity. Although the total antimony concentration is the basis for the hazard assessment in sediment and soil, and total antimony concentration is not a measure of its bioavailability, measured total antimony concentrations or analytically confirmed nominal doses are required for toxicity studies in sediments and soils. By preference, the dissolved antimony concentration in pore water is measured as a better indication for the toxic pressure during the test.

In addition, the ecotoxicological endpoint studied should be relevant at the population level. Preferred endpoints are mortality, growth, reproduction, hatching, development, and so on. Functional variables such as carbon or nitrogen transformation are also considered for effects on microbial processes. Results for endpoints such as biomarkers or changes in enzymatic processes, which do not have a clear link to the performance of an individual organism or population, are not considered. The exposure duration should be reported and should be representative for acute or (sub)chronic exposure or focusing on sensitive life stages. The test conditions should be suitable for the test organism and validity criteria for endpoints such as mortality, growth, and reproduction need to be fulfilled. Information on dose-response and statistics must be presented and statistically derived LC_{50} (i.e., concentration at which 50% mortality is found), EC_{50} (i.e., concentration at which 50% effect is found), EC_{10} (i.e., concentration at which 10% effect is found), or no observed effect concentration (NOEC, i.e., highest dose tested without a significant negative effect on the response) are retained.

14.3 Ecotoxicity data for antimony

14.3.1 Antimony toxicity in the aquatic environment

Reliable acute toxicity data identified for the effect of antimony on aquatic organisms are mainly derived from tests with $SbCl_3$, with also few data for Sb_2O_3 and $KSb(OH)_6$ (Tab. 14.1). Most reliable chronic toxicity studies are also based on $SbCl_3$ with some additional studies for Sb_2O_3 (Tab. 14.2). Reliable acute and chronic results were

Tab. 14.1: Acute antimony toxicity to aquatic organisms (all results based on measured antimony concentrations).

Test substance	Test organism	Endpoint	Exposure period / h	Effect parameter	Value / mg Sb L^{-1}	Reference
Fish						
$SbCl_3$	*Pimephales promelas*	Mortality	96	LC_{50}	14.4	[6]
$SbCl_3$	*Pimephales promelas*	Mortality	96	LC_{50}	21.9	[7]
$SbCl_3$	*Oncorhynchus mykiss*	Mortality	96	LC_{50}	>25.7	[6]
$SbCl_3$	*Ictalurus punctatus*	Mortality	96	LC_{50}	24.6	[8]
$KSb(OH)_6$	*Pargus major*, marine water	Mortality	96	LC_{50}	6.9	[9]
$SbCl_3$	*Pargus major*, marine water	Mortality	96	LC_{50}	12.4	[9]
Invertebrates						
$SbCl_3$	*Chlorohydra viridissimus*	Mortality	96	LC_{50}	1.77	[8]
$SbCl_3$	*Hyalella azteca*	Mortality	96	LC_{50}	21.6	[8]
$SbCl_3$	*Chironomus tentans*	Mortality	96	LC_{50}	4.1	[8]
$SbCl_3$	*Physa heterostropha*	Mortality	96	LC_{50}	14.2	[8]
$SbCl_3$	*Hydra oligactis*	Mortality	96	LC_{50}	1.95	[8]
$SbCl_3$	*Daphnia magna*	Mortality	48	LC_{50}	12.1	[7]
$SbCl_3$	*Gammarus* sp.	Mortality	96	LC_{50}	>25.7	[6]
$SbCl_3$	*Lumbriculus variegatus*	Mortality	96	LC_{50}	>25.7	[6]
$SbCl_3$	*Pycnopsyche* sp.	Mortality	96	LC_{50}	>25.7	[6]
$SbCl_3$	*Macrobrachium nipponense*	Mortality	96	LC_{50}	1.96	[10]
Algae and aquatic plants						
$SbCl_3$	*Pseudokirchnerella subcapitata*	Biomass; growth rate	72	EC_{50}	>36.6	[11]

Tab. 14.1 (continued)

Test substance	Test organism	Endpoint	Exposure period / h	Effect parameter	Value / mg Sb L^{-1}	Reference
Sb_2O_3	*Pseudokirchnerella subcapitata*	Biomass; growth rate	72	EC_{50}	>2.4	[12]
$KSb(OH)_6$	*Pseudokirchnerella subcapitata*	Inhibition of photosystem II	96	EC_{50}	>336	[13]
$SbCl_3$	*Lemna minor*	Growth (frond number)	96	EC_{50}	>25.5	[6]
$KSb(OH)_6$	*Lemna minor*	Growth (frond number)	168	EC_{50}	>219	[13]
$KSb(OH)_6$	*Wolffia arrhiza*	Growth (frond number)	168	EC_{50}	>219	[13]

Tab. 14.2: Chronic antimony toxicity to aquatic organisms (all results based on measured antimony concentrations).

Test substance	Test organism	Endpoint	Exposure period	Effect parameter	Value / mg Sb L^{-1}	Reference
Fish						
$SbCl_3$	*Pimephales promelas*, embryo stage	Length Weight Post-hatch survival	28 d	NOEC	1.13 2.31 4.5	[7]
Sb_2O_3	*Pimephales promelas*, embryo stage	Length Weight Post-hatch survival	30 d	NOEC	≥0.0075 ≥0.0075 ≥0.0075	[14]
Invertebrates						
$SbCl_3$	*Daphnia magna*	Reproduction	21 d	NOEC	1.74	[15]
$SbCl_3$	*Daphnia magna*	Reproduction	7 d	NOEC	3.9	[7]
Algae and aquatic plants						
$SbCl_3$	*Pseudokirchnerella subcapitata*	Biomass; growth rate	72 h	NOEC	2.11	[11]
Sb_2O_3	*Pseudokirchnerella subcapitata*	Biomass; growth rate	72 h	EC_{10}	2.4	[12]
$SbCl_3$	*Lemna minor*	Growth (frond number)	96 h	NOEC	12.5	[6]

identified for three trophic levels of aquatic organisms (primary producers, represented by algae or aquatic plants, primary consumers or herbivores, represented by invertebrates, and secondary consumers or predators, represented by fish).

The redox speciation of antimony was not measured during the exposure phase of these ecotoxicity tests, but the solubility and redox speciation of dissolved antimony from various antimony compounds, including sodium hexahydroxoantimonate ($NaSb(OH)_6$), antimony metal (Sb), antimony trioxide (Sb_2O_3), and antimony trichloride ($SbCl_3$), were studied for up to 28 days in standard OECD media at pH 6 and 8.5 [5]. Dissolved antimony from pentavalent antimony compounds remained in the pentavalent valence state after 28 days, whereas significant oxidation of Sb(III) to Sb(V) was evident for the trivalent inorganic antimony compounds. This transformation was not completed after 28 days, most likely because of the absence of any oxides in the test systems. Based on these findings, it can be concluded that acute/chronic toxicity tests with an Sb(V) substance as testing material reflect the toxicity of the pentavalent form, while toxicity tests with an Sb(III) substance as testing material reflect the toxicity of a mixture of both the trivalent and pentavalent form. Given the duration of the ecotoxicity tests (48 h – 30 d), equilibrium between both valence states is not expected to be reached during these tests. There is only one reliable test identified that studied the effect of different antimony substances in comparable conditions [9]. They observed 96-h LC_{50} values of 6.9, 12.4, and 0.93 mg Sb L^{-1} for toxicity of $KSb(OH)_6$, $SbCl_3$, and $SbCl_5$, respectively, to the marine fish *Pargus major* (red seabream), indicating a higher toxicity for pentavalent antimony substances compared to trivalent antimony substances. However, the results from using the pentavalent $SbCl_5$ were questionable and were not considered reliable. The higher toxicity of Sb(V) compared to Sb(III) could not be confirmed by the other reliable toxicity data identified and no evidence has been identified in scientific literature for Sb(V) being more toxic than Sb(III) for the environment. On the contrary, there seems to be more evidence that Sb(III) is more toxic than Sb(V) [16–18]. It can therefore be concluded that the use of test data that were generated with trivalent antimony compounds will be sufficiently protective for all antimony compounds: If equilibrium between Sb(III) and Sb(V) in the test medium is not reached at the time of exposure – and assuming that Sb(V) is less (or equally) ecotoxic than Sb(III) – then the outcome of such tests will not underestimate the toxicity of dissolved antimony at equilibrium.

The reliable acute toxicity data of Sb vary between 1.77 mg Sb L^{-1} (96-h LC_{50} for toxicity of $SbCl_3$ to the freshwater invertebrate *Chlorohydra viridissima* [8]) and >336 mg Sb L^{-1} (96-h LC_{50} for toxicity of $KSb(OH)_6$ to the freshwater alga *Pseudokirchnerella subcapitata* [13]) (Tab. 14.1). There are no clear differences in toxicity of antimony to different trophic levels. Chronic toxicity data for effect of antimony to aquatic organisms range from 1.13 mg Sb L^{-1} (28-d NOEC for adverse effect of $SbCl_3$ on growth and post-hatch survival of freshwater fish *Pimophelas promelas* embryos [7]) and 12.5 mg Sb L^{-1} (96-h NOEC for adverse effect of $SbCl_3$ on growth of the freshwater plant *Lemna minor* [6]) (Tab. 14.2). As for acute effects, there are no clear differences in toxicity of Sb to aquatic organisms belonging to different trophic levels.

Dissolution of Sb(III) salts such as $SbCl_3$ and subsequent hydrolysis of the Sb(III) ions released generates protons and increases the salt concentration by the anions released:

$$SbCl_3 + 3\,H_2O \rightarrow Sb(OH)_3 + 3\,H^+ + 3\,Cl^- \tag{14.1}$$

The oxidation of Sb(III) to Sb(V) only releases one additional proton per antimony atom:

$$Sb(OH)_3 + 2\,H_2O + 0.5\,O_2 \rightarrow Sb(OH)_6^- + H^+ \tag{14.2}$$

In the absence of any buffering reactions and not accounting for oxidation, an antimony concentration of up to 20 mg L^{-1} added as a Sb(III) salt would correspond to a theoretical pH value of 3.3. Such a decrease in pH was however not observed at these antimony concentrations in the test media where pH was measured and the pH decrease was limited up to maximum 0.5 units for antimony concentrations up to 20 mg L^{-1} [7, 19]. The chloride concentration at this antimony concentration (0.5 mM Cl) is still about a factor 10 smaller than the chloride concentration in standard reconstituted water recommended for toxicity tests (up to 4 mM Cl). It can therefore be concluded that none of the toxicity data for the aquatic compartment are confounded by the additions of counter ions (i.e., chloride) and/or protons resulting from the use of $SbCl_3$.

The rather limited dataset for antimony toxicity to aquatic organisms does not allow conclusions on the impact of water chemistry (pH, dissolved organic matter content, etc.) on the bioavailability and toxicity of antimony in the aquatic environment. Because the lowest reliable acute toxicity value is >1 but ≤10 mg L^{-1}, antimony substances are not classified as acute toxic for the aquatic environment under the European CLP Regulation (Regulation (EC) No. 1272/2008 on classification, labeling, and packaging of substances and mixtures) and as Acute Category 2 according to the Globally Harmonized System of Classification and Labelling of Chemicals (GHS, http://www.unece.org/trans/danger/publi/ghs/ghs_welcome_e.html). The lowest reliable chronic toxicity value is >1 mg L^{-1}, which would not result in a classification for chronic toxicity to aquatic organisms under GHS and CLP. Despite that the available toxicity data are conclusive and not sufficient for classification, antimony compounds have a harmonized classification as Aquatic Chronic 2 under CLP (see also chapter 16 on legislating antimony and its compounds).

According to the risk assessment practices under the European Union regulations, a Predicted No-Effect Concentration (PNEC) was based on the lowest chronic NOEC value and an assessment factor to take into account the remaining uncertainty due to, for example, interspecies variation or laboratory to field extrapolation. Reliable chronic toxicity data for antimony were available from fish, invertebrates, and algae, and the corresponding assessment factor for freshwater is 10, resulting in a PNEC for freshwater of 0.113 mg Sb L^{-1}. In the absence of chronic data for marine organisms, assessment factor of 100 must be applied on the lowest chronic NOEC value for freshwater organisms, resulting in a PNEC for marine water of 0.0113 mg Sb L^{-1}.

14.3.2 Antimony toxicity to sediment organisms

Three reliable and relevant chronic sediment toxicity tests with different invertebrate species are available [20–22] (Tab. 14.3). The test species all have different exposure routes, feeding habits, and ecological niches: (1) the bottom-dwelling *Hyalella azteca* (crustacean) is a surface deposit and filter feeder, (2) *Chironomus riparius* (insect) burrows within the sediment with a combined surface and subsurface feeding behaviour, and (3) *Lumbriculus variegatus* (oligochaete) is a head-down deposit feeder that feeds well below the sediment-water interface. Reliable NOEC values for the most sensitive endpoint per species were rather constant and varied between 112 (14-d NOEC for growth of *C. riparius* and 28-d NOEC for growth of *L. variegatus*) and 124 mg kg^{-1} dw (28-d NOEC for growth of *H. azteca*) (Tab. 14.3).

Tab. 14.3: Antimony toxicity to sediment organisms (all results based on measured antimony concentrations).

Test substance	Test organism	Endpoint	Exposure period / d	Effect parameter	Value / mg Sb kg^{-1} dw	Reference
$SbCl_3$	*Hyalella azteca*	Growth (weight	28	NOEC	124	[20]
		or length)	42	NOEC	272	
		Mortality	42	NOEC	272	
		Reproduction	42	NOEC	272	
$SbCl_3$	*Chironomus riparius* larvae	Growth (weight)	14	NOEC	112	[21]
		Mortality	14	NOEC	≥636	
		Emergence	28	NOEC	≥636	
$SbCl_3$	*Lumbriculus variegatus*	Growth (weight)	28	NOEC	112	[22]
		Mortality	28	NOEC	172	
		Reproduction	28	NOEC	172	

All tests were performed in artificial sediment spiked with $SbCl_3$. The sediment was equilibrated for one month before the start of the toxicity test. The overlying water was renewed at least three times a week during the test, removing an excess of chloride ions. The pH of the overlying water did not vary more than 0.3 units among treatments during the test. Hardness and oxygen concentration of the overlying water also did not show significant differences between treatments. It is therefore concluded that all effects observed could be attributed to the dissolved antimony concentrations and that results were not confounded by other changes in the test media due to the addition of $SbCl_3$.

According to the risk assessment practices under the European Union regulations, a PNEC was based on the lowest chronic NOEC value and an assessment factor of 10 and 50 for the freshwater and marine sediments, respectively. This resulted in PNEC values of 11.2 and 2.2 mg Sb kg^{-1} dw for freshwater and marine sediment, respectively.

14.3.3 Antimony toxicity in the terrestrial environment

Good quality toxicity data of antimony are available to various soil organisms belonging to the three trophic levels: decomposers, represented by microorganisms; primary producers, represented by plants; and primary consumers, represented by invertebrates. The data are based on soils spiked with $SbCl_3$, $Sb_2(SO_4)_3$, or Sb_2O_3 with mostly ≥1 week equilibration before the start of the toxicity tests (Tab. 14.4). Because of the rapid oxidation of Sb(III) to Sb(V) in soil [4], it can be expected that soil organisms were mainly exposed to Sb(V) forms during the toxicity tests. All results are based on measured or analytically confirmed nominal total antimony concentrations in the soil.

There is a discrepancy between toxicity data from soils amended with either soluble antimony salts, freshly amended Sb_2O_3, or equilibrated Sb_2O_3. The NOEC or EC_{10} values for the most sensitive endpoint per test ranged from 88 to 760 mg Sb kg^{-1} for soils amended with $SbCl_3$, from 60 to 100 mg Sb kg^{-1} for soils amended with $Sb_2(SO_4)_3$ and no toxicity was observed in soils freshly amended with Sb_2O_3 (NOEC values ≥823 or 1,804 mg Sb kg^{-1}). Testing antimony toxicity in soils amended with Sb_2O_3 and equilibrated for 31 weeks before the start of the toxicity tests resulted in NOEC values ranging between 999 and 2,930 mg Sb kg^{-1}. Care must be taken to interpret toxicity data of soluble Sb(III) forms due to potential confounding factors [4, 19]. As explained earlier, application of soluble Sb salts (e.g., $SbCl_3$) not only increases the antimony dose of the soil but also decreases soil pH, due to the hydrolysis of antimony and subsequent oxidation to $Sb(OH)_6^-$, and increases the salinity, due to the chloride or sulfate ions released (eqs. (14.1) and (14.2)). These changes in soil chemistry may also affect the response of the species tested and therefore it is not possible to attribute all observed toxic effects to antimony alone. Amending a soil with $CaCl_2$ to identical chloride concentrations as in an $SbCl_3$-amended soil resulted in identical effects on lettuce shoot yield (Fig. 14.1(a)), leaving no evidence that effects are only related to antimony itself [4, 19]. In addition, when expressing results based on the dissolved antimony

Tab. 14.4: Antimony toxicity to soil organisms (all results based on measured or confirmed nominal antimony concentrations).

Test substance	Equilibration period	Test organism	Endpoint	Exposure period / d	Effect parameter	Value /mg Sb kg^{-1} dw	Reference
Plants							
$SbCl_3$	1 week	*Lactuca sativa*	Shoot yield	24	EC_{10}	222	[23]
Sb_2O_3	1 week	*Lactuca sativa*	Shoot yield	24	NOEC	≥1,804	[23]
Sb_2O_3	31 weeks	*Lactuca sativa*	Shoot yield	17	NOEC	2,930 Pore water: 18.7 mg L^{-1} Estimated total conc. at complete dissolution: 710	[4]
Sb_2O_3	31 weeks	*Hordeum vulgare*	Root elongation	5	NOEC	999 Pore water: 9.7 mg L^{-1} Estimated total conc. at complete dissolution: 370	[4]
Invertebrates							
$Sb_2(SO_4)_3$	3 weeks	*Eisenia fetida*	Mortality Reproduction	21	NOEC NOEC	617 60	[24]
$Sb_2(SO_4)_3$	3 weeks	*Enchytraeus crypticus*	Mortality Reproduction	28	NOEC NOEC	384 100	[24]
$SbCl_3$	1 week	*Enchytraeus albidus*	Mortality Reproduction	42	NOEC NOEC	760 760	[25]

$Sb_2(SO_4)_3$	3 weeks	*Folsomia candida*	Mortality Reproduction	28	NOEC NOEC	100 <100	[26]
$SbCl_3$	1 week	*Folsomia candida*	Mortality Reproduction	28	EC_{10} EC_{10}	1,452 389	[23]
Sb_2O_3	1 week	*Folsomia candida*	Mortality Reproduction	28	NOEC NOEC	≥1,804 ≥1804	[23]
Sb_2O_3	31 weeks	*Folsomia candida*	Mortality Reproduction	28	NOEC NOEC	≥10,119 999 Pore water: 9.7 mg L^{-1} Estimated total conc. at complete dissolution: 370	[26]
Microorganisms							
$SbCl_3$	1 week	Native soil microorganisms	Nitrate formation Potential nitrification rate	28 7	EC_{10} EC_{10}	646 88	[23]
Sb_2O_3	1 week	Native soil microorganisms	Nitrate formation Potential nitrification rate	28 10	NOEC NOEC	≥1,804 ≥1,804	[23]
Sb_2O_3	no	Native soil microorganisms	Nitrate formation	28	NOEC	≥823	[27]
Sb_2O_3	31 weeks	Native soil microorganisms	Nitrate formation Potential nitrification rate	28 7	NOEC NOEC	≥10,119 2,930 Pore water: 18.7 mg L^{-1} Estimated total conc. at complete dissolution: 710	[4]

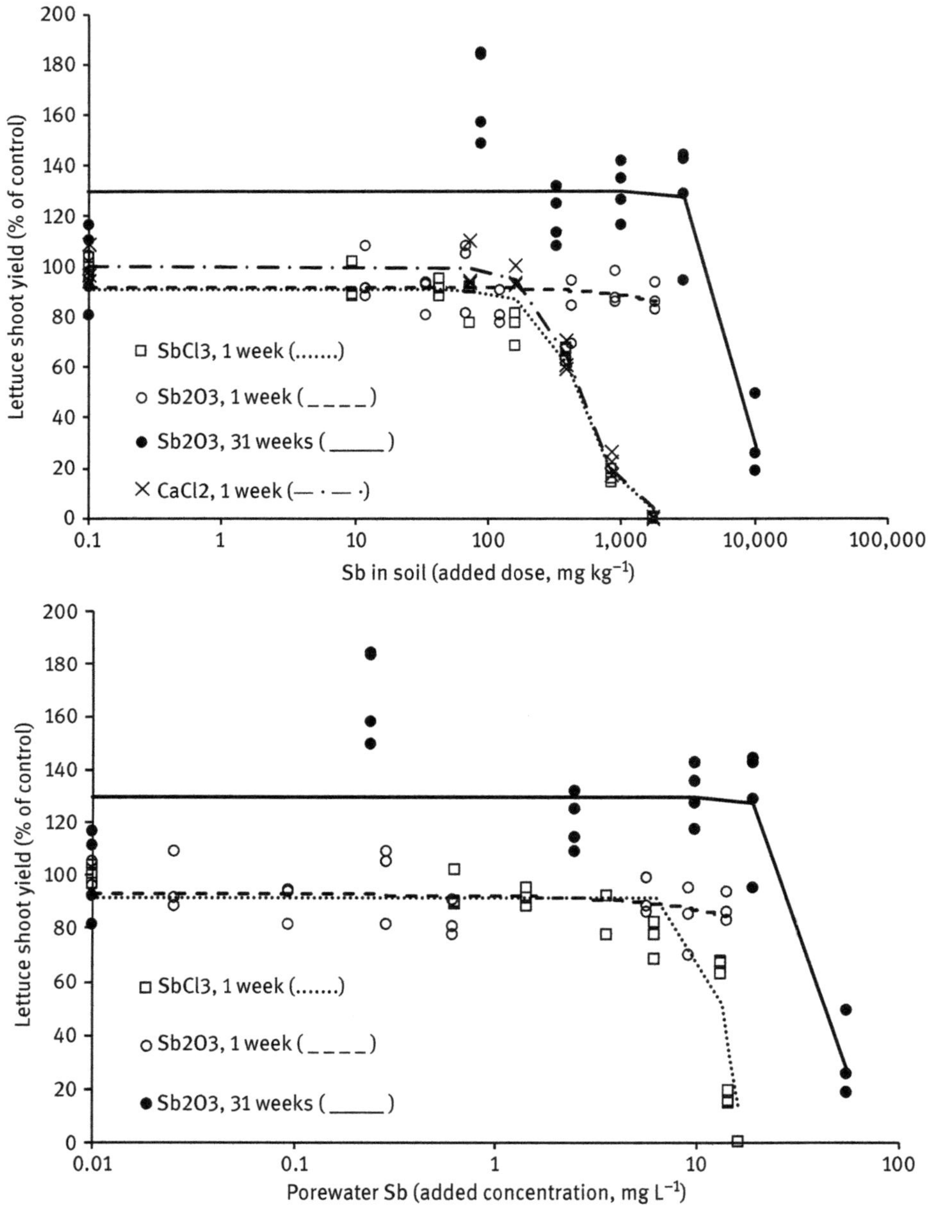

Fig. 14.1: Dose-response relationships of lettuce shoot yield in soils amended with $SbCl_3$ (1 week equilibration before testing), Sb_2O_3 (1 week or 31 weeks equilibration before testing), and $CaCl_2$ (1 week equilibration before testing). The lines are log-logistic dose-response curves fitted to the data. (a) dose expressed as added antimony concentration in soil ($CaCl_2$ was added at equivalent chloride concentrations as the $SbCl_3$ treatments); (b) dose expressed as dissolved antimony concentration in pore water. Data are derived from Oorts et al. [4].

concentration in pore water, which is for metal(loid)s, a better estimate of the bioavailable fraction, higher degree of toxicity, can be seen in the $SbCl_3$ amended soils compared to Sb_2O_3 amended soils, pointing to an additional cause of toxicity, next to the dissolved antimony ions (Fig. 14.1(b)). These confounding factors associated with the applications of soluble salts are also observed for other metals [28, 29] and are becoming more important when the metal ion is not strongly toxic, that is, when larger amounts of salts, and hence also of counter-ions, need to be added to generate toxic effects. Such confounding factors are also more pronounced in laboratory tests that generally use closed pots compared to field conditions, where the excess salts will be naturally leached out to deeper soil layers and groundwater by percolating rainwater [30]. It is therefore clear that toxicity tests with soluble antimony salts in closed systems have little relevance for predicting the toxicity of the antimony ion alone in soil.

The discrepancy between the artefacts related to the use of soluble antimony salts for soil toxicity testing and the accepted use of $SbCl_3$ data for derivation of toxicity thresholds for aquatic and sediment organisms is explained by the larger impacts of the counter-ions on the solution phase in soil compared to water and sediments [19]. This is due to the much smaller volume of solution in soil and hence the larger $SbCl_3$ concentration per volume solution required to observe toxic effect to soil organisms and to the difference in partitioning to the solid phase between antimony and chloride. Taking into account the moisture content of the soil used (28.9%) [4], the EC_{10} value of 222 mg Sb kg^{-1} soil for the effect of $SbCl_3$ on lettuce shoot yield corresponds to a theoretical total dose of 768 mg Sb L^{-1} pore water and 18.9 mM Cl. A large part of the antimony in soil is however adsorbed to the solid phase (the solid-liquid distribution coefficient (K_d) of antimony in soil ranges from 20 to 200 L kg^{-1} [4, 31]) and therefore is not present as dissolved antimony in soil solution. This is not the case for Cl^-, resulting in a higher chloride to antimony ratio in solution during terrestrial bioassays compared with aquatic toxicity tests.

The use of Sb_2O_3 amended soils for toxicity testing does overcome the confounding factors associated with the use of soluble antimony salts because the dissolution reaction does not generate protons and counter-ions:

$$Sb_2O_3 + 3\,H_2O \rightarrow 2\,Sb(OH)_3 \tag{14.3}$$

The difference in toxicity between freshly amended and equilibrated Sb_2O_3 in soil (Tab. 14.4, Fig. 14.1(a)) however indicates that total concentration-based toxicity data derived from soils freshly amended with Sb_2O_3 do also not result in a relevant estimate of antimony toxicity in soil. Sb_2O_3 will only slowly dissolve and transform to soluble antimony forms and the presence of a significant amount of undissolved, and therefore less bioavailable, antimony results in an underestimation of the potential antimony toxicity at full dissolution when based on total antimony concentrations in soil. Moreover, the variation of dissolved antimony concentrations during the exposure period of toxicity tests will result in a variable bioavailability and toxic pressure, complicating interpretation of the results. Results from tests with sufficient equilibration of Sb_2O_3 in soil are preferred as such tests ensure higher dissolved antimony concentrations and a more constant pore water antimony concentration during toxicity tests [4, 19]. Analysis of the pore water concentrations will yield a better estimate of the toxic pressure (Fig. 14.1(b)). Pore water–based toxicity thresholds can be recalculated to total soluble antimony concentrations in soil based on equilibrium solid-liquid distribution coefficient (K_d) derived for soluble Sb sources. The lowest NOEC for 31 weeks equilibrated Sb_2O_3 in soil is 999 mg total Sb kg^{-1} soil observed for reproduction of the springtail *Folsomia candida* [26] and root elongation of barley plants [4], corresponding to a measured Sb concentration in pore water of 9.7 mg Sb L^{-1}. Taking into account the K_d of 38 L kg^{-1} for the specific soil tested [4], this corresponds to a total soil antimony concentration at equilibrium of 370 mg Sb kg^{-1}. For further examples of K_d values in sediments and soils, see Chapter 10.

The data for antimony toxicity to soil organisms do not allow conclusions on the impact of soil chemistry on the bioavailability and toxicity of antimony in soil. Data for toxicity of other oxyanions (molybdenum, arsenic) to soil organisms indicate that their toxicity is lower in soils with higher clay or iron oxide content and with lower pH [32–34]. This is in line with the known binding of oxyanions on oxides and the larger number of positively charged sorption sites for anions at low pH, and therefore it can be assumed that the same trends also hold for antimony. This is confirmed by a multiple linear regression of K_d values for antimony with soil properties, which shows positive coefficients for iron and manganese oxides and a negative coefficient for pH [35].

According to the risk assessment practices under the European Union regulations, a PNEC was based on this lowest chronic NOEC value and an assessment factor of 10 (if chronic data are available for three trophic levels). This resulted in PNEC of 37 mg Sb kg^{-1} dw for soil.

14.4 Bioaccumulation and secondary poisoning

In addition to the antimony toxicity to aquatic, sediment, and soil organisms via direct exposure to antimony in water or solid phase, organisms can also be exposed to antimony via their food. Secondary poisoning is concerned with toxic effects in the higher members of the food chain, either living in the aquatic or terrestrial environment, which result from ingestion of organisms from lower trophic levels that contain accumulated substances. The assessment of the bioaccumulation or biomagnification potential is key in the evaluation of the risk for secondary poisoning.

Bioconcentration is the net result of the uptake, distribution, and elimination of a substance in an organism due to waterborne exposure, whereas bioaccumulation includes all routes, that is, air, water, soil, and food. Biomagnification is defined as accumulation and transfer of chemicals via the food chain, resulting in an increase of the internal concentration in organisms at higher levels in the trophic chain. There are no results of standard laboratory studies that measure the bioaccumulation of antimony identified. As for most natural elements, there are however sufficient measured data available for antimony concentrations in organisms and their environment, allowing to calculate bioaccumulation factors (BAF). Such field-based BAF results are preferred over laboratory-derived bioconcentration factors (BCF) or BAF because organisms can be assumed to be exposed during their entire life and hence biota antimony burdens are at steady state with the antimony concentrations measured in their environment, which is not necessarily the case in laboratory experiments.

There is substantial variation in the BAF values for aquatic organisms, which can be attributed to differences in species, trophic ecology, exposure concentrations, environmental conditions, and so on (Tab. 14.5). Antimony can accumulate (i.e., BCF or BAF values >1) in primary producers in the aquatic food chain (algae and plants) with BCF values ranging from 0.08 to 326 L kg^{-1} dw, but large differences were observed between species and studies [36–39]. Similarly, a strong variation was observed for antimony bioaccumulation in aquatic invertebrates, with reported BAF values between <0.0001 and 699.9 L kg^{-1} dw, indicating a potential bioaccumulation of antimony in these organisms [36, 37, 39–41]. No clear correlation between BAF and trophic ecology could be detected. Most studies also show that fish have lower BAF values compared to organisms from lower trophic levels [36, 37, 39, 42]. Therefore, it can be concluded that antimony does not biomagnify in the food chain [43]. Culioli et al. [36] found indeed that biomagnification factors, defined as ratios of antimony concentrations between consumers and diets, diminished with higher trophic levels in an aquatic ecosystem. It must be noted that all these studies only reported total dissolved antimony concentrations in water and hence no conclusions can be made on the relative bioaccumulation of Sb(III) versus Sb(V) species and on the importance of the dietary exposure for higher organisms.

As for aquatic organisms, there are no laboratory studies available on bioaccumulation of antimony in soil organisms. However, concentrations of antimony in soil and invertebrates at locations close to antimony smelter or mining activities have been reported (Tab. 14.5). All these studies show a bioaccumulation factor ≤1, indicating that antimony does not bioaccumulate in soil invertebrates, such as earthworms [39, 45, 46]. Several studies on antimony levels in soil organisms have been published. Many of these studies were conducted in contaminated mining areas. Natural occurring plants were collected onsite, and concentrations of antimony in the plants (whole plant, roots, shoots, leaves, aerial parts) were determined. All these studies confirm the

Tab. 14.5: Bioaccumulation of antimony in aquatic and soil organisms.

Test substance	Sb concentration	Test organism	Exposure period	BAF	Reference
Aquatic organisms (BAF in L kg^{-1} dw)					
Environmental Sb (mining area)	0.8 µg Sb L^{-1}	Invertebrates (*Hyalella Azteca*)	17 d	56	[40]
Environmental Sb (downstream of shooting range)	2.1–3.1 µg Sb L^{-1}	Fish (*Salmo trutta*)	23 d	19.4–28.6 (gill)* 4.5–6.7 (liver)*	[44]
Environmental Sb (Sb mining area)	0.015 µg Sb L^{-1}	Invertebrates: *Asellus aquaticus* *Hydropsyche pellucidula* *Leucorrhinia dubia* *Gammarus pulex*	Whole life (field)	 21.7* 13.3* 4.6* 2.9*	[41]
Environmental Sb (vicinity of ancient realgar mine)	149 µg Sb L^{-1}	Fish (*Salmo trutta*) Invertebrates: – Shredders – Scrapers – Collector-gatherers – Collector-filterers – Predators Plant (*Fontinalis antipyretica*)	Whole life (field)	3.0 699.9 420.5 230.3 332.0 168.0 326.4	[36]
Environmental Sb (vicinity of Sb mine)	381 µg Sb L^{-1}	Carnivores – Flat-head gudgeon (fish) – Dragonfly – Water Beetle Omnivores – Tadpole – Caddisfly Detrivores – Worm – Snail Herbivores – Mayfly – Water pennies Macroalgae	Whole life (field)	 0.0008 0.38 0.79 0.46 0.72 0.82 <0.0001 0.51 0.25 0.25	[37]

Tab. 14.5 (continued)

Test substance	Sb concentration	Test organism	Exposure period	BAF	Reference
Environmental Sb (As mining area)	Not reported	Fish Alga	Whole life (field)	10.8 0.25	[38]
Environmental Sb (As mining area)	Not reported	Fish Green alga River snail	Whole life (field)	2.34 78.1 34.4	[39]
Environmental Sb (vicinity of Sb mine)	<1.4–1,127.9 μg Sb L^{-1}	Algae Tadpole Trout	Whole life (field)	0.08–2.04 0–1.16 <detection limit	[42]
Soil organisms (BAF in kg kg^{-1} dw)					
Environmental Sb (vicinity of Sb smelter)	80–1,489 mg Sb kg^{-1} dw	Invertebrates	Whole life (field)	0.6–1.03	[45]
Environmental Sb (former mining and smelting area)	10.3–1,200 mg Sb kg^{-1} dw	Earthworms: *Lumbricus terrestris* *Octolasion cyaneum*	Whole life (field)	 0.034 0.063	[46]
Environmental Sb (near Sb smelter)	Not reported	Earthworms	Whole life (field)	0.19	[39]

* Wet weight-based results.

previous findings that the BAF for plants is typically lower than 1 and that the translocation factor within the plants is also lower than 1; that is, the highest levels of antimony are typically found in the roots (e.g., [47–49]). More detailed information on plant uptake, translocation, accumulation and toxicity can be found in Chapter 7.

It can be concluded that the bioaccumulation potential of antimony is low. For comparison, a substance fulfils the bioaccumulation criterion in a PBT (persistent, bioaccumulative, and toxic) assessment when the BCF in aquatic species is higher than 2,000. It is clear that antimony does not fulfill this criterion. Because of this low bioaccumulation potential of antimony and the fact that it is a natural element which is naturally present in the environment, it can be concluded that there is no risk for secondary poisoning of antimony in higher organisms.

14.5 Endocrine disrupting properties

Regulatory scrutiny on potential endocrine disrupting properties for environmental organisms is growing. Studies investigating the impact of antimony compounds on the endocrine system in aquatic organisms, amphibians, or aquatic organisms were however not identified. Most studies on health effects in rats state that antimony compounds do not disturb the reproductive function of males or females (e.g., [50]). There are however still some knowledge gaps on the effect of antimony on reproductive toxicity and on endocrine disruptive effects of antimony substances.

14.6 Environmental quality standards

Antimony is less commonly regulated compared to other metal(loid)s like arsenic, cadmium, copper, lead, and zinc. An overview of some published environmental quality standards is reported in Tab. 14.6.

When comparing such values, one should always check the protection goals associated with an environmental quality standard. Next to the aforementioned PNEC values, which are protective thresholds to be used for risk assessment purposes and aiming to avoid any adverse effects, other environmental quality standards may have lower protection levels as they may define thresholds for remedial actions in case of unacceptable effects (e.g., Serious Risk Concentration in the Netherlands [51]). Most no-effect concentrations for antimony in the aquatic environment vary between 5 and 300 µg Sb L^{-1} and thresholds predicting risk are above 1 mg Sb L^{-1}, all in agreement with the aquatic toxicity data available for antimony (Tabs 14.1 and 14.2). Antimony is not identified as a priority substance under the European Water Framework Directive (Directive 2013/39/EU). The quality standards for sediment (no-effect threshold at <15 mg Sb kg^{-1} dw and effect concentrations at ≥100 mg Sb kg^{-1} dw) are also well in line with the data reported in Tab. 14.3. There is however less consistency in the quality standards for soil. Published screening values for (unacceptable) risks in various European countries (Austria, Finland, Czech Republic) range between 2 and 40 mg Sb kg^{-1} dw [52], while no-effect concentrations vary between 37 and 100 mg Sb kg^{-1} dw [51, 53–54]. While the basis for these no-effect thresholds used by the European Union, Canada, Australia, New Zealand, and the Netherlands is well

Tab. 14.6: Environmental quality standards as defined under different regulatory frameworks.

Legislation	Limit	Note	Reference
European Union, Canada	Freshwater: 113 µg Sb L^{-1} Marine water: 11.3 µg Sb L^{-1} Freshwater sediment: 11.2 mg Sb kg^{-1} dw Marine sediment: 2.24 mg Sb kg^{-1} dw Soil: 37 mg Sb kg^{-1} dw	PNEC values: predicted safe concentration to be used for risk assessment purposes	[53, 54]
The Netherlands	MPC: Freshwater: 5.6 µg Sb L^{-1} Sediment: 14 mg Sb kg^{-1} dw Soil: 100 mg Sb kg^{-1} dw	Maximum Permissible Concentration (MPC): concentration in an environmental compartment at which no effect to be rated as negative is to be expected for ecosystems and humans.	[51]
	SRC_{eco}: Freshwater: 9,600 µg Sb L^{-1} Sediment: 110 mg Sb kg^{-1} dw Soil: 1,400 mg Sb kg^{-1} dw	Serious Risk Concentration (SRC_{eco}): concentration at which serious negative effects in an ecosystem may occur.	
Australia	Freshwater: 9 µg Sb(III) L^{-1} Marine water: 270 µg Sb(III) L^{-1}	Guideline values for protection of aquatic ecosystems. Only low-reliability trigger values were derived for Sb(III), only to be as an indicative interim working level.	[55]
USA	Freshwater: Acute effects: 9,000 µg Sb L^{-1} Chronic effects: 1,600 µg Sb L^{-1} Algae: 610 µg Sb L^{-1}	Lowest observed Sb toxicity	[56]

(continued)

Tab. 14.6 (continued)

Legislation	Limit	Note	Reference
USA	Soil: Invertebrates: 78 mg Sb kg^{-1} dw Mammals: 0.27 mg Sb kg^{-1} dw	Ecological soil screening level (ECO SSL)	[57]
Europe	Soil: 2–25 mg Sb kg^{-1} dw 5–40 mg Sb kg^{-1} dw	 Screening values for warning risk (residential use) Screening values for potential unacceptable risk (residential use)	[52]

documented and in line with the toxicity data reported here (Tab. 14.4), the basis for the screening values for potential risks is unknown, hindering a proper comparison of these values.

14.7 Conclusions

Information on the ecotoxicity of antimony in the environment is mainly based on experiments with trivalent antimony substances ($SbCl_3$, $Sb_2(SO_4)_3$, or Sb_2O_3) and only a few acute toxicity data to aquatic organisms are available for Sb(V) ($KSb(OH)_6$). Because of the expected oxidation of Sb(III) to Sb(V) in the oxic test media and based on the few comparative toxicity data for Sb(III) and Sb(V), it can be concluded that the results for Sb(III) substances are protective for all antimony substances. Therefore, all results are expressed based on total elemental antimony regardless of the substance tested. A good understanding of the environmental chemistry in the exposure media is critical for a proper evaluation of the ecotoxicity data to avoid toxicity due to confounding factors, such as acidification or increased ionic strength, from adding antimony salts being attributed to antimony.

Antimony toxicity to aquatic organisms is only observed at concentrations >1 mg Sb L^{-1}, which is a factor 1,000 above the natural background antimony concentration in water, which is typically below 1 µg Sb L^{-1} [58]. Toxicity in the sediment compartment occurs at concentrations ≥100 mg Sb kg^{-1} dw, again well above the 90th percentile of background concentrations in uncontaminated sediments in Europe of 2.1 mg Sb kg^{-1} [59]. Also toxicity in soil is only observed at elevated concentrations (≥370 mg Sb kg^{-1} dw), more than a factor 100 above the 90th percentile of antimony in agricultural soil in Europe (0.9 Sb kg^{-1} [60]). In none of the environmental compartments (water, sediment, soil), pronounced differences in sensitivity of organisms to antimony were identified, indicating a similar mode of action of antimony toxicity to all organisms. Because of the low bioaccumulation and biomagnification potential of antimony in aquatic or soil organisms, it is also concluded that there is no risk for secondary poisoning of higher organisms via transfer of antimony in the food chain. It can be concluded that antimony is not strongly toxic in the environment, especially when compared to other metal(loid)s such as silver, arsenic, cadmium, cobalt, copper, or selenium, which all show aquatic toxicity in the low µg L^{-1} range.

References

[1] OECD. Guidance on the incorporation of bioavailability concepts for assessing the chemical ecological risk and/or environmental threshold values of metals and inorganic metal compounds. OECD Environment, Health and Safety Publications, Series on Testing and Assessment, 259, 2016.

[2] Filella M, Belzile N, Chen YW. Antimony in the environment: A review focused on natural waters II. Relevant solution chemistry. Earth-Science Rev 2002, 59, 265–285.

[3] Belzile N, Chen YW, Wang ZJ. Oxidation of antimony (III) by amorphous iron and manganese oxyhydroxides. Chem Geol 2001, 174, 379–387.

[4] Oorts K, Smolders E, Degryse F, et al. Solubility and toxicity of antimony trioxide (Sb_2O_3) in soil. Environ Sci Technol 2008, 42, 4378–4383.

[5] Skeaff JM, Beaudoin R, Wang R, Joyce B. Transformation/dissolution examination of antimony and antimony compounds with speciation of the transformation/dissolution solutions. Integr Environ Assess Manag 2013, 9, 98–113.

[6] Brooke LT, Call DJ, Poirier SH, Lindberg CA, Markee TP. Acute toxicity of antimony III to several species of freshwater organisms. Report to batelle Memorial Research Institute, Columbus, OH 43201 IN partial fulfillment of work assignment No. 45, task order No. 4, subcontract No. F-4114(8834)-411. 1986.

[7] Kimball G The effect of lesser known metals and one organic to fathead minnows (*Pimephales promelas*) and *Daphnia magna*. 1978.

[8] TAI Environmental Sciences Inc. Results of acute toxicity testing of antimony trichloride using the freshwater species *Chironomus tentans, Physa heterostropha, Ictalurus punctatus, Hyalella azteca, Hydra oligactis* and *Chlorohydra viridissimus*. 1990.

[9] Takayanagi K. Acute Toxicity of Waterborne Se(IV), Se(VI), Sb(III), and Sb(V) on Red Seabream (*Pargus Major*). Bull Environ Contam Toxicol 2001, 66, 808–813.

[10] Yang JL. Comparative acute toxicity of gallium(III), antimony(III), indium(III), cadmium(II), and copper (II) on freshwater swamp shrimp (*Macrobrachium nipponense*). Biol Res 2014, 47, 2–5.

[11] Heijerick D, Vangheluwe M. Analysis of the results of a 72h-toxicity test with the freshwater green alga *Pseudokirchneriella subcapitata* using $SbCl_3$ as test substance. Study for the International Antimony Oxide Industry Association (IAOIA), 2004.

[12] Lisec. Screening and acute transformation/dissolution test with Sb_2O_3 in ecotox medium. LISEC study No.: WE-14-018. Study for the International Antimony Oxide Industry Association (IAOIA), 2001.

[13] Duester L, van der Geest HG, Moelleken S, Hirner AV, Kueppers K. Comparative phytotoxicity of methylated and inorganic arsenic- and antimony species to Lemna minor, *Wolffia arrhiza* and *Selenastrum capricornutum*. Microchem J 2011, 97, 30–37.

[14] Leblanc GA, Dean JW. Antimony and thallium toxicity to embryos and larvae of fathead minnows (*Pimephales-Promelas*). Bull Environ Contam Toxicol 1984, 32(5), 565–569.

[15] Heijerick D, Vangheluwe M. Analysis of the results of a 21 d chronic toxicity test with the freshwater flea *Daphnia magna* using $SbCl_3$ as test substance. Study for the International Antimony Oxide Industry Association (IAOIA), 2003.

[16] Cui XD, Wang YJ, Hockmann K, Zhou DM. Effect of iron plaque on antimony uptake by rice (*Oryza sativa* L.). Environ Pollut 2015, 204, 133–140.

[17] Flynn HC, Meharg AA, Bowyer PK, Paton GI. Antimony bioavailability in mine soils. Environ Pollut 2003, 124, 93–100.

[18] He MC, Yang JR. Effects of different forms of antimony on rice during the period of germination and growth and antimony concentration in rice tissue. Sci Total Environ 1999, 244, 149–155.
[19] Oorts K, Smolders E. Ecological threshold concentrations for antimony in water and soil. Environ Chem 2009, 6, 116.
[20] Heijerick D, Vangheluwe M. Analysis of the results of a 42-day chronic sediment test with *Hyalella azteca* using $SbCl_3$ as test substance. Study for the International Antimony Oxide Industry Association (IAOIA), 2003.
[21] Heijerick D, Vangheluwe M. Analysis of the results of a 28-day chronic sediment test with the midge *Chironomus riparius* using $SbCl_3$ as test substance. Study for the International Antimony Oxide Industry Association (IAOIA), 2005.
[22] Heijerick D, Vangheluwe M. Analysis of the results of a 28-day chronic sediment test with the oligochaete *Lumbriculus variegatus* using $SbCl_3$ as test substance. Study for the International Antimony Oxide Industry Association (IAOIA), 2005.
[23] Oorts K, Gasco G, Smolders E. Toxicity of antimony in the terrestrial environment. Research project for the International Antimony Oxide Industry Association (IAOIA), 2005.
[24] Kuperman RG, Checkai RT, Simini M, Phillips CT, Speicher JA, Barclift DJ. Toxicity benchmarks for antimony, barium, and beryllium determined using reproduction endpoints for *Folsomia candida, Eisenia fetida*, and *Enchytraeus crypticus*. Environ Toxicol Chem 2006, 25, 754–762.
[25] Heijerick D, Vangheluwe M. Analysis of the results of a 42-day soil toxicity test with *Enchytraeus albidus*, using $SbCl_3$ as a test substance. Study for the International Antimony Oxide Industry Association (IAOIA), 2003.
[26] Moser T. Acute and reproduction toxicity of Antimony Trioxide aged residues in contaminated soil according to ISO 11267 (1999) "Soil quality – Inhibition of reproduction of Collembola (Folsomia candida) by soil pollutants". Study for the International Antimony Oxide Industry Association (IAOIA), 2007.
[27] Lisec. The effect of Sb2O3 on the nitrogen transformation activity of soil microorganisms. LISEC study No. WE-12-004. Study for the International Antimony Oxide Industry Association (IAOIA). 2002.
[28] Speir TW, Kettles HA, Percival HJ, Parshotam A. Is soil acidification the cause of biochemical responses when soils are amended with heavy metal salts? Soil Biol Biochem 1999, 31, 1953–1961.
[29] Stevens DP, McLaughlin MJ, Heinrich T. Determining toxicity of lead and zinc runoff in soils: Salinity effects on metal partitioning and on phytotoxicity. Environ Toxicol Chem 2003, 22, 3017–3024.
[30] Smolders E, Oorts K, Van Sprang P, et al. Toxicity of trace metals in soil as affected by soil type and aging after contamination: Using calibrated bioavailability models to set ecological soil standards. Environ Toxicol Chem 2009, 28, 1633–1642.
[31] Nakamaru Y, Tagami K, Uchida S. Antimony mobility in Japanese agricultural soils and the factors affecting antimony sorption behavior. Environ Pollut 2006, 141, 321–326.
[32] Song J, Zhao FJ, McGrath SP, Luo YM. Influence of soil properties and aging on arsenic phytotoxicity. Environ Toxicol Chem 2006, 25, 1663–1670.
[33] Kader M, Lamb DT, Megharaj M, Naidu R. Sorption parameters as a predictor of arsenic phytotoxicity in Australian soils. Geoderma 2016, 265, 103–110.
[34] Oorts K, Smolders E, McGrath SP, Van Gestel CAM, McLaughlin MJ, Carey S. Derivation of ecological standards for risk assessment of molybdate in soil. Environ Chem 2016, 13, 168–180.

[35] Janik LJ, Forrester ST, Soriano-Disla JM, et al. GEMAS: Prediction of solid-solution phase partitioning coefficients (Kd) for oxoanions and boric acid in soils using mid-infrared diffuse reflectance spectroscopy. Environ Toxicol Chem 2015, 34, 235–246.
[36] Culioli JL, Fouquoire A, Calendini S, Mori C, Orsini A. Trophic transfer of arsenic and antimony in a freshwater ecosystem: A field study. Aquat Toxicol 2009, 94, 286–293.
[37] Telford K, Maher W, Krikowa F, et al. Bioaccumulation of antimony and arsenic in a highly contaminated stream adjacent to the Hillgrove Mine, NSW, Australia. Environ Chem 2009, 6, 133–143.
[38] Fu Z, Wu F, Amarasiriwardena D, et al. Antimony, arsenic and mercury in the aquatic environment and fish in a large antimony mining area in Hunan, China. Sci Total Environ 2010, 408(16), 3403–3410.
[39] Fu Z, Wu F, Mo C, et al. Bioaccumulation of antimony, arsenic, and mercury in the vicinities of a large antimony mine, China. Microchem J 2011, 97, 12–19.
[40] Couillard Y, Grapentine LC, Borgmann U, Doyle P, Masson S. The amphipod *Hyalella azteca* as a biomonitor in field deployment studies for metal mining. Environ Pollut 2008, 156, 1314–1324.
[41] Duran M, Kara Y, Akyildiz GK, Ozdemir A. Antimony and heavy metals accumulation in some macroinvertebrates in the Yesilirmak River (N Turkey) near the Sb-mining area. Bull Environ Contam Toxicol 2007, 78, 395–399.
[42] Dovick MA, Kulp TR, Arkle RS, Pilliod DS. Bioaccumulation trends of arsenic and antimony in a freshwater ecosystem affected by mine drainage. Environ Chem 2016, 13, 149–159.
[43] Obiakor MO, Tighe M, Pereg L, Wilson SC. Bioaccumulation, trophodynamics and ecotoxicity of antimony in environmental freshwater food webs. Crit Rev Environ Sci Technol 2017, 47, 2208–2258.
[44] Heier LS, Lien IB, Strømseng AE, et al. Speciation of lead, copper, zinc and antimony in water draining a shooting range-Time dependent metal accumulation and biomarker responses in brown trout (*Salmo trutta* L.). Sci Total Environ 2009, 407, 4047–4055.
[45] Ainsworth N, Cooke JA, Johnson MS. Distribution of antimony in contaminated grassland: 2. Small mammals and invertebrates. Environ Pollut 1990, 65, 79–87.
[46] Gál J, Hursthouse A, Cuthbert S. Bioavailability of arsenic and antimony in soils from an abandoned mining area, Glendinning (SW Scotland). J Environ Sci Heal – Part A Toxic/Hazardous Subst Environ Eng 2007, 42, 1263–1274.
[47] Vaculík M, Jurkovič L, Matejkovič P, Molnárová M, Lux A. Potential risk of arsenic and antimony accumulation by medicinal plants naturally growing on old mining sites. Water Air Soil Pollut 2013, 224, 1546.
[48] Cidu R, Biddau R, Dore E, Vacca A, Marini L. Antimony in the soil-water-plant system at the Su Suergiu abandoned mine (Sardinia, Italy): Strategies to mitigate contamination. Sci Total Environ 2014, 497–498, 319–331.
[49] Álvarez-Ayuso E, Otones V, Murciego A, García-Sánchez A, Regina IS. Antimony, arsenic and lead distribution in soils and plants of an agricultural area impacted by former mining activities. Sci Total Environ 2012, 439, 35–43.
[50] Coelho DR, De-Carvalho RR, Rocha RCC, Saint'Pierre TD, Paumgartten FJR. Effects of in utero and lactational exposure to SbV on rat neurobehavioral development and fertility. Reprod Toxicol 2014, 50, 98–107.
[51] van Leeuwen LC, Aldenberg T Environmental risk limits for antimony. National Institute for Public Health and the Environment, the Netherlands. 2012. http://edepot.wur.nl/66474
[52] Carlon C, D'Alessandro M, Swartjes F. Derivation Methods of Soil Screening Values in Europe. A review and Evaluation of National Procedures Towards Harmonisation. Ispra, European Commission, Joint Research Centre, 2007.

[53] Environment Canada, Health Canada. Screening assessment for the Challenge: Antimony trioxide (Antimony oxide): Chemical Abstracts Service Registry Number 1309-64-4. 2010. https://www.ec.gc.ca/ese-ees/9889ABB5-3396-435B-8428-F270074EA2A7/batch9_1309-64-4_en.pdf
[54] European Chemicals Agency (ECHA.) REACH Registration dossier for antimony (Accessed October 25, 2020, at https://echa.europa.eu/registration-dossier/-/registered-dossier/16124/1)
[55] Australian and New Zealand Environment and Conservation Council, Agriculture and Resource Management Council of Australia and New Zealand. Australian and New Zealand Guidelines for Fresh and Marine Water Quality. Volume 2 Aquatic Ecosystems – Rationale and Background Information. Canberra: 2000. https://www.waterquality.gov.au/sites/default/files/documents/anzecc-armcanz-2000-guidelines-vol2.pdf
[56] United States Environmental Protection Agency. Ambient water quality criteria for antimony. 1980. https://www.epa.gov/sites/production/files/2019-02/documents/ambient-wqc-antimony-1980.pdf
[57] United States Environmental Protection Agency. Ecological soil screening levels for antimony. 2005. https://www.epa.gov/sites/production/files/2015-09/documents/eco-ssl_antimony.pdf
[58] Filella M, Belzile N, Chen YW. Antimony in the environment: A review focused on natural waters I. Occurrence. Earth-Science Rev 2002, 57(l), 125–176.
[59] Salminen R, editor. Geochemical Atlas of Europe. Part 1: Background Information, Methodology and Maps. Espoo, Geological Survey of Finland, 2005.
[60] Reimann C, Birke M, Demetriades A, Filzmoser P, O'Connor P, editors. Chemistry of Europe's Agricultural soils – Part A: Methodology and Interpretation of the GEMAS data set – Geologisches Jahrbuch, Vol. B 102. Hannover, Schweizerbart, 2014.

Waldo Quiroz

Chapter 15
Antimony: human exposure

As with all chemical substances, the incorporation of such chemicals by human beings may be via a trophic route through the food chain or non-trophic route, such as breathing, dermic contact, or direct injection in the case of medical treatment. In the case of antimony, this metalloid presents itself in levels considered above normal in food, mainly in vegetables grown with contaminated water or soil [1] and also in the case of big cities, particulate atmospheric matter in which the enrichment factors containing this metalloid are high [2].

The degree of antimony exposure in humans must consider not only information related to concentrations but also comparative parameters to estimate whether or not this exposure is merely high or actually harmful. In that regard, in order to describe and evaluate the degree of antimony exposure in humans, we will use two references in this chapter. The first is the tolerable daily intake (TDI) of antimony. Second, we will use some values from international regulations in which values for antimony can be found.

15.1 Distribution and guidelines

Antimony levels on the earth crust are low. It presents itself in levels of 0.2 $\mu g\ g^{-1}$. From that point of view, we could think that antimony exposure has low plausibility. However antimony compounds present plenty of uses, mainly those associated to flame retardants and break pads in automobiles because of which there is an annual production more than 100,000 t [3]. In view of this it might be expected that the episodes of contamination by antimony are frequent.

However, contamination is a necessary requisite for evaluating its effects but is not enough to evaluate the degree of exposure by which tolerance and availability criteria must be analyzed for a more complete evaluation. Regarding tolerance, according to the World Health Organization's (WHO) "Guidelines for Drinking-Water Quality," the recommendation for antimony in water is 20 $\mu g\ L^{-1}$ [4] and the TDI for antimony in humans is 6 $\mu g\ kg^{-1}$ body weight, which considers a standard mass of a 60 kg adult and 2 L water consumption per day, which would provide a daily intake of 360 µg of antimony [4]. It is interesting to highlight that in the case of antimony and according to this reference, it has been reported that a daily intake value of antimony

Waldo Quiroz, Instituto de Química, Pontificia Universidad Católica de Valparaíso, Av. Universidad 330, Placilla, Valparaíso, Chile. waldo.quiroz@pucv.cl

https://doi.org/10.1515/9783110668711-015

through water consumption is very low, and its intake is mainly due to consumption of food containing antimony. Stricter guidelines for antimony in water can be found in the Netherlands or laxer in the Australian recommendations; we advise the reader to read chapter 14 and, in particular, Tab. 14.6 where more details can be found.

While it is true that the industrial use of antimony is high, its physicochemical properties make it an element with a low mobility and availability. For example, antimony oxide (Sb_2O_3) is a substance that presents a molar solubility of no more than 10 mg L^{-1} under neutral pH conditions, and it is a neutral species, $Sb(OH)_3$, at almost all pH values [5]. The previously mentioned findings lead us to anticipate that the availability of antimony when it exists as Sb(III) should be low. On the other hand, Sb(V) is a square bipyramid, a larger ion, and it is suspected that its uptake pathway to some cellular systems is through the AE-1 channel (HCO_3^-/Cl) [6]. Phytochelatins do not seem to play any role in phyto-availability [7]. In fact, the covalent bonds between Sb(V)/Sb(III) and sulfhydryl groups seem to be very labile, and possible complexes with bio-molecules through the formation of covalent bonds with –SH groups require no chromatographic methods for detection [8].

Tab. 15.1: Relevant parameters of antimony related to human exposure.

Parameter	Sb
Earth crust concentration (μg g^{-1})	0.2
Daily intake (tolerable, acceptable, μg)	360
Industrial production (t y^{-1})	10^5
WHO water guidelines (μg L^{-1})	20
Trivalent oxide solubility (mg L^{-1})	10
Main route of exposure	food
Affinity with biomolecules (–SH)	low

It is clear that because of its low solubility and low affinity for molecules, the mobility and bioavailability of antimony are low, and therefore, human exposure to antimony through the trophic route implausible. However, if we consider its industrial use, antimony should then present itself in the environment with a high degree of human exposure as a result of pollution, which would generate the production and use of such products. Nevertheless, evidence has the final word, and it is for this reason that in this chapter we will compare the baseline levels in addition to the reported cases of contamination in water, soil, food, and atmospheric particulate matter, which are the main routes of trophic and non-trophic access to humans. We will evaluate these routes by comparing them with toxicity criteria.

15.2 Human exposure to antimony through water

As much as for both intake reasons and irrigation reasons, antimony exposure through water must be considered. Regarding antimony levels in mineral water, two very different contexts can be identified in the literature. The first context is water that is polluted by antimony mining activities. This process has been described in Xikuangshan, China, in which antimony levels are reported ranging from 2 to 6,384 μg L^{-1} and 54 ± 47 μg L^{-1} in natural and drinking waters, respectively. It has also been reported that antimony levels in groundwater and surface and mine waters reach levels of up to 9,300 μg L^{-1} around abandoned antimony mines in which the suspended material of these waters is mainly in the form of Sb(III) [9]. Certainly these levels tend to surpass the ones recommended by WHO of 20 μg L^{-1} in water although the daily intake of 2 L of drinking water is below 360 μg of TDI.

The other context, in which antimony is found in high concentrations, ranging from 500 mg L^{-1} up to 10% wt, is in hot springs and geothermal water [10]. However, these types of water sources pose no real threat to human consumption or for irrigation because of their location in addition to their physical properties.

Outside the antimony mine context, the levels of this element in water are low. Antimony levels have been reported at 2.1 μg L^{-1} and of 0.6 μg L^{-1} for surface waters and groundwater, respectively [11], with typical concentrations in natural waters mainly found to be below 1 μg L^{-1} [10].

Recently, some concern has been generated due to the presence of antimony in water bottled in plastic polyethylene terephthalate (PET). This polymer is produced with antimony-, titanium-, or germanium-based catalysts. Studies in that regard have demonstrated that antimony levels found in bottled water range between 0.095 and 0.521 μg L^{-1}, which is far below a TDI of 360 μg if we consider a daily intake of 2 L. It is also below the United States Environmental Protection Agency (USEPA) guidelines of 6 μg L^{-1}. If we consider that this study was carried out with PET bottles with antimony levels of up to 213 ± 35 mg kg^{-1}, this finding shows that leaching from this plastic into water in normal storage conditions is very low and therefore human exposure to these types of products is quite minimal. Antimony leaching levels from PET bottles might become toxic only in extreme temperatures and time conditions [12].

Similar studies have reached similar conclusions; for instance, De Jesus et al. published a study in 2016 in which the antimony concentration was determined in PET containers and ranged between 194 and 323 mg kg^{-1}. Antimony leaching tests from these containers were carried out with a 3% acetic acid solution, red wine, and soft drinks and found that antimony content in all solutions after 10 days of storage was below 1 μg L^{-1} [13]. Similar studies have been carried out in citric juices, which hypothetically should dissolve a larger amount of antimony considering that citrate has a complexing capability for Sb(III) and Sb(V); however, in spite of these chemical considerations, total antimony levels found in citric juices, were around

1.1 µg L^{-1} [14]. Although antimony levels in waters or juices stored in PET bottles have been proven to be even 100 times higher compared with uncontaminated water or drinking water, the levels are still much lower than that established by USEPA, WHO regulations, or similar guidelines [15].

Studies about the leaching of antimony from PET bottles are still a matter of controversy because of the limited number of samples, and statistical treatments are often absent. This controversy is more complex if we consider the high number of variables on which this leaching depends, such as temperature, sunlight, time of contact, pH, bottle volume, and surface area. In Filella's review about antimony in PET bottles, it is shown that most of the published studies about the subject are based on analytical methods whose validations do not mention the use of Certified Reference Materials (CRMs) or they use CRMs with antimony levels far above those found in these samples [16]. However, in spite of this high number of variables, antimony concentration values reported in water in PET bottles are always a small percentage of the total antimony in PET bottles and are always below 2 µg L^{-1} and therefore below regulated values [16].

Based on the previous information, we can therefore state that human exposure to antimony through irrigation or consumption water is less frequent and mainly confined to water surrounding abandoned or active antimony mines. Demonstrated cases of antimony intoxication intake through water are almost nonexistent. In this context, a highly plausible hypothesis has been reported suggesting that some inhabitants in Pompeii (79 CE) could have died from antimony-contaminated water intake coming from the pipeline; this hypothesis is based on the analysis of the composition of a water pipe fragment excavated from House of Caecilius Jucundus giving that it was composed of 3,681 mg kg^{-1} of antimony and 7,872 mg kg^{-1} of tin. However, this eventual intake of antimony from dissolved pipeline material still remains to be confirmed by forensic studies for the determination of antimony in bones of these inhabitants not performed yet [17]. Another antimony-related intoxication case in the United States was reported in a review by de Belzile et al. in which more than 11.7 million people in 1,176 American communities drank antimony-contaminated water between 1998 and 2003. However, it was not possible to obtain the original reference to support it since the document link is no longer available. According to the WHO, the daily oral uptake of antimony ranges from 10 to 70 µg, which appears to be significantly higher than uptake via inhalation [18], but most of this daily oral uptake comes from food rather than water.

15.3 Human exposure to antimony through the vegetable–soil system

Having eliminated antimony intake through water as a harmful human exposure source to antimony, it is logical to evaluate the role of the soil since it is at the base of the terrestrial trophic chain. In contrast to its abundance in the terrestrial cortex, antimony levels in soils are very low compared to arsenic. On average, antimony concentration in soil is 0.2 mg kg^{-1}. Guideline value for antimony in soils considering health risks according to Finland government is 10 mg kg^{-1} [19].

Antimony can reach higher concentrations in soils due to mining/smelting activities from 50 to more than 10,000 mg kg^{-1} [20, 21] increasing up to 700 mg kg^{-1} [22, 23]. Vehicle traffic activities give concentrations close to 10 mg kg^{-1}.

Given that the presence of antimony-contaminated soils is not exclusive of heavily-contaminated areas, it can be asked "How much of the antimony present in the soil enters the trophic chain?" Several studies determined antimony concentrations in vegetables that grow in these contaminated soils. It has been reported that some plants, such as *Dryopteris erythrosora, Fagopyrum dibotrys, Rumex patientia L.*, and *Poecilotheria vittata*, accumulate antimony from mining activity-contaminated soils [9]. In 2011 Qi et al. report antimony concentrations ranging from 3.92 to 143.69 mg kg^{-1} in 34 plants in the Xikuangshan antimony deposit areas [24].

Antimony concentration levels in plants can originate from two sources: (1) contaminated soil and (2) contaminated particulate matter deposited on the aerial parts of the vegetables. In 2015, Pierart et al. wrote a review in which they summarized antimony reported levels in edible plants to date [25]. This review described that antimony can be found in every plant tissue from roots to fruits in a wide range of concentrations ranging from 0.004 to 1,400 mg kg^{-1}. However, antimony contamination in these vegetables does not always come from contaminated soil. As a matter of fact, concentrations of antimony in both soil and vegetables are very low, and it has been reported that low levels of metalloid are transferred from soil to *Rosemary* edible aerial parts [26]. Hammel and collaborators conducted a study in which a historical mining area residual material was used in lands for agricultural purposes. Nineteen vegetable and crop species cultivated in soils with 500 mg kg^{-1} of antimony were investigated. Despite the high antimony concentration in the soil, antimony levels in those plants were close to normal with concentrations in grain and storage organs up to 0.09 mg kg^{-1} and 0.34 and 2.2 mg kg^{-1} in shoots and leaves, respectively [27].

If we only consider the soil variable, it has been reported that antimony enters vegetables through the growth process in contaminated soil, and that antimony is mainly retained in the roots via a strong bond between antimony and iron, which leads to an antimony decrease in the plant. This phenomenon has been described in rice in which antimony was found to form part of an iron plaque that limits its access to the plant [28]. It has been reported that antimony accumulates from 10 to 12 times more in the root and from 28 to 54 times more in the iron plaque than in the rest of the rice plant, which shows that human exposure to antimony from the soil–rice system is very unlikely [29]. Similar patterns have been reported on herbal crops as described by Affholder and collaborators in which the wild Rosemary (*Rosmarinus officinalis L.*) crop grown in contaminated soils with 309 ± 230 mg kg^{-1} of antimony had antimony levels on stems and leaves below 0.6 mg kg^{-1} [26]. The previously mentioned cases are just some representative cases that reflect a general tendency. In a review by Pierart, edible vegetables, such as barley, cabbage, carrots, potatoes, spinach, tomatoes, wheat, endive, corn salad, dill, and oats, grown on highly contaminated soil were reported to present antimony levels below 3 mg kg^{-1}. The only cases in which contaminated soil increased antimony levels in vegetables were cucumbers, colza, peanuts, rye, and corn although, on several of these crops, it is not clear which part of the plant accumulated antimony or, in the case of corn, antimony was mainly accumulated in the root and not in the edible parts [25]. Other published cases in which these vegetables exceeded these values are very limited and mainly concern experimental crops with artificially added antimony standards that accumulated mainly in the roots [30].

Phyto-availability of antimony from soils depends on the redox form of antimony. It is known that phyto-availability depends in part on the solubility of a substance, and in that sense, phyto-availability of Sb(III) in soils should be far less than Sb(V) because it is known that Sb_2O_3 solubility is very low compared to Sb(V) [5]. It has been suggested that atmospheric particulate matter could be a more important pathway given that penetration through stomata openings is a major mechanism involved in foliar uptake of particulate matter [31].

15.4 Human exposure to antimony through aquatic ecosystems

In aquatic ecosystems of fresh or saltwater, antimony access to the trophic chain strongly depends on the sediment–water interface and the type of biota developed and whether this biota is autotrophic or heterotrophic. As we saw in previous chapters, water content in non-contaminated systems is very low. Concentrations in unpolluted waters were shown to be below 1 µg L^{-1}, and antimony concentrations in surface marine waters were 184 ± 45 ng L^{-1} [10].

Regarding sediments, these are considered contaminant sinks and this consideration does not seem to be the exception in the case of antimony. Non-contaminated sediments have antimony levels of a few mg kg^{-1} in contrast to the abundance of this element in the terrestrial cortex [10], and in the case of contaminated environments, antimony concentration levels have been reported to be from 57 to 7,316 mg kg^{-1} in river sediments, which decrease as the distance from the world's largest antimony mine increases [32].

Human exposure from aquatic ecosystems implies that antimony enters the trophic chain. Chemical variables that have an influence on this process are pH, the presence of counter ions, and redox conditions [33] of the water–sediment system in which reductive conditions tend to favor antimony immobilization in the form of Sb(III), species that has more affinity for the particulate phase [34] mainly due to the low solubility of its compounds.

Given that antimony mobility conditions on aquatic systems depend on many variables, it is impossible to make any a priori judgement about the degree of human exposure to antimony only considering the total antimony concentrations of sediments or contaminated water. In order to conduct an evaluation, antimony concentrations in the biota present in the fields of contaminated ecosystems need to be evaluated. In that regard, one of the more useful reviews about bioaccumulation of antimony in environmental freshwater and food chains was published by Obiakor et al. In this work, it was concluded that field studies indicate that antimony is not biomagnified in food webs consisting of autotrophs, macroinvertebrate consumers, tadpoles, and fish. However, some authors claim that limited data are available for antimony bioaccumulation in different geographic regions across spatiotemporal scales [35]. The reported concentrations in this study for organisms ranged from non-detectable to 674.9 mg kg^{-1}, with higher concentrations associated with mining and smelting activities. However, the association between contamination of sediment or water with high antimony levels in aquatic organisms is very low, and contradictory results are presented. For instance, in the study by Dovick et al., the bioaccumulation trends of antimony in a freshwater ecosystem affected by mine drainage was studied and showed a high correlation between water contamination and concentration in microorganisms with a bioaccumulation factor (BAF) of 100 to 10,000. However, biomagnification was not observed; on the contrary, a decrease through the trophic levels primary producers was observed to be > tadpoles > macroinvertebrates > trout [36].

Other studies that have tried to establish similar relationships between antimony levels in water and sediment with microorganisms concluded that the dynamic performance of antimony in the ecological environment is still not clear [37]. From the studies that were analyzed, one that demonstrates a clear antimony transfer in the water–sediment system to the aquatic biota is the work of Culioli et al. in which different sites were studied with antimony levels in waters and sediments of 2 to 149 μg L^{-1} and 31 to 1,057 mg kg^{-1}, respectively. In this work, antimony levels are reported on bryophytes between 0.34 and 49 μg g^{-1} and between 8.7 and 198 μg g^{-1} in benthic macroinvertebrates. However, these high levels of antimony were not reflected in the *Salmo trutta* (Salmonidae) samples whose highest antimony concentration was 0.45 μg g^{-1}. Similar to the work of Dovick, BAF factors between consumer and diet diminish with higher trophic levels [38]. Based on these data, it can be concluded that the bioaccumulation of antimony in terrestrial and aquatic ecosystems is low. More details about BAF in aquatic and terrestrial isms can be found in chapter 14 and Tab. 14.5.

15.5 Human exposure to antimony through atmospheric contamination

In the atmosphere, antimony presence is due to natural emissions, such as volcanic activity and rock erosion, and anthropogenic emissions in which fossil fuel combustion, metallurgic activities, and combustion waste have been described [39]. Recently, heavy-weight vehicle braking pads, which consist of copper and antimony alloys, have also been described as one of the main sources of antimony emissions [40].

After elimination of lead from gasoline in many parts of the world and due to the extensive use that antimony has in braking pads in heavy-weight vehicle traffic, antimony has now become one of the elements with greater enrichment in particulate matter in big cities. Studies carried out in urban cities of Argentina [41], Sweden [42], Japan [43], and Chile [44] among others have established that antimony in atmospheric particulate matter in urban zones presents concentrations above 100 mg kg^{-1} in particulate matter PM_{10}. In a study in Japan, antimony presented the highest enrichment factor, 10,000, out of all of the analyzed elements. Studies that show that vehicle traffic is a source of antimony contamination in the city airs provide sufficient evidence to suggest that antimony contamination is a common phenomenon on a global scale.

In a 1980's report from the USEPA, a summary of the standards of air quality on work places was provided, showing that most countries have adopted the value of 0.5 mg m^{-3} [45] assuming that workers complete an 8-h shift and 40-h work week [46]. If we use a daily air breathing of 10 m^3 as a standard we must accept that this regulation is clearly above the 360 µg TDI; therefore, we believe that this value needs to be considered as a doubtful reference either because it is outdated or based on occupational exposure. In this book, more information about limitations of the regulations on occupational exposure guidelines can be found in chapter 16. Modern legislation that considers the size and the concentrations of the aerosol has proposed more rigorous guidelines.

In the absence of a more stringent regulation and if, we use the above EPA value as a reference, we see that antimony levels are several order of magnitude below. For instance, in previous studies from Canada, United States, and Europe from the 1980s, in which particulate matter levels were reported to be higher than current ones in most of the big cities, antimony concentrations in the atmosphere ranged from 0.5 to 470 ng m^{-3} [47] and from 0.9 to 15.3 ng m^{-3} [48]. Much higher concentrations have been reported in the air of an indoor shooting range with peak antimony concentrations at 119 μg m^{-3} [49]. If we consider a daily volume of breathing air of 10 m^3, we find that antimony levels in cities are quite different from a TDI of 360 μg Sb. In the case of shooting ranges, these values could be exceeded. However, shooting ranges concern a limited occupational case. Exposure to antimony-contaminated atmospheres above the reference value of 0.5 mg m^{-3} was found in the Tyneside Factory in the United Kingdom in 1960 in which antimony values in the air were in the range of 0.53 to 36.70 mg m^{-3}. This particular factory was closed in 2002 [50]. A more recent case is presented in workers of glass and engineering plastic industries involving Sb_2O_3 in Taiwan, where reported levels in the air reached 2.5 ± 0.6 mg m^{-3} [51].

It is evident that human exposure to atmospheric particulate matter is most likely to occur in big cities. However, studies that have provided some evidence in which inspired antimony enters the human body in harmful amounts is scarce. For instance, it has been reported that harbour workers exposed to heavy vehicle traffic presented antimony levels in their blood (27 ± 9 ng g^{-1}) up to nine times the values in control groups [52]. On the other hand, workers in the mining sector presented anomalous antimony levels in urine [53]. The atmospheric route has been described as the main source of antimony for pregnant women in urban areas in Sabadell, Spain. This finding suggests that, despite higher dietary intake of antimony than that found by the inhalation route, atmospheric uptake of antimony may be significant in humans living in high vehicle traffic areas [54]. In the case of workers of Sb_2O_3 industry, antimony levels in blood of 4 ± 1 μg L^{-1} have been reported. Antimony concentrations 0.8 ± 0.2 μg L^{-1} were found in glass workers and 0.21 ± 0.06 μg L^{-1} in other workers [51].

Antimony levels in urine of people that work or attend shooting ranges are higher. Compare, for instance, values of those in the special forces (0.34 μg L^{-1}), maintenance staff (0.13 μg L^{-1}), and shooting instructors (0.32 μg L^{-1}) with the police officers before discharging their weapons (0.06 μg L^{-1}) and a reference value (0.04 μg L^{-1}) [55]. It is difficult to evaluate the degree of antimony-related toxicity at this level of enrichment since there are no recommended limits for antimony in urine. Reported concentrations of antimony in urine are in the range of 0.10 to 1.8 μg L^{-1} [56] and for the case of people exposed to antimony at shooting ranges, antimony levels are within this range or even lower. Lacking regulated values for antimony in urine, recommendations for urinary arsenic levels could be used as a reference (50 μg L^{-1}) [57]. If we consider that the TDI of arsenic is much lower than antimony, and if we take this difference as a basis to estimate a limit value for antimony in urine, it is easy to realize that reported values for antimony in urine are far below any criteria for toxic levels. Some studies have reported that the kidney tissue starts to deteriorate when exposed to levels of Sb(III) of 100 μg L^{-1} [58].

Despite the previous cases, human exposure to antimony through atmospheric particulate matter has not been studied on a global scale and the scarce information available remains anecdotals until more evidence is gathered. In conclusion, we can claim that the atmospheric route is a human exposure pathway at the global scale particularly important in cities with high vehicle traffic. We can also conclude that exposure to a contaminated atmosphere generates access for this metalloid into the human body, thus causing an increase in urinary and blood antimony levels. Fortunately, the degree of human exposure to antimony through this route has not yet reached levels that are harmful enough to our health so as to generate world alarm. Instead, the episodes in which these levels are seen occur in occupational niches; therefore, decisions to minimize the exposure degree of antimony on these specific types of working environments depend more on occupational regulations than on environmental ones.

15.6 Antimony in food

On average, approximately 5 µg of antimony is consumed in food on a daily basis, with 0.2 to 1.1 ng g^{-1} being found in meats, vegetables, and seafood [59]. Outside of the merely environmental context, processed food is a trophic route for human exposure to any contaminant. We have seen that bottled water, which could be considered within this category, is a typical case of a food product that is contaminated by the industrial process for its sale, specifically its container. However, more sophisticated products, that might be a potential source of contamination from the industrial production chain, need to be considered.

Ruiz-de-Cenzano et al. determined the antimony content in commercially baby foods where concentrations ranging from 0.6 to 6.9 ng g^{-1} are values far from reaching a TDI limit, especially for babies [60].

In the work of Hansen et al., it was reported that antimony levels in 7 out 26 commercial juice samples were above the established safe limits for drinking water (5 µg L^{-1} established by the Commission of the European Communities) with the highest value of 13.6 µg L^{-1} in glass bottle of sour cherry juice [61]. In a similar work by Savić et al., juices from seven manufacturers were analyzed, and in all samples the antimony levels were below the detection limit (LOD) of the analytical method (2 µg L^{-1}) [62]. Kilic et al. analysed 13 brands of energy drinks and reported antimony concentrations ranging from none detected to 0.42 µg L^{-1} [63]. Lachenmeier et al. studied the about chemical composition of vintage pre-ban absinthe in which antimony concentrations in 32 samples were below the LOD of the method of (0.001 mg L^{-1}) [64]. Antimony levels have been reported to be from 67 to 76 ng L^{-1} in red wine from Castelao [65]. The impact of processing wine was studied by Hopfer et al., and showed that it did not have a significant influence on the content of antimony in 65 red wine samples, some of which contained up to 0.033 µg L^{-1} [66]. Also, low antimony concentrations were found in 50 Nigerian fruit juices and soft drinks; all of them were between 0.21 and 2.77 µg L^{-1} [67].

In the work of Zand et al., a multielement analysis of weight loss–related foods in Spain was reported. Seventy-four samples of cereal bars, cereal cakes, and meals replacement shakes were analyzed and all antimony concentrations found were below 0.01 mg kg^{-1}, the authors calculated the daily intake for 100 g of these products and it was far below any tolerance parameter for antimony [68]. Zand et al. published another work analyzing eight samples from four popular brands of commercial infant foods in the United Kingdom, and all of the results demonstrated concentrations below the LOD of 0.6 µg 100 g^{-1} of commercial product (6 µg kg^{-1}) [69].

Karabagias et al. performed an analysis of 37 citrus honey samples in four countries and found concentrations levels between 0.45 and 0.73 mg kg^{-1} in these samples [70], which was far from a normal daily intake of these types of products after considering the TDI upper limit.

Overmeire et al. analyzed samples from 19 commercial egg production farms and 22 privately owned free-range egg farms in Belgium. Antimony concentrations in the products ranged from 0.89 to 9.02 µg kg^{-1} [71], which was far below the possible maximum tolerable intake.

In conclusion, none of the studies on processed food showed antimony levels that generated a risk of exceeding the TDI. Furthermore, it is important to note that, from all food, including those non-processed, vegetables are the only ones that presented high antimony levels. These levels could make them somewhat unsafe for human consumption. However, these cases in general are very few, and they are mainly grown around antimony mines as indicated in the review by Belzile et al [72].

15.7 Antimony in humans: absence of evidence or evidence of absence?

Perhaps the absence of conclusive evidence about antimony exposure points to the absence of human exposure to antimony beyond occupational-related exposures or those related to being in the vicinity of an antimony mine. However, with the purpose of discarding that this absence of evidence is not due to a lack of studies, it is pertinent to review the published literature about the presence of antimony in human beings.

In environmental toxicology, body fluids are the best samples to evaluate human exposure to a chemical substance because of their non-invasive nature, easy sampling and being related to recent exposure. Hair and nails are also useful. They are excretions of our body whose sampling is more complex due to their growth speed. They provide exposure information on larger time frames especially on chemical substances which tend to accumulate on them such as lead [73] or mercury [74].

Since antimony levels in some of these fluids and human excretions reflect such exposure (effect), they can complement or substitute for the lack of information on environmental exposure (cause). However, published literature does not support this hypothesis. Extensive reviews about human exposure to antimony with information about antimony concentrations in human tissues (hair, nails, teeth) [75], human blood [76], and human-excreted biofluids (urine, milk, saliva) [77] have been published. Regarding the presence of antimony in human tissues, Filella et al. showed that much of the published information might not be reliable because, in addition to lacking adequate certified reference materials, often antimony concentrations were very close to the detection limits of the analytical techniques used. These constrains have impeded the establishment of reliable background concentrations of antimony for human hair, nails, and teeth [75].

The situation is similar in the case of human excreted biofluids. Likewise, a great portion of the published information shows a lack of validated reference materials and concentrations that were close to the LOD. However, in the case of urine, the authors suggested an upper limit of 0.1 μg L^{-1} as a reference value in healthy unexposed individuals. Besides, the authors concluded that the urine analysis appears justified when examining occupational and environmental exposure to nearby localized sources [77].

Finally, in the case of human blood, the problem of lack of adequate validated reference materials and concentrations close to the LOD still persists. However, the authors suggest a reference value for nonexposed people of 1 μg L^{-1} although its use as an indicator of antimony exposure is kept in a plausible status for future studies [76].

Based on discussed data, we can confirm that human exposure to Sb is very unlikely and very infrequent. Human exposure to antimony at the food intake level is mainly present in the case of non-processed foods, such as vegetables, which originate from terrestrial ecosystems, or in the case of food coming from marine ecosystems, when contaminated mainly by mining activity. In the case of human exposure to antimony via atmospheric contamination in cities with vehicle traffic, it is not completely clear whether this human exposure to antimony is harmful because it is below current regulations although such regulations in our judgement are underestimated if we compare them to the TDI. Despite the previously mentioned issues, given that this atmospheric contamination has not reflected frequent cases of antimony enrichment in body fluids, it is believed that there is not enough evidence to consider this path of human exposure alarming. Only occupational antimony exposure at shooting ranges and mining facilities have yielded high antimony levels in human urine.

This type of exposure is mainly of an occupational character, and based on our judgement, this type of exposure deserves concern and policies involving occupational safety but does not justify global environmental concern. Due to all of the reasons mentioned above, we can confirm that human exposure to antimony is still low and limited to specific occupational circumstances or mining environments that are clearly identified and avoidable. In this case, it can be stated that for human exposure to antimony, the absence of evidence is evidence of absence. However, these cases of occupational exposure are sufficient to apply the ethical precautionary principle to avoid or at least minimize human exposure to antimony coming from

Tab. 15.2: Risky exposure to antimony from environmental compartments.

Environmental compartment	Risky human exposure
Water	Nearby Sb mines [9], hot springs [10]
Air	APM from heavy weight vehicle traffic[a] [41–44]
Soil	Nearby Sb mines [20–23]
Sediment	Nearby Sb mines [32]
Vegetables	Vegetables around Sb mines[b] [28–30]
Marine biota	Nearby mining and smelting activities [36]
Processed food	–
Occupational	Shooting ranges [49, 55], Sb mines facilities [50–51]

[a]High enrichment factors but concentration still remains below guidelines.
[b]Few species were capable to accumulate antimony.

occupational conditions and also to avoid the use of antimony compounds in braking pads, ideally looking for replacements for flame retardants, catalysts, and metal alloys.

References

[1] Bagherifam S, Brown TC, Fellows CM, Naidu R. Bioavailability of arsenic and antimony in terrestrial ecosystems: A review. Pedosphere 2019, 29, 681–720.

[2] Smichowski P. Antimony in the environment as a global pollutant: A review on analytical methodologies for its determination in atmospheric aerosols. Talanta 2008, 75, 2–14.

[3] Anderson C. Antimony production and commodities. In: Dunne RC, Kawatra SK, Young CA., ed. SME Mineral Processing & Extractive Metallurgy Handbook. Society for Mining, Metallurgy, and Exploration (SME), 2019.

[4] World Health Organization (WHO). Chemical fact sheets-Antimony. In: Guidelines for Drinking-Water Quality, 4th edition, incorporating the 1st addendum, 2017 (Accessed February 02, 2021, at http://www.who.int/publications/i/item/9789241549950).

[5] Oorts K, Smolders E, Degryse F, Buekers J, Gascó G, Cornelis G, Mertens J Solubility and toxicity of antimony trioxide (Sb_2O_3) in soil. Soil. Environ Sci Technol 2008, 42, 4378–4383.

[6] Barrera C, Lopez S, Aguilar L, Mercado L, Bravo M, Quiroz W. Pentavalent antimony uptake pathway through erythrocyte membranes: Molecular and atomic fluorescence approaches. Anal Bioanal Chem 2016, 408, 2937–2944.

[7] Corrales I, Barceló J, Bech J, Poschenrieder C. Antimony accumulation and toxicity tolerance mechanisms in Trifolium species. J Geochem Explor 2014, 147, 167–172.

[8] Verdugo M, Encinar JR, Costa-Fernández JM, Menendez-Miranda M, Bouzas-Ramos D, Bravo M, Quiroz W. Study of conformational changes and protein aggregation of bovine serum albumin in presence of Sb(III) and Sb(V). PLoS ONE 2017, 12, e0170869.

[9] He M, Wang N, Long X, Zhang C, Ma C, Zhong Q, Wang A, Wang Y, Pervaiz A, Shan J. Antimony speciation in the environment: Recent advances in understanding the biogeochemical processes and ecological effects. J Environ Sci 2019, 75, 14–39.
[10] Filella M, Belzile N, Chen YW. Antimony in the environment: A review focused on natural waters I. Occurrence. Earth Sci Rev 2002, 57, 125–176.
[11] Alderton D, Serafimovski T, Burns L, Tasev G. Distribution and mobility of arsenic and antimony at mine sites in FYR Macedonia. Carpath J Earth Environ Sci 2014, 9, 43–56.
[12] Westerhoff P, Prapaipong P, Shock E, Hillaireau A. Antimony leaching from polyethylene terephthalate (PET) plastic used for bottled drinking water. Water Res 2008, 42, 551–556.
[13] de Jesus A, Dessuy MB, Huber CS, et al. Determination of antimony in pet containers by direct analysis of solid samples using graphite furnace atomic absorption spectrometry and leaching studies. Microchem J 2016, 124, 222–227.
[14] Rüsz Hansen H, Pergantis SA. Detection of antimony species in citrus juices and drinking water stored in PET containers. J Anal At Spectrom 2006, 21, 731–733.
[15] Shotyk W, Krachler M, Chen B. Contamination of Canadian and European bottled waters with antimony from PET containers. J Environ Monit 2006, 8, 288–292.
[16] Filella M. Antimony and PET bottles: Checking facts. Chemosphere 2020, 261, 127732.
[17] Charlier P, Bou Abdallah F, Bruneau R, Jacqueline S, Augias A, Bianucci R, Perciaccante A, Lippi D, Appenzeller O, Rasmussen KL. Did the Romans die of antimony poisoning? The case of a Pompeii water pipe (79 CE). Toxicol Lett 2017, 281, 184–186.
[18] World Health Organization (WHO). Antimony in Drinking-water. Background document for development of WHO Guidelines for Drinking-water Quality, WHO/SDE/WSH/03.04/74, 2003. (Accessed February 09, 2021, at https://apps.who.int/iris/handle/10665/75375).
[19] Ministry of the Environment, Finland. Government Decree on the Assessment of Soil Contamination and Remediation Needs. 214/2007 (Accessed February 02, 2021, at www.ecolex.org/details/legislation/government-decree-on-the-assessment-of-soil-contamination-and-remediation-needs-no-214-of-2007-lex-faoc113198)
[20] Pérez-Sirvent C, Martínez-Sánchez MJ, Martínez-López S, Hernández-Córdoba M. Antimony distribution in soils and plants near an abandoned mining site. Microchem J 2011, 97, 52–56.
[21] Okkenhaug G, Zhu YG, Luo L, Lei M, Li X, Mulder J. Distribution, speciation and availability of antimony (Sb) in soils and terrestrial plants from an active Sb mining area. Environ Pollut 2011, 159, 2427–2434.
[22] Okkenhaug G, Grasshorn Gebhardt K-A, Amstaetter K, Lassen Bue H, Herzel H, Mariussen E, Rossebø Almås Å, Cornelissen G, Breedveld GD, Rasmussen G, Mulder J. Antimony (Sb) and lead (Pb) in contaminated shooting range soils: Sb and Pb mobility and immobilization by iron based sorbents, a field study. J Hazard Mater 2016, 307, 336–343.
[23] Lafond S, Blais JF, Martel R, Mercier G. Chemical leaching of antimony and other metals from small arms shooting range soil. Water Air Soil Pollut 2013, 224, 1–15.
[24] Qi C, Wu F, Deng Q, Liu G, Mo C, Liu B, Zhu J. Distribution and accumulation of antimony in plants in the superlarge Sb deposit areas, China. Microchem J 2011, 97, 44–51.
[25] Pierart A, Shahid M, Séjalon-Delmas N, Dumat C. Antimony bioavailability: Knowledge and research perspectives for sustainable agricultures. J Hazard Mater 2015, 289, 219–234.
[26] Affholder M-C, Prudent P, Masotti V, Coulomb B, Rabier J, Nguyen-The B, Laffont-Schwob I. Transfer of metals and metalloids from soil to shoots in wild rosemary (*Rosmarinus officinalis* L.) growing on a former lead smelter site: Human exposure risk. Sci Total Environ 2013, 454–455, 219–229.
[27] Hammel W, Debus R, Steubing L. Mobility of antimony in soil and its availability to plants. Chemosphere 2000, 41, 1791–1798.

[28] Huang Y, Chen Z, Liu W. Influence of iron plaque and cultivars on antimony uptake by and translocation in rice (Oryza sativa L.) seedlings exposed to Sb(III) or Sb(V). Plant Soil 2012, 352, 41–49.
[29] Ren JH, Ma LQ, Sun HJ, Cai F, Luo J. Antimony uptake, translocation and speciation in rice plants exposed to antimonite and antimonate. Sci Total Environ 2014, 475, 83–89.
[30] Feng R, Wei C, Tu S, Ding Y, Wang R, Guo J. The uptake and detoxification of antimony by plants: A review. Environ Exp Bot 2013, 96, 28–34.
[31] Schreck E, Foucault Y, Sarret G, Sobanska S, Cécillon L, Castrec-Rouelle M, Uzu G, Dumat C. Metal and metalloid foliar uptake by various plant species exposed to atmospheric industrial fallout: Mechanisms involved for lead. Sci Total Environ 2012, 427–428, 253–262.
[32] Wang X, He M, Xi J, Lu X. Antimony distribution and mobility in rivers around the world's largest antimony mine of Xikuangshan, Hunan Province, China. Microchem J 2011, 97, 4–11.
[33] Herath I, Vithanage M, Bundschuh J. Antimony as a global dilemma: Geochemistry, mobility, fate and transport. Environ Pollut 2017, 223, 545–559.
[34] Wu XD, Song JM, Li XG, Yuan HM, Li N. Behaviors of dissolved antimony in the Yangtze River Estuary and its adjacent waters. J Environ Monit 2011, 13, 2292–2303.
[35] Obiakor MO, Tighe M, Pereg L, Wilson SC. Bioaccumulation, trophodynamics and ecotoxicity of antimony in environmental freshwater food webs. Crit Rev Environ Sci Technol 2017, 47, 2208–2258.
[36] Dovick MA, Kulp TR, Arkle RS, Pilliod DS. Bioaccumulation trends of arsenic and antimony in a freshwater ecosystem affected by mine drainage. Environ Chem 2016, 13, 149–159.
[37] Kawamoto Y, Morisawa S. The distribution and speciation of antimony in river water, sediment and biota in Yodo River, Japan. Environ Technol 2003, 24, 1349–1356.
[38] Culioli J-L, Fouquoire A, Calendini S, Mori C, Orsini A. Trophic transfer of arsenic and antimony in a freshwater ecosystem: A field study. Aquatic Toxicology 2009, 94, 286–293.
[39] Shotyk W, Krachler M, Chen B. Anthropogenic impacts on the biogeochemistry and cycling of antimony. Met Ions Biol Syst 2005, 44, 171–203.
[40] Von Uexküll O, Skerfving S, Doyle R, Braungart M. Antimony in brake pads-a carcinogenic component? J Clean Product 2005, 13, 19–31.
[41] Gómez DR, Giné MF, Bellato ACS, Smichowski P. Antimony: A traffic-related element in the atmosphere of Buenos Aires, Argentina. J Environ Monit 2005, 7, 1162–1168.
[42] Sternbeck J, Sjödin Å, Andréasson K. Metal emissions from road traffic and the influence of resuspension – Results from two tunnel studies. Atmos Environ 2002, 36, 4735–4744.
[43] Furuta N, Iijima A, Kambe A, Sakai K, Sato K. Concentrations, enrichment and predominant sources of Sb and other trace elements in size classified airborne particulate matter collected in Tokyo from 1995 to 2004. J Environ Monit 2005, 7, 1155–1161.
[44] Quiroz W, Cortés M, Astudillo F, et al. Antimony speciation in road dust and urban particulate matter in Valparaiso, Chile: Analytical and environmental considerations. Microchem J 2013, 110, 266–272.
[45] United States Environmental Protection Agency (US EPA). Ambient Water Quality Criteria for Antimony. Office of Water Regulations and Standards. Washington DC, USA, 1980 (Accessed February 02, 2021, at www.epa.gov/wqc/ambient-water-quality-criteria-antimony).
[46] Agency for Toxic Substances and Disease Registry (ATSDR). Toxicological profile for arsenic. Atlanta, GA: U.S. Department of Health and Human Services, Public Health Service. 2007. (Accessed February 02, 2021, at www.atsdr.cdc.gov/toxprofiles/tp.asp?id=22&tid=3)
[47] Schroeder WH, Dobson M, Kane DM, Johnson ND. Toxic Trace Elements Associated with Airborne Particulate Matter: A Review. JAPCA 1987, 37, 1267–1285.

[48] Smichowski P, Gómez DR, Dawidowski LE, Fernanda Giné M, Claudia Sánchez Bellato A, Reich SL. Monitoring trace metals in urban aerosols from Buenos Aires city. Determination by plasma-based techniques. J Environ Monit 2004, 6, 286–294.
[49] Dams R, Vandecasteele C, Desmet B, et al. Element concentrations in the air of an indoor shooting range. Sci Total Environ 1988, 77, 1–13.
[50] McCallum RI. Occupational exposure to antimony compounds. J Environ Monit 2005, 7, 1245–1250.
[51] Wu CC, Chen YC. Assessment of industrial antimony exposure and immunologic function for workers in Taiwan. Int J Environ Res Public Health 2017, 14, 689–677.
[52] Quiroz W, De Gregori I, Basilio P, Bravo M, Pinto M, Lobos MG. Heavy weight vehicle traffic and its relationship with antimony content in human blood. J Environ Monit 2009, 11, 1051–1055.
[53] Quiroz W, Arias H, Bravo M, Pinto M, Lobos MG, Cortés M. Development of analytical method for determination of Sb(V), Sb(III) and TMSb(V) in occupationally exposed human urine samples by HPLC-HG-AFS. Microchem J 2011, 97, 78–84.
[54] Fort M, Grimalt JO, Querol X, Casas M, Sunyer J. Evaluation of atmospheric inputs as possible sources of antimony in pregnant women from urban areas. Sci Total Environ 2016, 544, 391–399.
[55] Vandebroek E, Haufroid V, Smolders E, Hons L, Nemery B. Occupational exposure to metals in shooting ranges: A biomonitoring study. Saf Health Work 2019, 10, 87–94.
[56] Caroli AA S, Coni E, Petrucci F, Senofonte O, Violante N. The assessment of reference values for elements in human biological tissues and fluids: A systematic review. Crit Rev Anal Chem 1994, 24, 363–398.
[57] Agency for Toxic Substances and Disease Registry (ATSDR). Toxicological profile for arsenic. Atlanta, GA: U.S. Department of Health and Human Services, Public Health Service. 2007. (Accessed February 02, 2021, at www.atsdr.cdc.gov/toxprofiles/tp.asp?id=22&tid=3)
[58] Roldán N, Pizarro D, Verdugo M, Salinas-Parra N, Quiroz W, Reyes-Martinez C, Figueroa S, Quiroz C, Gonzalez AA. Antimony(III) induces fibroblast-like phenotype, profibrotic factors and reactive oxygen species in mouse renal cells. Environ Chem 2020, 17, 182–190.
[59] Cooper R, Harrison A. The exposure to and health effects of antimony. Indian J Occup Environ Med 2009, 13, 3–10.
[60] Ruiz-de-Cenzano M, Rochina-Marco A, Cervera ML, de la Guardia M. Evaluation of the content of antimony, arsenic, bismuth, selenium, tellurium and their inorganic forms in commercially baby foods. Biol Trace Elem Res 2017, 180, 355–365.
[61] Hansen C, Tsirigotaki A, Bak SA, et al. Elevated antimony concentrations in commercial juices. J Environ Monit 2010, 12, 822–824.
[62] Savić SR, Petrović SM, Stamenković JJ, Petronijević BZ. The presence of minerals in clear orange juices. Advanced Technologies 2015, 4, 71–78.
[63] Kilic S, Cengiz MF, Kilic M. Monitoring of metallic contaminants in energy drinks using ICP-MS. Environ Monit Assess 2018, 190, 202–210.
[64] Lachenmeier DW, Nathan-Maister D, Breaux TA, Sohnius EM, Schoeberl K, Kuballa T. Chemical composition of vintage preban absinthe with special reference to thujone, fenchone, pinocamphone, methanol, copper, and antimony concentrations. J Agric Food Chem 2008, 56, 3073–3081.
[65] Kaya AD, Bruno De Sousa R, Curvelo-Garcia AS, Ricardo-Da-Silva JM, Catarino S. Effect of Wood Aging on Wine Mineral Composition and $^{87}Sr/^{86}Sr$ Isotopic Ratio. J Agric Food Chem 2017, 65, 4766–4776.
[66] Hopfer H, Nelson J, Collins TS, Heymann H, Ebeler SE. The combined impact of vineyard origin and processing winery on the elemental profile of red wines. Food Chem 2015, 172, 486–496.

[67] Roberts II, Orisakwe OE. Evaluation of potential dietary toxicity of heavy metals in some common Nigerian beverages: A look at antimony, tin and mercury. QScience Connect 2011, 2, 1–10.
[68] Zurita-Ortega A, Cervera-Mata A, Delgado G, Zurita-Ortega F, Rufián-Henares JÁ, Pastoriza S. Mineral profile of weight loss related foods marketed in Spain. Food Chem 2020, 313, 126156.
[69] Zand N, Chowdhry BZ, Wray DS, Pullen FS, Snowden MJ. Elemental content of commercial 'ready to-feed' poultry and fish based infant foods in the UK. Food Chem 2012, 135, 2796–2801.
[70] Karabagias IK, Louppis AP, Karabournioti S, Kontakos S, Papastephanou C, Kontominas MG. Characterization and geographical discrimination of commercial Citrus spp. honeys produced in different Mediterranean countries based on minerals, volatile compounds and physicochemical parameters, using chemometrics. Food Chem 2017, 217, 445–455.
[71] Van Overmeire I, Pussemier L, Hanot V, De Temmerman L, Hoenig M, Goeyens L. Chemical contamination of free-range eggs from Belgium. Food Addit Contam 2006, 23, 1109–1122.
[72] Belzile N, Chen YW, Filella M. Human exposure to antimony. I. Sources and intake. Crit Rev Environ Sci Technol 2011, 41, 1309–1373.
[73] Niculescu T, Dumitru R, Botha V, Alexandrescu R, Manolescu N. Relationship between the lead concentration in hair and occupational exposure. Br J Ind Med 1983, 40, 67–70.
[74] Airey D. Mercury in human hair due to environment and diet: A review. Environ Health Perspect 1983, 52, 303–316.
[75] Filella M, Belzile N, Chen Y-W. Human exposure to antimony. II. Contents in some human tissues often used in biomonitoring (Hair, Nails, Teeth). Crit Rev Environ Sci Technol 2012, 42, 1058–1115.
[76] Filella M, Belzile N, Chen Y-W. Human exposure to antimony. IV. Contents in human blood. Crit Rev Environ Sci Technol 2013, 43, 2071–2105.
[77] Filella M, Belzile N, Chen Y-W. Human exposure to antimony. III. Contents in some human excreted biofluids (urine, milk, saliva). Crit Rev Environ Sci Technol 2013, 43, 162–214.

Caroline Braibant

Chapter 16
Legislating antimony and its compounds

16.1 Introduction

This chapter aims to address the most recent legislative status of antimony and its compounds in terms of environmental and health legislation, also known as EHS or HSE legislation. While EHS legislation constitutes the big-picture thinking, regulations are one sort of legislative instrument made and maintained by an authority, justifying regulatory action.

Regulatory scrutiny on antimony is first and foremost the result of an unfortunate "guilty by association" approach. Indeed, metals and metalloids are typically known as "low hanging fruits" in the world of legislations, and what other reason would be needed to scrutinize such an inorganic chemical when it is on top used as synergist for and with some flame-retardant compounds carrying a very stigmatized "halogen" label.

With time and increasing scientific evidence becoming available, the proportionality and the efficiency of the current regulatory pressure applied on antimony and its compounds may be questioned. Indeed, despite having intrinsic toxicity properties as all naturally occurring or synthetic/man-made chemicals, antimony's toxicity is expressed at excessively high concentrations, which are very uncommon in current production and use practices. For regulatory control over the production and use of antimony and antimony compounds to be of added value, the very low potency of antimony's hazards and exposures to antimony should be considered.

This brings us to the topic of hazard and exposure, and ultimately, risk and safety assessments. Regulating chemicals involves understanding their intrinsic potential to cause adverse impacts on health or the environment (their toxicity or hazard) as well as the conditions under which a human being or the environment may come into contact with that chemical (the exposure). It is only when the exposure conditions meet or exceed the hazard levels that there is a risk, and hence, the potential need to manage it. Managing risks can be done via so-called voluntary initiatives, taken by each producer and user to minimize or control the exposure; or via regulatory measures, imposed on all producers and users by a competent authority.

The following sections summarize the main regulatory measures applicable to the production and use of antimony and antimony compounds in particular in the EU, which is one of the leading regions in setting EHS standards for chemicals.

Caroline Braibant, Secretary-General 2016–2020, International Antimony Association Belgium.
caroline.braibant@antimony.com

https://doi.org/10.1515/9783110668711-016

The reference contact point to obtain information about EHS legislation is the International Antimony Association (i2a), a nonprofit industry association established in Brussels, Belgium. The most important information and contact details of i2a can be found on their website: www.antimony.com.

16.2 Criticality and recycling

Antimony is recognized as a critical raw material/mineral in the EU [1], the United States [2], Japan, and Australia [3]. These regions have defined antimony as being crucial to their economies while at the same time, as being subject to frequent or possibly unreliable and hindered access to. This means that the regions are highly dependent on imports (with the related risk of supply shortage), confronted with limited substitution options to replace antimony in its applications, and in high need of the material which provides an essential function in the manufacture of products, which are key to the regional economy and/or national security.

Before addressing any EHS regulatory measures, it is important to understand what this criticality entails. Indeed, when a state realizes that it is in critical need of a material which is subject to possible shortage, two solutions may be proposed:

i. Develop local sourcing of that material, and/or
ii. Reduce the dependence on it, via substitution.

While the first solution stimulates local exploration, extraction, and recycling projects (produce more), the second solution pushes research and development initiatives to find alternative chemicals, techniques, or technologies (use less). This "produce more vs use less" adds confusion to the future of antimony, which is already unpredictable due to the battle that exists between fire safety, sustainability/circularity, and EHS regulatory objectives, among others.

16.3 Classification

In EHS legislation terms, classifying a chemical involves putting that chemical in a given "hazard class" based on the nature of the physical, health, or environmental hazard that can be caused by that chemical. Within each "hazard class," a "hazard category" can be further assigned, specifying the severity of the hazard. The purpose of the classification is to inform about the intrinsic hazards of the chemical, which is the first step of risk management: awareness and communication. The label and safety data sheet (SDS) accompanying classified chemicals further assists with the communication in order to minimize the likelihood of unsafe handling or practices with the chemical.

The United Nations (UN) have developed a Globally Harmonized System (GHS) of classification and labeling of chemicals aimed to harmonize the classification, labeling, and SDS of chemical worldwide, beyond the end points that were already harmonized under the UN Transport of Dangerous Goods. The GHS contains a number of blocks that are progressively taken up by nations worldwide, towards a full harmonization.

Whereas in some countries and regions, classifications are national or regional recommendations, in others, classifications are established by authorities. This is the case of the European Classification and Labelling Regulation (CLP) [4], which contains, in Part 3 of its Annex VI, a table containing mandatory classifications for a number of chemicals, including the ones shown in Tab. 16.1 for antimony and antimony compounds.

The classifications included in Annex VI of the EU CLP Regulation are updated on a regular basis to take into account the updates to the GHS as well as new information and harmonized classification decisions made by European authorities for each substance.

Human health hazards are those caused by the "toxicity" of the chemical, while environmental hazards are those caused by the "eco-toxicity" of the chemical. These hazards are investigated with laboratory animals and conditions. (Eco-)toxicity is intrinsic to each and every chemical. The only difference between chemicals is the concentration needed to cause that (eco-)toxicity and the mode of action leading to the expression of that (eco-)toxicity. The smaller the concentration needed, the lower the category of the hazard class will be. Lower classification category numbers can hence be associated with a high potency and severity of effects.

In terms of mode of action, this is particularly important for hazard classes such as carcinogenicity, mutagenicity, and reproductive toxicity. The GHS for these classes is however established on the basis of available evidence rather than on the severity of that observed evidence. The European Chemicals Agency (ECHA) in charge of implementing the CLP Regulation in the EU has developed comprehensive guidance [5] on the implementation of the GHS classification criteria, which can assist authorities, producers, and users in deriving the applicable classification for each chemical.

The main hazards and GHS classifications of antimony substances are the following:

- Skin corrosive 1B: Chlorides, no matter the nature of the counter-ion, are known to be corrosive chemicals. As such, antimony chlorides, due to the very low pH of the compounds, have a harmonized classification as skin corrosive. The category 1B indicates that compared to other corrosive chemicals, the antimony chlorides may require more quantity to cause skin irritation or burns.
- Specific Target Organ Toxicity Single Exposure category 3: Similar to the effect chlorides can have on "external" skin, irritations and burns can also be caused inside the respiratory tract, when the chloride vapors are inhaled. Due to this, antimony chlorides have a harmonized classification as STOT SE 3, which is the GHS classification assigned to all respiratory irritants.

Tab. 16.1: Harmonized and recommended classifications for antimony compounds under the CLP Regulation [4].

CAS number	Name	Recommended classification and labeling			
		Phys-chem	**Human health**	**Environment**	**Transport**
7440-36-0	Antimony metal	Not classified	Not classified	Not classified	Not dangerous
7440-36-0	Antimony metal powder	Not classified	Carc. 2, H351 STOT Rep. Exp. 2, H373	Not classified	UN No.: 2871 class: 6.1
1309-64-4	Diantimony trioxide	Not classified	Carc. 2, H351 STOT Rep. Exp. 2, H373	Not classified	Not dangerous
13145-04-6	Diantimony trisulfide	Not classified	Carc. 2, H351 STOT Rep. Exp. 2, H373	Not classified	Not dangerous
1314-60-9	Antimony trichloride	Not classified	Skin corr. 1B, H314, STOT SE 3 H335 – C ≥ 5%	Aquatic chronic 2, H411	UN No.: 1733, class 8, corrosive
29736-75-2	Diantimony tris ethylene glycolate	Not classified	Acute tox. 4, H302, H332	Aquatic chronic 2, H411	UN No.: 3077, class: 9
1314-60-9	Diantimony pentoxide	No classified	Not classified	Not classified	Not dangerous
7647-18-9	Antimony pentachloride	Oxid. liquid 2, H272	Skin corr. 1B, H314, STOT SE 3 H335 – C ≥ 5%	Aquatic chronic 2, H411	UN No.: 1730, class 8, corrosive
15432-85-6	Sodium antimonate	Not classified	Acute tox. 4, H302, H332	Aquatic chronic 2, H411	UN No.: 3077, class: 9
33908-66-6	Sodium hexahydroxoantimonate	Not classified	Acute tox. 4, H302, H332	Aquatic chronic 2, H411	UN No.: 3077, class: 9
12208-13-8	Potassium hexahydroxoantimonate	Not classified	Acute tox. 4, H302, H332	Aquatic chronic 2, H411	UN No.: 3077, class: 9

– Acute oral toxicity category 3 or 4: Antimony and antimony compounds are known to cause gastric irritation. As such, some of them carry a harmonized classification as having a low to medium acute oral toxicity. Indeed, the categories 3 and 4 indicate a relative mild oral toxicity. Quantitatively speaking, this means that antimony compounds carrying an acute oral toxicity category 3 have a lethal dose or LD_{50}[1] between 50 and 300 mg kg^{-1} body weight of the tested animal. Those carrying an acute oral toxicity category 4 have a LD_{50} between 300 and 2,000 mg kg^{-1} body weight.

- Carcinogenicity category 2: Diantimony trioxide has a mandatory classification of "Suspected human carcinogen." This classification is also that retained by the International Agency for Research on Cancer (IARC) [6]. The category 2 is assigned because the available carcinogenicity evidence is not sufficiently convincing to classify diantimony trioxide as category 1A or 1B. Instead, only limited evidence exists, and questions remain about the human relevance of the available animal data. It is important to note that this classification is based on carcinogenicity evidence observed in inhalation studies only, and as such is associated to that specific exposure route by producers and users of diantimony trioxide. In addition to this mandatory classification, the available evidence suggests that a STOT Repeated Exposure (RE) category 2 for lung also applies to antimony substances in respirable (<4 µm) particle-sized powders. Although this is not included in Annex VI of CLP, it is a self-classification recommended by authorities and implemented on a "voluntary" basis by most antimony and antimony compounds' powders suppliers.
- Aquatic chronic toxicity category 2: The toxicity of antimony and antimony compounds to aquatic organisms is relatively mild in that the no observed concentration (NOEC) on fish, crustacea, and algae or other aquatic plants is >1 mg L^{-1}.

More information about the hazards covered by the classification mentioned earlier can be found in chapters 14 and 15. It is important to note that by the time of writing the three chapters, ECHA has requested the generation of additional mutagenicity, reproductive toxicity, repeated dose inhalation, and oral toxicities about antimony metal and four antimony trivalent compounds. Once these study results become available, they will be utilized to review the classifications applicable to each substance.

16.4 Risk assessment

The first risk assessments on antimony substances were performed by the United States Environmental Protection Agency (USEPA) in the late 1990s [7]. In 2008, the EU risk-assessed diantimony trioxide under the former Existing Substances Regulation 793/93 and concluded that there was no need for further information and/or

1 Dose required to kill half the members of a tested population after a specified test duration.

testing and no need for risk reduction measures beyond those which were already in place [8].

Risk assessments are performed by many authorities aiming to understand the need to regulate chemicals. Indeed, only if a risk can be expected, would a protective regulatory measure be justified. The most comprehensive regulatory system requiring the risk assessment of chemicals to ensure their safe production, trade, and use is the EU-REACH Regulation [9]. REACH stands for Registration, Evaluation and Authorisation of CHemicals. In the Registration, industry is obliged to risk assess each chemical it places on the EU market in volumes of 1 t y^{-1} or more per legal entity. Some of the Registration Dossiers (risk assessment files) submitted by industry to ECHA are then scrutinized as part of the Evaluation, which may result in a call for more evidence or for risk management measures (such as restriction, authorization, or other measures outside REACH), as described in the ECHA scheme mentioned later (Fig. 16.1).

ECHA‘s decision-making is based on three elements: the hazard assessment, the exposure assessment, and the risk characterization ratio (RCR) calculation.

During the hazard assessment, all the available toxicity and ecotoxicity evidence is reviewed in order to understand the hazard in its nature and manifestation and to establish a level below which the hazard is not expected to express itself. This is, in terms of toxicity: the Derived No Effect Level (DNEL) and in terms of ecotoxicity: the Predicted No Effect Concentration (PNEC). The DNEL and PNEC provide an indication of the safe level above which human exposure and environmental release should be avoided or controlled, respectively.

The exposure assessment is performed to determine whether the production or use of the chemical can be performed under conditions that would meet and not exceed that DNEL and PNEC. The exposure assessment is performed by measuring actual exposure to workers or environmental releases, modeling these, or through a combination of measured evidence and modeling. This results in workplace exposure levels or predicted environmental concentrations (PEC) that can be compared with the DNEL and PNEC, respectively, in order to calculate the risk.

According to ECHA‘s Guidance on Information Requirements and Chemical Safety Assessment, Part E: Risk Characterisation [11], the mathematical formula to calculate the RCR is

$$\text{RCR} = \frac{\text{PEC}}{\text{PNEC}} \text{ or } \frac{\text{Exposure}}{\text{DNEL}}$$

Only if the DNEL is far inferior to the exposure levels, and the PNEC far inferior to the PEC, would the resulting RCR be below 1, and provide an indication of safety. The smaller the ratio is (below 1), the more confidence this brings in the likely safety of the use for which the RCR has been calculated.

In the EU-REACH Dossiers submitted for antimony metal and nine antimony compounds, RCRs below 1 have been obtained for all substances and uses. The

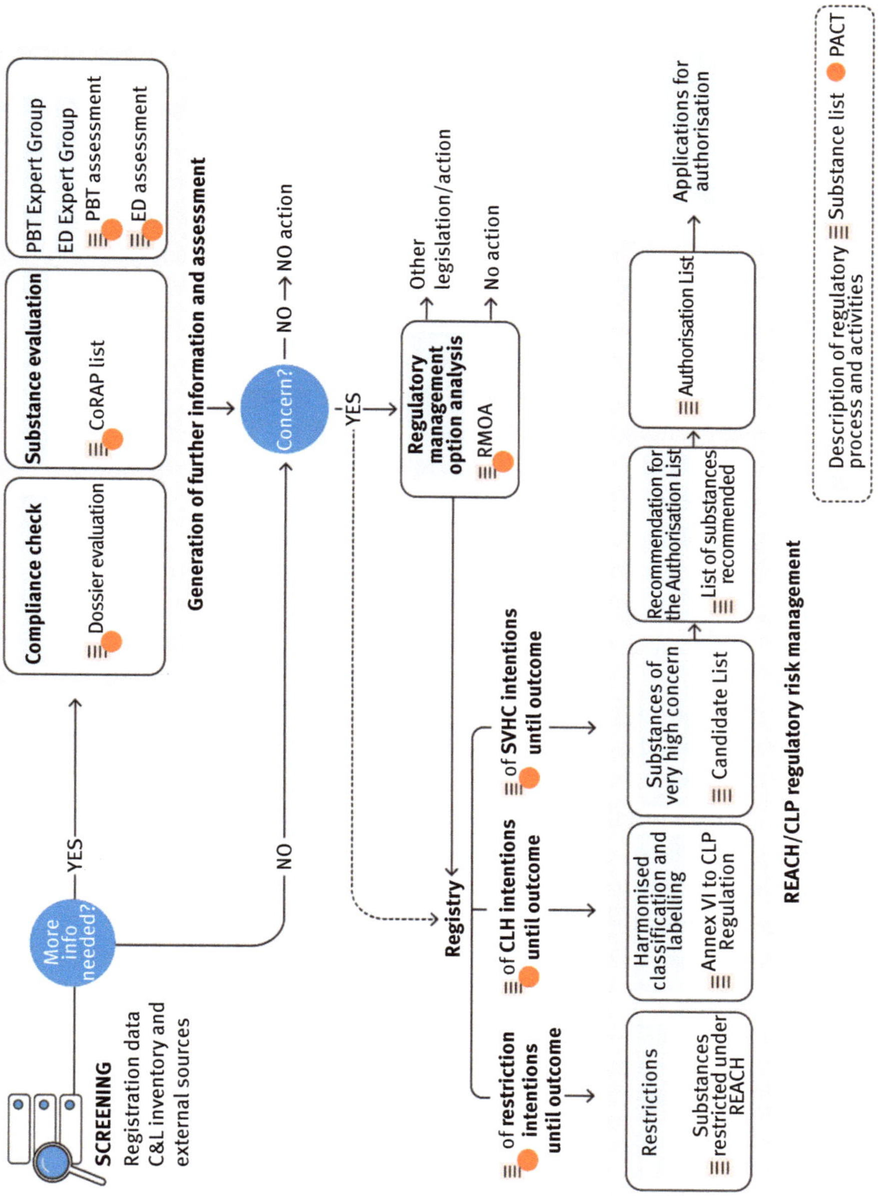

Fig. 16.1: REACH and CLP machinery serving ECHA's Integrated Regulatory Strategy and the SVHC Roadmap [10].

DNELs and PNECs have been calculated for each substance using the available evidence, weight of evidence, and read-across/grouping, as well as the applicable assessment or uncertainty factors.

PNECs for antimony substances have been calculated for fresh- and marine water, as well as for the sediment and terrestrial compartments. For example, the PNECs for fresh- and marine water are, respectively, 0.113 mg L^{-1} and 0.0113 mg L^{-1}. This means that provided there is not more than these concentrations released to each type of water basin, there is no risk for the aquatic environment. Environment and Climate Change Canada also concluded in 2020 that the 11 antimony compounds that they risk assessed do not meet the environmental toxicity criteria of the Canadian Environmental Protection Act [12].

DNELs in particular, have been calculated for the following:
- Long-term, inhalation respirable,[2] local effects (mg m^{-3}) for workers
- Long-term, inhalation respirable, local effects (mg m^{-3}) for consumers
- Long-term, inhalation inhalable,[3] local effects (mg m^{-3}) for workers
- Long-term, inhalation inhalable, local effects (mg m^{-3}) for consumers
- Long-term, oral, systemic effects (mg kg^{-1} bw d^{-1}) for consumers
- Long-term, dermal, systemic effects (mg kg^{-1} bw d^{-1}) for workers
- Long-term, dermal, systemic effects (mg kg^{-1} bw d^{-1}) for consumers

The lower DNELs are typically the inhalation respirable ones,[4] where exposure should be prevented the most, considering the (inhalation) carcinogenicity and lung toxicity that may result from prolonged and excessive breathing of fine particles of antimony substances. The highest DNELs are those associated with the long-term dermal effects, as antimony is known to have a very low dermal absorption rate, and the exposure via the skin would therefore require higher quantities before exceeding a safe threshold.

It should be noted that both hazard and exposure assessments evolve over time, as new evidence, approaches and requirements come into play. The EU-REACH Dossiers in particular are subject to regular updates by the registering industry, which will communicate the resulting DNELs, PNECs, and RCRs per use in their extended SDS, that is, SDS containing so-called exposure scenarios, which

2 Fraction that can be inhaled and reach the deep lung, typically measuring less than 4 µm in diameter.

3 Fraction that can be inhaled but will be retained in the upper lung and coughed out or swallowed, measuring between 4 and 100 µm.

4 For example, the lowest DNEL long-term, inhalation *respirable*, local effects for workers is that for antimony metal, with a value close to 0.05 mg m^{-3}. The same DNEL for the *inhalable* fraction is 0.263 mg m^{-3}. These DNELs take into account the 2017 US NTP carcinogenicity studies on diantimony trioxide [16].

provide information about the risk assessment of each production step and use the chemical is involved in.

More information about the hazards triggering the above DNELs and PNECs can be found in the chapters 14 and 15. By the time of writing these three chapters, 5 of the 10 antimony REACH Registration Dossiers are undergoing a group evaluation by ECHA [10, 13]. The final evaluation decisions issued by ECHA in March 2020 have requested industry to generate new evidence about the mutagenicity, the reproductive toxicity, and the repeated dose inhalation and oral toxicities of the evaluated substances. Once these study results become available, they will be utilized to review the hazard and risk assessments conducted for these substances.

The risk assessment conducted so far by industry is resulting in a set of safety measures to be implemented during the handling and use of each substance. These measures are identified through the risk assessment and the refinement of the RCR calculation. They ensure the safe use of antimony substances. They include maximum quantities to be handled at a given time, maximum times and daily frequencies of handling the antimony substance, as well as engineering and personal protective controls to prevent oral, dermal, and particularly inhalation exposures of workers.

Beyond the set of workplace measures recommended by industry in the SDS, authorities have historically performed their own risk assessments and come up with other risk management measures. These may be site restrictions, to be applied at the level of the workplace where the chemical is extracted, produced or transformed; and product restrictions, aimed to protect the consumer who is using an article containing the chemical. Finally, in anticipation of articles reaching their end of life status and becoming a waste, antimony-containing articles are also subject to risk management measures. These are addressed in the forthcoming chapters.

16.5 Site restrictions

Site restrictions can be workplace or environment related. They are established for industrial sites based on the known hazards and risks of the substance (cf. list of classification end points in Section 16.2, "Classification").

Bearing in mind the lung hazard and classification of some antimony substances, it is foremost important to limit the generation of respirable powders, and then the potential to inhale these. Industry will typically achieve this via three approaches: developing alternatives to dry powders (e.g., wetted powder, paste, and masterbatch preparations), installing local exhaust ventilation systems, and providing workers with appropriate respiratory protective equipment. The standard that is used to determine the efficacy of these measures is one imposed by authorities: "Occupational Exposure Limit values" (OEL) or "Threshold Limit Values" (TLV). OELs and TLVs are derived by industrial occupational hygiene and epidemiology specialists on behalf of authorities or the broader scientific community. The American Conference of Governmental Industrial Hygienists (ACGIH) was set up in 1938 and is the author of the current OEL or TLV of 0.5 mg m^{-3} in place for antimony and antimony compounds in most jurisdictions [14]. The OELs in place to date are summarized by i2a on their website (Tab. 16.2).

Tab. 16.2: Mandatory workplace levels for antimony and antimony compounds in various jurisdictions [15].

Country	Occupational exposure limit / mg m^{-3}	Maximum exposure time	Document date
UK	0.5	8 h TWA	Second edition – 2011
German MAK-Commission	0.006 (expressed as respirable antimony)		2018
Finland	0.5	8 h TWA	2016
Belgium	0.5	8 h TWA	2014
France	0.5	8 h TWA	2016
Sweden	0.25		2011
US NIOSH and OSHA	0.5	8 h TWA	2007
Australia	0.5	8 h TWA	2018
Japan	0.1	8 h TWA	2013
China	0.5	8 h TWA	2007

The 0.5 mg m^{-3} value recommended by ACGIH predates the release of the 2017 US NTP Carcinogenicity studies on diantimony trioxide [16], which has observed adverse effects on the lung at lower air concentrations than those observed before 2017. The US NTP results have inspired the revised Japanese OEL and the MAK level applied in Germany. The current 0.5 mg m^{-3} level applied by the US NIOSH and OSHA, as well as many other jurisdictions, is also being reconsidered in light of ACGIH's review of the US NTP studies. In reviewing this value, there is a need to decide, first, on the size of the workplace aerosol that should be targeted (respirable or inhalable, cf. Section 16.3, "Risk assessment," for definitions), and then on the concentration of antimony in that aerosol fraction that should not be exceeded. This is how the German MAK arrived at the first respirable-fraction value of 0.006 mg m^{-3} for diantimony trioxide and trisulphide.

It is likely that one of the outcomes of the evaluation of antimony substances under REACH may result in the identification of the need of an EU-wide OEL. In that case, ECHA's Risk Assessment Committee (RAC) will assess all available data, interpret it, and produce the applicable calculations to arrive at a respirable or an inhalable OEL. In any case, the German MAK value and potential revised ACGIH TLV may inform ECHA in determining that OEL. In addition, the setting of an OEL or TLV should take into account the actual workplace exposure measurements that are regularly collected by industry to (i) verify compliance with existing limit values in place and (ii) contribute to value-setting attempts by, for example, ACGIH and ECHA.

Although the potential of antimony to cause environmental toxicity is rather low (aquatic PNECs are high, and the environmental classification is among the least severe ones), best practice foresees that air or water emissions of chemicals in general, and metals in particular, are avoided or limited. Such limits are established depending on the main industrial activity of the site, the local conditions, and the best available technologies (BAT) available. In the EU, the Industrial Emissions Directive or IED [17] sets the minimum conditions under which operating permits can be granted by local competent authorities to industrial sites. Specific conditions are established per sector, with each one, its own associated emission limit values (AELs). The BAT AELs for antimony are prescribed in BAT AEL Conclusion documents[5] in sectors such as ceramic manufacturing [18], glass manufacturing [19], nonferrous metals industries [20], and the incineration of waste [21]. These BAT AELs are generally imposing a maximum emission limit of 0.01–0.5 mg m^{-3} to the air, which is very much inspired from the workplace levels in place.

16.6 Product restrictions

Antimony substances can be used for and therefore be present in lead articles, flame-retardant polymers and textiles, polyethylene terephthalate matrices, and glass, among others. Where such products are foreseen to come into contact with consumers, and in particular with so-called vulnerable populations such as children, a number of restrictions on the content or the release of antimony have been imposed by authorities. These are either related to the known or suspected hazards of antimony, or triggered by the "precautionary principle" set out in Principle 15 of the 1992 Rio Declaration on Environment and Development.

Contrary to the "preventive principle," which aims to avoid and penalize "serious consequences" on the basis of "clear and convincing evidence," the role of the

5 BAT Reference or BREf documents are available from: https://eippcb.jrc.ec.europa.eu/reference/.

"precautionary principle" is to intervene to avoid such same consequences, but on the basis of a "threat of serious or irreversible damage" [22].

Due to its use as polymerization catalysts in the production of PET, which is one of the most common food contact materials, a number of investigations have led to the limitation of the amount of antimony allowed in the plastic matrix used by so many consumers. As such, the European Food Safety Administration (EFSA) has set, in 2004, a migration limit at 0.04 mg kg^{-1} of food (as antimony) [23] (allowing for 10% of the Tolerable Daily Intake (TDI) being allocated to food contact materials). This migration limit will be typically taken up in food contact material legislation, such as the EU Plastic Food Contact Materials Regulation 10/2011 [24].

As regards the concentration of antimony allowed in the drinking water contained in some of these PET bottles, for example, the World Health Organization (WHO) concluded in 2004 on a guideline value of 20 μg L^{-1} [25].

As antimony can be used as a synergist in flame retardant plastics necessary for electronic and electric applications, and other polymers as functional pigment, it is may be present in plastic articles used for childcare, or toys, for example. Again, regulations exist to avoid excessive exposures to antimony and other chemicals in such products.

In the EU, Directive 2009/48 relating to the safety of toys [26] foresees three limits for antimony in toys:
- 45 mg kg^{-1} in dry, brittle, powder-like or pliable toy material
- 3 mg kg^{-1} in liquid or sticky toy material
- 560 mg kg^{-1} in scraped-off toy material

In the United States, a number of states hold lists of chemicals which are subject to reporting obligations when they are present in products for children, such as clothing and products that "facilitate sleep" (e.g., Washington State list of high priority chemicals). Such a listing triggers a number of administrative obligations including, the requirement to submit notices regarding the use of antimony in children's products, their manufacture, or their import/distribution. The lists are typically influenced by Proposition 65 [27], which originally listed antimony based on its existing IARC classification [6] as suspected carcinogen. Hazard-based only, Proposition 65 and other similar state lists included antimony substances without the performance of a use-specific risk assessment. The listing is solely justified by the presence of antimony in the product and the fact that the product is used by "vulnerable populations," that is, a precautionary principle approach.

16.7 Waste and recycling

Antimony is used in several applications and can be found in a number of end-of-life or waste materials, including more particularly:

- Lead products (in which antimony had been added to strengthen this naturally soft metal), and
- Polymers (in which traces of antimony can remain following its use as polymerization catalyst, or in which it was added in relatively higher concentrations to serve as flame retardant synergist or technical pigment).

As regards the classification and transport of wastes containing antimony, these are subject to the Basel Convention, and the UN and EU waste legislations. The EU Commission Decision establishing a List of Wastes [28] categorizes antimony as a "heavy metal" which triggers an almost default hazardous classification of the wastes they are contained in. This means that antimony-containing wastes cannot be shipped to non-OECD countries and have to be handled as locally as possible (good for the environment and for local economies), but also triggers a number of burdensome requirements to access such wastes from other lands.

The default categorization of antimony-containing wastes as hazardous wastes also constitutes one of the various obstacles to the economic viability of recycling projects (which are being stimulated for critical raw materials such as antimony), as explained later.

The recycling of antimony contained in Pb-based products takes place as part of the existing highly efficient recycling loop for such Pb-based products. The continued safe use and recycling of lead hence also ensures that the critical antimony is retained on the market for future use in that value chain.

The recycling of antimony in polymers is less established to date. As regards polyethylene terephthalate (PET), the traces of antimony contained in that polymer following its function as catalyst remain in the PET recycling loop, which constitutes the most circular polymer type to date, thanks to the existence of comparably accessible and affordable collection, sorting and recycling techniques. The main regulatory condition recycled PET should meet in terms of antimony safety is to ensure that the resulting antimony content does not exceed that applicable to food contact materials.

As regards the antimony contained in other thermoplastics, typically used in electronic and electric equipment (EEE), the process to recover antimony can follow several routes. Their development has started in the late 2010s and a number of projects are now moving to pilot plant phases. At this point in time in the projects, the economic viability of the recycling needs to be assessed in terms of the regulatory compliance that the output materials should meet, and the cost of meeting those.

16.8 Conclusion

The need to decrease the dependency on imports of antimony is clearly contradictory with the sometimes excessive or precautionary EHS regulatory requirements imposed on the substance and the products and wastes it is contained in.

The objective of EHS legislation is safety. The Oxford Languages [29] definition of safety is "the condition of being protected from or unlikely to cause danger, risk, or injury." EHS legislation is hence there to protect society and the environment from the danger, risk or injury that may be caused by chemicals. This is done by a number of legislative tools or actions: (i) classifying, labeling, and communicating about the safe use conditions of antimony; (ii) properly considering the hazards, exposure, and risk to identify the more efficient risk management measures; and (iii) implementing the necessary site or product restrictions to reinforce the protection. In the case of antimony substances, the current classifications, risk assessment conclusions, workplace exposure limit values, environmental emission limit values, and maximum product concentrations or migration limits seem to address all hazard and exposure possibilities.

The excessive regulatory pressure put on antimony constitutes an additional burden to the promising recycling avenues that are being developed. Instead, regulations should target the integrated concept of "safe circularity" and "industrial symbiosis" that would render recycling of critical raw materials such as antimony safe and more attractive to technical experts and investors.

References

[1] Critical raw materials. DG GROW, European Commission. 2020. (Accessed October 20, 2020, at https://ec.europa.eu/growth/sectors/raw-materials/specific-interest/critical_en)

[2] Final List of Critical Minerals 2018: A Notice by the Interior Department on 05/18/2018. US Federal Register. 2018. (Accessed October 20, 2020 at https://www.federalregister.gov/documents/2018/05/18/2018-10667/final-list-of-critical-minerals–2018)

[3] Australia's Critical Minerals Strategy 2019. Department of Industry, Innovation and Science and Austrade, Australian Government. 2019. (Accessed October 20, 2020 at https://www.industry.gov.au/sites/default/files/2019-03/australias-critical-minerals-strategy-2019.pdf)

[4] Regulation (EC) No 1272/2008 on the classification, labelling and packaging of substances and mixtures. European Commission. 2008. (Accessed October 20, 2020 at https://eur-lex.europa.eu/legal-content/EN/TXT/PDF/?uri=CELEX:02008R1272-20200501&from=EN)

[5] Guidance on the Application of the CLP Criteria. Guidance to Regulation (EC) No 1272/2008 on classification, labelling and packaging (CLP) of substances and mixtures – Version 5.0. European Chemicals Agency. 2017. (Accessed October 20, 2020 at https://echa.europa.eu/documents/10162/23036412/clp_en.pdf/58b5dc6d-ac2a-4910-9702-e9e1f5051cc5)

[6] International Agency for Research on Cancer (IARC) – Summaries & Evaluations: VOL.: 47 (1989) (p. 291) Antimony trioxide and antimony trisulfide. International Agency for Research on Cancer. 1989. (Accessed October 20, 2020 at http://www.inchem.org/documents/iarc/vol47/47-11.html)

[7] Integrated Risk Information System (IRIS) – Chemical Assessment Summary: Antimony; CASRN 7440-36-0. U.S. Environmental Protection Agency. National Center for Environmental Assessment. 2002. (Accessed October 20, 2020 at https://cfpub.epa.gov/ncea/iris/iris_documents/documents/subst/0006_summary.pdfandhttps://cfpub.epa.gov/ncea/iris2/chemicalLanding.cfm?substance_nmbr=676)

[8] European Union Risk Assessment Report: Diantimony trioxide (CAS No: 1309-64-4; EINECS No: 215-175-0). European Commission. 2008. (Accessed October 20, 2020 at https://echa.europa.eu/documents/10162/553c71a9-5b5c-488b-9666-adc3af5cdf5f)

[9] Regulation (EC) No 1907/2006 of the European Parliament and of the Council on the Registration, Evaluation, Authorisation and Restriction of Chemicals (REACH). European Commission. 2006. (Accessed October 20, 2020 at https://eur-lex.europa.eu/legal-content/EN/TXT/PDF/?uri=CELEX:02006R1907-20200428&from=en)

[10] Grouping speeds up regulatory action Integrated Regulatory Strategy: Annual Report. European Chemicals Agency. May 2020. (Accessed October 20, 2020 at https://echa.europa.eu/documents/10162/27467748/irs_annual_report_2019_en.pdf/bd23e8cb-a55a-24af-4be3-7a29828ebb09)

[11] Guidance on Information Requirements and Chemical Safety Assessment – Part E: Risk Characterisation. Version 3.0. European Chemicals Agency. 2016. (Accessed October 20, 2020 at https://echa.europa.eu/documents/10162/13632/information_requirements_part_e_en.pdf)

[12] Screening Assessment Antimony-containing Substances. Environment and Climate Change Canada + Health Canada. 2020. (Accessed October 20, 2020 at https://www.canada.ca/en/environment-climate-change/services/evaluating-existing-substances/screening-assessment-antimony-containing-substances.html#toc17)

[13] Substance evaluation – CoRAP. European Chemicals Agency. 2020. (Accessed October 20, 2020 at https://www.echa.europa.eu/information-on-chemicals/evaluation/community-rolling-action-plan/corap-table/-/dislist/name/-/ecNumber/-/casNumber/-/lec_submitter/-/cse_public_lifecycle/Not+started/haz_detailed_concern/-/)

[14] Antimony and Compounds: TLV(R) Chemical Substances 7th Edition Documentation. American Conference of Governmental Industrial Hygienists. 2001. (Accessed October 20, 2020 at https://www.acgih.org/forms/store/ProductFormPublic/antimony-and-compounds-tlv-r-chemical-substances-7th-edition-documentation)

[15] Workplace Restrictions. International Antimony Association. 2020. (Accessed October 2020, 20 at https://www.antimony.com/regulations-compliance/workplace-restrictions/)

[16] NTP Technical report on the toxicology and carcinogenesis studies of antimony trioxide (CAS RN 1309-64-4) in Wistar Han [CRL:WI(Han)] rats and B6C3F1/N mice (inhalation studies) – NTP TR 590. United States National Toxicology Program. 2017. (Accessed October 20, 2020 at https://ntp.niehs.nih.gov/ntp/htdocs/lt_rpts/tr590_508.pdf)

[17] Directive 2010/75/EU of the European Parliament and the Council on industrial emissions. European Commission. 2010. (Accessed October 20, 2020 at https://eur-lex.europa.eu/legal-content/EN/TXT/PDF/?uri=CELEX:32010L0075&from=EN)

[18] Reference Document on Best Available Techniques in the Ceramic Manufacturing Industry. IPPC Bureau, European Commission. 2007. (Accessed October 20, 2020 at https://eippcb.jrc.ec.europa.eu/sites/default/files/2019-11/cer_bref_0807.pdf)

[19] Commission Implementing Decision establishing the best available techniques (BAT) conclusions for the manufacture of glass. European Commission. 2012. (Accessed October 20, 2020 at https://eur-lex.europa.eu/legal-content/EN/TXT/PDF/?uri=CELEX:32012D0134&from=EN)

[20] Commission Implementing Decision establishing the best available techniques (BAT) conclusions for the non-ferrous metals industries. European Commission. 2016. (Accessed October 20, 2020 at https://eur-lex.europa.eu/legal-content/EN/TXT/PDF/?uri=CELEX:32016D1032&from=EN)
[21] Commission Implementing Decision establishing the best available techniques (BAT) conclusions for waste incineration. European Commission. 2019. (Accessed October 20, 2020 at https://eur-lex.europa.eu/legal-content/EN/TXT/PDF/?uri=CELEX:32019D2010&from=EN)
[22] Workshop on EU legislation. Principles of EU Environmental Law – The Preventive and Precautionary Principles. European Commission. 2012. (Accessed October 20, 2020 at https://ec.europa.eu/environment/legal/law/pdf/principles/9%20Preventive%20and%20Precautionary%20Principles_revised.pdf)
[23] Opinion of the Scientific Panel on food additives, flavourings, processing aids and materials in contact with food (AFC) on a request from the Commission related to a 2nd list of substances for food contact materials (Ethylene carbonate, REF. No 16955, Methacrylic acid, 2-sulphoethyl ester, REF. No 21370, Antimony trioxide, REF. No 35760, cis-endo-bicyclo[2.2.1] heptane-2,3-dicarboxylic acid, disodium salt, REF. No 38505, Pimelic acid, calcium salt, REF. No 76415). European Food Safety Authority. 2004. (Accessed October 20, 2020 at https://efsa.onlinelibrary.wiley.com/doi/epdf/10.2903/j.efsa.2004.24a)
[24] Regulation (EU) No 10/2011 on plastic materials and articles intended to come into contact with food. European Commission. 2011. (Accessed October 20, 2020 at https://eur-lex.europa.eu/legal-content/EN/TXT/PDF/?uri=CELEX:32011R0010&from=EN)
[25] Antimony in Drinking-water – Background document for development of WHO Guidelines for Drinking-water Quality. WHO/SDE/WSH/03.04/74. World Health Organization. 2003. (Accessed October 20, 2020 at http://www.who.int/water_sanitation_health/water-quality/guidelines/chemicals/antimony.pdf)
[26] Directive 2009/48/EC on the safety of toys. European Commission. 2009. (Accessed October 20, 2020 at https://eur-lex.europa.eu/legal-content/EN/TXT/PDF/?uri=CELEX:32009L0048&from=EN)
[27] Chemicals considered or listed under Proposition 65: Antimony Oxide (Antimony trioxide). California Office of Environmental Health Hazard Assessment. 1990. (Accessed October 20, 2020 at https://oehha.ca.gov/proposition-65/chemicals/antimony-oxide-antimony-trioxide)
[28] Commission Decision of 3 May 2000 establishing a list of wastes and a list of hazardous waste. European Commission. 2000. (Accessed October 20, 2020 at https://eur-lex.europa.eu/legal-content/EN/TXT/PDF/?uri=CELEX:02000D0532-20150601&from=EN)
[29] Oxford Languages for Google. Oxford University Press. 2020. (Accessed October 20, 2020 at https://languages.oup.com/google-dictionary-en/)

Appendix

Tab. 1: Properties of the isolated atom.

Property (unit)	Value	Reference
General		
Atomic number	51	
Periodic table group	15	
Periodic table group name	pnictogen	
Periodic table period	5	
Block	p-block	
Electronic configuration		
Ground state electron configuration	[Kr] $4d^{10}\ 5s^2\ 5p^3$	
Shell structure	2.8.18.18.5	
Atomic weight[a]	121.760(1)[b]	[1]
Naturally occurring isotopes		
Mole fraction	^{121}Sb: 0.5721(5)[c] ^{123}Sb: 0.4279(5)[c]	[2]
Atomic mass (μu)	^{121}Sb: 120903810.1(2.8)[c] ^{123}Sb: 122904214.0(1.6)[c]	[3]
Spin	^{121}Sb: 5/2+ ^{123}Sb: 7/2+	[4]
Nuclear magnetic moment (nuclear magnetons)	^{121}Sb: +3.3592 ^{123}Sb: +2.5466	[4]
Electron affinity (eV)	1.047401(19)	[5]
Ionization energies (eV)		
$Sb \rightarrow Sb^{+}$	8.608389(12)[c]	[6]
$Sb^{+} \rightarrow Sb^{2+}$	16.626(25)[c]	
$Sb^{2+} \rightarrow Sb^{3+}$	25.3235	
$Sb^{3+} \rightarrow Sb^{4+}$	43.804(25)[c]	
$Sb^{4+} \rightarrow Sb^{5+}$	[55.00](10)[c]	
$Sb^{5+} \rightarrow Sb^{6+}$	99.51(5)[c]	

[a]Scaled to an atomic weight of 12 for carbon-12 (^{12}C), where ^{12}C is a neutral atom in its nuclear and electronic ground state, having the result that atomic weight values are dimensionless.
[b]Uncertainty is a decisional uncertainty; it is given in parentheses following the last significant figure to which it is attributed. Geological materials are known in which the element has an isotopic composition outside the limits for normal material. The difference between the atomic weight of the element in such materials and that given in the table may exceed the stated uncertainty.
[c]Uncertainties in parentheses, following the last significant digits to which they are attributed.

https://doi.org/10.1515/9783110668711-017

Tab. 2: Properties of the bound atom.

Property (unit)	Value	Reference
Covalent radii (pm)[a]		
Empirical atomic radius	145	[7]
Cordero	139	[8]
Pyykkö:		[9]
molecular single bond	140	
molecular double bond	133	
molecular triple bond	127	
Crystal radii (pm)[b]		
Sb(III), 4-coordinate	90	[10]
Sb(III), 5-coordinate	94	
Sb(III), 6-coordinate	90	
Sb(V), 6-coordinate	74	
Effective ionic radii (pm)[b]		
Sb(III), 4-coordinate	76	[10]
Sb(III), 5-coordinate	80	
Sb(III), 6-coordinate	76	
Sb(V), 6-coordinate	60	
van der Waals radius (pm)		
Batsanov	220	[11]
Alvarez	247[c]	[12]
Electronegativity (Pauling units)		
Allred Rochow	1.82	[13]
Pauling	2.05	[14]
Allen	1.984	[15]

[a]Various methods for calculating additive covalent radii exist. Values for three of them are given: values in Slater [7] are general-purpose atomic radii covering covalent, ionic, and metallic molecules: Refer to the original publications for details.
[b]See Gibbs et al. [16] for a discussion of the meaning of these parameters.
[c]Uncertainty qualified as large; polynuclear excluded.

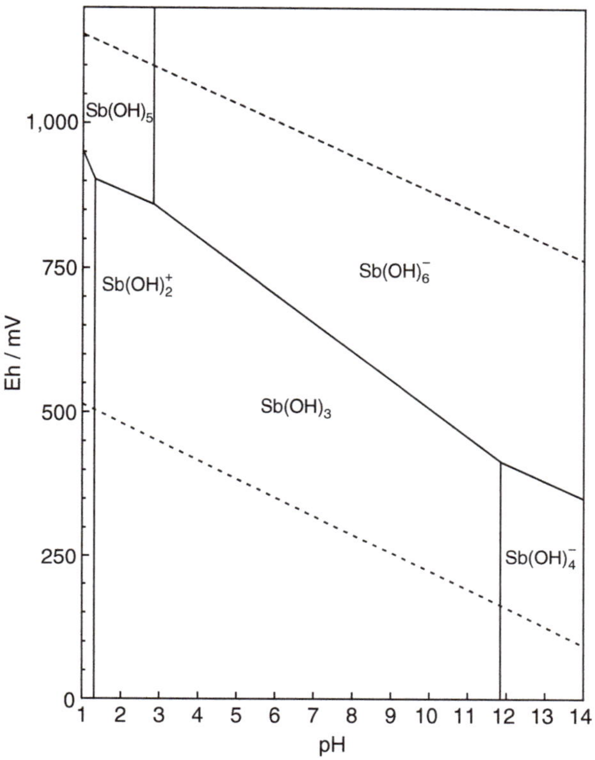

Fig. 1a: Eh – pH diagram of antimony in the Sb–H_2O system at a dissolved antimony concentration of 1×10^{-8} mol L^{-1} at 25 °C and 0 ionic strength.

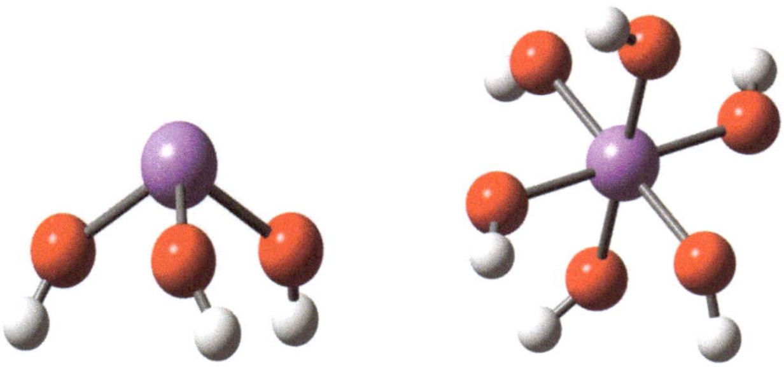

Fig. 1b: Main chemical species of (left) Sb(III) – tetrahedral $Sb(OH)_3$ – and (right) Sb(V) – octahedral $Sb(OH)_6^-$ – in aqueous solution. Ball and stick models with oxygen in red, hydrogen in white and antimony in purple.

References

[1] Meija J, Coplen TB, Berglund M, Brand WA, De Bièvre P, Gröning M, Holden NE, Irrgeher J, Loss RD, Walczyk T, Prohaska T. Atomic weights of the elements 2013 (IUPAC Technical Report). Pure Appl Chem 2016, 88, 265–291.

[2] Meija J, Coplen TB, Berglund M, Brand WA, De Bièvre P, Gröning M, Holden NE, Irrgeher J, Loss RD, Walczyk T, Prohaska T. Isotopic compositions of the elements 2013 (IUPAC Technical Report). Pure Appl Chem 2016, 88, 293–306.

[3] Wang M, Audi G, Kondev FG, Huang WJ, Naimi S, Xu X. The AME2016 atomic mass evaluation (II). Tables, graphs and references. Chin Phys C 2017, 41, 030003.

[4] Sansonetti JE, Martin WC. Handbook of basic atomic spectroscopic data. J Phys Chem Ref Data 2005, 34, 1559–2259.

[5] Scheer M, Haugen HK, Beck DR. Single- and multiphoton infrared laser spectroscopy of Sb^-: A case study. Phys Rev Lett 1997, 79, 4104–4107.

[6] Kramida A, Ralchenko Y, Reader J, NIST ASD Team NIST. Atomic Spectra Database (ver. 5.8), [Online]. Available: https://physics.nist.gov/asd [2020, November 16]. National Institute of Standards and Technology, Gaithersburg, MD. 2020.

[7] Slater JC. Atomic radii in crystals. J Chem Phys 1964, 41, 3199–3204.

[8] Cordero B, Gómez V, Platero-Prats AE, Revés M, Echeverría J, Cremades E, Barragán F, Alvarez S Covalent radii revisited, Dalton Trans 2008, 2832–2838.

[9] Pyykkö P. Additive covalent radii for single-, double-, and triple-bonded molecules and tetrahedrally bonded crystals: A summary. J Phys Chem A 2015, 119, 2326–2337.

[10] Shannon RD. Revised effective ionic radii and systematic studies of interatomic distances in halides and chalcogenides. Acta Cryst A 1976, 32, 751–767.

[11] Batsanov SS. Van der Waals radii of elements. Inorg Mater 2001, 37, 871–885.

[12] Alvarez S. A cartography of the van der Waals territories. Dalton Trans 2013, 42, 8617–8636.

[13] Allred AL, Rochow EG. A scale of electronegativity based on electrostatic force. J Inorg Nucl Chem 1958, 5, 264–268.

[14] Pauling L. The Nature of the Chemical Bond. An Introduction to Modern Structural Chemistry, Third edition. Ithaca, US, Cornell University Press, 1961.

[15] Allen LC. Electronegativity is the average one-electron energy of the valence-shell electrons in ground-state free atoms. J Am Chem Soc 1989, 111, 9003–9014.

[16] Gibbs GV, Ross NL, Cox DF, Rosso KM, Iversen BB, Spackman MA. Bonded radii and the contraction of the electron density of the oxygen atom by bonded interactions. J Phys Chem A 2013, 117, 1632–1640.

Index

https://doi.org/10.1515/9783110668711-018

Bei Fragen zur Produktsicherheit wenden Sie sich bitte an:
If you have any questions regarding product safety,
please contact:

Walter de Gruyter GmbH
Genthiner Straße 13
10785 Berlin
productsafety@degruyterbrill.com